Monographiae Biologicae

Series Editor

Henri J. Dumont, Department of Biology, Ghent University, Ghent, Belgium

The Monographiae Biologicae provide a forum for top-level, rounded-off monographs dealing with the biogeography of continents or major parts of continents, and the ecology of well individualized ecosystems such as islands, island groups, mountains or mountain chains. Aquatic ecosystems may include marine environments such as coastal ecosystems (mangroves, coral reefs) but also pelagic, abyssal and benthic ecosystems, and freshwater environments such as major river basins, lakes, and groups of lakes.

In-depth, state-of-the-art taxonomic treatments of major groups of animals (including protists), plants and fungi are also elegible for publication, as well as studies on the comparative ecology of major biomes. Volumes in the series may include single-author monographs, but also multi-author, edited volumes.

Prospective authors and/or editors should consult the Series Editors for more details. Any comments or suggestions for future volumes are welcomed;

Series Editor:
Henri J. Dumont, State University of Ghent, Institute of Animal Ecology, Belgium
E-mail: henri.dumont@ugent.be

Tamara A. Shiganova • Henri J. Dumont
Alexander S. Kazmin

Non-Native Species in the Ponto-Caspian Seas

Introduction, Impact, Geographical Distribution, Consequences

Springer

Tamara A. Shiganova
Shirshov Institute of Oceanology
Russian Academy of Sciences
Moscow, Russia

Henri J. Dumont
Department of Animal Ecology
Ghent University
Ghent, Belgium

Alexander S. Kazmin
Shirshov Institute of Oceanology
Russian Academy of Sciences
Moscow, Russia

ISSN 0077-0639 ISSN 2215-1729 (electronic)
Monographiae Biologicae
ISBN 978-3-032-18639-3 ISBN 978-3-032-18640-9 (eBook)
https://doi.org/10.1007/978-3-032-18640-9

This Springer imprint is published by the registered company Springer Nature Switzerland AG
The registered company address is: Gewerbestrasse 11, 6330 Cham, Switzerland

Acknowledgements

T.A. Shiganova performed this study in the framework of state assignment of the Ministry of Education and Science of the Russian Federation (FMWE-2024-0023 "Bio-resources, biodiversity and the structure of pelagic communities of the World Ocean, their spatial and temporal dynamics").

A.S. Kazmin performed this study in the framework of state assignment of the Ministry of Education and Science of the Russian Federation (FMWE-2024-0016 "Multi-scale hydro-physical processes in the World Ocean and its boundary layers: their investigation by methods of operational oceanography, ship observations, remote sensing, theoretical, numerical and laboratory modeling").

Competing Interests The authors have no competing interests to declare that are relevant to the content of this manuscript.

Contents

Chapter 1
Introduction

Abstract The wide range of issues associated with the study of aquatic species dispersal from one marine or river basin to another are briefly formulated. It is noted that since the beginning of the eighteenth century, a sharp increase in the number of non-native animal and plant species introduced into recipient ecosystems has been observed. The geographical location of the seas considered in the monograph (Black Sea, Sea of Azov, Caspian Sea, Aral Sea, Sea of Marmara) and adjacent marine basins and land areas is presented. The topics that will be covered in the monograph are listed.

Dispersal of aquatic species from one marine or river basin to another may occur both naturally and as a man-mediated process. Initially, it used to happen under natural conditions. Such crossings of natural boundaries took place in all historical epochs and on different time scales. It was related to and caused by the geological processes that favored the merging or splitting of basins, or to climate variations and accompanying changes in the environmental conditions. Geographical isolation of basins established in the past, often sufficiently long time ago, allowed aboriginal flora and fauna to evolve and build up a specific biological diversity. Until recently, that diversity reflected the historical evolution of each given region and was characterized by a number of species and sometimes higher taxa produced in the process of local evolution, being endemic to the site of origination. If plant or animal species were somehow introduced, by whatever rare natural event, to a new basin, this could cause a temporary disturbance or not, but such events may be considered to have been so spaced in time that they usually did not cause sensitive changes to marine of freshwater ecosystems as a whole.

Gaining some momentum in the early eighteenth century, accelerated in the nineteenth century, and culminating in the second half of the twentieth century, a sharp increase in the number of non-native animal and plant species introduced to foreign ecosystems, terrestrial and aquatic alike, occurred. In marine systems, they appeared in the pelagic as well as in near-shore zones. Currently, the "spontaneous" appearance of invasive, non-native species in various taxonomic groups, ranging from algae and protozoa to fishes, and their naturalization to new biotopes has

T. A. Shiganova et al., *Non-Native Species in the Ponto-Caspian Seas*,
Monographiae Biologicae 96, https://doi.org/10.1007/978-3-032-18640-9_1

become like an epidemic to coastal regions of the World Ocean, inland seas, and to marine, brackish and freshwater systems (Carlton 1985).

At present, the international scientific community fully realizes the global character of this ecological problem: day after day, a displacement of thousands of representatives of flora and fauna occurs, as a by-product of commercial and non-commercial anthropogenic activities. This causes serious ecological and economic impacts. Some invaders, like *Vibrio cholerae,* are extremely dangerous to human health. In the 1990s, scientific bodies and programs were set up to synthesize the available data on non-native species, to recommend measures of control over the propagation of such species and to protect the natural biological diversity of the host basins. Among them, the most prominent are the International Global Invasive Species Program (Mooney 1999) and the Globallast Program. The latter is aimed at developing procedures for control of ballast water, e.g., conventions were accepted for monitoring of ballast water release and special coatings for ships to prevent fouling species from attaching. The 2005 IMO International Convention for the Control and Management of Ships' Ballast Water and Sediments, "Ballast Water Convention" or BWC (adopted in February 2004 (not in force). The full text of the draft Ballast Water Convention is available on line at http://globallast.imo.org/index.asp?page=mepc.htm) addresses the prevention and reduction of the risks to environment, human health, property, and resources arising from the transfer of harmful aquatic organisms and pathogens via management of ballast waters and sediments.

In this book, we explore the ecological effects of the invasive species currently established in the southern seas of Eurasia (Black Sea, Sea of Azov, Caspian Sea, Aral Sea, Sea of Marmara, Mediterranean Sea; Fig. 1.1), with special attention to the pathways and vectors of their penetration, expansion, establishment, and effect upon the recipient ecosystems. The problem of the appearance of non-native species in these seas became especially critical after the invasion of the ctenophore *Mnemiopsis leidyi*, which had and continues to have an enormous effect on the ecosystems concerned, their functioning, and in particular, their commercial fish resources (Shiganova 1997, 1998; Shiganova et al. 2001, 2004a, b, c).

In our monograph, we intend to cover the following aspects:

1. Verification of non-native species in the Black Sea, Sea of Azov, Caspian Sea, Aral Sea, Sea of Marmara, and Aegean Sea (Fig. 1.1) by means of the genetic analyses of target species
2. Identification of the main pathways and vectors of invasions
3. Determination of the factors that render a basin suitable for the establishment of invaders
4. Revealing of species capable for successful establishment in the new basin and further expansion, and qualities allowing this species to establish in new biotopes
5. Assessment of non-native species role in the ecosystem and determination of effects upon the ecosystems of the recipient basins

Fig. 1.1 Study area and major artificially constructed canals (The Suez Canal and Volga-Don Canal)

6. Analysis of the after-effects of the invasion of the new species: why some species cause an absolute degradation of the ecosystem, while others may provide its restoration; why some species replace valuable native species, while the other may lead to an enrichment of the ecosystem
7. Impact of abiotic environmental factors and climate change on non-native species

Increasing awareness of the problems associated with invasions of non-native species in Ponto-Caspian seas has led to rapid growth in research on this topic. This book provides updated summary of this fast-growing field of research, and is an essential text for established researchers, undergraduate, and graduate students.

References

Carlton JT (1985) Transoceanic and interoceanic dispersal of coastal marine organisms: the biology of Ballast water. Oceang Mar Biol Ann Rev. Margaret B (ed). Aberdeen University Press 23: 313–371

Mooney HA (1999) The Global Invasive Species Program (GISP). Biol Invasions 1:97–98. https://doi.org/10.1023/A:1010023129637

Shiganova TA (1997) *Mnemiopsis leidyi* abundance in the Black Sea and its impact on the pelagic community. In: Ozsoy E, Mikaelyan A (eds) Sensivity of the North, Baltic Sea and Black Sea to antropogenic and climatic changes, vol Sp. Kluwer Academic Publishers, pp 117–130

Shiganova TA (1998) Invasion of the Black Sea by the ctenophore *Mnemiopsis leidyi* and recent changes in pelagic community structure. Fish Oceanogr. Globec Special Issue Ed. Steeve Coombs 7(3–4):305–310

Shiganova TA, Bulgakova YV, Volovik SP, Mirzoyan ZA, Dudkin SI (2001) A new invader, *Beroe ovata* Mayer 1912 and its effect on the ecosystems of the Black and Azov Seas. Hydrobiologia. Purcel JE, Graham WM, Dumont HJ. Kluwer Academic Publishers 451:187–197

Shiganova TA, Dumont HJD, Mikaelyan A, Glazov DM, Bulgakova YV, Musaeva EI, Sorokin PY, Pautova LA, Mirzoyan ZA, Studenikina EI (2004a) Interaction between the Invading Ctenophores *Mnemiopsis leidyi* (A. Agassiz) and *Beroe ovata* Mayer 1912, and their Influence on the Pelagic Ecosystem of the Northeastern Black Sea. In: Dumont H, Shiganova T, Niermann U (eds) The Ctenophore Mnemiopsis leidyi in the Black, Caspian and Mediterranean Seas and other aquatic invasions, NATO ASI Series, 2. Environment. Kluwer Academic Publishers, pp 33–70

Shiganova TA, Dumont H, Sokolsky AF, Kamakin AM, Tinenkova D, Kurasheva EK (2004b) Population dynamics of *Mnemiopsis leidyi* in the Caspian Sea, and effects on the Caspian ecosystem. In: Dumont H, Shiganova T, Niermann U (eds) The Ctenophore Mnemiopsis leidyi in the Black, Caspian and Mediterranean Seas and other aquatic invasions, NATO ASI Series, 2. Environment. Kluwer Academic Publishers, pp 71–111

Shiganova TA, Christou ED, Bulgakova JV, Siokou-Frangou I, Zervoudaki S, Siapatis A (2004c) Study on the distribution and biology of the invader *M.leidyi* in the northern Aegean Sea, comparison with indigenous species *Bolinopsis vitrea*. In: Dumont H, Shiganova T, Niermann U (eds) The Ctenophore Mnemiopsis leidyi in the Black, Caspian and Mediterranean Seas and other aquatic invasions, NATO ASI Series, 2. Environment. Kluwer Academic Publishers, pp 113–135

Chapter 2
Brief Insight in Geological History of the Ponto-Caspian Seas Basins

Abstract Brief insight in the paleogeographic evolution of the Mediterranean Sea, the Black Sea, the Sea of Azov, the Caspian Sea, and the Aral Sea basins is presented. All of them are remains of the Tethys Sea, a prehistoric ocean, existed during much of the Mesozoic Era and early-mid Cenozoic Era, which was the predecessor to the modern Indian Ocean, the Mediterranean Sea, and the Eurasian inland marine basins. The geological history of tectonic processes and the associated repeated disconnections and mergers of basins in various geological epochs has been traced back to the present state. Geologically speaking, hyperkinetic development process ended when the present-day biota of the basins were formed. During the Late Quaternary, each of the southern basins was subject to its own evolutionary history. Regardless of differences in physical-geographic factors, hydrological regimes, and resident fauna, all modern Ponto-Caspian seas have significant common features: salinity different from standard ocean value (either exceeding normal oceanic salinity like the Mediterranean Sea, or below it like the Black Sea, Sea of Azov, Caspian and Aral Seas) and isolation from the world ocean.

The Mediterranean Sea, the Black Sea, the Sea of Azov, the Caspian Sea, and the Aral Sea (Lake) form a system of inland basins, extending as a chain for more than 6000 km, from the North Atlantic to the deserts of Middle Asia. Currently, the total area and volume of these reservoirs is approximately 3,400,000 km^2, and 4,400,000 km^3 correspondingly. The coasts of these sea-system were the cradle of our modern civilization and today represent an attracting area with an advanced economic development and significant recreational value.

The paleogeographic evolution of these basins associated with the geological structure of the region, which belongs to the Alpine geosyncline belt. It was formed in the Cenozoic era from a system of fracture zones in the earth's crust. Despite a number of uncertainties in our understanding of the details of the processes involved, most geologists agree in their description of the genesis of these seas (Unified International Geochronological Scale of 2004; regional scales by Fedorov (1978) and Svitoch and Yanina (1997)).

All of them are remains of the *Tethys Sea* or the *Neo-Tethys*, a prehistoric ocean existed during much of the Mesozoic Era and early-mid Cenozoic Era (Fig. 2.1). It

T. A. Shiganova et al., *Non-Native Species in the Ponto-Caspian Seas*,
Monographiae Biologicae 96, https://doi.org/10.1007/978-3-032-18640-9_2

Fig. 2.1 The Tethys Sea

was the predecessor to the modern Indian Ocean, the Mediterranean Sea, and the Eurasian inland marine basins (primarily represented today by the Black Sea and Caspian Sea), a broad sea that formerly extended between the present Indian and Atlantic Oceans.

The Tethys Sea (Ocean) existed during the most of the Mesozoic, yet began to shrink at the end of the Eocene, when a major continental plate, the Indian subcontinent, and the Arabian Peninsula collided with Asia. Close to the contact zone of the two plates, one of which slide under the other (subduction), orogenic (mountain-forming) processes became active, accompanied by volcanic activity. In its eastern part, the enormous chain of the Himalayan mountains emerged. The mountain systems of the Lesser Caucasus, Asia Minor, and the Balkans subdivided its western part into a Southern and a Northern basin. The southern basin, which included the area of the present Mediterranean Sea, retained its connection with the ocean. It also occupied the southern periphery of the Scythian-Turanian Plate and the margin of the Crimean-Caucasian geosyncline system. The Northern Basin or Paratethys (Fig. 2.2a) extended from the pre-Alpine and pre-Carpathian foothills to Middle Asia and included the basins of the present-day Black and Caspian seas. In the west, it was periodically connected with the Mediterranean Sea.

The Paratethys was a vast expanse of water that across its geological history went through a complicated evolution, at times forming a united large marine or brackish-water body and sometimes a series of isolated saline or desalinated pools.

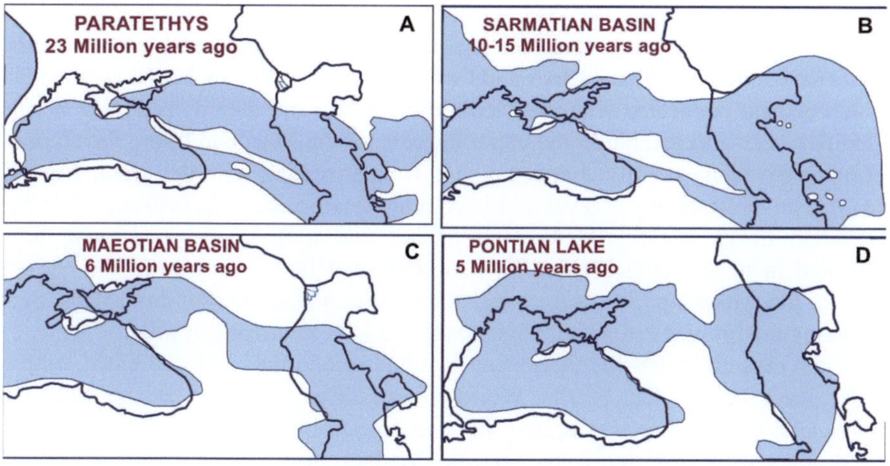

Fig. 2.2 (**a**) Paratethys at the end of the Oligocene (Rogl and Steininger 1983) (**b**) Sarmatian basin (Kolesnikov 1941); (**c**) Maeotian basin (Kolesnikov 1941); (**d**) Pontian Lake (Andrusov 1988)

In the Neogene, uplifting of the Alpine, Carpathian, Balkan, and Transcaucasian areas caused significant changes in the geomorphology of land and sea. In the Middle Miocene (15–10 million years ago), simultaneously with the uplifting of the Alps and Carpathians, the Mediterranean Basin arise. It occupied not only the southern plains but also the Volyn' and Podol highlands and, in the east, approached the Turanian Plate. In its northern part, a giant closed waterbody, the Sarmatian Lake was formed (Fig. 2.2b). It engulfed the Pannonian, Black, Caspian, and Aral Seas, extended from Vienna to the Aral Sea, and its northern boundary approached modern Krivoi Rog and Dnepropetrovsk cities. It was the largest inland brackish-water that ever existed. In its centre, the Caucasus rose in the form of a mountainous island.

The Sarmatian Sea rapidly desalinates by the numerous rivers that emptied into it; it was probably fresher than the present-day south basin of the Caspian Sea (12‰). Due to this, typical representatives of the mediterranean marine fauna disappeared, and a specific vertebrate fauna evolved, that included fishes looking like mullet, cod, and herring, "marine" mammals such as dolphins and other cetaceans (*Cetotherium*), and seals similar to the present-day Caspian seal. The Sarmatian basin continued to desalinate further, and its salinity fell to values significantly lower than those in the present-day main body of Black Sea. However, like in the Black Sea, significant differences in salinity between the upper and lower layers formed, which led to the development of a deep hydrogen sulfide zone. It is believed that desalination coupled with a deep hydrogen sulfide zone formation contributed to the extinction of most of the Sarmatian fauna; a specific fauna developed instead (Zenkevich 1963). At the end of the Miocene, a regression occurred that forced the division of this previously fully flooded area into the three separate basins, i.e., the Mediterranean, Black, and Caspian Seas. Their sizes changed repeatedly, but finally the previous Sarmatian lake shrunk and became known as the Maeotian Sea (now the Sea of Azov) (Fig. 2.1c). The Caucasus stuck out as a large peninsula and occupied almost the entire land zone between the Black and Caspian Sea areas. For a while, this Maeotian lake remained connected with the Mediterranean Sea and thereby acquired marine Mediterranean species, but in the upper deposits of the Maeotian basin, the composition of the fauna sharply changed again. The connection with the Mediterranean Sea, presumably falling in level and becoming hypersaline, was broken, while the Maeotian Lake, in contrast, became fresher and a brackish-water fauna now appeared in it. Arkhangelskyi (1935) suggests that the conditions in the eastern part of the Maeotian Sea were similar to those in the present-day Black Sea, including the presence of a deep layer saturated with hydrogen sulfide.

In the meantime, the Mediterranean Sea had turned into a closed basin, with a water level below that of the World Ocean. The Gibraltar, Bosphorus, and Dardanelle straits were closed, and the Aegean Sea had fallen dry. Progressively, the drought worsened, and in the end, the Mediterranean almost completely dried up. During this so-called Messinian crisis, only some salt pools remained on the floor of the basin. The lower Nile River, for example, had carved out a canyon deeper than the Colorado Grand Canyon. At about 5 Ma, the blockage of the Gibraltar Strait was breached, and the waters of the Atlantic Ocean rapidly filled up the Mediterranean Sea (inter alia, flooding the Nile valley as far south as Aswan).

During the Pliocene, the Maeotian Sea was replaced by the almost freshwater Pontian Lake (Fig. 2.2d). The (future) Black, Azov, and Caspian Seas (including the Aral Lake) were connected with each other across the zone of the current territories of the Stavropol' and Krasnodar districts, and the northern Caucasus. In the Pontian Lake, the pre-existing marine fauna went extinct and a brackish-water fauna of freshwater origin evolved. Its representatives survive to date in the Caspian Sea, the Sea of Azov and in the low-salinity regions of the northwestern and western Black Sea. These elements of the present-day fauna are typically the ones referred to in the scientific literature as Pontian relics or Ponto–Caspian species. In the Caspian Sea, they are considered as autochthonous or endemic, since the Caspian Sea (especially its low-salinity northern part) is the place where they are conserved best.

At the end of its existence (late Pontian period), a tectonic uplift in the region of the northern Caucasus subdivided the Pontian Lake into a Caspian and a Black Sea part. Each of these later evolved independently. For details on the geological evolution of the seas, their (dis)connection to the World Ocean, and their changes in salinity and biota, the reader may refer to Andrusov (1988), Bogachev (1933), Arkhangelskyi (1935), Derzhavin (1951), Zenkevich (1959, 1963), Leont'ev et al. (1977), and Starobogatov (1994).

For the purpose of this book, we were mainly interested in the final period of this, geologically speaking, hyperkinetic development process that ended when the present-day biota of the basins were formed. During the Late Quaternary, each of the southern basins was subject to its own evolutionary history. The largest temperature, salinity, and faunal variations and the smallest sea level oscillations occurred in the Black Sea. The Caspian Sea experienced the largest sea level oscillations yet least intensive salinity changes.

Regardless of differences in physical-geographic factors, hydrological regimes, and resident fauna, all Ponto-Caspian seas have significant common features. All are characterized by salinity values different from standard ocean value. They either exceed normal oceanic salinity (Mediterranean Sea) or are below normal values (Black Sea, Sea of Azov, Caspian and Aral seas). All basins are isolated from the world ocean and have abyssal water temperatures higher than oceanic ones (except the Sea of Azov and the Aral Sea): 8–9 °C in the Black Sea, 5–6 °C in the Caspian Sea, and 13.5–13.7 °C in the Mediterranean Sea (Zenkevich 1963). These temperatures result from the winter convection and correspond to the mean winter temperature of their surface water.

At present, the Black and Caspian and, especially, the Sea of Marmara are characterized by significant salinity differences between the surface and deep waters. Salinity stratification is reinforced by temperature stratification in the warm season of the year, when surface water is heated to 27–30 °C. In contrast, in winter, the surface layer is extensively cooled and, at selected places with the lowest salinity, an ice cover is formed. The salinity and, sometimes, temperature stratification cause deep water to stagnate long enough to develop hydrogen sulfide in the near-bottom layer. Throughout the year or seasonally, hydrogen sulfide saturation extends over thicker or thinner layers of water. Arkhangelskyi and Strakhov (1938) thinks that hydrogen sulfide contamination is characteristic not only of the present-day Black Sea but represents a phenomenon characteristic of the entire southern basins of the Neogene geosyncline.

The waters of the Black Sea, Sea of Azov, Caspian, and Aral seas differ from the waters of the ocean in salt composition; these differences increase in eastward direction.

The physical evolution of the marine environments on the southern flanks of Europe and Asia during the Tertiary and Quaternary affected their faunas in many ways. The paramount variable causing biotic change was salinity, a factor that fluctuated strongly on various timescales. Precisely this turbulent history caused the non-homogeneity of its fauna and flora. The present-day biota of the southern seas are largely relicts, often referred to under the name "ancient autochthonous assemblage." Fundamentally, they represent what remained of an ancient Tethys fauna (including Sarmatian, Pontian, and Caspian species) after it had been exposed to a repeated alternation of phases of desalination and salination. To some extent, this fauna is supplemented by comparatively recent invaders of freshwater origin and by immigrants (pseudo-relics) from the Arctic. The latter worked their way down south via the Volga River, and are therefore mainly restricted to the Caspian Sea. In addition, Atlantic elements advanced from the west, through the Mediterranean; a few such forms ventured as far as East as the Aral Lake.

The fauna of the basins differs in the proportion of the groups considered. The fauna of the Mediterranean Sea consists of south-boreal and subtropical species derived from the Atlantic Ocean. The Caspian Sea best preserved the remarkable relict "Caspian" fauna, while the Sea of Azov and the Black Sea are inhabited by a mixed Ponto-Caspian and marine Mediterranean fauna.

References

Andrusov NI (1988) An outline of the history of the development of the Caspian Sea and its inhabitants. Reproduction of the original author's publication of the 1889 edition. St. Petersburg Publishing House, HER Media

Arkhangelskyi AD (1935) The geological structure of the USSR. The western part. Issue 2. United Scientific and Technical Publishing House, 427 pp

Arkhangelskyi AD, Strakhov NM (1938) The geological structure and history of the development of the Black Sea. USSR Academy of Sciences, Moscow. 228 pp

Bogachev VV (1933) Materials on the study of the tertiary ichthyofauna of the Caucasus. In: Proceedings of Azerbaijan Oil Research Institute, vol 15. 62 p

Derzhavin AN (1951) Study in the history of the fauna of the Caspian and freshwater basins of Azerbaijan. In: Alizade AN, Asadov SM, Derzhavin AN (eds) Animal World of Azerbaijan. Izdatelstvo Akademii Nauk Azerbaidzhanskoi SSR, Baku, pp 34–83. (in Russian)

Fedorov PV (1978) Pleistocene of the Ponto-Caspian Sea. Nauka, Moscow. 165 pp. (in Russian)

Kolesnikov VN (1941) The development of the fauna of the Caspian Sea. Rep USSR Acad Sci 30:1

Leont'ev OK, Maev EG, Rychagov GI (1977) Geomorphology of the coasts and bottom of the Caspian Sea. Moscow State University, Moscow. 208 pp

Rogl, Steininger (1983) Evolution of the Indian Ocean and the drift of India, July 1987. Hydrobiologia 150(3):203–222. https://doi.org/10.1007/BF00008704

Starobogatov YI (1994) The Systematization and Paleontology Dreissena polymorpha (Pall) (Bivalvia, Dreissenidae). Nauka Publication, pp 18–46

Svitoch AA, Yanina TA (1997) Quaternary deposits of the Caspian Sea Coasts. Moscow, MSU Publ. 286 pp. (in Russian)

Zenkevich LA (1959) Classification of brackish-water reservoirs on the example of the seas of the USSR. Izv USSR Acad Sci Geogr 2:2–11

Zenkevich LA (1963) Biology of the Seas of USSR. USSR Academy of Sciences, Moscow. 739 pp. (in Russian)

Chapter 3
Pathways and Vectors of Non-native Species Introduction in the Ponto-Caspian Seas

Abstract Pathways and vectors of non-native species introduction in the Ponto-Caspian seas are reviewed. Appearance of new species in the aquatic ecosystems may be the result of a spontaneous penetration via the sea straits and rivers (natural range expansion, which is a frequent phenomenon) or, most commonly, as a result of human activity. The following root causes (vectors) for the appearance of non-native species are considered: natural expansion via existing waterways; aquariumistics; aquaculture; intentional introduction of commercially valuable species; accompanied releases; construction of canals and reservoirs, that opens up new highways to the penetration of non-native species into new basins, fully related to unintentional human intervention; shipping, including transportation of invasive species with the ballast waters or as fouling communities. Consequences of the most prominent canals construction (i.e., Volga–Don Canal and Suez Canal) are discussed. Main invasive corridors for the species invasion in Europe (Southern, Central, and Northern) are distinguished. It is pointed out that the species invasion processes accelerated dramatically in the past few decades due to new technologies, shipping intensity increase, global expanding of shipping routes, increased basin connectivity, and global warming.

Appearance of new species in the aquatic ecosystems, referred to as non-native species, invaders, exotics, or non-indigenous species (NIS), may be the result of a spontaneous penetration via sea straits and rivers (natural range expansion, which is a frequent phenomenon) or, most commonly, as a result of human activity. Humans, intentionally or occasionally, are surprisingly efficient at opening up pathways for the introduction of new aquatic species.

According to *International Convention on Biological Diversity (CBD)*, the following terms are in use:

- The region from which a taxon is originated is referred to as the *native* or the *donor area*.
- The region to which a taxon is transferred is referred to as the *recipient area*.
- The modes of the invasion of non-native species are referred to as the *vectors*; among them, natural and anthropogenic ones are distinguished.

T. A. Shiganova et al., *Non-Native Species in the Ponto-Caspian Seas*,
Monographiae Biologicae 96, https://doi.org/10.1007/978-3-032-18640-9_3

- The routes along which invasion of non-native taxa occurs are referred to as *invasion pathways*. In case taxon penetration via these pathways is repeated or systematic, they are named *invasive corridors*.

The species that appear in a recipient region and produce an independent reproductive population here are regarded as *established* non-native (briefly, non-natives) species. The term *invader* is applied to a species that exert a negative effect on other species or on the entire ecosystem of the recipient basin.

To refer to a species whose home range is uncertain, the term "*cryptogenic*" is used. Commonly, these include cosmopolitan species, which are universally spread and for which it is difficult to ascertain the region of origin.

3.1 Vectors of Non-native Species Introduction

At present, following root causes (vectors) for the appearance of non-native species are most widespread.

Natural expansion The presence of a given species in a given new location may be the result of a natural penetration via existing waterways. This mode of expansion was probably rare before the twentieth century, but recently became prominent because of global warming, giving thermophile species a chance to extend northward, and making cold-water species retreat further poleward. Due to the slow but steady temperature increase in the twentieth century, environmental conditions in the temperate zone of the northern hemisphere became sufficiently favorable to host populations of a number of warm-water loving taxa on a permanent basis.

Aquariumistics Aquarium animals and plants often travel huge distances from their natural range to final destination. There are often few or no restrictions on trading such biota, and even if such regulations exist, there is often a lack of enforcement. Aquarium owners are often sloppy and/or lack ecological insight, so many of them occasionally dispose their aquarium pets while these are still alive or viable, allowing them to escape in an environment that is essentially alien to them. In the twentieth century, aquarium trade became so popular that it acquired an industrial character. In this manner, a series of marine species, originally destined for aquaria but later "liberated," found their way to the sea. Some of them deeply influenced the recipient marine ecosystem.

Among the best-known cases of this kind is that of the benthic green alga *Caulerpa taxifolia*, cultivated since 1982 in the State Aquarium of Monaco (Oceanographic Museum of Monaco) (Meinesz and Hesse 1991). At some point, it accidentally found its way to the near-shore waters of the western part of the Mediterranean Sea and spread with yacht anchors and fishing nets. After few years, it occupied more than 6000 hectares of near-shore marine bottom. It rapidly and deeply changed the substrate and the benthic community living in and on it (Thibaut and Meinesz 2004).

The aquarium sunfish *Lepomis gibbosus* was imported as an ornamental species, stocked in aquaria and small ponds, from the North America to Europe as early as in the eighteenth century (Svetovidov 1964). From these ponds, sunfish penetrated the basins of major rivers such as the Rhine, Oder, and Danube and associated inland basins. At present, sunfish dwells in the basins of the Tissa and Bug Rivers, in the lower reaches of the Danube, Dniester, and Dnieper Rivers, and in Odessa Bay and Berezan Lagoon; single specimens are even encountered in the fresher areas of the Black Sea (Boltachev et al. 2003). Sunfish juveniles feed on crustaceans, while adult individuals prefer insects and small fishes. It inflicts damage to fisheries in both natural and artificial basins, consuming to extinction the eggs, larvae, and small juveniles of various commercial fishes and even competing for food with them (Svetovidov 1964). An analysis of the dynamics of its distribution allows to expect its further expansion eastward, and penetration in the nearest future in the basins of the rivers, emptying in the Sea of Azov (Don and Kuban'.

Aquaculture Since a long time, mariculture and marine and freshwater aquaculture provided an intentional introduction and breeding of new commercial species in ponds, lagoons, bights, and near-shore regions, either free or in suspended cages being a major vector of invasive taxa.

Intentional introduction of commercially valuable species Many animal species, mainly crustaceans and fish, have been intentionally introduced to new aquatic environments in order to create new or additional food resources.

Accompanied releases Along with a species destined for intentional release, undesirable animal and plant species may occasionally be introduced. Among them, in selected cases, even parasites can be released, become established, and change host afterwards, potentially or in the facts causing major damage to the new host.

Construction of canals and reservoirs A formidable vector that opens up new highways to the penetration of non-native species into new basins, fully related to unintentional human intervention, is the construction of canals, connecting previously isolated seas, lakes, and rivers. Through them, species from one basin may find an easy boulevard to penetrate to others, either in a natural way, or attached to or inside of ships. A most prominent example was caused by the construction of the *Volga–Don Canal* (Fig. 3.1, Introduction).

Soon after it opened (in 1952), species from the Black Sea and the Sea of Azov, together with invaders that had previously established in these seas, started to penetrate the Caspian Sea. Some of these invaders went further, taking advantage of the Volga–Baltic system to end up in the Baltic and White seas. Thus, the so-called *northern* invasive corridor for aquatic Ponto–Caspian species appeared (Panov et al. 2007) (Fig. 3.2).

From there, selected species were transported by ships across the Atlantic, establishing themselves in the Great Lakes of North America (Jażdżewski et al. 2004; Mills et al. 1993).

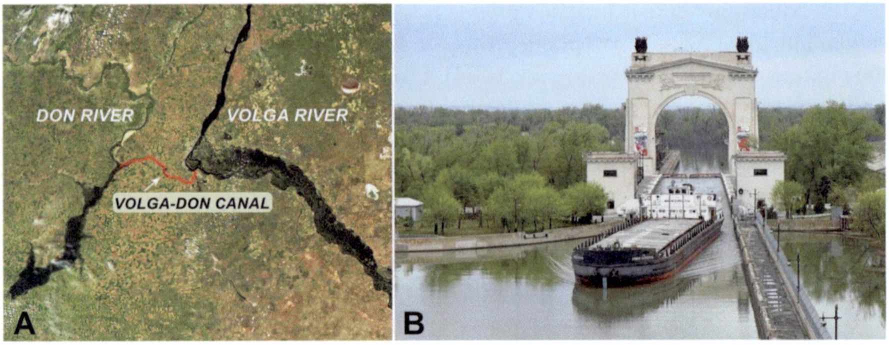

Fig. 3.1 (**a**) Volga-Don Canal (satellite view); (**b**) a dry-cargo ship passes a gateway on the Volga-Don Canal

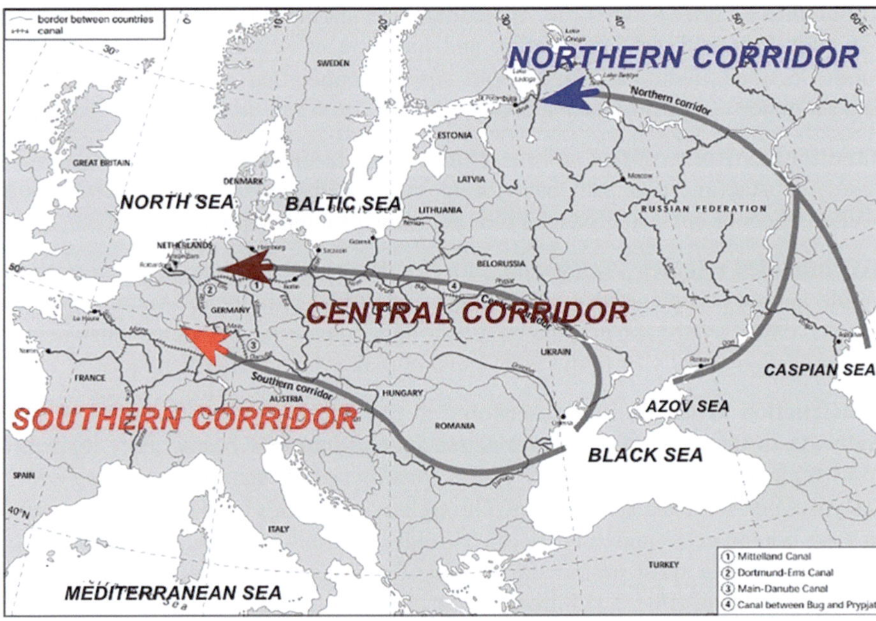

Fig. 3.2 Main invasive corridors for the species invasion in Europe (Panov et al. 2007; modified)

After the construction of a system of canals and reservoirs in the Danube region, the *southern* invasive corridor appeared, which was used by brackish-water and fresh-water Ponto–Caspian species to expand into Western Europe (Ojaveer et al. 2002) (Fig. 3.2).

As early as in the eighteenth century, drainage systems inside the basins of the Visla and Oder rivers started to be created; as a result, by the twentieth century, the Black and the North seas became connected via this network of drainage canals. A third and main invasive corridor for the "escape" of aquatic, first of all, Ponto–

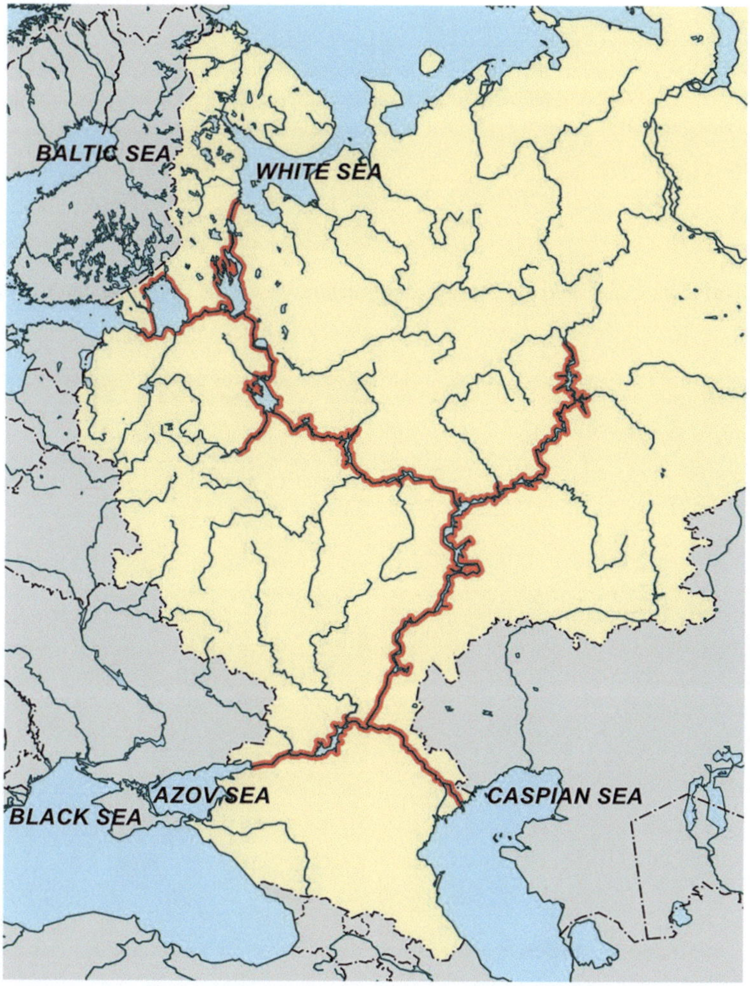

Fig. 3.3 The united system of waterways in the European part of Russia

Caspian animal species from their previously closed basins—the *central corridor*—was thereby created (Fig. 3.2).

The united system of waterways in the European part of Russia that has developed to date, connects the Black Sea, the Sea of Azov, the Caspian Sea, the Baltic Sea, and the White Sea (Fig. 3.3), providing pathways for invasive species exchange.

With the commissioning of the *Suez Canal* in 1869, which linked the Red and the Mediterranean seas (Fig. 1, Introduction; Fig. 3.4), a marine connection between the Atlantic, Indian, and Pacific oceans became established, but together with the new navigation route it had intended to create, a new invasion corridor appeared. Via this surface connection, many species, native to the Red Sea, started penetrating the eastern shores of the Mediterranean Sea, slowly making their way to the north and

Fig. 3.4 (**a**) Suez Canal before reconstruction (satellite view); (**b**) ships pass through the Suez Canal

Fig. 3.5 Suez Canal before (left) and after (right) reconstruction in 2015 (satellite view)

west, eventually reaching the Aegean Sea. These are collectively known under the name Lessepsian immigrants (Por 1978) after Ferdinand De Lesseps, the "father" of the canal. Also, many species from the Red Sea, Indian, and Pacific oceans have been conveyed with ships to various regions of the Atlantic Ocean, Mediterranean and Black seas. At present, the Mediterranean Sea has become the largest recipient basin for invasive species in the world.

Since the very beginning it was recognized that the canal was inadequate for coping with the increasing volume of traffic and size of vessels, and recurrent expansions were carried out, to deepen and widen it. The most recent enlargement of the canal was celebrated in August 2015. This latest expansion, dubbed the "New Suez Canal," added a 72.4 km lane parallel to the existing canal (Fig. 3.5).

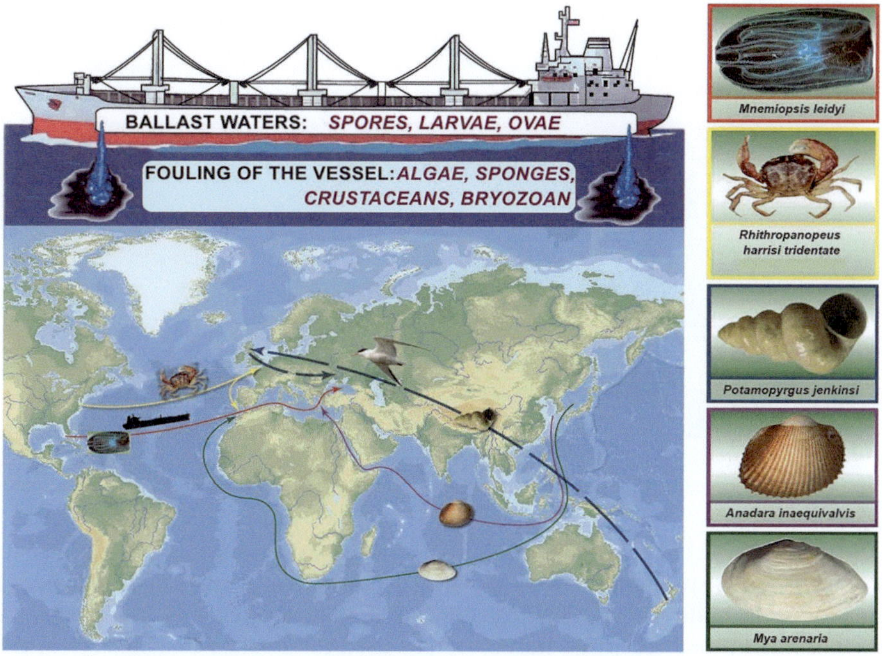

Fig. 3.6 Examples of long-distance transfer of marine species with ships

Among nearly 700 non-native species, currently inhabiting the Mediterranean Sea, about 50% were introduced through the Suez Canal since 1869 (Galil et al. 2015). This is one of the most prominent corridors for invasions of marine species in the world. Molecular methods demonstrated high levels of gene flow between the Red Sea and the Mediterranean seas.

Shipping Currently, the most important vector of aquatic plants and animals invasions is transport via water, inside or on the hull of ships due to the global intensification of navigation. Organisms are transferred as elements of hull fouling communities and in ballast water or sediment in the ballast chambers and tanks. Some examples of global long-distance transfer of organisms to European seas by this way are shown in Fig. 3.6.

Owing to the present-day rates and scale of the cargo traffic, this pathway enables a rapid and virtually global dispersal (Fig. 3.7) of representatives of flora and fauna, including parasites, pathogenic organisms, and toxic algae.

Ship ballast systems, destined for the stabilization of the balance of ships during cargo operations and refueling, appeared in the second half of the nineteenth century. At first, solid ballast was used. In the twentieth century, liquid (water) ballast rapidly became preferred because it is cheaper and more readily available. Ballast is of vital importance for tankers and cargo ships, and some other types of ships unable to navigate when empty. The rapidly growing global demand in oil products in the

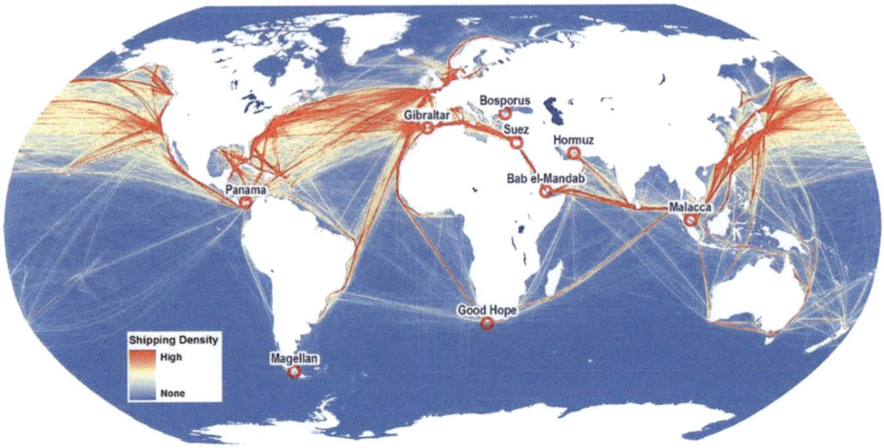

Fig. 3.7 Global maritime shipping routes (https://image.slideserve.com/1095489/maritime-shipping-routes-and-strategic-locations-control-l.jpg)

second half of the twentieth century provided a tremendous increase in the number of tankers being built. Their individual tonnage rose to the point of giving them a leading role in total sea cargo traffic. Ballast used to be taken immediately in the tanks and its volume might comprise up to 40% of cargo capacity. After ballast water release, the tanks may contain about 0.4% (according to some data, up to 1%) of the total mass of carried oil products, mainly heavy fractions; their concentration in the ballast water might reach up to 100 mg l^{-1} and higher, which normally represents a lethal dose to biota (Boltachev et al. 2003). Thus, the ballast water of tankers, which comprise the greater part of the annual water discharge dumped by commercial ships, do not represent a serious danger of biological contamination, but on the other hand, the pollution of the marine environment with oil products derived from the dumping of ballast waters of tankers became a global phenomenon and caused serious ecological problems. In 1954, an International Convention was accepted, accompanied by numerous supplements (OILPOL-54/69) aimed at the prevention of sea pollution with oil. In 1973, it was followed by the stricter International Convention on the prevention of sea pollution from ships, and a Protocol (MARPOL-73/78) supplemented by numerous regional and national regulations. According to MARPOL-73/78, it was required that tankers built after the adoption of this Convention should have their ballast facilities isolated from any oil products, i.e., they should be equipped with separate tanks and ballast systems. This type of ship could exchange its ballast water without any restriction. Soon, this led to an ever-growing number of catastrophic invasions of non-native species of plants and animals, and regulations on intake, replacement, and dumping of ballast waters was adopted. In the inland waters such as the Black and Caspian seas, special restrictions on all operations with ballast waters were enforced. Unfortunately, most of the conventions on ballast waters cause practical problems with navigation safety, technical implementation, and economic expense. For example, according to the International

Convention, any ship entering the Black Sea, has to replace all ballast water with the domestic Black Sea water. According to rules, the removal and replacement should be performed not closer than 30 miles offshore and at the depths no less than 20 m, with an obligatory registration of the time and place of this operation on the ship's log-book. However, with an average velocity of 16 knots, the distance from the Bosporus to say Sevastopol (280 miles) will take about 18 h for modern tankers. Depending of the ship's tonnage, a complete replacement of the ballast water would require more than a day, which is a significant financial expense, especially if the ship have to stop or slow down to perform the operation (Boltachev et al. 2003). Therefore, the ships' owners frequently resort to a partial replacement of the water in the open sea, which provides only a minor avoiding of the species release danger in the recipient basin. About 10 billion tons of ballast water is annually carried across the World Ocean, along with about 3000 different species of hydrobionts (International Marine Organization—IMO 1998).

Inland and semi-enclosed seas of Eurasia are especially vulnerable. Their species structure is conservative because of the long-term isolation from the major oceanic basins and are characterized by low species diversity. Thus, an introduction of even a single alien species may cause a major restructuring of the ecosystem and even its degradation. The semi-enclosed Mediterranean Sea with a salinity, exceeding oceanic, possess the highest number of non-native species (Galil 2007; Zenetos et al. 2005), followed by the Baltic Sea. Among these non-natives, benthic species are overwhelmingly dominating (Gollasch 2006). However, during recent decades, with the above-mentioned introduction of separate tanks for ballast waters, the number of pelagic species has increased in a spectacular way.

Thus, unfavorable conclusion is that marine ecosystems, particularly that of inland seas, are changing at an alarming rate. These changes have accelerated dramatically in the past few decades due to new technologies, shipping intensity increase, global expanding of shipping routes, and increased basin connectivity. These events have accelerated invasions of aquatic species from their native ranges to distant locations. Every day, many non-native species are being transported around the world (Carlton et al. 1999; Carlton 1996).

This process keeps intensifying; new invaders are regularly registered in various marine and freshwater basins. Starting with the 1980s, ballast waters became the top cause for unintentional invasions, with fouling and aquaculture in second rank. According to calculations by Gollasch (2006), yet another non-native species appears in Europe on average every 11 months; in the Mediterranean Sea, this interval is as short as 6 weeks.

Below, the list of ten most harmful invasive species and their impact on ecosystem according to the Global Invasive Species Program evaluation is provided (Table 3.1). One species from the list, i.e., *Vibrio cholerae* (El Tor biovar, which is particularly dangerous for human health), was somehow transported to the Black Sea and the Sea of Azov. Another species—the ctenophore *Mnemiopsis leidyi*—invaded the Black Sea and caused enormous damage there. The Chinese mitten crab *Eriocheir sinensis* penetrated the Black Sea, the Sea of Azov, and, recently, the Caspian Sea.

Table 3.1 Ten most destructive invasive species

Species	Native to	Introduced to	Impacts
Cholera *Vibrio cholerae* (various strains)	Various strains with broad ranges	South America, Gulf of Mexico and other areas	Some cholera epidemics appear to be directly associated with ballast waters. E.g., an epidemic began simultaneously at three separate ports in Peru in 1991, sweeping across South America, affecting more than a million people and killing more than ten thousand by 1994. This strain had been previously reported only in Bangladesh.
Cladocera *Cercopagis pengoi*	Black and Caspian Sea	Baltic Sea, American Lakes	Reproduces rapidly to form large populations that dominate the zooplankton community, clogging fishing nets and trawls, with associated economic impacts
Mitten Crab *Eriocheir sinensis*	Yellow Sea	Western Europe, Baltic Sea, West Coast North America	Undergoes mass migrations for reproduction. Burrows into river banks and dykes causing erosion and siltation. Preys on native fish and invertebrate species, causing local extinctions. Interferes with fishing.
Toxic Algae (Red/Brown/ Green tides) Various species	Various species with broad ranges	Several species have been transferred to new areas in ships' ballast water.	May form harmful algal blooms. Depending on the species, can cause massive kills of marine life through oxygen depletion, release of toxins and/or mucus. Can foul beaches and impact tourism and recreation. Some species may contaminate filter-feeding shellfish and cause fisheries to be closed. Consumption of contaminated shellfish by humans may cause severe illness and death.
Round Goby *Neogobius melanostomus*	Black, Azov and Caspian Seas	Baltic Sea and North America	Highly adaptable and invasive. Increases in numbers and spreads quickly. Competes for food and habitat with native fishes including commercially important species, and preys on their eggs and larvae. Spawns multiple times per season and survives in poor water quality.

(continued)

Table 3.1 (continued)

Species	Native to	Introduced to	Impacts
North American comb jelly *Mnemiopsis leidyi* Aggassiz,1865	Eastern seaboard of the Americas	Black, Azov, Caspian Seas	Reproduces rapidly (self-fertilizing hermaphrodite) under favorable conditions. Feeds excessively on zooplankton and fish eggs. Depletes zooplankton stocks; alters food web and ecosystem function. Contributed significantly to collapse of Black and Azov Sea fisheries in 1990s, with massive economic and social impact. Currently has similar impact in the Caspian Sea.
North Pacific Seastar *Asterias amurensis*	Northern Pacific	Southern Australia	Reproduces in large numbers, reaching "plague" proportions in invaded environments. Feeds on shellfish, including commercially valuable scallop, oyster and clam species.
Zebra Mussel *Dreissena polymorpha*	Eastern Europe (Black Sea)	Western and Northern Europe, including Ireland and Baltic Sea; eastern half of Northern America.	Fouls all available hard surfaces in mass numbers. Displaces native aquatic life. Alters habitat, ecosystem and food web. Causes fouling problems on infrastructure and vessels. Blocks water intake pipes, sluices and irrigation ditches. Economic costs to USA alone of around US$750 mln between 1989 and 2000.
Asian Kelp *Undaria pinnatifida*	Northern Asia	Southern Australia, New Zealand, West Coast of USA, Europe and Argentina	Grows and spreads rapidly, both vegetatively and through dispersal of spores. Displaces native algae and marine life. Alters habitat, ecosystem and food web. May affect commercial shellfish stocks through space competition and alteration of habitat.
European Green Crab *Carcinus maenas*	European Atlantic Coast	Southern Australia, South Africa, USA, Japan	Highly adaptable and invasive. Resistant to predation due to hard shell. Competes with and displaces native crabs and becomes a dominant species in invaded areas. Consumes and depletes wide range of prey species. Alters inter-tidal rocky shore ecosystem.

References

Boltachev AP, Danilyuk ON, Pakhorukov NP (2003) On the expansion of the range and possible consequences of the introduction of the sunfish Lepomis macrochirus (Perciformes, Centrarchidae) in the Black Sea countries // Evolution of marine ecosystems under the influence of intruders and artificial mortality of fauna. Abstracts of the reports. Rostov-on-Don, pp 51–53

Carlton JT (1996) Pattern, process, and prediction in marine invasion ecology. Biol Conserv 78:97–106

Carlton JT, Geller JB, Reaka-Kudla M (1999) Historical extinctions in the sea. Annu Rev Ecol Syst 30:515–538

Galil BS (2007) Loss or gain? Invasive aliens and biodiversity in the Mediterranean Sea. Mar Pollut Bull. https://doi.org/10.1016/j.marpolbul.2006.11.008

Galil BS, Ferdinando Boero Marnie L, Campbell James T, Carlton EC, Fraschetti S, Gollasch S, Hewitt CL, Jelmert A, Macpherson E, Marchini A, McKenzie C, Minchin D, Occhipinti-Ambrogi A, Ojaveer H, Olenin S, Piraino S, Ruiz GM (2015) "Double trouble": the expansion of the Suez Canal and marine bioinvasions in the Mediterranean Sea. Biol Invasions 17:973–976. https://doi.org/10.1007/s10530-014-0778-y

Gollasch S (2006) Overview on introduced aquatic species in European navigational and adjacent waters. Helgol Mar Res 60:84–89

IMO (1998) International Marine Organization Assembly Resolution A.868(20). Guidelines for the Control and Management of Ships' Ballast Water to Minimize the Transfer of Harmful Aquatic Organisms and Pathogens

Jażdżewski K, Konopacka A, Grabowski M (2004) Recent drastic changes in the gammarid fauna (Crustacea, Amphipoda) of the Vistula River deltaic system in Poland caused by alien invaders. Divers Distrib 10(2):81–87. aquaticinvasions/2004/AI_2004_2_1

Meinesz A, Hesse B (1991) « Introduction et invasion de l'algue tropicale Caulerpa taxifolia en Méditerranée nord-occidentale ». link.springer.com›article/. https://doi.org/10.1023/A:1014549500678

Mills EL, Leach JH, Carlton JT, Secor CL (1993) Exotic species in the great lakes: history of biotic crises and anthropogenic introductions. J Great Lakes Res 19(1):1–54

Ojaveer H, Leppäkoski E, Olenin S, Ricciardi A (2002) Ecological impact of Ponto-Caspian invaders in the Baltic Sea, European inland waters and the Great Lakes: an inter-ecosystem. In: Leppäkoski E, Gollasch S, Olenin S (eds) Invasive aquatic species of Europe. distribution, impacts and management. Springer, Dordrecht. https://doi.org/10.1007/978-94-015-9956-6_41

Panov VE, Dgebuadze YY, Shiganova TA, Filippov AA, Minchin D (2007) A risk assessment of biological invasions in the inland waterways of Europe: The Northern Invasion Corridor case study. In: Francesca Gherardi, biological invaders in inland waters: profiles, distribution, and threats. Springer, pp 639–656

Por FD (1978) Lessepsian migration. The influx of Red Sea biota into the Mediterranean by way of the Suez Canal. Springer-Verlag, Berlin Heidelberg New York Publ. New York, pp 88–194

Svetovidov AN (1964) In: Pavlovsky EN (ed) Fishes of the Black Sea. Nauka, Moscow, p 550

Thibaut T, Meinesz A (2004) Cauler-pa taxifolia: 18 years of infestation in the Mediterranean Sea. In: Aquatic invasions in the Black, Caspian, and Mediterranean Seas: The Ctenophores Mnemiopsis leidyi and Beroe in the Ponto-Caspian and other Aquatic Invasions. Springer Netherlands, Dordrecht, pp 287–298

Zenetos A, Çinar ME, Pancucci-Papadopoulou MA, Harmelin JC, Furnari G, Andaloro F, Bellou N, Streftaris N, Zibrowius H (2005) Annotated list of marine alien species in the Mediterranean with records of the worst invasive species. Mediterr Mar Sci 6(2):63–118

Chapter 4
The Black Sea

Abstract Various aspects of original ecosystem state and non-native species invasions in the Black Sea are discussed in details, including history of the Black Sea basin and its biota, main features of the Black Sea physical geography, hydrology and hydrochemistry, ecosystem state before the non-native species invasion, non-native species introduction in the Black Sea, and non-native biota composition. The Black Sea is considered as the main recipient region and, at the same time, a donor of non-native species. Special attention was paid to invasive ctenophores *Mnemiopsis leidyi* and *Beroe ovata* and the role of the latter as a bio-controller of *M. leiduyi* population. Also, the Marmara and Aegean Seas as the nearest neighbors of the Black Sea and the connection route between the Black and Mediterranean Seas were distinguished. Their water exchange with the Black Sea, biota composition, non-native species, and ecosystem state have been considered.

4.1 Historical Insight in the Black Sea Basin and Its Biota

In the mid-Miocene (15–10 million years ago) movements of the earth's crust created the Alps, Carpathians, Balkans, and Caucasus, simultaneously shrank the Tethys Ocean and split it into a series of brackish waters, severed from the ocean. One of them, called the Sarmatian Sea, included the modern Black, Azov, Caspian and Aral Seas (Lakes) (Fig. 4.1a). In the late Miocene-early Pliocene, it get connection with the ocean; salinity increased and marine biota appeared. This basin is called the Meotic Sea (Fig. 4.1b). In the Pliocene, connection with the ocean again completely stopped and an almost fresh enormous Pontic Lake-Sea appeared in place of the saline Meotic Sea (Fig. 4.1c). At that time, the future Black, Azov, and Caspian Seas were still connected with each other in the area of the modern territory of Stavropol, Krasnodar and the North Caucasus. In the Pontic Lake-Sea, marine fauna died out and brackish-water flora and fauna formed. Several scientists (Zenkevich 1963; Sovinskii 1904) supposed that Pontian relicts survive in the Caspian, the Sea of Azov and in low-salinity regions of the western and northwestern Black Sea. Representatives of this fauna today are united under the name "Pontian relics." In the Caspian Sea, especially in the Northern Caspian, they were preserved

T. A. Shiganova et al., *Non-Native Species in the Ponto-Caspian Seas*,
Monographiae Biologicae 96, https://doi.org/10.1007/978-3-032-18640-9_4

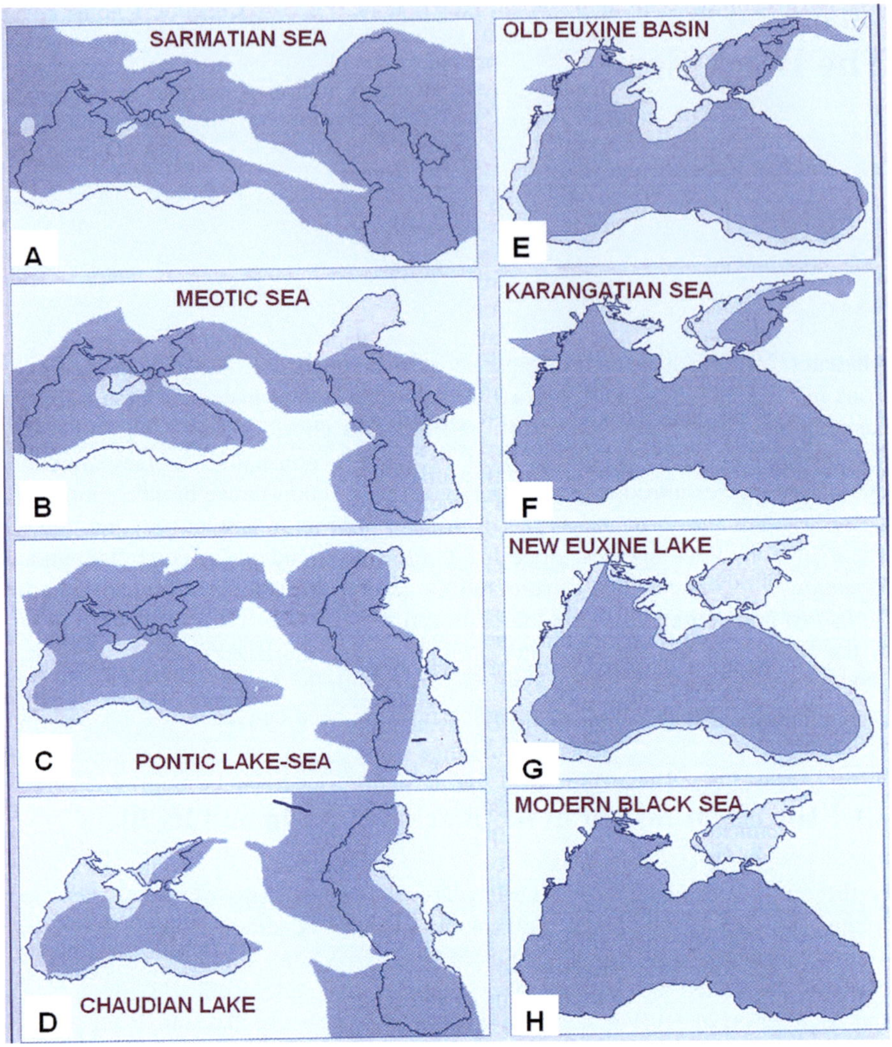

Fig. 4.1 Geological history of the Black Sea. (**a**) Sarmatian Sea, (**b**) Meotic Sea, (**c**) Pontic Lake-Sea, (**d**) Chaudian Lake, (**e**) Old Euxine basin, (**f**) Karangatian Sea, (**g**) New Euxine Lake, (**h**) modern Black Sea

in the best way and were the prototype of the modern Caspian autochthonous fauna (Andrusov 1988). At the end of the Pontian period, the uplift of the North Caucasus separated the Caspian from the Black Sea and Azov basins, although short-term connections still occurred. Next, the warmer and less saline Kuyal'nik Lake was relatively similar to the Pontian Lake.

The Chaudian Lake, the last Pliocene basin in the area (Fig. 4.1d), strongly differed from the Kuyal'nik; its fauna had much in common with the present

Caspian. The Sea of Azov probably did not exist. All this suggests that, almost until the end of the Pleistocene, the Black Sea was low in salinity and separated from the Mediterranean. The Sea of Marmara, linked with the Black Sea via the Bosporus, had no connection with the Mediterranean (Arkhangelskii and Strakhov 1938).

At the beginning of the Quaternary period, salinity and habitats in the Black Sea region began to change. The outline of the basin changed as well. Climate changes became more important than tectonics: a suite of glaciations and warm interglacial (Mindel-Riss, Riss-Wurm, and Holocene) occurred (Muratov et al. 1980). At the end of the Mindel (400000–500,000 years before present—B.P.) the Chaudian Lake, filled with melted water, transformed into the Old Euxine basin (Fig. 4.1e). Its outlines were reminiscent of those of the present-day Black Sea and Sea of Azov and it contained a Pontian fauna. In the southwest, it was connected with the Sea of Marmara, but still isolated from the ocean. In the northeast, the Kuma-Manych depression provided a link with the fresh-water Caspian Sea.

During the Riss-Wurm interglacial (100000–150,000 years B.P.), the Dardanelles first connected with the Mediterranean, creating the Karangatian Sea (Fig. 4.1f) with salinity higher than that of the present-day Black Sea, allowing marine biota to invade. Pontian species were marginalized in river mouths and coastal lagoons. About 18,000–20,000 years B.P., the New Euxine Lake (Fig. 4.1g) again became isolated from the ocean, filled with melted water and freshened. Its marine fauna receded and Pontian biota thrived and penetrated the Ponto-Azov region (Mordukhai-Boltovskoi 1960; Fedorov 1978). This period lasted for about 10,000 years and ended with the formation of the present-day Black Sea (Fig. 4.1h). Around 8000 years B.P. (according to other sources 5000 years B.P.), the Bosporus and Dardanelles opened (Fedorov 1978) resulting in salinization of the Black Sea. Mediterranean species penetrated and dominated the new basin; the Pontian fauna once again retreated to lagoons and river mouths (Mordukhai-Boltovskoi 1960). Table 4.1 summarizes the qualitative variations of the basins' salinity in the Quaternary.

The fauna composition in the modern Black and Azov seas is the result of long-term evolution and is due to both large-scale geological processes and the introduction of species through a system of interconnected seas (via natural straits and

Table 4.1 Salinity transformation in the area of modern Black Sea basin in the Quaternary (after Arkhangelskii and Strakhov 1938; modified)

Basin	Salinity	Glaciation stage
Chaudian Lake-Sea	Low	Mindel glaciation
Old Euxine Lake-Sea	Low	
Karangatian Sea (end of connection with the Mediterranean)	High	Riss-Wurm interglacial Riss glaciation
New Euxine Lake	Low	Wurm glaciation
Ancient Black Sea basin (new connection with the Mediterranean)	Mid	Holocene warming
Modern Black Sea	12‰–22‰	Anthropocene

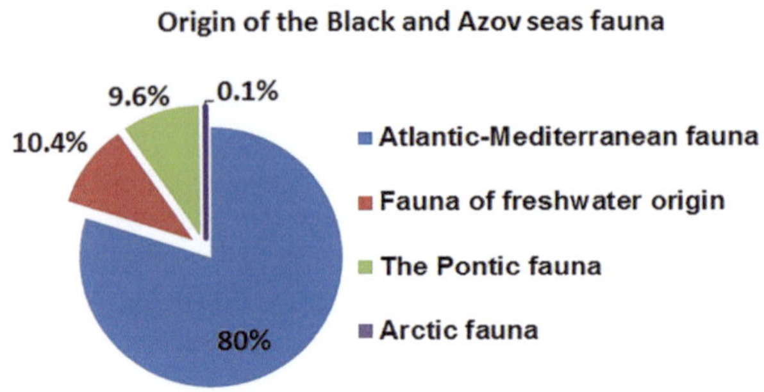

Fig. 4.2 The origin of the modern Black and Azov seas fauna. (After Mordukhai-Boltovskoi 1960, 1969a, b, c; with additions)

artificial canals). The origin of the modern Black and Azov seas free-living multicellular fauna is shown at Fig. 4.2. It is noteworthy that the major part (80%) constitutes of species of marine Atlantic-Mediterranean origin, whereas the contribution of genetically fresh-water species and "Pontian relics" is about 10% each. The proportion of Arctic species is quite insignificant (0.1%).

Pontic relics (Ponto-Caspian species) The presence of representatives of a specific fauna in the waters of the Black and Azov seas basin, which have a common origin with the Caspian fauna, has been known for a long time, but in the twentieth century they attract the attention of biologists and ecologists due to the wide distribution of its individual representatives. Pontic relics or Ponto-Caspian species (according to international terminology adopted in recent years) are of Caspian origin and have settled in Ponto-Azov (the Azov-Black Sea basin, the term by Mordukhai-Boltovsky (1960) relatively recently, most likely in the post-glacial period through the Kumo-Manych depression after significant rise of the Caspian Sea level and its desalination. Mordukhai-boltovskoy (1960) calls representatives of these species "Caspian fauna," meaning a specific complex of species characteristic of the modern Caspian Sea. It consists of crustaceans (mainly amphipods, mysids, Cumans, cladocerans, and some other groups), mollusks (bivalves and gastropods), fish (primarily bovine, herring, and sturgeon), and some other groups, and is considered autochthonous for the Caspian (i.e., originated in it). Species included in this complex according to their systematic position is sharply different from the usual marine (ocean) life. They form special groups of species, or genera, or even families. This is, undoubtedly, due to the antiquity of the Pontic or "Caspian" fauna (in the terminology of Mordukhai-Boltovsky 1960), which for a long time developed in isolation from the ocean. The Caspian fauna, as mentioned above, is endemic to the Ponto-Caspian basin, since it does not live outside of it at all (except for species that have spread in the modern era as a result of human activity). A large number of forms, endemic to the entire Ponto-Caspian basin, is explained by their long-time genetic isolation (lack of kinship with forms of other seas).

In the Black Sea, as a result of increasing salinity at the last stage of geological history, the Caspian fauna or Pontic relics retreated to the lower reaches of rivers and estuaries of the North-Western and Western parts of the Black Sea (from the Danube to the Dnieper), where 95 species belonging to 50 genera habits. In the rivers and estuaries of the Sea of Azov, 54 species belonging to 32 genera resides (Mordukhai-Boltovskoi 1960). There are five main rivers, flowing into Black Sea with their estuaries: the Danube, the Dniester, the Dnieper with the Bug, the Don and the Kuban. All other rivers of the basin are either much smaller or do not form deltas and estuaries. In the Sea of Azov basin, relics are concentrated mostly in the estuaries of rivers, where the water is almost fresh. In each of these areas the Ponto-Caspian inhabiting fauna has some specific species composition. Some species are endemic to certain areas, but not endemic to Ponto-Azov as a whole, but more significant differences in the quantitative ratios of species and groups. Table 4.2 shows the distribution of Ponto-Caspian species by regions (Mordukhai-Boltovskoi 1960).

Table 4.2 Composition and distribution of Ponto-Caspian species in the main areas of the Azov-Black Sea basin (according to Mordukhai-Boltovsky 1960)

	Estuarine systems						
Species	Danube	Dniester	Dnieper -Bug	Don	Kuban	Sea of Azov	Black Sea
Invertebrates							
Hydrozoa							
Cordylophora caspia	+	+	+	+	+	+	–
Moerisia maeotica	+	–	+	+	+	+	+
Polypodium hydriforme	+	–	+	+	+	–	–
Tricladida sp.	+?	–	+	–	–	–	–
Thalassioplanina geniculata	–	–	–	–	–	–	+
Polychaeta							
Hypania invalida	+	+	+	+	–	–	–
Psammoryctes deserticola	–	+	+	–	–	–	–
Archaebdella esmonti	–	+	+	+	–	–	–
Victorella pavida	+	–	–	–	–	–	–?
Bivalvia							
Dreissena polymorpha	+	+	+	+	+	–	–
D. rostriformis bugensis	–	–	+	–	–	–	–
Monodacna pontica	+	+	+	+	–	–	–
M. colorata	–?	–	+	+	+	+	–
Adacna plicata relicta	+	+	+	–	–	–	–
A. laeviuscura fragilis	+	+	+	–	–	–	–
Theodoxus pallasi	+?	–?	+	+	+	+	–
Pyrgulina (=Micromelania) lincta	+	+	+	+	–	–	–
P.(=Caspia) gmelini	–	+	+	+	–	–	–
P.(=Clessiniola) variabilis	–?	+	+	–	–	–	–

(continued)

Table 4.2 (continued)

Species	Estuarine systems					Sea of Azov	Black Sea
	Danube	Dniester	Dnieper -Bug	Don	Kuban		
P.(=C.) triton	–	–	+	–	–	–	–
Cladocera							
Cercopages pengoi	–	+?	+	+	–	+	–
Evadne trigona	–	+	+	+	–	+	–
E.maeotica	+	–	+	+	–	+	–
E. anonyx	–	–	–	–	–	+	–
Copepoda							
Heterocopa caspia	+	+	+	+	+	–	–
Eurytemora grimmi	+	–	+	–	–	–	–
Limnocletodes behningi	+	–	–	+	–	–	–
Amphipoda							
Dikerogammarus haemobaphes	+	+	+	+	+	+?	–
D.villosus	+	+	+	+	–	–	–
D.palmatus	–	–	+	–	–	–	–
Pontogammarus maeoticus	+	+	+	+	+	+	+
P.abbreviatus	+	+	–	–	–	–	–
P.sarsi	+	+	+	+	+	–	–
P.weidemanni	+	–	+	+	–	+?	–
P.robustoides	+	+	+	+	+	–	–
P.aralensis	–	+	+	+	–	–?	–
P.subnudus	–	+	–	–	–	–	–
P.crassus	+	+	+	+?	+	+?	–
P.obesus	+	+	+	+	+	–	–
Niphargoides carpulentus	+	–	+	–	–	–	–
N.compactus	+	–	+	–	–	–	–
N.motasi	+	–	–	–	–	–	–
N.spinicaudatus	+	–	–	–	–	–	–
N.intermedius (=N. borodini intermedius?)	+	+	+	–	–	–	–
Stenogammarus compressus	+	+	+	–	–	–	–
S.similes	+	+	+	+?	–	–	–
S.compresso-similis	+	–	–	–	–	–	–
S. macrurus	+	+	+	+	–	–	–
Chaetogammarus ischnus	+	+	+	+	+	–	–
Ch.placidus	+	–	+	–	–	–	–
Ch.deminutus	+	+	+	–	–	–	–
Ch.warpachchowskyi	+	+	+	–	–	–	–

(continued)

Table 4.2 (continued)

Species	Estuarine systems					Sea of Azov	Black Sea
	Danube	Dniester	Dnieper -Bug	Don	Kuban		
Iphigenella acanthopoda	+	+	–	–	–	–	–
I.andrussowi	+	+	+	–	–	–	–
Gammarus (Iph?) shablensis	+	–	+	–	–	–	–
Cardiophilus baeri	+	–	+	+	+	+	+
Gmelina pusilla	+	+	+	+	–	–	–
G.kusnetzowi	–	–	+	+	+	–	–
G.costata	+	+	–	–	–	–	–
Gmelinopsis tuberculata	–	+	+	+	–	–	–
Amathillina cristata	–	–	+	+	–	–	–
Pandorites podoceroides	–	–	–	+	–	–	–
Corophium nobile	+	+	+	–	–	–	–
C.robustum	+	+	+	+?	–	–	–
C.chelicorne	+	+	+	+	–	–	–
C.maeoticum	+	+	+	+	–	–	–
C.curvispinum (C. devium)	+	–?	+	–	–	–	–
C.sowinskyi (=C. curvispinum)	+	+	+	+	+	–	–
C.mucronatum	–	+	–	–	–	–	–
Mysidacea							
Paramysis baeri	+	+	+	+	+	–	–
P.kessleri sarsi	+	+	+	–	–	–	–
P.ullskyi	+	+	+	+	+	–	–
P.kowalewskyi	+	+	+	+	+	+	–
P.intermedia	+	+	+	+	+	–	–
Katamysis warpachowskyi	+	+	+?	–	–	–	–
Limnomysis benedeni	+	+	+	+	+	–	–
Hemimysis anomala	+	+	+	+	–	+	+
H.serrata	+	–	–	–	–	+	+
Cumacea							
Pterocuma pectinata	+	+	+	+	+	+	–
P.sowinskyi	–	+	–	+	+	–	–
P.rostrata	+	+	+	+	–	–	–
Pseudocuma cercaroides	+	+	+	+	–	–	–
P.laevis	+	–	–	–	–	–	–
Stenocuma tenuicauda	+	–	+	+	–	–	–
S.graciloides	+	+	+	+	–	–	–
Schizorhynchus eudorelloides	+	+	+	+	–	–	–

(continued)

Table 4.2 (continued)

Species	Estuarine systems					Sea of Azov	Black Sea
	Danube	Dniester	Dnieper -Bug	Don	Kuban		
Sch.abbreviatus	+	+	+	+	–	–	–
Caspiocuma campylaspoides	–	+	+	+	–	–	–
Volgocuma telmatophora	–	+	+	+	–	–	–
Iaera sarsi	+	+	+	+	–	–	–
Decapoda							
Astacus leptodacdactylus	+	+	+	+	+	–	–
A.colchicus	–	–	–	–	–	–	–
A.pachypus	–	–	+	+	+	+	+?
Caspihalacarus hyrcanus	+	–	+	+	–	–	–
Total invertebrates	**69 (73?)**	**61 (63?)**	**76 (77?)**	**56 (59?)**	**27**	**16 (20?)**	**6 (7?)**
Pisces							
Acipenser ruthenus	+	+	+	+	–	–	–
A.nudiventris	+	–	+	–?	–?	+	+
A.guldenstadti colchicus	+	+	+	+	+	+	+
A.stellatus	+	+	+	+	+	+	+
Huso huso	+	+	+	+	+	+	+
Alosa (Caspialosa)caspia	+	+	+	+	+	+	+
A.(C)kessleri pontica	+	+	+	+	+	+	+
A.(C)brachnikovi maeotica	+	+	+	+	+	+	+
Clupeonella delicatula	+	+	+	+	+	+	+
C.abrau	–	–	–	–	–	–	–
Pungitus platygaster	+	+	+	+	+	+	+
Lucioperca marina	–	–	+	–	–	–	–
L.volgensis	+	+	+	+	–	–	–
Abramis sapa	+	+	+	+	+	–	–
Gobitis caspia	+	–	–	–	–	–	–
Chalcalburnus chalcoides	+	–?	+	+	+	+	+
Ruttilus frisii	+	+	+	+	+	+	+
Knipowitschia longicaudata	–	–	+	+	+	–	–
Neogobius ratan	–	–	+	+	+	+	+
N.platyrostris	–	–	–	–	+	+	+
N.cephalarges	–	+	–	+	+	+	+
N.kessleri	+	+	+	–	–	–	–
N.syrman	–	+	+	+	+	+	+
N.fluviatilis	+	+	+	+	+	+	+
N.melanostomus	+	+	+	+	+	+	+

(continued)

Table 4.2 (continued)

Species	Estuarine systems					Sea of Azov	Black Sea
	Danube	Dniester	Dnieper -Bug	Don	Kuban		
Mesogobius gymnotrachelus	+	+	+	+	–	–	+
M.batrachocephalus	+	+	+	+	+	+	+
Proterorhinus marmoratus	+	+	+	+	+	+	+
Caspiosoma caspium	–	–	+	+	–?	+	–
Benthophiloides brauneri	–	–	+	–	–	–	–
Bentophilus stellatus	+	+	+	+	+	+	–
B.ctenolepidus magistri	–	–	–	+	+	+	–
Total Pisces	**22**	**21 (22?)**	**27**	**25 (26?)**	**23 (25?)**	**22**	**21 (22?)**
In Total	**91 (95?)**	**82 (85?)**	**103 (104?)**	**81 (85?)**	**50 (52?)**	**38 (42?)**	**27**

In general, the distribution of Ponto-Caspian fauna within the basin resembles the distribution of freshwater fauna: most species inhabit pre-estuarine areas and oligohaline and lower mesohaline zones. However, the Ponto-Caspian fauna is more resistant to salinization than freshwater one, and several dozens of Caspian species live in the upper mesohaline and even polyhaline waters of the Black Sea.

Arctic species The Arctic complex, which penetrated from the north into the southern seas in the postglacial period, is very poorly represented (0.1%, Fig. 4.2) in the Black Sea, mainly by flora species. The reason for the Arctic species depletion, along with the unsuitable salinity, is the absence of deep-sea species because of anoxic conditions below 80–150 m in the entire sea basin.

Atlantic-Mediterranean species The Mediterranean fauna, which inhabits the Black and partially the Sea of Azov, penetrated through the Kumo-Manych system and further east, up to the Aral Sea, despite the weak connection that existed between the Black and Caspian Seas.

The Mediterranean fauna has become significantly poorer in composition as it moves eastward. Already upon reaching the Black Sea, the Mediterranean fauna becomes fourfold poorer, and barely 2.5% of it remains in the Sea of Azov (Zenkevich 1963). The qualitative depletion of biota in the Black Sea primarily captures the most stenohaline systematic groups of the Mediterranean Sea—radiolarians, siphonophores, ctenophores, corals, many groups of crustaceans, especially decapods, all groups of mollusks, especially cephalopods and gastropods, echinoderms, tunicates, and fish. Currently, the total number of described species in the Black Sea is relatively low—3774. Of these, 1.619 species of fungi, algae and higher plants, 1983 species of invertebrates, 168 species of fish, and 4 species of marine mammals (Mordukhai-Boltovsky 1969a, b, c).

4.2 Main Features of the Black Sea Physical Geography, Hydrology, and Hydrochemistry

The main physical and geographical features The modern Black Sea extends between 46°32′–40°35′ N and 27°27′–41°42′ E for 1149 km in zonal and 610 km in meridional direction. Its surface area is 413,488 km^2 and volume is 529,954 km^3. The maximum and average depth is 2243 m and 1197 m, correspondingly. The deep-water part makes up the majority of the basin; shelf is very narrow (except the northwestern part) and the continental slope is very steep (Fig. 4.3). Black Sea almost completely lacks islands and large Crimean Peninsula stands out in the north. On the north, the Kerch and Taman' peninsulas separate the Black Sea from the Sea of Azov via the narrow Kerch Strait. In the southwest, it has access to the Marmara Sea through the Bosphorus and Dardanelles Straits and further into the Mediterranean Sea of the Atlantic Ocean.
The inland position, elongation along the parallel with a small difference in the latitudes of the northern and southern coasts, and a relatively small area determine generally homogeneous climatic conditions over the Black Sea. The main features of the sea climate as a whole are formed under the influence of atmospheric macro-circulation over the Mediterranean region. At the same time, orography and outlines of the coast may produce noticeable local climatic differences at some areas. In general, the Black Sea climate is similar to the Mediterranean one (warm, humid winters, hot and dry summers). Its southeastern part, protected by mountains, is characterized by a humid subtropical climate (abundance of precipitation, warm winters and hot summers).

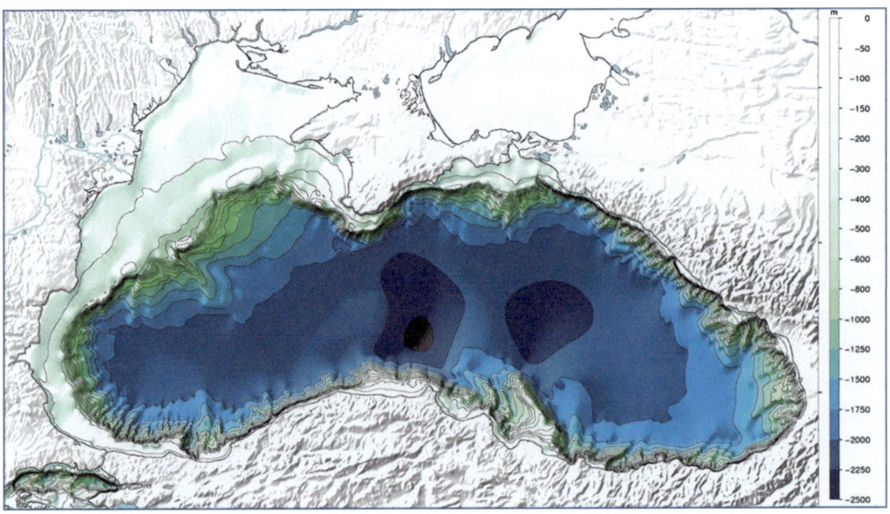

Fig. 4.3 Bathymetric map of the modern Black Sea (https://anapacity.com/content/images/glubina-chernogo-morya-01.jpg)

The main seasonal weather features in the Mediterranean climatic region, and in particular over the Black Sea, are governed by the interaction of the Siberian and The Azores maxima, the Asian minimum and the Mediterranean winter cyclone, the named due to the intensification of cyclonic activity here in winter and its weakening in summer. In winter, northeasterly winds with an average speed of 7–8 m/s predominate over almost the entire sea. Only in the southeastern part of the sea easterly winds with average 5–7 m/s prevail. In general, the air temperature in winter decreases from the open sea to the coast to −2 °C in the northwest, 0 °C in the northeast and to +4–5 °C in the southeast. The minimum air temperature over the open sea is rarely below 0 °C, but in the northwestern part it may reach −4–5 °C.

In summer, due to large-scale pressure pattern (positioning of the Azores maximum and the Asian minimum), northwesterly winds prevail over the entire sea with the average speed over the open sea 3–5 m/s, and over the coast—2–5 m/s. The air temperature above the sea at this time is rather uniform. The average air temperature in August varies from 22 °C in the northwest to 23–24 °C in the west and in the center and up to 24–25 °C in the east of the sea.

The annual river flow into the Black Sea basin is about 346 km^3. Its main part (270 km^3, or 80%) is provided by rivers flowing into the northwestern part of the sea (the biggest ones are Danube, Dniester, Dnieper, Bug).

Hydrology The temperature of the Black Sea's upper layer has a marked seasonality. In winter, water temperature ranges from −0.5 °C in the northwest to about 9–10 °C in the southeast. In summer the surface layer is warmed to between 23 and 26 °C. Spatially, the temperature in general, increases from the northwest to the southeast. However, high resolution satellite data indicate significant spatial inhomogeneity of SST distribution (Fig. 4.4). Thermic frontal zones are apparent in the winter season between the intensively cooling shallow northwestern area and warmer waters of central part, and in spring-summer season between the actively warming waters of the southeastern part and cooler central area (Kazmin 2016).

The vertical distribution of temperature for most of the year is characterized by highest values on the surface, some decrease to the horizons of 60–75 m, from where it rises very slowly with depth and reaches 9.2 °C at the bottom at depths of 2000 m, which is explained by the geothermal influx of heat from the bottom. At the horizons of 75–100 m there is a cold intermediate layer (CIL), where water temperature throughout the year retain values 7–8 °C in open areas of the sea. Seasonal changes in water temperature are manifested up to the horizons of 150–200 m, but most clearly, they are expressed in the upper 50–60-meter layer. The winter cooling and related convection forms an upper mixed layer extending to depths of about 50–100 m, with temperatures at the lower boundary of about 6.5–8 °C.

An average salinity of the Black Sea is rather high (21.90‰). Still, it is approximately half that of the oceans. Currently, the salt budget is relatively stable due to the balance between salt intake with the bottom Bosphorus current intrusion from the Sea of Marmara and removal by the upper Bosphorus current. The distribution of salinity on the sea surface is characterized by its increase from the northwest to the southeast (Fig. 4.5). This is explained by significant river discharge into the

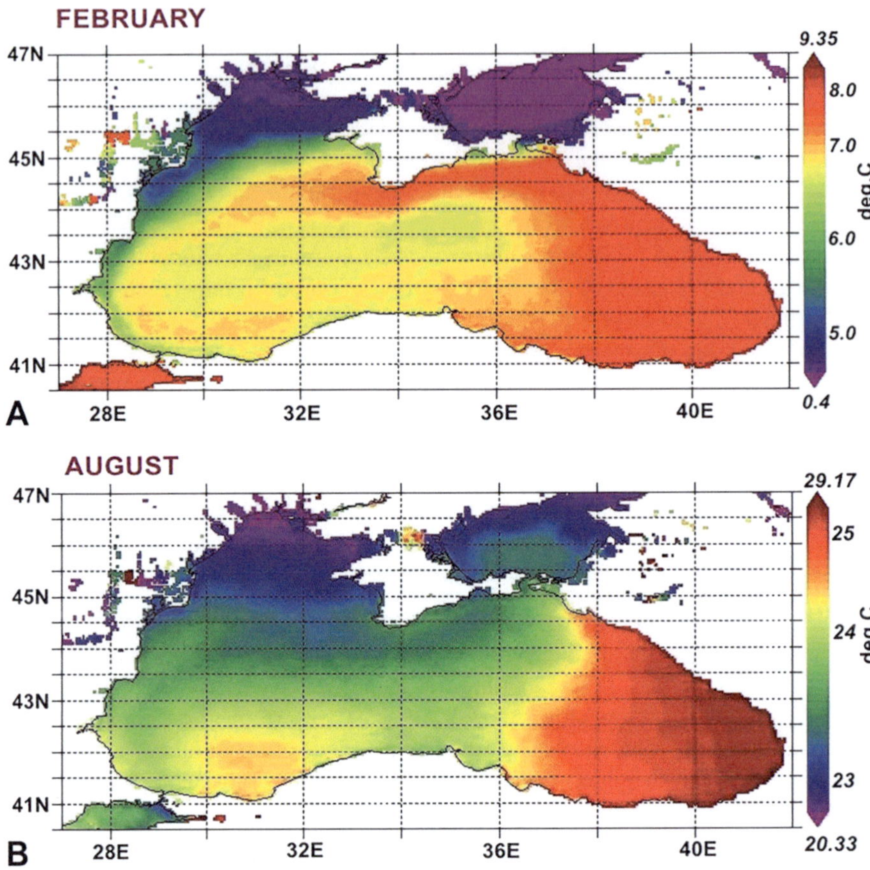

Fig. 4.4 Climatological (1982–2008) sea surface temperature (°C) of the Black Sea in February (**a**) and August (**b**). High resolution AVHRR Pathfinder v.5.0 data (NOAA/NODC)

northwestern part of the sea. Salinity also reduces to 5–10‰ in a narrow coastal zone, near the mouths of large rivers. A slight desalination near the Kerch Strait and the eastern coast of Crimea is caused by the penetration of less salty waters of the Sea of Azov here.

Surface currents in the Black Sea are wind-generated, with the main cyclonic (counterclockwise) stream encircling the entire sea near the coast (Rim current) where its speed may reach 40–50 cm/s. Two large-scale cyclonic gyres (eddies) with flow velocities up to 10 cm/s in the central and up to 25 cm/s in the peripheral regions exist inside this rim (Fig. 4.6). Flows in the Bosphorus are complex, with surface Black Sea water going out and deep saltier water coming in from the Sea of Marmara. Water exchange through the Bosphorus is relatively slow, and a complete recycling of Black Sea waters takes about 2500 years.

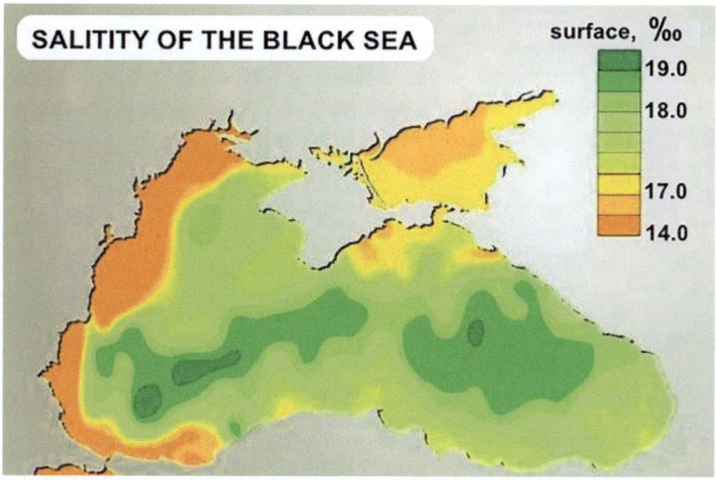

Fig. 4.5 Sea surface salinity of the Black Sea (summer). (Credit Philippe Rekacewicz, UNEP/GRID-Arendal (https://www.grida.no/resources/6539). Modified)

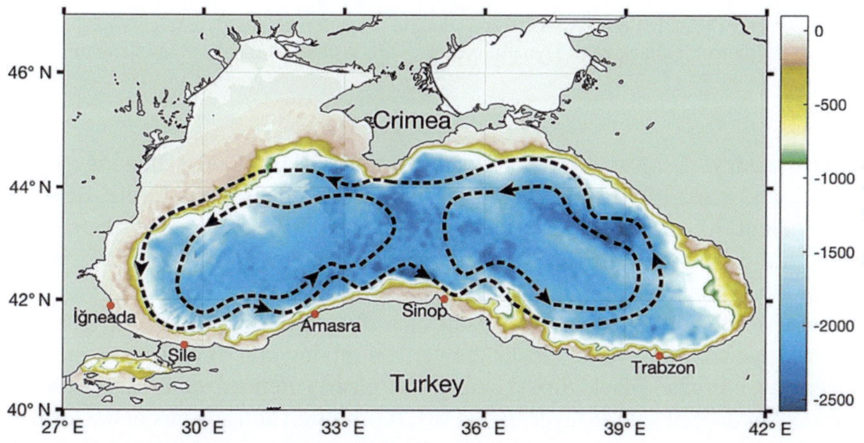

Fig. 4.6 General circulation at the surface of the Black Sea. (https://os.copernicus.org/articles/18/1081/2022/os-18-1081-2022-f01-web.png. Modified)

While climatologically averaged general circulation in the Black Sea looks rather smooth (Fig. 4.6), synoptical situation is quite different and demonstrated pronounced eddy formation activity both in the coastal zone and in the open sea (Fig. 4.7). Intensive eddy formation in the coastal current zone promotes exchange between coastal waters and open sea.

Fig. 4.7 Black Sea satellite optical image. (Captured by the Moderate Resolution Imaging Spectroradiometer (MODIS) on NASA's Aqua satellite (https://visibleearth.nasa.gov/images/90318/turquoise-swirls-in-the-black-sea). Dynamic features are visualized due to phytoplankton spring bloom)

Hydrochemistry A most important feature of the Black Sea is that oxygen is dissolved (and a sea life is made possible) only in the upper water levels. Below a depth of about 70–100 m at the center and 100–150 m near the edge, there is no oxygen; in those reaches the sea is contaminated by hydrogen sulfide (H_2S), which results in a dead zone inhabited only by adapted bacteria (Fig. 4.8).

These specific conditions are the result of a sharp difference in salinity/density (halocline/ pycnocline), which blocks the vertical mixing and oxygen transport into the deep layer of the Black Sea. There is no clearly defined boundary between oxygen and hydrogen sulfide layers in the sea. It is blurred and appears in the form of transitional zone, the position of which depends on the variability of dynamic processes. In the hydrogen sulfide zone of the Black Sea, the content of this gas increases from 0.15 mg/l at the horizon of 150 m to 8.0–10.0 mg/l at the horizon of 1500 m, and deeper it almost stable.

The Black Sea is rich in nutrients. The concentration and distribution of nitrites and nitrates varies widely depending on their consumption by phytoplankton. In autumn and winter, the concentration of these substances is highest in the upper layer; at horizons of 50–75 m it can reach 15 micrograms/l. In summer, they are found only in areas affected by river runoff.

The sea is characterized by a relatively high phosphate content compared to other seas, especially in deep waters. The largest amount of these substances in the upper layer with a thickness of up to 50–100 m is observed in January–February, when

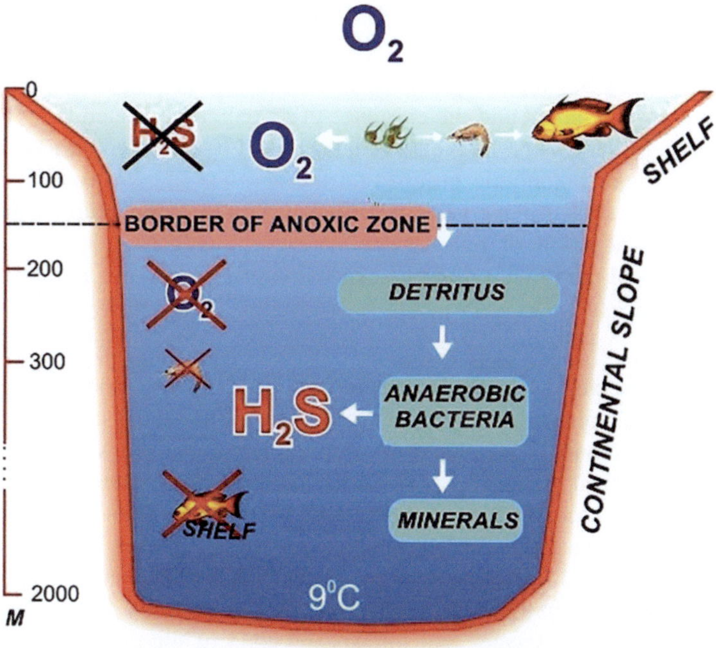

Fig. 4.8 Two-layer vertical structure of the Black Sea basin with the upper life-sustaining oxygenated layer (O_2) and hydrogen sulfide (H_2S) contaminated dead zone

intensive mixing occurs and phosphates enter here from the underlying layers. In May–June, their minimum content is noted, up to complete disappearance, which is associated with the maximum development of phytoplankton. The deep waters of the sea are a zone of accumulation of phosphates, from where they come to the surface. The first maximum locates at the horizon of 130 m, and the highest concentration of phosphates (up to 200–250 mcg/l) is observed at the bottom layers.

The amount of dissolved silicium in the Black Sea water is high throughout the entire thickness of the sea. Its concentration in the surface layer is 900–1300 mcg/l, and in coastal areas it is always higher than in the open sea. The silicium content increases with depth, except for horizons of 10–25 m, where the bulk of the diatoms, consuming this substance, is concentrated. The silicium concentration at the bottom reaches 7000–8000 and even 9000 mcg/l.

The noted hydrological and hydrochemical features of the Black Sea make it one of the most peculiar seas of our planet.

In this section, information from Dobrovolsky and Zalogin (1982), https://www.britannica.com/place/Black-Sea and https://en.wikipedia.org/wiki/Black_Sea have been partially used.

4.3 Ecosystem State Before the Non-native Species Invasion

The contemporary Black Sea's plant and animal life developed under conditions of relatively low salt content and the existence of an oxygen-depleted layer below the surface oxygen-rich waters. Its biota is characterized by a limited variety of species within most major taxonomic groups and the absence of numerous others. This section provides a concise overview of the primary biological conditions prior to the widespread introduction of non-indigenous species, which is linked to climate change, increased maritime traffic, the expansion of aquaculture, and other anthropogenic influences.

Phytoplankton

Phytoplankton serves as a crucial trophic connection within any ecosystem; it influences overall productivity, reflects the health of the ecosystem, and supports the growth and activity of zooplankton communities. In the Black Sea, phytoplankton consists of typical marine, brackish-water, and freshwater species, accounting for up to 40% of total species diversity, along with over 100 benthic–planktonic diatom species. The abundance of brackish-water and freshwater species varies according to the salinity levels in different areas of the Black Sea. The highest concentration of these species is found in the western and northwestern regions. According to Pitsyk (1979), a total of 746 species and intraspecies taxa of phytoplankton have been recorded for the Black Sea (Table 4.3), comprising 525 marine, brackish-water–marine, and brackish-water species; 211 freshwater and freshwater–brackish-water species; and 10 species with unclear salinity preferences.

As documented by Proshkina-Lavrenko (1971), the Black Sea's planktonic algoflora in the early 1970s comprised approximately 342 diatom species (Chrysophyta), around 200 dinoflagellate (Dinoflagellata) species, 90 green algae species, 50 chrizomonad species, along with 32 and 16 species of blue–green and euglena algae, respectively. Such species distribution is typical for inland seas. Diatom algae play the dominant role (Proshkina-Lavrenko 1971), which aligns with the expected balanced functioning of an inland basin.

Within the freshwater species composition, additional assemblages are observed, resulting from the riverine inflow of the Danube, Dnieper, and Dniester rivers in the sea's northwestern region. There, the species inventory is primarily enriched by new Cyanophyta and Euglenophyta species (Nesterova 2006).

Diatoms, a prevalent form of phytoplankton, are extensively distributed and have consistently held significant importance within the Black Sea's trophic system. Their greatest species diversity is found in the shallow waters of coastal areas, as the majority are neritic species. The most commonly represented genera include *Chaetoceros*, *Coscinodiscus*, *Thalassiosira*, *Nitzschia*, and *Rhizosolenia* (Georgieva 1993).

During the early 1990s, the genus *Chaetoceros* contributed the greatest diversity of diatom species, encompassing 40 species and 8 varieties. Following in terms of species count was the genus *Coscinodiscus*, represented in the Black Sea by 23 species, among which *C. janischii*, *C. jonesianus*, and *C. perforates* were the most

Table 4.3 Phytoplankton taxonomic composition (species count) (following Pitsyk 1979)

Division	Marine, brackish-water–marine, and brackish-water species			Freshwater and freshwater–brackish-water species			Total phytoplankton	
	Number of genera	Number of species	%	Number of genera	Number of species	%	Number of genera	Number of species
Bacillariophyta	55	245	46,7	24	93	44,1	64	342
Pyrrophyta	34	193	36,8	6	12	5,7	36	205
Chlorophyta	11	23	4,4	26	66	31,3	36	91
Chrysophyta	25	47	8, 9	2	3	1,4	27	51
Cyanophyta	5	8	1,5	10	24	11,4	12	34
Euglenophyta	3	3	0,6	6	13	6,1	7	17
Xantophyta	3	6	1,1	–	–	–	3	6
Total	136	525	100	74	211	100	185	746

frequently observed. In the northwestern region, common occurrences included members of the genus *Thalassiosira* (*T. parva*, *T. corsuda*, *T. decipiens*, *T. excentrica*), as well as *Cerataulina pelagica*, *Sceletonema costatum*, and *Leptocylindrus danicus*. The family Rhizosoleniaceae was represented by *Proboscia alata* (=*R. alata*), *Dactyliosolen fragilissimus* (*R. fragissima*), and the invasive species *Pseudosolenia calcar-avis* (*R. calcar-avis*). In more desalinated coastal areas, representatives of the genus *Melosira*, such as *M. granulata* and *M. italica*, achieved their peak biomass during spring. The pennate diatom *Thalassionema nitzschioides* was widespread in coastal and neritic waters, alongside various benthic–pelagic species from the genera *Rhabdonema*, *Navicula*, *Amphora*, *Pleurosigma*, *Cymbella*, and *Synedra*.

Dinoflagellata were always the second significant group of planktonic algae with respect to both species composition and abundance. In the 1960–1970s, they were represented by 205 species and intraspecies taxa. Dinoflagellata occur in plankton throughout the year, though in the summer period they reach their highest diversity and the peak of their development is observed in the late spring and the summer, when they produce an intensive outburst of their abundance sometimes causing water "blooming."

The principal genera identified were *Protoperidinium* (36 species), *Ceratium* (15 species), *Dinophysis* (16 species), *Gymnodinium* (16 species), *Amphidinium* (13 species), *Gonyaulax* (12 species), *Gyrodinium* (10 species), *Prorocentrum* (8 species), and *Glenodinium* (species). Neritic species were typically dominant. The most frequently observed large algal species, with cell dimensions exceeding 50 μm, included *Ceratium fusus*, *C. tripos*, *C. inflatus*, *C. furca*, *Peridinium crassipes*, *P. divergens*, *Dinophysis caudata*, *Gonyaulax polyedra*, and *Pyrophacus horologicum*. In contrast, most dinoflagellate species exhibited average sizes ranging from 20 to 40 μm. Prominent among these medium-sized dinoflagellates were *Prorocentrum cordatum*, *Prorocentrum compressum*, *Scrippsiella trochoidea*, *Protoperidinium steinii*, *Gonyaulax spinifera*, *Protoceratium reticulatum*, and *Prorocentrum micans*. The nannoplankton group (5–20 μm) comprised a comparatively smaller number of species (Morozova-Vodyanitskaya 1954; Georgieva 1993).

Throughout the region, beginning as early as the 1980s and 1990s, substantial quantities of small phytoplankton forms such as Prymnesiophyta algae (including Coccolithophora and Silicoflagellata) were present. This group is not characterized by high species diversity (51 species), but it is extensively distributed, and large populations of particular species emerge (Georgieva 1993). In more recent years, the contribution of Prymnesiophyta to the overall phytoplankton abundance has grown considerably; a shift from the Peridiniales algal community to a Prymnesiophyta-dominated one is evident. Currently, intense "blooms" of Coccolithophora in the open areas of the sea are routinely observed during the late spring to early summer months (Mikaelyan 1997).

In the sea's northwestern section, elevated eutrophication levels and alterations in the phytoplankton community structure were documented in the 1970s and 1980s. This is attributed to the increasing discharge of dissolved mineral phosphates and nitrates from major river systems (Bodeanu 1993). The structural changes in phytoplankton within Romanian waters are linked not only to the influx of phosphates and nitrates but also to the rising quantities of organic matter delivered by the Danube River. This trend has continued over the last 20 years, leading to the prolific growth of mixotrophic algae belonging to the Dinoflagellata and Euglenophyta species.

***Noctiluca scintillans* (Macartney) Kofoid & Swezy (Dinophyceae)**
A key contributor to the wet plankton biomass in the Black Sea is the heterotrophic dinoflagellate group, notably the colorless pyrophyte alga *Noctiluca scintillans*. This spherical organism, measuring under 1 mm in diameter and possessing a single flagellum, can enlarge its size by ingesting substantial prey. Functioning as both a saprophyte and a predator, it aligns more closely with zooplankton in its ecological role, trophic position, and dimensions (Sorokin 2002).

In assessments of zooplankton abundance, *N. scintillans* is typically categorized as non-trophic zooplankton. While widely distributed across the World Ocean, its presence is more concentrated in boreal waters (HElCOM 2012).

N. scintillans serves as the primary source of biological light in the Black Sea, meaning research into the sea's bioluminescence is inherently connected to the study of this organism's ecology. The species derives its name from its ability to emit light when stimulated. The brightness of this luminescence is influenced by temperature, with peak intensity recorded between 12 and 15 °C. In contrast, under very low experimental temperatures, the proportion of glowing individuals drops significantly; for instance, in one experiment at 1.6 °C, only 6% of specimens exhibited luminescence (HELCOM 2012).

Reproduction and Seasonal Dynamics of *N. scintillans*
Based on long-term published and original data, the rise in *Noctiluca* abundance following the winter decline begins in April, coinciding with the start of water warming at temperatures around 8 °C (HElCOM 2012).

N. scintillans exhibits both sexual (Fig. 4.9) and vegetative reproduction (Fig. 4.10); the latter method—binary fission of the parent into two daughter organisms—is the most frequent and occurs throughout the year (Harrison et al. 2011).

During asexual reproduction in *N. scintillans*, cell division peaks around midnight and is least active during daylight hours. An alternative reproductive strategy involving zoospore formation occurs simultaneously in no more than 5% of cells, and only from spring to early summer (Harrison et al. 2011), when it achieves its highest quantitative levels. In sexual reproduction, individuals emerge from spores that have rested on the seabed, which accounts for the species' preference for shallow shelf areas.

Variations in the rate of water column warming across years lead to differences in the timing and speed of population growth; consequently, the abundance peak occurs in different calendar periods, though at similar temperature ranges. The highest *N. scintillans* abundance is typically recorded in May or June, when over 50% of the population is found in the upper water layer at temperatures between 12 °C and 15 °C (Zauka 2004). For instance, our data suggest the spring abundance peak likely coincides with the most active phase of sexual reproduction, observed when surface layer temperatures reach 15–20 °C. It appears that the development intensity of *N. scintillans* is influenced by water temperature. If surface layer

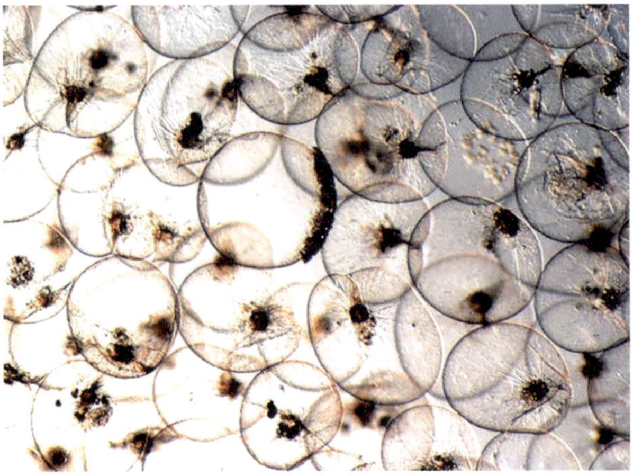

Fig. 4.9 Sexual reproduction of *N. scintillans* (formation of zoospores)

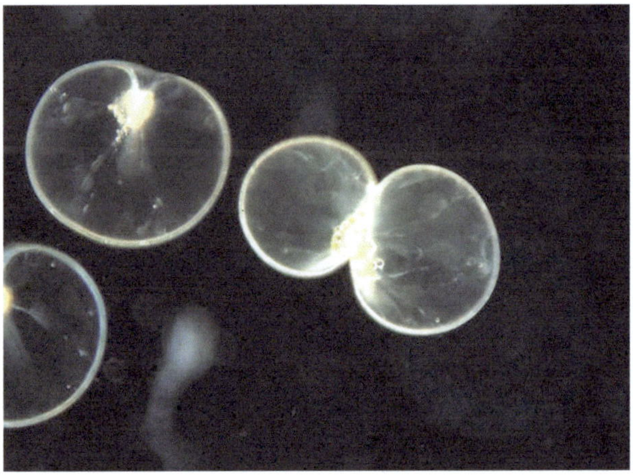

Fig. 4.10 Asexual reproduction of *N. scintillans*

temperatures during the spring reproductive period (May–June) fall below 16 °C or rise above 20 °C, *N. scintillans* abundance remains low. The highest abundances occur at temperatures between 16 and 18 °C.

When surface temperatures rise above 20 °C, the abundance of *N. scintillans* typically declines. For instance, in August, it accounts for just 6% of the total plankton abundance. However, in the coastal waters of the sea's northwestern region, a second surge in *N. scintillans* numbers occurs specifically in August; in certain years, it can represent up to 99% of the wet plankton biomass in these near-shore areas (Petranu 1997). This phenomenon was likely linked to elevated concentrations of its food sources during periods of eutrophication. Based on our observations, during colder years when August surface-layer temperatures were lower, *N. scintillans* also attains high abundance and biomass levels in the northeastern part of the sea, particularly in coastal zones, where it formed the majority of the plankton.

Vertical Distribution of *N. scintillans*

The vertical positioning of *N. scintillans* is closely tied to the temperature conditions and the seasonal characteristics of water column stratification in the Black Sea's open waters. Near coasts and within bays, various local variations emerge in both the vertical placement and the peak values of cell densities. The vertical distribution of *N. scintillans* becomes particularly distinctive during periods of coastal "blooms." The most substantial aggregations of *N. scintillans* are typically found in narrow layers; influenced by wave action, stratification, and temperature, these layers can be situated right at the surface or at depths of 15, 20, and even 30 m. Below these depths, the abundance of *N. scintillans* declines sharply, though in the lower section of the oxygenated zone, within the daytime zooplankton maximum layer, its concentration consistently—yet modestly—increases once more (Vinogradov et al. 1992).

Feeding of *N. scintillans*

N. scintillans possesses a broad dietary range, though its feeding habits within the Black Sea remain inadequately researched. Its primary diet consists of detritus and phytoplankton; however, its intake occasionally incorporates small zooplankton organisms (Mironov 1954; Greze and Fedorina 1979) and microplankton, particularly bacteria. Data from other areas indicate that *N. scintillans* can also ingest copepods, their eggs, nauplii, and fish larvae and eggs (Kiørboe and Titelman 1998).

Interannual Variability of *N. scintillans*

Prior to the 1970s, the overall population of *N. scintillans* in the sea was relatively low. Following the onset of eutrophication in the early 1970s, its numbers in the northwestern region surged sevenfold, reaching up to 9087 individuals per cubic meter. In the coastal waters of the northeastern area (Blue Bay) during April and May, the abundance of *N. scintillans* peaked at 1400 milligrams per cubic meter, accounting for over 60% of the mesoplankton biomass (Pasternak 1983). In the years that followed, substantial year-to-year fluctuations in both its abundance and biomass were recorded.

Mesozooplankton

Mesozooplankton serves as a vital trophic connection within marine ecosystems, shaping the dynamics of all other trophic levels, the overall functioning of the pelagic system, and the movement of carbon across the food web. The effectiveness with which matter and energy are transferred from autotrophic to initial heterotrophic stages of production is influenced by the abundance of zooplankton.

As a consumer of phytoplankton and micro-plankton, zooplankton regulates their populations; simultaneously, it provides the primary food source for small pelagic fish and all pelagic fish larvae, thereby influencing the state of fishery resources.

Lowered water salinity and the presence of a hydrogen sulfide layer below 150–200 m depths have led to a notably reduced diversity of zooplankton species in the Black Sea compared to the Mediterranean Sea. Overall, Black Sea plankton comprises approximately 150 zooplankton species (including 70 brackish-water and freshwater inhabitants from desalinated zones and estuaries) and about

50 meroplankton species (Kovalev et al. 1983). Several plankton groups common in the Mediterranean, such as Doliolids, Salps, Pteropods, Siphonophors, and Euphausiids, are either absent or seldom found in the Black Sea (Mordukhai-Boltovskoi 1969a, b, c). Zooplankton species in the Black Sea vary ecologically and in origin: the most abundant marine types include euryhaline and thermophilic species of Mediterranean origin alongside relatively cold-water species of North Atlantic boreal origin; brackish-water and freshwater species are largely confined to river deltas and estuaries.

Species Composition
Rotifers, the tiniest members of the mesozooplankton, constitute a group with relatively low population numbers. The majority of rotifer species are freshwater organisms; they are found in brackish-water zones, with only a small number being marine species. Within the Black Sea, planktonic rotifers live in the brackish, desalinated regions influenced by river discharge.

Brackish-water species are primarily observed in the northwestern section of the sea and are largely composed of members from the Bosminidae, Daphniidae, and Cercopagidae families.

The dominant species in the coastal community is the neritic warm-water *Pseudevadne tergestina* Claus, 1877.

Cladocera (Diplostraca) exhibit a clear succession of species. The cold-water *Podon leuckartii* (G.O. Sars, 1862) and *Podon intermedius* Lilljeborg, 1853 are present from January to May. Starting in May, a warm-water species community begins to emerge, featuring the most abundant *Penilia avirostris*, *Pleopis polyphemoides*, and *Evadne spinifera* (Table 4.4) (Mordukhai-Boltovskoi 1969a, b, c).

Members of the Cladocera (Diplostraca) are epipelagic organisms found in surface layers down to depths of 10–25 m, both in coastal zones and offshore. Most warm-water species are active from May through November, with peak abundance typically occurring between July and September. During the summer, and more specifically from late summer into early autumn, these warm-water Cladocera contribute substantially to the total biomass of epipelagic plankton, particularly in nearshore regions. The most abundant species is *Penilia avirostris*, which dominates in both offshore and coastal waters. In July, or more commonly in August–September, *Penilia avirostris* significantly contributes to the abundance and biomass of coastal zooplankton (Mordukhai-Boltovskoi 1969a, b, c).

Copepoda
Apart from a few outliers, copepods consistently form the dominant portion of plankton in any given basin, typically accounting for at least 70% of the total planktonic fauna.

Based on their origins, the copepods of the Black Sea can be categorized into several groups: Mediterranean warm-water species like *Centropages ponticus* and hyponeuston species from the Pontellidae family (including *Pontella mediterranea*, *Anomalocera pattersoni*, and the less common *Labidocera brunescens*); Mediterranean eurythermic species, primarily represented by *Paracalanus parvus*, *Acartia*

Table 4.4 Main Cladocera (Diplostraca) species inhabiting the Black Sea

Species	Warm-water	Cold-water	Brackish-water
Fam. Sididae (Baird, 1850)			
Genus Penilia Dana			
P. avirostris Dana	+		
Genus *Diaphanosoma* Fischer 1850			
D. brachyurum (Lievin)			+
Fam. Daphnidae Straus 1820			
Genus *Daphnia* O.F. Muller 1785			
D. longispina O.F. Muller			+
Genus *Moina* Baird			
M. micrura Kurz			+
Fam. Bosminidae G.O.Sars			
Genus *Bosmina* Baird 1845			
B. longirostris (O.F. Muller)			+
Fam. Cercopagidae Giesbrecht, 1892			
Genus *Cercopagis* Krøyer, 1849			
C. pengoi (Ostroumov)			+
Fam. Podonidae M. Boltovskoi			
Genus *Pleopis*			
P. poliphemoides (Leuckart)	+		
P .tergestina (Claus)	+		
Genus *Podon* Lilijeborg			
P .intermedius Lilljeborg		+	
P .leuckarti (G.O. Sars)		+	
Genus *Evadne* Loven			
E. anonyx G.O. Sars			+
E. spinifera P.E. Muller	+		
E. normanni Loven	+		
Genus *Podonevadne* Gibitz			
P. trigona G.O. Sars			+
Genus *Cornigerius* M. Boltovskoi			
C.maeoticus maeoticus (Pengo)			+
Fam. Leptodoridae Lilljeborg 1861			
Genus *Leptodora* Lilljeborg			
L. kindtii (Focke)			+

clausi, and *Oithona nana*; Atlantic boreal cold-water species, such as *Calanus euxinus*, *Pseudocalanus elongatus*, and *Oithona similis*; and brackish-water species, which include the Pontian–Caspian *Heterocopa caspia* and *Eurytemora grimmi* (Mordukhai-Boltovskoi 1969a, b, c).

Furthermore, Black Sea copepods include neritic species. These can be subdivided into those inhabiting the surface layer—largely part of the warm-water assemblage that flourishes in summer above the seasonal thermocline—and those found in deeper layers of the open Black Sea. These different groups may exchange species, particularly during winter and spring (Table 4.5).

The nearshore community includes various copepod species from the Temoridae family, such as *Eurytemora velox* and *E. affinis*, along with members of the Harpacticoida and Tisbidae families. In coastal waters, *Centropages ponticus* and *Paracalanus parvus* are frequently observed, though they can also be found in open-sea environments. Along the Crimean shoreline, the smaller species *Acartia clausi* and *A. latisetosa* are commonly recorded (Greze et al. 1979; Kovalev 1993).

Among Cyclopoidae, the most abundant species are *Oithona nana* and *O. similis*. During the 1980s, *Oithona nana* was a dominant form in epipelagic plankton and was also plentiful in nearshore areas, serving as a valuable food source for fish

Table 4.5 Main species of Copepoda in the Black Sea

Species	Warm-water	Cold-water	Eurythermal	Near-surface	Deep-water	Brackish-water
Order Calanoida G.O. Sars						
Fam. Pontellidae Dana						
Genus *Anomalocera* Templeton						
A. patersoni Templeton			+	+		
Genus *Pontelia* Dana						
P. mediterranea Claus	+			+		
Genus *Labidocera* Lubbock						
L. brunescens Czernjavsky	+			+		
Fam. Calanidae Dana						
Genus *Calanus*						
C. euxinus Hulsemann		+			+	
Fam. Centropagidae Giesbrecht						
Genus *Centropages* Kroyer						
C. ponticus Karaw	+			+		
Fam. Paracalanidae Giesbrecht						
Genus *Paracalanus* Boeck						
P. parvus			+			
Fam. Diaptomidae						
Genus *Diaptomus* Westwood						
D. gracilis Sars						+
D. salinus E.Daday						+

(continued)

Table 4.5 (continued)

Species	Warm-water	Cold-water	Eurythermal	Near-surface	Deep-water	Brackish-water
Fam. Pseudocalanidae Sars G.O.						
Genus *Pseudocalanus* Boeck						
P. elongatus Boeck		+				
Fam. Acartiidae Sars						
Genus *Acartia* Dana						
A. clausi Giesbrecht			+			
Genus Paracartia T. Scott						
P. latisectosa (Kricz)	+					
Genus *Acanthacartia* Steuer						
A. italica Steuer	+					
Fam. Temoridae Giesbrecht, 1893						
Genus *Eurytemora* Giesbrecht						
E. velox Lilljeborg						+
E. affinis (Poppe)						+
Genus *Heterocope* Sars						
H. caspia Sars						+
Order Cyclopoida Sars						
Genus *Oithona* Baird						
Oithona nana Giesbr			+			
O. similis Claus		+				
Fam. Cyclopinidae Sars G.O.						
Genus *Cyclopina* Claus						
C. gracilis Claus						
Fam.Cyclopidae Rafinesque						
Genus *Euryte* Philippi						
E.longicanda Philippi						
Fam. Monstrillidae Dana						
Genus *Monstrilla* Dana						
M. grandis Giesbr.				+		
M. helgolandica Claus				+		
M. ostroumowi Karaw				+		
Order Harpacticoida						
Fam.Tisbidae (Stebbing)						
Tisbe furcata (Baird)			+			

larvae. In recent decades, this species has nearly vanished, while *O. similis* has grown significantly more common (Kovalev et al. 1993; Petranu 1997).

Class Appendicularia
***Oikopleura (Vexillaria) dioica* Fol, 1872**

Within the Black Sea, the Appendicularia group is represented by a solitary species -*Oikopleura dioica*. It frequently attains significant population densities both in coastal zones, particularly during the spring reproductive season, and in open waters within the thermocline layer (Zaika and Andryushchenko 1969). Along the Crimean shores, it exhibits two distinct abundance peaks—one in May–June and another in August; in the coastal waters of the sea's northeastern region, these peaks are noted in April–May and again in August (Pasternak 1983).

Phylum Chaethognata
***Parasagitta setosa*, (J. Müller, 1847)**

A key element among predatory zooplankton is the holoplanktonic species *Parasagitta setosa* (J. Müller, 1847), a member of the Chaetognatha. The small, nearshore benthic form *Spadella cephaloptera*, which inhabits macrophyte beds, is of lesser importance: prior to 1960, it was frequently observed in Black Sea bays from May through November (Mordukhai-Boltovskoi 1969a, b, c).

Parasagitta setosa, representing a eurythermal assemblage, occupies open waters. During daylight hours, it remains below the cold intermediate layer, ascending to surface layers at night. The vertical distribution of selected chaetognaths correlates with the vertical oxygen profile (Greze 1989; Vinogradov et al. 1992; Kovalev 1993). Its reproductive period extends from June to November, with juveniles favoring surface waters above the thermocline (Vinogradov et al. 1992).

Seasonal Variability of Mesozooplankton

During the 1980s in the Black Sea, the seasonal variation in zooplankton biomass exhibited two distinct peaks, as documented by Vinogradov et al. (1989) and Shiganova et al. (2014a, b). The initial peak, linked to the spring phytoplankton bloom, occurred from April to May and was initiated by the development of Copepoda. The reproduction of cold-water and eurythermic Copepoda species—such as *Calanus euxinus*, *Pseudocalanus elongatus*, *Oithona similis*, and *Paracalanus parvus*—along with the emergence of their eggs, nauplii, and copepodites, contributed to the early rise in zooplankton biomass during spring and early summer. Concurrently, eurythermic species like *Acartia clausi* and *Oikopleura dioica*, as well as warm-water species such as the Cladocera (Diplostraca) *Pleopis polyphemoides*, were also developing. The biomass of these species typically reached its first maximum by April, influenced by water temperature and the availability of phytoplankton as food. In coastal waters by May or early June—depending on the preceding months' temperatures—the reproduction of cold-water zooplankton species declined or ceased, while eurythermic copepods (*Acartia clausi*, *Paracalanus parvus*, and *Oithona nana*) continued to reproduce, as noted by Pasternak (1983), Kovalev et al. (1993), and Vinogradov et al. (1992).

In the Black Sea during the 1980s, seasonal changes in zooplankton biomass showed two clear peaks, as reported by Vinogradov et al. (1989) and Shiganova et al. (2014a, b). The first peak, associated with the spring phytoplankton bloom, took place from April to May and was driven by the growth of Copepoda. The reproduction of cold-water and eurythermic Copepoda species—including *Calanus euxinus*, *Pseudocalanus elongatus*, *Oithona similis*, and *Paracalanus parvus*—together with the hatching of their eggs, nauplii, and copepodites, led to the initial increase in zooplankton biomass in spring and early summer. At the same time, eurythermic species such as *Acartia clausi* and *Oikopleura dioica*, along with warm-water species like the Cladocera (Diplostraca) *Pleopis polyphemoides*, were also developing. Biomass for these groups generally reached its first high point by April, affected by water temperature and phytoplankton food supply. In coastal areas by May or early June—depending on temperatures in the preceding months—reproduction of cold-water zooplankton species slowed or stopped, while eurythermic copepods (*Acartia clausi*, *Paracalanus parvus*, and *Oithona nana*) kept reproducing, according to Pasternak (1983), Kovalev et al. (1993), and Vinogradov et al. (1992).

The Cladocera (Diplostraca) assemblage in the Black Sea is composed of broadly distributed marine species and demonstrates a distinct seasonal pattern. Between January and May, cold-water species like *Podon leuckarti* and *Podon intermedius* are observed. Following this, a warm-water species complex begins to appear in May, with *Penilia avirostris*, *Pleopis polyphemoides*, and *Evadne spinifera* being the predominant members (Mordukhai-Boltovskoi 1969a, b, c).

A principal predatory zooplankton species is the holoplanktonic Chaetognatha *Parasagitta setosa*. This organism belongs to the eurythermic complex and is frequently encountered in pelagic waters. It stays below the cold intermediate layer during daylight hours and migrates upward to the surface layer at night. Its breeding season spans from June to November. Juvenile specimens reside in the surface layer, above the thermocline (Vinogradov et al. 1992; Kovalev et al. 1993). *P. setosa* fulfills a significant role in the trophic network, functioning as a predator that ingests roughly 85% of its body mass daily. Its prey consists of copepods, *Oikopleura dioica*, and larger organisms like small pelagic fish (Steele 1970). Simultaneously, *P. setosa* acts as a crucial dietary component for pelagic fish (Shiganova 2009).

A secondary peak in zooplankton biomass occurred from August to September. This interval was characterized by the expansion of warm-water, surface-dwelling species, including *Penilia avirostris*, various other Cladocera species, and the copepod *Centopages ponticus*. It also aligned with the continued growth of eurythermic copepods such as *Paracalanus parvus*, *Acartia clausi*, *Oithona nana*, and *O. similis*, alongside meroplanktonic larvae derived from benthic species. Before the incursion of *M. leidyi* into the Black Sea, the late summer zooplankton maximum often exceeded the spring peak in scale, particularly in nearshore areas (Pasternak 1983; Kovalev et al. 1993; Vinogradov et al. 1992).

Table 4.6 Inventory of Hydrozoa species present in the Black Sea that exhibit a medusa life stage

Species	Marine	Brackish-water
Rathkea octopunctata (Sars)	+	
Corymorpha nutans Sars	+	
Cladonema radiatum Dujardin	+	
Coryne tubulosa Sars	+	
Obelia longissima (Pallas)	+	
Blackfordia virginica Mayer (non-native species)		+
Olindias inexpectata (Ostroumov)		+
Moerisia maeotica (Ostroumov)		+

Gelatinous Macroplankton

Class Hydrozoa

Hydrozoa, the most basic Coelenterata, consist of both solitary and colonial polyps and medusas. The majority of Hydrozoa exhibit metagenesis within their life cycles. Medusas are part of the plankton and reproduce sexually. Table 4.6 enumerates species that possess a planktonic medusae stage. Hydromedusae are predominantly found in the northwestern section of the Black Sea. In areas with reduced salinity, population surges may be linked to the invasive *Blackfordia virginica* and the Pontian-Caspian species *Moerisia maeotica*. During spring, *Obelia longissima* and *Coryne tubulosa* can occasionally be plentiful (based on authors' observations from May 1997).

Class Scyphozoa

Fam. Ulmaridae

Aurelia aurita (L., 1758)

Two indigenous species of Scyphomedusae inhabit the Black Sea: *Rhizostoma pulmo* (Macri, 1778) and *Aurelia aurita* (L., 1758). Nevertheless, recent genetic studies indicate that the Black Sea *Aurelia aurita* is part of the clade Borealis (Ramsak et al. 2012) and should be designated as *Aurelia* sp. pending the completion of further molecular analyses in other regions.

Aurelia sp. exhibits an alternation between sexual and asexual reproduction, known as Metagenesis, and progresses through four distinct developmental stages—medusa, planula, scyphistoma, and ephyra—each with unique morphological characteristics and lifestyles (Mironov 1967) (Fig. 4.11).

The planula larva forms within specialized oral lobes of the female before being released into the water, where it remains for approximately 1 week before settling on the seabed.

From the planula develops the scyphistoma, a polyp-like stage commonly found in shallow zones. This phase can persist for more than 3 years, influenced by food availability and temperature. The scyphistoma subsequently gives rise to ephyrae. In the Black Sea, ephyrae emerge during spring and late autumn. Consequently, *Aurelia* sp. in this region exhibits two distinct generations: a winter generation, which develops from November–December through May, and a spring generation,

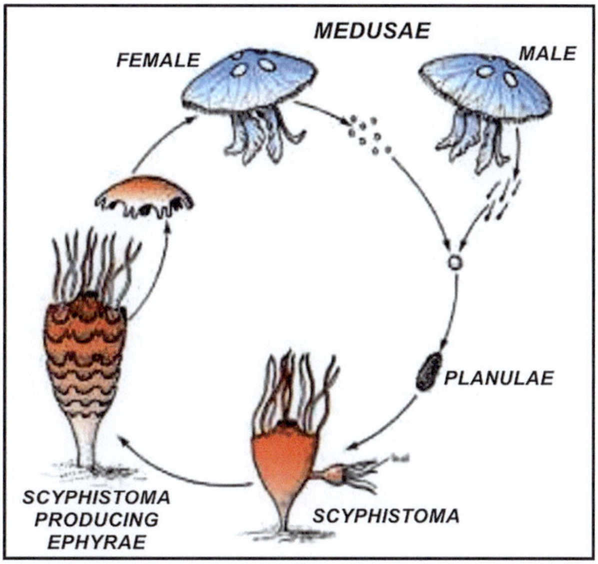

Fig. 4.11 Life cycle of *Aurelia aurita*. (http://jellieszone.com/scyphomedusae.html; modified by authors)

developing from April–May until August–September. According to our observations, the autumn generation continues developing until April, while the spring generation initiates in May.

During spring, very small medusae (3–8 mm) develop from the ephyrae. By May, the average biomass of these juvenile medusae remains relatively low (Fig. 4.12), with larger individuals (exceeding 100 mm) from the previous generation accounting for only 3–6%. The biomass of the spring medusa generation rises steadily until July, though smaller medusae still constitute a minor portion of the population. Growth of *Aurelia* sp. continues into August, with average sizes reaching 70 ± 29 mm and biomass increasing, while small individuals gradually disappear. However, by October, the winter generation of medusae begins to develop, with both size and biomass rising across all areas (Fig. 4.12), and scyphistomae start producing ephyrae once more (Shiganova 2009).

Aurelia sp. ranks among the most widespread and numerous Scyphozoa species, occupying near-shore marine environments across all oceans from 70° N to 40° S. This jellyfish can endure temperature ranges from −0.6 to 31 °C and survive salinity fluctuations from levels exceeding typical oceanic conditions—such as 43‰ in the Red Sea—down to 4–5‰, as observed in the Gulf of Finland and Bothnian Bay within the Baltic Sea. In the Black Sea, *Aurelia* sp. is found at temperatures between 4 and 27 °C and salinities of 10–19‰ (Annensky 2009). This species also contributed substantially to adverse ecosystem shifts prior to the arrival of *M. leidyi*; accordingly, it will be examined more closely in the following discussion.

The substantial increase in the *Aurelia* sp. population emerged as a factor contributing to the degradation of the Black Sea ecosystem, even prior to the arrival of the ctenophore *Mnemiopsis leidyi* (Mironov 1971; Gomoyu and Kupriyanov

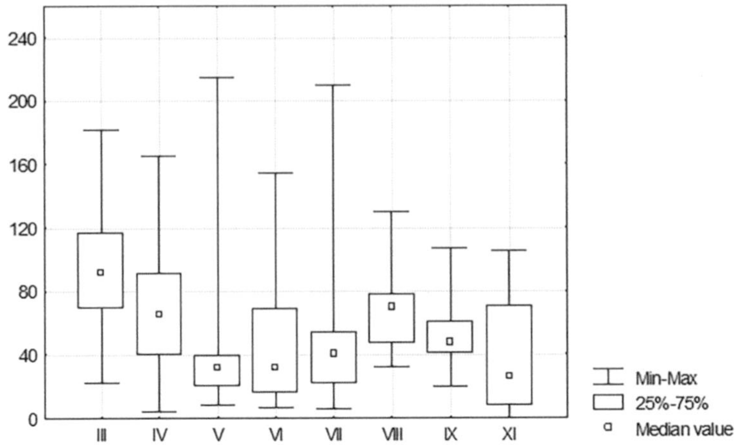

Fig. 4.12 Seasonal changes in the bell diameter (mm) of *Aurelia* sp. medusae

1980; Annensky 2009; Zaitsev and Polishchuk 1984; Lebedeva and Shushkina 1991). Their preference for cooler waters can be understood by noting that *Aurelia aurita* is more prevalent in boreal regions and is thus considered a cryophilic species, despite its broad geographic range across various climatic zones.

Assessment of the ecological function of medusae in the Black Sea relied on balance data and information regarding medusa standing stock. The food requirements of medusae account for 9–13% of total primary production and can be met at a rate of 95–100% annually; in this process, medusae consume 34–67% of the total mesozooplankton production or 47–90% of copepod production (Annensky 2009). These estimates indicate a significant adverse effect of the enlarged *Aurelia* sp. population on edible zooplankton.

Given that this species does not rely exclusively on predation, it can also be classified as sestofagous. In this context, however, the role of medusae in regulating the primary constituent of particulate matter—detritus—is minimal. This is because the amount they ingest is roughly equivalent to the portion of their diet that is not assimilated.

Fam. Rhizostomidae
Gen. Rhizostoma Cuvier
***Rhizostoma pulmo* (Macri)**

Rhizostoma pulmo (Macri 1778) is another member of the Class Scyphozoa. This species is relatively abundant in the Black Sea, primarily occupying coastal zones and sometimes drifting into open waters via currents (Zaitztev et al. 2007).

Similar to most scyphozoans, the life cycle of *R. pulmo* (Fig. 4.13) features a free-swimming planula stage, a sessile and asexually reproducing polyp (scyphistoma), and a free-swimming ephyra that matures into a medusa. In mature females, the brown-colored gonads hold ripe eggs with a distinct egg membrane. Male gonads appear whitish-blue and contain motile sperm. The sex of adult medusae can be

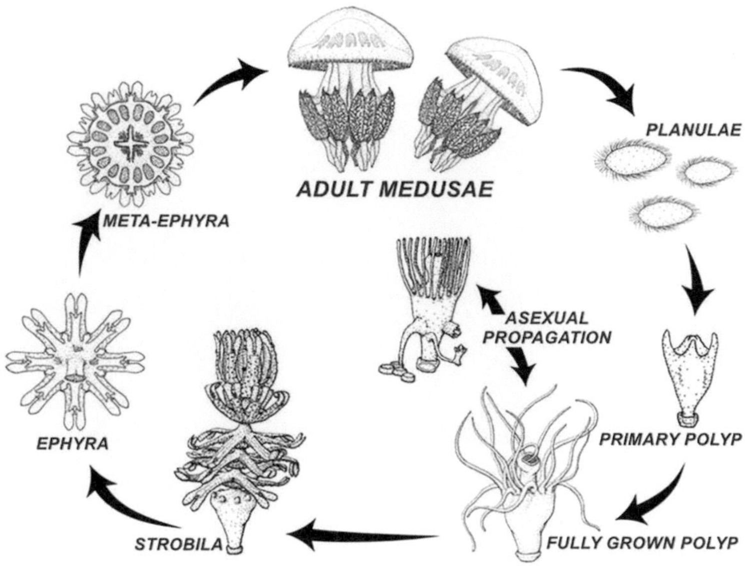

Fig. 4.13 Life cycle schematic of the *Rhizostoma pulmo* jellyfish. (After Fuentes et al. 2011; modified by authors)

identified in free-swimming individuals because the color of the gonads, visible through the bell, shows sexual dimorphism. Once released, the eggs sink to the seafloor and develop into planulae within 2 days.

The free-swimming phase continues for 1–5 days, concluding when the planula descends to the seabed and attaches itself. Within 2 days of settling, scyphistomae develop from the planula. The lowest temperatures suitable for scyphistomae growth are 15 °C in summer and 10 °C in winter. Over 2 years, it can grow to a maximum height of 2.3 mm and a width of 1.6 mm.

Asexual reproduction involves forming podocysts, which then give rise to scyphistoma, undergo strobilation, and release viable ephyra scyphistomae. Scyphistoma generates the highest number of podocysts when water temperatures remain at or above 20 °C. If the water is cooler, production rates drop considerably.

Type Ctenophore
Order Cydippidea
Fam.Pleurobrachiidae
***Pleurobrachia pileus* O. Muller**

Pleurobrachia represents the most ancient group within the ctenophore order Cydippidea. Until the 1980s, *Pleurobrachia pileus* was recognized as the sole ctenophore species inhabiting the Black Sea (Fig. 4.14). These organisms are relatively small; within the Black Sea, their body length typically varies between 5 and 25 mm.

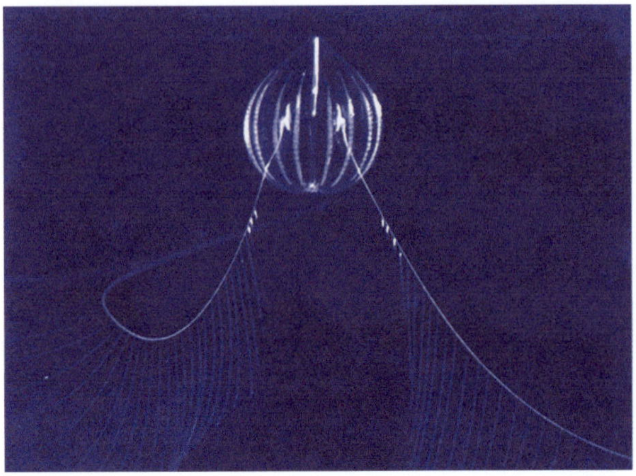

Fig. 4.14 *Pleurobrachia pileus* with extended tentacles

Ctenophores possess transparent bodies with oval or spherical forms. Positioned at the body's far end is a narrow, slit-like mouth, which typically remains shut. During feeding, the mouth of *Pleurobrachia* opens widely. Like all ctenophores, *Pleurobrachia pileus* has eight rows of swimming combs that start some distance from the aboral pole. These comb rows can extend meridionally up to two-thirds of the creature's total length. *P. pileus* propels itself using these swimming combs, moving with its oral end forward. The aboral organ is found at the opposite end. Flanking it on both sides are two tentacular bulbs, into which the ctenophore can withdraw its tentacles. *P. pileus* is capable of moving with its tentacles fully retracted. When extended, the tentacles can stretch to roughly twenty times the animal's own length. The outstretched tentacles display a series of many long lateral filaments. Their surface is covered with sticky cells called colloblasts, which assist the animal in capturing prey.

Spatial Distribution of *P. pileus*
Pleurobrachia pileus thrives in cold-water environments; within the Black Sea, it is classified as a bathypelagic species and primarily inhabits the open ocean beneath the thermocline layer. Prior to the emergence of *M. leidyi*, the significance of *P. pileus* as a predator of zooplankton was notably substantial within the planktonic ecosystem.

Seasonal Distribution of *P. pileus*
Based on our field observations, the highest seasonal abundance of *P. pileus* occurs during its reproductive periods in spring (late March–April) and again in August–September (Fig. 4.15), aligning with findings from other studies (Gruzov et al. 1994; Mutlu and Bingel 1999). The average individual size is smallest during the peak spring reproduction, increases over the summer, and reaches a maximum in July. By August, a new cohort of small individuals appears while larger ones persist, resulting in a population composed of two generations and a corresponding two-size structure. The biomass of *P. pileus* rises through the summer months as the ctenophores grow in size.

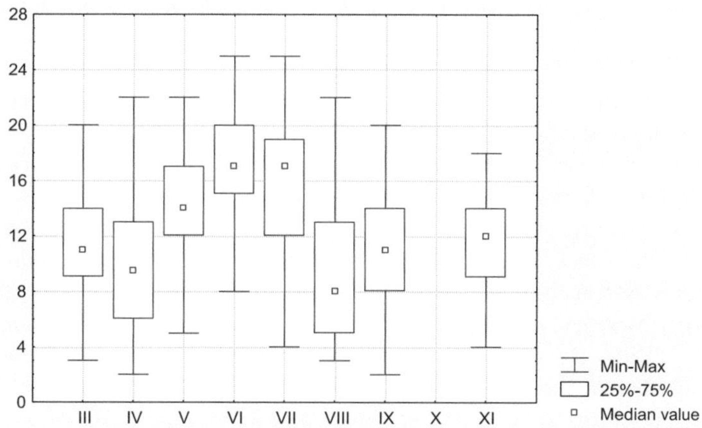

Fig. 4.15 Seasonal variations of *Pleurobrachia pileus* size. (After Shiganova 2009)

Feeding and Functional Role of *P. pileus*
P. pileus is classified as a zooplankton predator. In the cold intermediate layer, this species, together with *P. setosa*, establishes powerful predatory connections that can substantially adjust material flows across the ecosystem's trophic levels (Gruzov et al. 1994).

In the Black Sea, the diet of *P. pileus* is composed of Copepoda (90%), Cladocera (1%), Mollusca (1%), and fish eggs and larvae (1%). The five principal copepod species found in the gut of *P. pileus* are *Calanus euxinus* (39%), *Pseudocalanus elongatus* (30%), *Acartia clausi* (28%), *Oithona similis* (2%), and *Paracalanus parvus* (1%) (Mutlu and Bingel 1999).

Ichthyofauna
The Black Sea's ichthyofauna comprises approximately 200 species and subspecies, a number three times lower than that found in the Mediterranean Sea (Svetovidov 1964; Rass 1993a, b; Boltachev 2003). Presently, four principal groups of fish are recognized based on their origin. The majority of the fish population, around 140 species and subspecies, consists of true marine fishes. This group itself is divided into two subgroups: Mediterranean immigrants, which include subtropical and tropical euryhaline species, and boreal Atlantic relics.

The warm-water community of species originating from the Mediterranean primarily occupies the upper, warmer layer. Pelagic Mediterranean fish that permanently reside in the Sea of Marmara and the Aegean Sea include large, commercially important predatory species such as the Atlantic horse mackerel *Trachurus trachurus* (Linnaeus, 1758), the Mediterranean horse mackerel *Trachurus mediterraneus* (Steindachner, 1868), the Atlantic bonito *Sarda sarda* (Bloch, 1793), and the bluefish *Pomatomus saltatrix* (Linnaeus, 1766). These fish species spawn within the Black Sea (Svetovidov 1964). The Atlantic mackerel *Scomber scombrus* (Linnaeus, 1758) and the Atlantic chub mackerel *Scomber colias* Gmelin,

1789 move into the Black Sea to feed. The swordfish *Xiphias gladius* Linnaeus, 1758 occasionally migrates into the northwestern and western regions of the sea; sporadic spawning events have even been recorded (Gordina and Bagnyukova 1992). The Atlantic bluefin tuna *Thunnus thynnus* (Linnaeus, 1758) is a regular presence in the Black Sea (Rass 1994).

Over 60 species are permanent residents of the Black Sea (Vodyaniysky 1930; Zenkevich 1963; Svetovidov 1964). The primary commercial species are small pelagic planktivorous fish. The most prevalent among these are two anchovy subspecies: the Black Sea anchovy (*Engraulis encrasicolus* Alexandrov, 1927) and the Azov Sea anchovy (*E. encrasicolus maeoticus* Pusanov, 1936). Currently, these two subspecies are classified as a single species, *Engraulis encrasicolus* (Linnaeus, 1758). However, the Black Sea anchovy resides permanently in the Black Sea, whereas the Azov anchovy migrates to the Sea of Azov in spring for spawning and returns to the Black Sea in autumn for wintering. Based on registered catch volumes, the Black Sea anchovy is the foundational commercial fish of the region. During the 1980s, anchovy constituted the largest portion of the total catch for all countries, according to data from the General Fisheries Commission for the Mediterranean (GFCM).

The Black Sea horse mackerel *Trachurus mediterraneus* (Steindachner, 1868), a commercially significant species, preys on zooplankton and small fish. It comprises three distinct local populations—northern, eastern, and southwestern—which vary in their geographic distribution, spawning seasons, growth rates, and migratory patterns. The northern population inhabits the vicinity of the Kerch Strait. The western population is found along the southwestern Black Sea coasts and the Bosporus area, migrating in winter to the Sea of Marmara or the northern Turkish shores. The eastern population's range stretches from the Kerch region westward, following the Caucasian coastline to Turkey's western coast. During winter, mackerel stay near the shore at depths of up to 100 m. The largest individuals, reaching about 30 cm, belong to the southwestern population (Zuev and Mel'nikova 2003).

Within the group of fish that reside year-round in the Black Sea, several notable species include *Scomber australasicus* Cuvier, 1832, the gray mullet *Chelon saliens* (Risso, 1810), the golden mullet *Chelon auratus* (Risso, 1810), the red mullet *Mullus ponticus* Essipov, 1927, the garfish *Belone euxini* Günther, 1866, and various members of the Sparidae and Scienidae families, *Dasyatus pastinaca* (Svetovidov 1964; Rass 1992, 1993a, b).

Grey mullets reproduce during the middle of the summer. They are among the uncommon species that travel to open waters to spawn. As autumn arrives and temperatures drop, the developing young depart from the sea and migrate southward to overwinter near the shores of Bulgaria, Turkey, and the Caucasus. In the spring, the juveniles make their way back to the northern Black Sea coast (Zaitsev 2006). Grey mullets feed on epiphytic organisms gathered from the seabed.

Among the smaller species found near the seabed, the red mullet *Mullus ponticus* holds the greatest commercial importance. It lives in coastal waters and gathers in large groups over soft sediment bottoms. In terms of its life cycle characteristics, four distinct subpopulations can be identified: the Caucasian, the Crimean, the Balkan, and the Turkish groups. The primary fishing activity for red mullet occurs during spring and autumn along the Crimean and Caucasian shores (Zuev and Mel'nikova 2003).

Cold-water fish originating from the boreal Atlantic region comprise 13 species. These mainly occupy the cool intermediate water layer and migrate upward to shallower depths in the autumn and winter months. The most common among them are the sprat *Sprattus sprattus* (Linnaeus, 1758) and the Black Sea whiting *Merlangius merlangus* (Linnaeus, 1758) (Rass 1993a, b; Svetovidov 1964).

The Black Sea sprat, *Sprattus sprattus*, is another widely distributed species that has become a significant component of the fishery, particularly in recent decades. This increase is attributed both to the adoption of new fishing methods and the decline in populations of predatory fish and dolphins that prey on sprats. The majority of sprat catches, approximately 50 thousand tons, occur along the Crimean coasts.

Demersal, or bottom-dwelling, fish comprise several species: the brill *Scophtalmus rhombus* (Linnaeus, 1758), the European flounder *Platichthys flesus* (Linnaeus, 1758), the turbot *Scophthalmus maximus* (Linnaeus, 1758), the thornback ray *Raja clavata* Linnaeus, 1758, the spiny dogfish *Squalus acanthias* (Linnaeus, 1758), and the Mediterranean sand eel *Gymnammodytes cicerelus* (Rafinesque 1810). Among these bottom-living fish, the most commercially valuable is the turbot *Scophthalmus maximus* (Linnaeus, 1758), which inhabits the continental shelf. Its primary concentration is found in the northwestern part of the sea. Turbots favor locations with depths of 50–100 meters and a flat seabed of sand or silt-sand. They venture into shallower areas (5–40 meters) only during the spring spawning season. Their diet consists of small benthic fish and benthos. The average turbot measures 30–50 cm in length and weighs around 4 kg.

The Black Sea's fish community also contains 22 species of brackish-water Pontian–Caspian relicts. Of these, only the kilka Clupeonella cultriventris (von Nordmann, 1840) holds commercial importance, whereas gobies from the Gobiidae family and the percarina *Percarina maeotica* Kuznetsov, 1888 are not commercially targeted (Rass 1993a, b; Boltachev 2003).

Among the Pontian–Caspian species, the round goby *N. melanostomus* has become an invasive organism, spreading extensively across Eastern and Western Europe and entering the Baltic Sea (Jazdzewski and Konopasca 2002; Slynko et al. 2000). In 1990, it was introduced via ballast water into the St. Clair River near Lake Ontario (Canada) and subsequently expanded further into the North American Great Lakes (Mills et al. 1993).

Anadromous and semi-anadromous fishes are represented by 25 species. All are of ancient origin, dating back at least 1.5–2 million years. The distinctive life cycle of these species is historically linked to their feeding migrations into the freshened Pontian Lake and their subsequent return migrations for spawning (Rass 1993a, b). Among this group, sturgeons have consistently been the most prized commercial fish.

Lastly, the Black Sea also hosts typical freshwater species that may occasionally move from rivers into the near-estuary zones of the sea, typically during flood periods (e.g., carps, breams, pike-perches, basses, and others) (Rass 2001).

Change of the Ichthyofauna Species Diversity and Commercial Fish Stocks Initial signs of declining species diversity within the fish community were observed as far back as the 1970s. These changes were linked to multiple factors, such as the regulation of major river flows, eutrophication in the sea's northwestern sector, intensified fishing activity, degradation of spawning and feeding habitats, and the obstruction of traditional fish migration pathways, among others (Zaitstev 1993; Rass 1992, 2001; Prodanov et al. 1997; Shiganova and Bulgakova 2000).

Furthermore, concurrent with rising human populations, greater recreational use of coastal areas, and increased discharge of agricultural and industrial waste, spawning conditions were severely degraded. This affected both fish species that deposit eggs on sediment or aquatic vegetation and those that spawn in nearshore waters.

During the 1970s, purse seines targeting pelagic fish and various trawls—including pelagic, bathypelagic, and bottom types—were introduced to catch pelagic and demersal species. This gear was employed to harvest bonito (*Sarda sarda*), Atlantic mackerel (*Scomber scombrus* and *S. colias*), Mediterranean horse mackerel (*Trachurus mediterraneus*) (Ivanov and Beverton 1985), and later also bluefish (*Pomatomus saltatrix*) (Ivanov and Beverton 1985). Turkish fishing vessels captured these species even in the Bosporus, where they gathered in large, dense schools during their migration into the Black Sea (Gucu 2002). Between 1970 and 1975, a decline in the numbers of migrating fish was noted—most notably the once-abundant bonito (*Sarda sarda*), which entered the Black Sea in May to feed and spawn before returning to the Sea of Marmara and the Aegean Sea at the end of summer to overwinter. Catches of bluefish (*Pomatomus saltatrix*) in the northern part of the sea also dropped after 1971 (Fig. 4.16) (Caddy and Griffiths 1990).

Two potential explanations can be offered for this occurrence. Firstly, the extensive Turkish fishing fleet captured the vast majority of fish migrating through the Bosphorus (Gucu 2002), a fact supported by GFCM data (Fig. 4.16). A second possible reason is the dramatic alteration of conditions within the spawning and feeding grounds of the Black Sea's northwestern region. Prior to 1970, the highly productive, shallow northwestern area served as the primary zone for feeding migrations and, for certain species, for the spawning of seasonal migrants. Due to reduced river inflow, eutrophication, and related hypoxia in the western and northwestern sections, the habitat quality for demersal fish species degraded. Consequently, the stocks of several species, and their corresponding catches, saw a substantial decrease. Along the Bulgarian coast, the turbot population declined because of high mortality rates among larvae and juveniles, linked to deteriorating environmental conditions in the pelagic zone and near-bottom layer caused by widespread algal blooms.

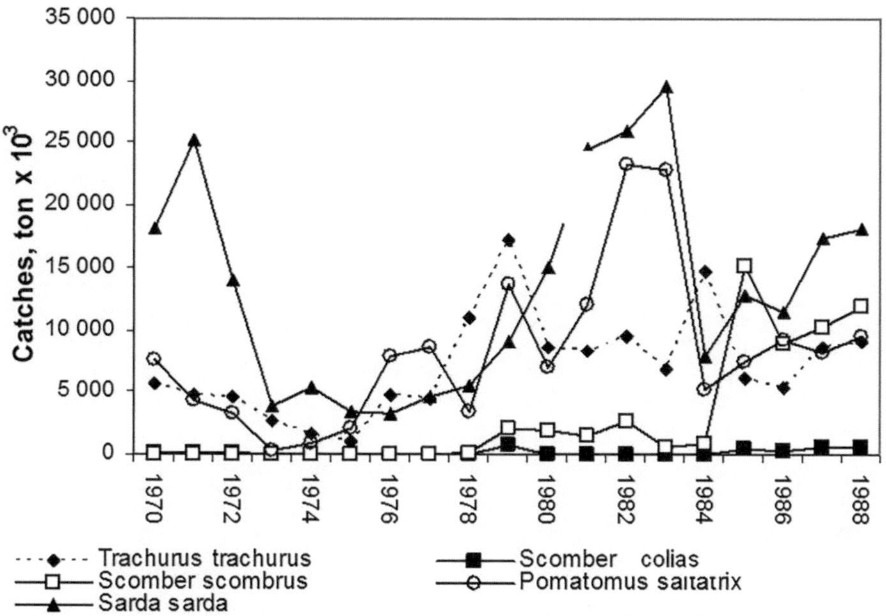

Fig. 4.16 Catches of major pelagic fish species (seasonal migrants to the Black Sea). (GFCM data; authors' visualization)

However, catches of species such as Black Sea whiting (*Merlangus merlangus*), flounders, and red mullets rose. From the late 1970s to the early 1980s, whiting constituted the largest portion of demersal fish catches (Fig. 4.17).

During the 1970s, harvests of migratory and freshwater-migratory predatory species, including sturgeons (beluga and stellate sturgeon), anadromous herrings, and pikeperch (Rass 1992), declined sharply in the northwestern Black Sea as a result of river runoff regulation.

Several fish species, such as the beluga *Huso huso* (L.), the Black Sea salmon *Salmo trutta labrax* Pallas, the common seahorse *Hippocampus ramulosus* Leach, the European seabass *Dicentrarchus labrax* (L), *Labrus viridis* L., and others, neared extinction and are now included in red lists (Boltachev 2003).

In the early 1970s, over half of the total catch consisted of large pelagic fish: mackerels, Atlantic horse mackerel, Atlantic bonito, bluefish, grey mullets, shads, beluga, and sturgeons. Species such as anchovies, flatfish, red mullet, Black Sea horse mackerel, stellate sturgeon, and kilka made up only around 20% of the catch (Rass 1992).

In the 1970s, eutrophication in coastal areas led to profound shifts throughout the marine ecosystem, altering species composition substantially. Large pelagic predators vanished from fishery hauls, while short-lived species such as sprat and anchovy expanded dramatically (Figs. 4.18 and 4.19). From the 1970s until 1988—the year *Mnemiopsis leidyi* appeared—the number of species declined by a factor of 4–7.

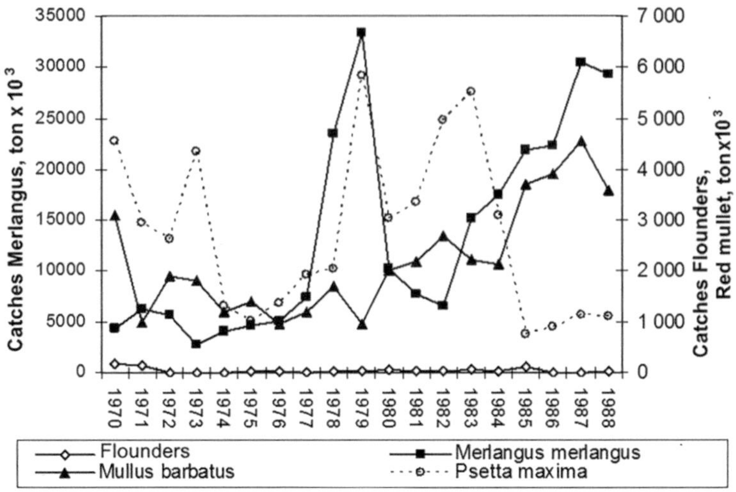

Fig. 4.17 Overall catches of bottom-dwelling fish species in the Black Sea. (Data from GFCM; chart created by the authors)

During the 1980s, reduced predation pressure from fish and dolphins allowed small pelagic stocks, including anchovy, sprat, and Black Sea horse mackerel, to grow considerably. Consequently, fisheries increasingly focused on these small pelagic species, and the share of whiting (*Merlangius merlangus*) in total catches also rose. Throughout the 1970s and early 1980s, small pelagic fish dominated the catch. By the 1980s, total annual landings reached 600–800 thousand tons.

The combined haul of anchovies and sprats accounted for over 90% of the total catch, with anchovies representing the largest portion. The remainder was dominated by the highest Turkish catch of *Trachurus mediterraneus*, while species including flatfishes, grey mullets, red mullets, and horse mackerels contributed only 1–7% of the total catches (Figs. 4.19 and 4.20).

Following the economic challenges that emerged after 1985, Russia's fishing fleet was largely depleted, numerous operations ceased, and the total catch volume previously achieved by the Soviet Union saw a major decline. As previously mentioned, this period also coincided with the disappearance of large pelagic fish from the hauls. Catches became dominated by small pelagic species, primarily anchovy and sprat, with smaller contributions from Black Sea whiting, Black Sea horse mackerel, and red mullet; anchovy also remained a component of the former USSR's catches. The majority of the commercially valuable fish species that were once harvested were effectively lost to the fishing industry. Consequently, well before the arrival of *M. leidyi*, the diversity of species in the catches had already deteriorated substantially.

Current literature frequently cites overfishing as a key factor for certain species, notably anchovy and Mediterranean horse mackerel, and even Black Sea whiting (Prodanov et al. 1997). This overfishing is presented as the rationale for the drop in

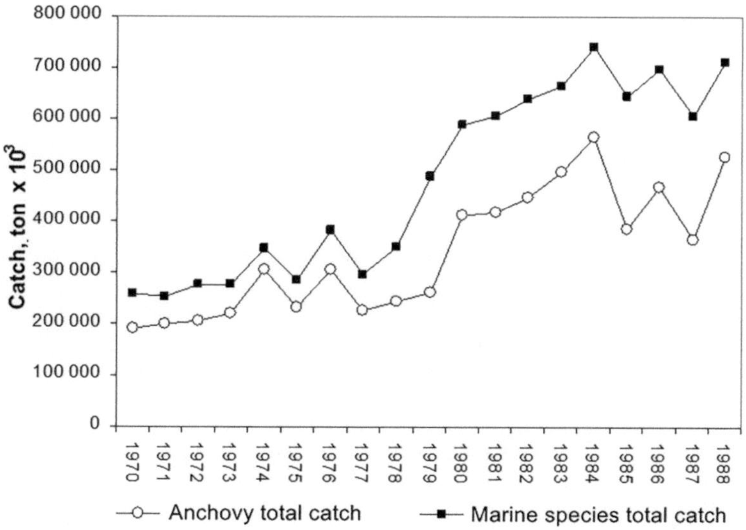

Fig. 4.18 Anchovy catch and overall marine species catch in the Black Sea. (GFCM data; chart created by the authors)

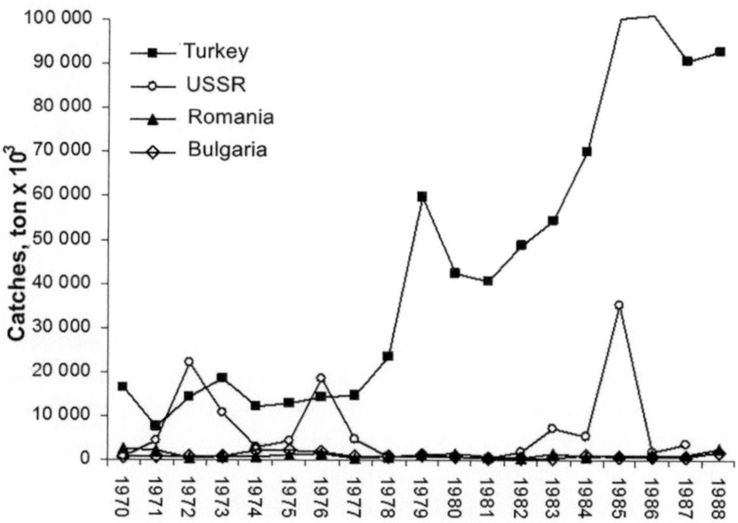

Fig. 4.19 Catches of the Black Sea horse mackerel, *Trachurus mediterraneus*, by different countries. (Based on GFCM data; chart created by the authors)

catches after 1988 and the subsequent massive population explosion of the invasive comb jelly *Mnemiopsis leidyi* (Niermann 2004; Gucu 2002; Daskarov et al. 2007). However, there is no evidence of overfishing affecting Azov Sea anchovy in the years leading up to the appearance of *M. leidyi* in that region. Both the fish stocks and the catches only began their decline after 1988.

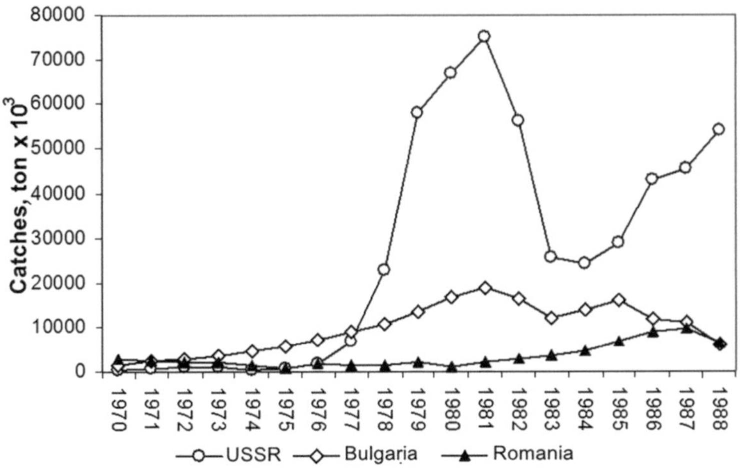

Fig. 4.20 Catches of *Spratus spratus* (Linnaeus, 1758) by different countries in the Black Sea. (GFCM data; authors' plot)

The Sprat is a pelagic schooling fish that feeds on plankton and stands as the most abundant cold-water species within the Black Sea's ichthyofauna. It has a short life cycle, with a maximum lifespan of 5–6 years, and the majority of the population consists of individuals under 4 years of age. Spawning begins at 1 year old and occurs annually. The primary concentrations of Sprat are found in the Danube region and along the Crimean coast (Fashchuk et al. 1995). As a cold-water species, Sprat avoids temperatures exceeding 15 °C, which leads it to typically inhabit depths greater than 35–50 m. During spring, schools of Sprat that move into shelf waters can be located above the thermocline (Nadolinskii 2004). Spawning takes place almost year-round but is most intense in colder periods. At this time, reproduction occurs in the sea's surface layer, whereas in the warmer season it shifts to depths of 40–80 m.

Consequently, alterations to the Black Sea's fish populations commenced as early as the 1970s and were driven by human-induced factors. These included modifications to river runoff, the introduction of pollutants via major rivers, eutrophication, the degradation of spawning and feeding grounds, and the obstruction of traditional fish migration routes, rather than by overfishing. It is clear that during this era, the decline in small pelagic fish stocks was not attributable to excessive fishing. It is plausible that Turkey engaged in overfishing of large pelagic species, but this cannot be definitively confirmed due to a lack of adequate data (Gugu 2002). Nonetheless, these larger fish evidently stopped migrating to the northern regions of the sea. This cessation likely resulted from their capture in the Bosporus area and from shifts in their migration paths to spawning and feeding locations. Regarding demersal fish, turbot harvests declined because of eutrophication, pollution, and hypoxia in the seabed layers. Catches of other bottom-dwelling species remained stable, while those of certain types, such as whiting and red mullet, actually rose. During the

1980s, the population of small pelagic fish expanded substantially. This growth was facilitated by the reduced pressure from predators, namely large pelagic fish and dolphins. As a result, their commercial importance and catch volumes increased significantly. The subsequent reduction in their stocks and catches only occurred following the explosive proliferation of *M. leidyi*, not as a consequence of overfishing.

Macrophytobenthos

The Black Sea's macrophyte community comprises 304 species, falling into three algal divisions (with 85, 77, and 142 species of green, brown, and red algae, respectively) and one division of higher flowering plants (Kalugina-Gutnik 1979; Mil'chakova 2003). The taxonomic structure, ecological forms, and distribution of the algoflora are primarily governed by two environmental factors: temperature and salinity. The absence of endemic algal forms points to a relatively recent origin for this flora. It consists of both cold-water and warm-water assemblages. In the northwestern section of the sea, particularly around the Zernov phyllophore field and Novorossiysk Bay, the flora resembles that of boreal regions. Warm-water Mediterranean species that migrated into the Black Sea encountered suitable conditions, allowing them to establish dense populations. In this context, the southeastern coastal area and Turkey's near-shore waters are particularly notable, showing the closest similarity in species composition to Mediterranean coastal waters off France (Kalugina-Gutnik 1979; Minicheva 2003). Red algae from the Rhodophyta division are the dominant group, whereas brown and green algae (Phaeophyta and Chlorophyta) are represented by fewer species. The smallest floristic group, Phaeophyta, includes key species that act as principal and regional edificators (Maksimova and Luchina 2002).

The seasonal progression of species within these taxonomic groups is not strongly evident. Chlorophyceae, Rhodophyceae, and Fucophycae emerge first in February, March, and April, respectively. The highest species count is recorded during the summer months. Below depths of 25 meters, the qualitative makeup of the macrophyto-benthos remains consistent year-round (Kalugina-Gutnik 1979).

Historically, a defining characteristic of the Black Sea's benthic vegetation was the presence of phyllophore associations, which featured a layered structure in the plant cover, a consistent species composition, and elevated productivity.

Within the Black Sea, five distinct species of phyllophore are documented: *Phyllophora nervosa* (DC) Grev., *Ph. brodiaei* (Turn) J. Ag., *Ph. preudoceranoides* (Gmel.), *Ph. membranifolia* (G. et W.) J. Ag., and *Ph. trailii* Holm ex Batt. (Kalugina 1954). *Phyllophora nervosa* is an endemic species of the Mediterranean basin, widely distributed throughout both the Black and Mediterranean seas. The remaining three species are characteristic of North Atlantic flora and are absent from the Mediterranean Sea. While these species inhabit shallower depths in northern seas, they have adapted to deeper waters (20–50 m) in the Black Sea. Their survival as cold-water forms is facilitated by the region's consistently low year-round temperatures of 6–10 °C. Consequently, the developmental timing of these Black Sea populations remains aligned with that of their northern counterparts (Kalugina-

Gutnik 1979). Prior to the 1980s, the most abundant species was the free-living form *Phyllophora nervosa*. It was predominantly found in the northwestern Black Sea, forming extensive layers over sandy and silty sediments mixed with shell fragments. Its biomass stock not only surpassed that of other phyllophore species but exceeded the combined biomass of all other phytobenthos, ranging from 0.5 to 10.8 kg m^{-2}.

During the peak period of human-induced environmental stress, particularly from eutrophication in the northwestern section of the sea between 1960 and 1975, the composition of phytoplankton communities experienced substantial alterations. These changes were marked by the loss of certain algal species and their subsequent replacement by others. For instance, the algae *Cystoseira barbata f. Hoppii* (Ag.), known for its high sensitivity to pollution and nutrient enrichment, either deteriorated or vanished from coastal plant communities. Concurrently, complex ecosystems supporting a wide array of algae, invertebrates, and fish also disappeared, giving way to communities dominated by *Cladophora*, *Enteromorpha*, and *Porphyra* (Minicheva 1993). The extent of seabed covered by *Phyllophora nervosa* declined dramatically. The most significant aggregation of free-living phyllophora was found at depths of 25–60 meters in the northwestern region. In the mid-1900s, the Zernov Phyllophora Field spanned approximately 11,000 km^2. By the mid-1980s, both the field's area and the algal biomass had plummeted, shrinking by factors of 20 and 50, respectively (Zaitsev 2006). Within the Zernov field, *Phyllophora nervosa* played a dual role as a crucial oxygen producer and a habitat for 118 invertebrate species and 47 fish species. Its decline triggered a sharp reduction in the diversity of invertebrates and fish associated with these underwater meadows. This degradation is believed to have been driven by eutrophication, leading not only to significant economic damage but also to severe, cascading consequences for the entire northwestern shelf ecosystem. Furthermore, nutrient pollution contributed to a rise in the populations of algae such as *Enteromorpha* and *Ceramium* (Zaitsev 2006).

Consequently, among the distinct characteristics of macrophytobenthos that have emerged in recent years due to eutrophication in the Black Sea, the most significant include (1) the contraction of the phytal zone, primarily driven by the upward shift of its lower boundary to shallower depths; (2) a diminished role for large, perennial species with complex structures; (3) an increased prominence of short-lived, highly productive opportunistic species (r-strategists) featuring filamentous thalli, which are either facultative or obligatory epiphytes; and (4) alterations in the ecological and morphological traits of certain abundant species (Maksimova and Luchina 2002).

Benthos

Within the Black Sea, benthic organisms outnumber zooplanktonic species, with 1518 benthic animal species documented by the mid-1970s (Kiseleva 1979). The majority of this benthic fauna, approximately 1200 species, originates from the Mediterranean-Atlantic region, while around 300 species are derived from freshwater and Pontian–Caspian sources (Kiseleva 1979).

The benthic animal community in the Black Sea can be categorized into the following groups:

Porifera. Sponge fauna of the Black Sea is represented by 29 species, of which 5 are endemic species (Kaminskaya 1968; Mordukhai-Boltovskoi 1972).

Nemertini. Nemertin fauna of the Black Sea includes 31 species (Muller 1968).

Polychaeta. The Black Sea hosts 195 species of Polychaeta worms (Kiseleva 1979). The habitat of 29 Polychaeta are restricted to the near-Bosphorus region and the rest 164 species could be considered as the Black Sea ones (Revkov 2003).

Within the Pontian-Caspian Polychaeta worm community, *Hypania invalida* has emerged as an invasive species, extensively colonizing the Tsimlyansk and Manych reservoirs in the Don River basin, the Zaporozh'e reservoir of the Dnieper basin, and Lake Balkhash in Kazakhstan. Through the Southern corridor, it entered the upper Danube River and its reservoirs, subsequently reaching the Rhine River and other European waterways via canal networks (Mordukhai-Boltovskoi 1979; Ketelaars 2004).

Crustacea This class is represented in the Azov-Black Sea basin by more than 530 species (Mordukhai-Boltovskoi 1969a, b, c). Of them, 507 species are referred to the Black Sea bottom fauna, including species of Ostracoda (109), Harpacticoida (154), Cirripedia (6), Amphpoda (103), Isopoda (29), Tanaidacea (6), Cumacea (23), Decapoda (32), Mysidacea (19), Acarina (27), Pantopoda (8) and Insecta (11) (Kiseleva 1979; Revkov 2003).

Cirripedia The most common mass species is the non-native *Balanus improvisus,* while another non-native species of the same genus *B. eburneus* are relatively rare. In 1995, one more representative of Cirripedia—*Chthamalus montagui*, was found for the first time off the coasts of the Crimea (Zevina and Poltarukha 1999). Among the Black Sea barnacle widely spread is *Chthamalus stellatus* (Revkov 2003).

Decapoda The Black Sea is home to between 32 and 37 species of decapods (Revkov 2003). These crustaceans typically inhabit seagrass beds, sandy bottoms in shallow zones, and underwater rocky outcrops and cliffs. Among the most prevalent and widespread species are *Carcinus mediterraneus*, *Pachygrapsus marmoratus*, *Eriphia verrucosa*, *Diogenes pugilator*, *Pilumnus hirtellus*, and *Pisidia longimana*. In recent years, populations of the first three species have declined markedly due to unregulated harvesting and habitat disruption. Concurrently, the invasive Dutch crab *Rhithropanopeus harrisii* has been expanding its range (Revkov 2003). The large stone crab, *Eriphia verrucosa*, is found along rocky shorelines. The lobster *Homarus gammarus*, which can exceed 30 cm in length, is the largest decapod in the Black Sea. It is commercially fished in Turkey and is occasionally observed off the coasts of Bulgaria, the Caucasus, and Crimea (Kobyakova and Dolgopol'skaya 1969). Certain shrimp species, such as the grass shrimp *Palaemon adspersus* and the stone shrimp *P. elegans* (Zaitsev 2006), also hold local commercial importance. The Pontian–Caspian crayfish *Astacus leptodactylus* has been successfully introduced to water bodies in Europe (Ketelaars 2004).

Mysidacea Within the Pontian-Azov basin, mysids are documented as comprising 19 species, with 8 of these inhabiting fresh or low-salinity waters (Mordukhai-Boltovskoy 1960). Several have become invasive. Following the development of canals and reservoirs, they have extensively colonized freshwater basins across Europe, even reaching the Baltic Sea. These species are *Hemimysis anomala*, *Limnomysis benedeni*, *Katamysis warpachowskyi*, *Paramysis lacustris*, *P. baeri*, *P. intermedia*, *P. kessleri*, and *P. ullskyi*; notably, *Hemimysis anomala* and *Limnomysis benedeni* have also entered the Baltic Sea. *Hemimysis anomala* has been introduced into the Great Lakes of North America (Ojaveer et al. 2002; Ketelaars 2004; Leppakoski 2004). Certain species, including *Limnomysis benedeni*, *Paramysis lacustris*, *P. baeri*, *P. intermedia*, *P. kessleri*, and *P. ullskyi*, were deliberately introduced into freshwater systems (Ketelaars 2004).

Cumacea Twenty-three species of Cumacea have been documented in the Black Sea (Kiseleva 1979). The most frequently observed species across all studied regions are *Cumella pygmaaea euxinica*, *Eudorella truncatula*, *Iphinoe elisae*, and *Iphinoe maeotica*; the last of these is not found in the area north of Cape Tarkhankut (Revkov 2003). Pontian-Caspian cumacean species that have become naturalized in other basins include *Caspiacuma campylaspoides*, *Pseudocuma cercaroides*, *Pterocuma pectinata*, *P. sowinskyi*, *P. rostrata*, and *Stenocuma tenuicauda* (Mordukhai-Boltovskoy 1960). The majority of these species were deliberately introduced to enhance biodiversity and serve as supplementary food sources for fish (Ketelaars 2004).

Anisopoda Of six anisopod species in the Black Sea, two of them, i.e., *Leptochelia mergellinae* Smith, 1960 and *Pontotanais borceai* Bacescu, 1960 restricted to the near-Bosphorus area (Makkaveeva 1969). The rest four species are genuine Black Sea inhabitants and encountered off the Crimean coasts. Three of them (*Apseudopsis ostroumovi, Leptochelia savignyi,* and *Tanais cavolini*) forms specific near-shore biocenosis.

Isopoda Isopods includes 29 species (Kusakin 1969). Among them is *Idotea baltica basteri*, which lives in macrophytes clusters and the near-shore thickets and always was one of the widely spread species (Zaitsev 2006). The representative of Isopoda *Gnathia bacescoi* was found only in the Bosphorus region; meanwhile, after 1969, it was also observed near the Crimea (Zaitsev and Aleksandrov 1998).

Amphipoda This is one of the most numerous (along with Ostracoda and Harpacticoida) groups of crustaceans which includs 111 species (Mordukhai-Boltovskoi 1969a, b, c, 1972). The most invasive are the Pontian–Caspian species and 15 of them propagated via the northern invasive corridor to other basins: *Corophium mucronatum, C. maeoticum, C. sowinskyi, C. curvispinum Echinogammarus ischnus, Chaetogammarus warpachowskyi, Obesogammarus crassus, Pontogammarus robustoides, P. sarsi, P. obesus, P. subnudus(?), Stenogammarus macrurus, Dikerogammarus villosus, D.haemobaphes, Gmelina pusilla, G. kusnetzowi, G. costat,* and *Echinogammarus ischnus*. Introduction of the latter two species resulted in substrate changes and *C. curvispinum* became a species-edificator of the entire biotope. Some Pontian–Caspian amphipods were intentionally introduced in other basins, among them are *Corophium nobile, C. robustum, Amathillina cristata, Pontogammarus aralensis, Pontogammarus robustoides, P. maeoticus, P. subnudus(?), Dikerogammarus villosus,* and *D. haemobaphes* (Ketelaars 2004).

Pantopoda According to Kiseleva (1979), there are 7 species of sea spiders in the Black Sea; meanwhile Becesku (1969) reported about 8 species. Off the coasts of the Crimea, 5 species were registered, including *Anoplodactylus petiolatus*, which has earlier been noted only in the Bosphorus region. In 1986, it was encountered off the Crimean coasts near Yalta at a depth of 10–20 m within the *Chamelea gallina* assemblage (Sergeeva 1992).

Loricata Testaceous mollusks in the Black Sea are represented only by three species: *Middendorffia caprearm, Lepidochitona cinerea,* and *Acanthochitona fascicularis* (Starobogatov 1972).

Gastropoda Among 100 gasteropod species, 18 were noted exclusively in the Bosphorus area. The rest are represented by 82 species, of which 71 species are also encountered in the Mediterranean Sea. Regarding the origin, most species (46) are of Mediterranean–Boreal origin, 15 are Mediterranean endemics, 8 species are of a Mediterranean–Lusitanian origin, and 5 species belong to the brackish-water Pontian-Caspian species (Chukhchin 1984). The most common species in the near-shore biocoenoses are *Bittium reticulatum, Calyptraea chinensis, Tritia reticulata, Gibbila adriatica, Cyclope neritea, Hydrobia acuta, Rissoa parva, R. splendida, Tricolia pullus,* and the invader *Rapana venosa* (Revkov 2003).

Bivalvia Bivalve mollusks represent a key element of the shallow-water biocoenoses in the Black Sea. They constitute 12 out of 16 biocoenoses found on unconsolidated sediments, where they act as the dominant organisms (Kiseleva 1981). A total of 89 bivalve species have been recorded. The distribution of 22 species is confined to the Bosporus region. Fourteen species are native to the lagoons, estuaries, and coastal lakes of the northwestern Black Sea. Among the marine species prevalent in the Black Sea are the mussel *Mytilus galloprovincialis*, the cockle *Cerastoderma lamarcki*, the donacilla *Donacilla cornea*, and the phaseolina *Modiolula phaseolina* (R. A. Philippi, 1844) (Revkov 2003).

Echinodermata Fourteen species of echinoderms are represented in the Black Sea. Among them, there are 8 species of holothurians and 4 ophiura species; starfishes and sea urchins are represented by a single species each. Eight species are encountered only in the Bosphorus region and five ones may be referred to typical Black Sea forms (Revkov 2003).

Chordata In the Black Sea, this group is formed by 8 species of ascidia and a lancelet (Kiseleva and Zaika 1972).

Meroplankton

A key element shaping the zooplankton community is benthic fauna, whose larval stages inhabit the pelagic zone. These organisms form a distinct ecological category known as meroplankton. This group represents a significant, yet insufficiently researched, part of coastal plankton. The occurrence of benthic invertebrate larvae in the plankton is influenced by a variety of factors: the reproductive season of adults, water temperature, currents, the length of the developmental period, and the available food sources (such as phytoplankton, microzooplankton, and detritus) (Murina 2003).

In certain biocoenoses, the proportion of animals that include a pelagic larval stage in their reproduction can account for 75–90% of the total benthic abundance (Kiseleva 1979). Recent findings (Murina 2003) indicate that 162 out of 442 species (36.4%) belonging to the five major macrozoobenthos taxa in the Black Sea—Polychaeta, Bivalvia, Gastropoda, Cirripedia, and Decapoda—produce pelagic larvae. The majority of biocoenoses found over both soft and hard substrates, artificial reefs, and fouling communities are dominated by species whose development includes a pelagic larval phase. Consequently, the abundance of widespread species like *M. galloprovincialis*, *M. lineatus*, and *B. improvisus* is exceptionally high.

To forecast the population levels of bottom-dwelling invertebrate larvae, three key elements must be taken into account: (1) the state of the adult populations, particularly the reproductive capacity of species found in the shelf region; (2) the temperature regimes that determine spawning timing; and (3) the larvae's ability to survive amid anthropogenic pollution (Murina 2003). Additionally, the duration since the arrival of the invasive ctenophore *M. leidyi* is highly significant.

Mammals

The marine mammal population in the Black Sea consists of one seal species and three dolphin species. The Mediterranean monk seal, *Monachus monachus* (Hermann, 1779), is now extremely rare in the Black Sea, with only occasional individuals entering from the Sea of Marmara into the Anatolian coastal waters and the Bosphorus area. However, in the past (specifically between 1992 and 1997), groups of these seals were sighted along rocky sections of the Turkish coastline, where they located shelter and even established breeding grounds (Kirac et al. 1998; Johnson 1998). The seal's virtual disappearance from the region can likely be attributed to two primary factors: the degradation of their traditional haul-out sites and their incidental killing during historical dolphin hunts (Kirac and Savas 1996).

Within the dolphin community, the most prevalent species is the common dolphin, *Delphinus delphis* L. Two additional species—the harbor porpoise *Phocoena phocoena* L. and the bottlenose dolphin *Tursiops truncatus* Mont. (specifically the Black Sea subspecies *T.t. ponticus* Barabash)—are less common and have historically existed in smaller numbers (Geptner et al. 1976). Systematic research into the distribution, population size, and seasonal movements of Black Sea dolphins began in the 1930s.

Due to unregulated hunting of dolphins in the Black Sea throughout the twentieth century—driven by various commercial and medicinal uses—their population experienced a severe decline. Estimates placed their numbers at 1–2.5 million individuals

Table 4.7 Changes in dolphin feeding habits and variations in fish catches in the Azov–Black Sea area (thousand tons)

Feeding object	Consumption by dolphins		Fish catches[a]			
	Initial level (mid of twentieth century)	1980s	1949–1953	1960	1988	1991
Total	2400–3600	215	57–102	414	993	
Anchovy	1500–2200	120	12.7–57.5	60	551	86
Sprat	700–1100	60	0.07–2.7	4.5	68	15
Whiting	52–70	20	<0.01	0.01	35	27
Gobies	60–80	5	5–23	38	3	0,4

Prepared by the authors
[a]After Averkiev (1960), Danilevskii and Maiorova (1979), Moiseev (1989), Rass (1992), and Chashchin (1995)

prior to the 1940s, but by 1950–1966 the population had fallen to roughly 500,000. By the period 1980–1984, the count had dropped further to only 50,000–100,000 individuals (Caddy and Griffiths 1990). Assessments conducted in the 1980s suggested a potential recovery in dolphin numbers; however, after 1988, a sharp decrease in their primary food source—anchovy—led to another decline in dolphin abundance.

Shifts in the dietary patterns of Black Sea dolphins and the trends in commercial fish harvests (Table 4.7) reveal that, relative to the mid-twentieth century, dolphin fish consumption in the mid-1980s had decreased by a factor of 7–11. Concurrently, total fish catches grew more than tenfold from the late 1940s through 1988. The relationship between these two metrics also shifted significantly. Whereas dolphins originally consumed fish quantities roughly ten times greater than human catches, by the 1980s the commercial harvest exceeded total natural consumption by a factor of 4–5. It is likely that fluctuations in fish stocks linked to the dolphin population decrease may also have impacted lower trophic levels of the ecosystem, due to altered consumption of phytoplankton and zooplankton by fish.

4.4 Non-native Species Introduction in the Black Sea

The Black Sea with its thin active layer is vulnerable to external influences, including climate changes and anthropogenic impacts. Conditions, favorable for the vital activity of organisms are existing only in the upper oxygenated layer with a thickness of 100–150 m. Those specific conditions are the result of a pronounced salinity/density stratification (halocline/ pycnocline), which prevents vertical mixing and oxygen transport into the deep layers (see Sect. 4.2).

The Black Sea upper layer temperature rise by about 2 °C during the last three decades due to global warming along with significant increase of water salinity in the past 15 years resulted in decrease of the oxygenated layer thickness and, accordingly, to the rise of the hydrogen sulfide contaminated zone. In addition, there is an increase in wind cyclonic vorticity and the intensity of geostrophic circulation, as well as a decrease in riverine runoff and precipitation over the Black Sea. The upper layer warming increased the risks of the introduction of thermophilic species, first of all Mediterranean ones into the Black Sea ecosystem, which may result in its significant changes.

In the second half of the twentieth century, there was a sharp increase in the number of non-native species of plants and animals in the ecosystems of the seas and coastal areas of the oceans. The reasons for this, as well as pathways and vectors of non-native species introduction have been discussed in Chap. 2.

The non-native invaders cause considerable changes in the structure and dynamics of marine ecosystems and the food web by outcompeting original inhabitants. Thus, an understanding the causes and impacts of non-native species invasions as well as development of effective responses to existing and future threats is a crucial issue.

Analysis of the pathways and vectors of species invasion and their establishment indicates that the Black Sea in the second half of the twentieth century became a region-recipient of a large number of species, and at the same time a donor area, primarily for the seas of the Ponto-Caspian basin. A number of circumstances contributed to that. Among the seas of the Ponto-Caspian basin, the Black Sea features the most intensive navigation with the various areas of the World Ocean. As a result, since the second half of the twentieth century considerable number of marine and brackish-water species was brought with ships to the port areas (Shiganova 2010). An important is the geographical location of the Black Sea between the Mediterranean Sea with its high species diversity and the Sea of Azov and the Caspian Sea with low species diversity. The Black Sea also occupies an intermediate position in terms of the abiotic environmental factors: it is significantly less saline than the Mediterranean Sea, and more saline than the Azov and Caspian seas. However, the main factor contributing to the establishment of the non-native species in the Black Sea is significant anthropogenic eutrophication and disruption of the functioning of the ecosystem as a whole. In addition, the Black Sea became the most sensitive to species invasions due to natural conditions. It is a non-tidal, meromictic basin with a thin (60–200 m depending on vertical stratification) surface oxygenated layer. The present flora and fauna, which were formed under conditions of relatively low salinity (12–22‰) and the existence of an anoxic zone beneath the upper oxygen-containing layer, are distinguished by low species diversity of most taxonomic groups and complete absence of many of them. However, the biota is highly productive, especially in near-shore regions, because of a high abundance of key planktonic and benthic species and large populations of commercial fish species (Svetovidov 1964; Greze 1979; Kiseleva 1979).

The Black Sea biota is 79% of Atlantic-Mediterranean origin, 10% of freshwater origin, and 10% of Ponto-Caspian origin. Minor component is an Arctic assemblage (1%) containing mainly flora (Fig. 4.21).

The biota of Atlantic-Mediterranean origin is composed of species of the Lusitanian province and of the boreal zone of the Atlantic Ocean. The Lusitanian origin species are warm-water species that inhabit the upper layer of the Black Sea. The

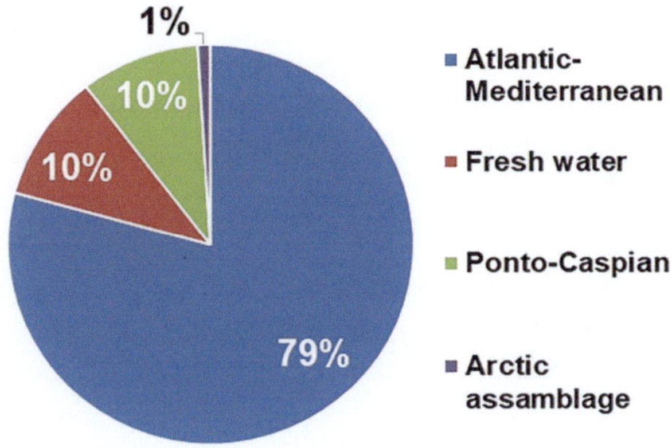

Fig. 4.21 Original composition of the Black Sea biota

Atlantic boreal origin species are found in moderately cold water and have features of cold-water relicts. They are mainly benthic, demersal, and pelagic species that inhabit the cold intermediate layer and lower depths. In addition to salinity, impoverishment of the Black Sea biota is associated with the absence of deep-water species below the oxic layer. The number of species in the Black Sea is currently relatively low and amount to 3786. Of these, 1619 are fungi, algae, and higher plants, 1983 are invertebrates, 180 are fish, and 4 are mammals (Morduhai-Boltovskoi 1969a, b, c; Zaitsev and Mamaev 1997; Shiganova 2009).

The near-Bosporus region of the Black Sea represents a sort of an intermediate adaptation area at the route of invasion of Mediterranean species into the Black Sea. Among those species there are representatives of various systematic groups like phytoplankton, macrophytes, trophic holozooplankton, and benthos (Chuhchin 1984; Kiseleva 1979; Greze 1966; Georgieva 1993; Shiganova et al. 2012). Meanwhile, selected Mediterranean species of phyto- and zooplankton have been found off the Crimean and northwestern coasts, i.e., beyond the near-Bosporus region, as early as the 1960s and the beginning of the 1970s (Kuzmenko 1966; Senichkina 1993; Kovalev and Finenko 1993). This process continued in the following years (Senicheva 2001; Zagorodnaya and Kolesnikova 2003) and the number of penetrated and even established species increasing each year (Table 4.8); full detail on non-native species in the Black Sea is provided in the next section). Among the seasonal migrants from the Mediterranean Sea, there are several species of fishes that commit seasonal feeding or spawning migrations to the Black Sea. These species refer to as migrants; meanwhile, they play or at least played a significant role in the trophic dynamics of the Black Sea ecosystem.

In the second half of the twentieth century, a number of events occurred that had a significant impact on the intensity of the introduction of non-native species into the Black Sea (Fig. 4.22). Since the early 1950s, active anthropogenic interference began (including regulation of riverine runoff, dumping of large amounts of pollutants and nutrients, etc.), which led to serious disturbances of the aquatic environment. Biotic changes have also been caused by other human activities: deliberate

Table 4.8 Number of the non-native species recorded in the Black Sea in various systematic groups (?—uncertain status)

Systematic groups	Number of established species	Number of Mediterranean species (with different status) in vicinity of Bosphorus
Fungi	**4**	
Pathogenic bacteria	**1**	
Parasites of fishes	**3**	
Microplankton (Tintinnidae)	5	
Phytoplankton	**24**	33
Fern Azollaceae	**2**	
Macrophytes	**14**	36
Kamptozoa	**1**	
Hydrozoa	6	
Ctenophora	**2**	
Oligochaeta	**1**	
Polychaeta	**11**	
Copepoda	**3**	59
Echinodermata	**4 (?)**	
Asteroidea		3
Order Decapoda	**9**	2
Order Amphipoda	**1**	1
Order Balanomorpha	4	
Class Bivalvia	**9**	6
Class Gastropoda	**6**	
Ascidiacea	**1**	
Polychaeta		3
Pisces (intentional introduction)	**12**	
Pisces (natural penetration)	**14**	
Pisces (shipping)	**4**	
Pisces (aquariumists)	**3**	
Total 232	**126**	143

stocking of commercial species and release of accompanying, penetration of species from previously isolated basins connected by the canal systems, careless releases by aquarists.

In turn, those disturbances facilitated the processes of non-native species establishment. In 1980s, ballast tanks were introduced on the ships. In the first years of this innovation, tanks were discharging directly in the harbors. As a result, considerable numbers of marine and brackish water species have arrived with ballast water or on fouled hulls. In 1980s, upper layer temperature of the Black Sea rise associated with the global warming, also began. It resulted in intensified expansion of warm-water species from the Sea of Marmara and the Mediterranean Sea (Shiganova et al. 2012).

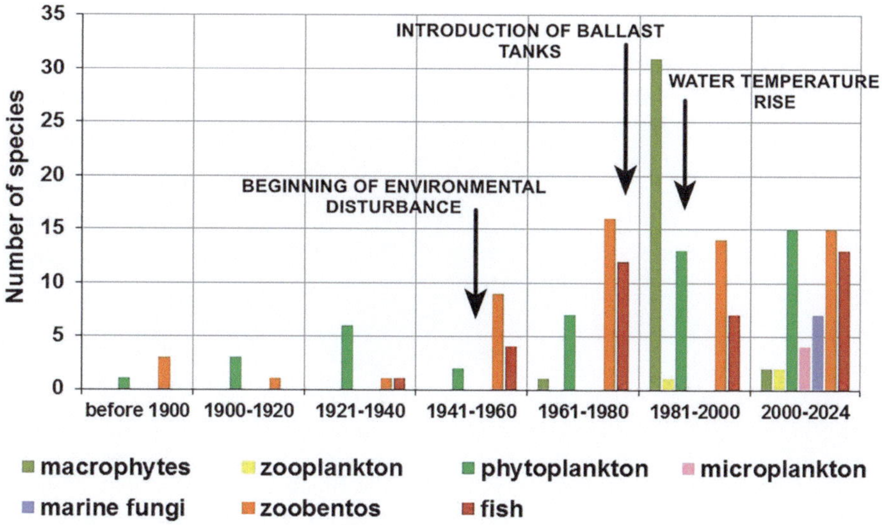

Fig. 4.22 Changes of species composition and chronology of disturbing events since 1900s

Donor regions of non-native species for the Black Sea with their relative share are shown at Fig. 4.23. The largest number of species (42%) arrived from the Mediterranean Sea as a result of natural expansion or with the ballast waters. Their share continues to increase due to the warming of the upper layer of the Black Sea. Only a few of these species have become abundant and almost none have reached other Ponto-Caspian seas. Among them are the representatives of micro-, phyto- and zooplankton, demersal plankton, macrophytes, zoobenthos and fishes. These species usually have subtropical or even tropical origins; most still occur only in the southern parts of the Black Sea and near the Bosporus Strait where salinity is higher. They are not considered as established in the Black Sea. Certain species have penetrated to the nearshore regions off Bulgaria, Romania and Ukraine in the course of their migrations or carried with currents or ballast waters.

About 8% of established non-native species arrived in the Black Sea from the waters of North Atlantic coast of America, mainly in the 1980s–1990s. Most of those species habit in temperate climate and are eurythermal and euryhaline and widely distributed in coastal waters of the World Ocean. Invaders from this region have greatly affected the Black Sea ecosystem. Among them, the most prominent are the ctenophores *Mnemiopsis leidyi* A.Agassiz 1865 and *Beroe ovata* Bruguiere 1789, and the copepod *Acartia tonsa* (Dana 1849). The barnacles *Amphibalanus eburneus* (Gould, 1841) and *Amphibalanus improvisus* (Darwin, 1854) were introduced from the same area much earlier. Through this pathway, brackish-water species inhabiting in brackish bays and estuaries also were accidentally introduced. The most successful of these are the hydromedusas *Blackfordia virginica* (Mayer 1910) and *Bougainvillia* (*Perigonimus*) *megas* (Kinne 1956).

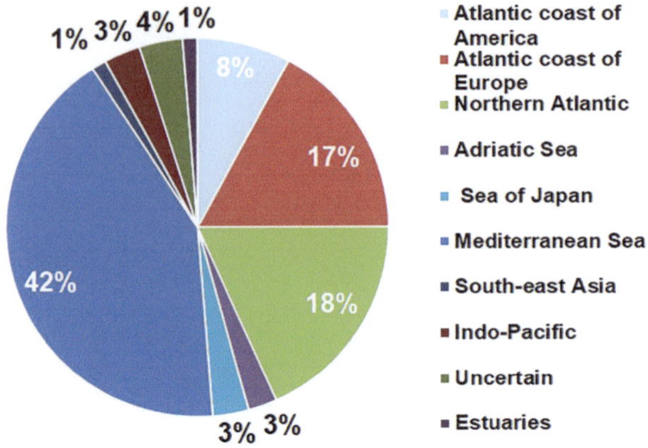

Fig. 4.23 Donor regions of non-native species for the Black Sea with their relative share (%)

One more source of non-native species for the Black Sea is the waters of North Atlantic coast of Europe (17%); most of these species being brought by ships. The most successful is the crab *Rhithropanopeus harrisii* (Maitland 1874). Also, some of these species dispersed from coastal estuaries of Europe via rivers and canals to deltas and brackish areas of the Black Sea. Among them are the Chinese mitten crab *Eriocheir sinensis* (Milne-Edwards 1853) and the pumpkinseed fish *Lepomis gibbosus* (Linne 1758).

About 18% of non-native species arrived in the Black Sea from the unknown part of the North Atlantic; most notable are the diatom *Rhizosolenia calcar-avis* (Schultze 1858) = *Pseudosolenia calcar-avis* (M.Schultze) and bivalve *Mya arenaria* (Linne 1758).

The Japan Sea is another source of invaders for the Black Sea (3%), including the Rapa whelk *Rapana venosa* (Valenciennes 1846) and the intentionally introduced mullet *Planiliza haematocheila* (Temminck & Schlegel, 1845) = (*Lisa haematocheila*). With this mullet, three species of fish's parasites were introduced. Aquarists occasionally releasing medaka *Oryzias latipes* (Temminck and Schlegel), originally from Japanese freshwater.

Of interest are Indo-Pacific species (3%) that might have arrived with ballast water, or recently as Lessepsian migrants (see Chap. 2). In the last few years, this group has expanded substantially. Other species from the coastal waters of the Pacific or Indian oceans first established in coastal waters of Europe and in the Adriatic Sea and then, apparently, colonized the Black Sea from there. The species include the polychaetes *Capitellethus dispar* (Ehlers 1907) and *Glycera capitata* (Örsted 1843), the gastropod *Potamopyrgus jenkinsi* (Smith 1889), and the Chinese mitten crab *Eriocheir sinensis* (H. Milne-Edwards 1853). None of these have become abundant and widely distributed. The attempted introduction of five fish species from estuaries of the Japan Sea, the Amur River, and other rivers of northeast Asia did not succeed. However, two fern species and strains of *Vibrio cholerae* were brought from these regions.

Some of considered species itself are non-native for the regions, from which they reached the Black Sea. Thus, adaptation to the Black Sea conditions was secondary and followed adaptation to initial recipient area conditions. For instance, crab *Rhithropanopeus harrisii* (Gould 1841) first moved to Europe from coastal waters of North America; the Chinese mitten crab first came from eastern Asia to the North Sea; the aquarium fish pumpkinseed was first brought to Europe from North America; the bivalve *Anadara inaequivalvis* (Bruguière 1789) from the coastal waters of Philippines was first released into the Adriatic Sea; the mosquito fish *Gambusia holbrooki* (Girard 1859) from Central America was first transferred to the Adriatic Sea, the polychaete tubeworm *Ficopomatus enigmaticus* (Fauvel 1923) = *Mercierella enigmatica* first came from the coastal India to coastal Europe, and the gastropod *Potamopyrgus antipodarum* (Gray 1843) = *Potamopyrgus jenkinsi* (E. A. Smith 1889) from coastal New Zealand first invaded Europe (Shiganova 2009).

In total, 232 non-native species have been established (Table 4.8). The ratio of the freshwater, brackish water, and euryhaline species in the non-native biota is almost equal to those in the native community. Many established marine euryhaline species are able to inhabit in the whole sea basin. Brackish-water species have established in the less saline northwestern part of the sea, while freshwater species established in the river mouths, bays and estuaries. Most of euryhaline, eurythermal planktonic species have established and occur in the upper layer; several of them may temporarily penetrate below the thermocline. Warm-water non-native planktonic species may develop only during the warm seasons and inhabit the warmer, upper layer above the thermocline. Benthic non-native species are also represented by species with high tolerances for various abiotic factors. The similarities in the ratios of the various types of organisms in native and non-native biotas might be due to the fact that, when there are stressful environmental factors (low salinity, strongly continental climate, etc.), only ecologically similar species can establish but those are more competitive the natives.

The most devastating event in terms of its consequences for the local biota was the invasion of *Mnemiopsis leidyi*, which caused cascading effects at higher trophic levels, from a decrease in zooplankton to collapsing planktivorous fish and drastic declines among large pelagic fish and dolphins. Similar effects occurred at lower trophic levels: (1) decrease of zooplankton, resulted in increase in phytoplankton, which was released from zooplankton grazing pressure and (2) increase of bacterioplankton, which led to increase of its predators, zooflagellates, and ciliates (Shiganova et al. 2004a, b, c, 2019a, b). Ten years later another warm-water ctenophore, *Beroe ovata*, an obligate predator of *M. leidyi*, was introduced with the ballast water from North Atlantic coast of America (Konsulov and Kamburska 1998). *B. ovata* invasion reduced *M. leidyi* populations and the ecosystem began to recover at all trophic levels (Shiganova et al. 2014a, b).

Diatom *Rhizosolenia calcar-avis* (Schultze 1858) is an additional food item for large zooplankton species, but when its populations explode, it suppress more valuable native phytoplankton species. While in general the role of the fouling species *Amphibalanus improvisus* is negative, its larvae are consumed by small pelagic fishes.

An introduction of other organisms may have positive effect in a certain sense. For instance, bivalves *Anadara inaequivalvis* (Bruguiere 1789) and *Mya arenaria* (Linnaeus 1758), though replacing native species, constitutes valuable food for zoobenthic species and benthophagous fishes, while their larvae are consumed by small pelagic fishes. Crab *Rhithropanopeus harrisii* (Gould 1841) has become an additional food for benthophagous fishes. The mullet *Planiliza haematocheila* (Temminck and Schlegel 1845) competes with native mullets but is itself a valuable commercial species.

Most of established non-native species are euryhaline and eurythermal. Species with these traits not only established but also begin to expand, first reaching the Sea of Azov and the Caspian Sea and then invaded the Sea of Marmara, the Aegean and Adriatic seas and the other areas of Mediterranean Sea. Some species, e.g., the Rapa whelk, have been widely dispersed from the Black Sea with the ballast waters into the Mediterranean Sea and Atlantic coastal waters of the North and South America. Thus, the Black Sea has become a region-recipient of a large number of non-native species and in turn, a secondary donor area for the expansion of non-native species in adjacent seas and further in the World Ocean (Shiganova 2009, 2010).

Formerly, species of Mediterranean origin have been recorded either temporarily or permanently in the Black Sea, mainly near the Bosphorus area and were rarely considered as established non-native species. In recent times, Mediterranean species of phytoplankton, zooplankton, benthic and fish species have been encountered also in the northwestern and northeastern parts of the Black Sea, far from the Bosphorus. This process of adaption of Mediterranean biota was most probably facilitated by rising of water temperatures and shipping intensification. Mediterranean species have reached the center, southwestern, southeastern and northeastern regions being transported by the currents, Mediterranean water intrusion and by vessels. The number of species delivered with ballast waters has increased, especially in the harbor's areas. Some species have been represented by only a few recorded specimens, while others developed and reached a high abundance (Alexandrov 2004; Boltachev 2006; Selifonova 2009). The process of the establishment of Mediterranean species already invaded the sea is also continues. Selected species of this kind may temporarily become subdominant species.

Usually, most of phyto- and zooplankton species remain rare or are abundant only in some specific years (Alexandrov 2004; Vershinin et al. 2005). It means that conditions of the Black Sea with low salinity and low winter temperature do not favor maintenance of self-sustaining population of the most of these species. During the past few years, a new tendency of gelatinous species expanding from the Mediterranean, first into the Sea of Marmara and then into the Black Sea has appeared (Shiganova and Ozturk 2010).

Detailed information on non-native species in the Black Sea, including systematic, numbers in various taxonomic groups, origins, ways of introduction, status etc., is provided in the following Sect. 4.5. Special attention will be paid to both to the most harmful invaders and commercial valuable species, suitable for aquaculture.

4.5 Non-native Biota Composition

In this section currently available information on non-native species composition of various systematic groups, including microplankton, phytoplankton, ferns and macrophytes, zooplankton, gelatinous plankton, benthos and fish, which arrived and established in the Black Sea is compiled (Tables 4.9, 4.10, 4.12, 4.14, 4.16, 4.17, and 4.19).

Microplankton

Among the representatives of microplankton, *Vibrio cholerae* (El Tor) poses the greatest danger to humans. It can persist in coastal waters and infect people on adjacent land areas. Epidemic outbreaks of cholera caused by El Tor strain were first recorded in 1970 in the coastal zones of the Black Sea adjusted to the Kerch and Odessa regions (Burgasov et al. 1986). Since then, outbreaks of El Tor have become quite common in a number of Azov-Black Sea cities (Shadrin 2000).

Three non-native species of marine fungi were found in the bottom samples collected in the Odessa Bay (Alexandrov 2004) (Table 4.9).

Tintinnidae were first recorded in the open part of the Black Sea and at the coast of Novorossiysk in 2001–2004 (Gavrilova 2005). Tintinnid ciliates are a vital microplanktonic link in aquatic food webs. They feed on planktonic algae and cyanobacteria and can consume up to 27% of annual primary production in the coastal waters. At the same time, tintinnids constitute an important seasonal food component for meso- and macrozooplankton, such as copepods (Capriulo and Carpenter 1983), large zooflagellates and fish larvae.

The first non-native Tintinnidae species were registered in the Black Sea in August and September of 2001: *Eutintinnus lususundae* Entz, 1885, *E. tubulosus* Ostenfeld, 1899, *E. apertus* Kofoid et Campbell, 1929 and *Eutintinus* sp. (Polikarpov et al. 2003; Gavrilova 2001, 2005; Kurilov 2005). The representative of another genus, *Salpingella decurtata* Jörgensen, 1924, was recorded for the first time in 2002 and since then it has been regularly observed in the samples during summer and autumn (Gavrilova 2010). The other findings of new Tintinnidae ciliate species in the Black Sea dated back to 2009 (*Rhizodomus tagatzi* Strelkov et Wirketis, 1950; *Tintinnopsis tocantinensis* Kofoid et Campbell, 1929), and *Amphorellopsis acuta* Schmidt, 1902 was first recorded in 2010 (Gavrilova 2010; Selifonova 2011). Thus, nine species new for the studied region of the Black Sea have been recorded in 1997–2014, namely *E. lususundae* (Fig. 4.24a), *E. tubulosus* (Fig. 4.24b), *E. apertus* (Fig. 4.24c), *Eutintinnus* sp. (Fig. 4.24d), *S. decurtata* (Fig. 4.24e), *T. tocantinensis* (Fig. 4.24g), *R. tagatzi* (Fig. 4.24f), *A. acuta* (Fig. 4.24h), and *Nolaclusilis* sp. (Fig. 4.24i).

Currently, *E. lusus-undae* has spread widely throughout the Black Sea (Polikarpov et al. 2003) and penetrated into the Sea of Azov (Kreneva 2003).

Table 4.9 Non-native species of microplankton in the Black Sea

Species	Donor region	Vector	First record	Status of establishment	References
Bacteria *Vibrio cholerae* (El Tor strain)	Southeast Asia	Bacillus-carrying	1970 Kerch, Odessa areas	Established	Burgasov et al. (1986)
Marine fungi Ascomycota *Savoryella lignicola*	Southeast Asia	Shipping	2004 Odessa Bay	Unknown	Alexandrov (2004)
Ascomycota *Cirrenalia basiminuta*	Indo-Pacifica	Shipping	2006 Odessa Bay	Unknown	Alexandrov (2006)
Sordariomycetes *Savoryella lignicola*	unknown	Shipping?	2006 Odessa Bay	Unknown	Alexandrov (2006)
Phylum Ciliophora Doflein, 1901 **Class Spirotrichea Butschli**					
Fam. Tintinnidae Claparede & Lachmann, 1858 *Eutintinnus* sp.	Mediterranean Sea		2001–2004 Sevastopol		Gavrilova (2005)
E. tubulosus (Ostenfeld, 1899) Kofoid & Campbell, 1939	?		Novorossiysk areas, Azov Sea		Selifonova (2011)
Eutintinnus haslae Taniguchi and Hada, 1981	?	Shipping		Established	Gavrilova and Dovgal (2016)
Eutintinnus apertus Kofoid & Campbell, 1929	?				Gavrilova and Dovgal (2016)
Atys angustatus E. A. Smith, 1872	?				Gavrilova and Dovgal (2016)
Salpingella decurlata Jorgensen, 1924	Mediterranean Sea	Shipping	2001–2004 Sevastopol, Novorossiysk, Tuapse areas	Established	Gavrilova (2005), Selifonova (2011)
Tintinnopsis directa Hada, 1932 *Tintinnopsis campanula* (Ehrenberg, 1840)	Mediterranean Sea	Shipping	2002 Novorossiysk, Tuapse areas	Established	Gavrilova (2005, 2010)
Eutintinnus lusus-unda (Entz, 1885) Kofoid & Campbell, 1939	Mediterranean Sea	Shipping	2001–2004 Novorossiysk	Established	Polikarpov et al. (2003)

Table 4.10 Non-native species of phytoplankton in the Black Sea

Species	Donor region	Vector	First record	Status of establishment	References
Phylum Ochrophyta **Class Dictyochophyceae** Fam. Dictyochaceae *Octactis speculum* (Ehrenberg) F.H. Chang, J.M. Grieve & J.E. Sutherland, 2017	Marine	Shipping	Georgian waters	Established	Chang (2024) https://www.algaebase.org
Infraphylum Dinoflagellata **Class Dinophyceae** **Order Gonyaulacales** Fam. Pyrocystaceae *Alexandrium monilatum* (Howell) Balech, 1995	Atlantic coast of North America	Shipping (ballast waters); currents advection	1991 Varna Bay	Established	Moncheva (1995)
1. *Alexandrium pseudogonyaulax* 2. (Biecheler) Horiguchi ex K. Yuki & Y. Fukuyo, 1992	Eastern Atlantic, Mediterranean	Shipping (ballast waters)	2001–2002 Varna Bay	Established	Terenko (2003)
Alexandrium minutum Halim, 1960	Mediterranean, Sub-tropical Atlantic	Shipping (ballast waters)	1991 Varna Bay	Established (?)	Moncheva (1995)
Fam.Actinomonadaceae *Apedinella radians* (Lohmann) P.H. Campbell, 1973	marine	Shipping?	Georgian waters	Established	https://www.algaebase.org
Gymnodinium aureolum (E.M. Hulburt) Gert Hansen, 2000	Mediterranean	Shipping (ballast waters)	2002 Varna Bay	Established	Terenko (2003)
Class Dinophyceae **Order Gymnodiniales** Fam. Gymnodiniaceae *Gymnodinium uberrimum* (Allman) (Kofoid et Swezy, 1921)	Fresh waters, ponds, reservoirs	Shipping (ballast waters)	1999 Varna Bay	Established	Moncheva et al. (2000), Terenko (2003)

(continued)

Table 4.10 (continued)

Species	Donor region	Vector	First record	Status of establishment	References
Fam.Gymnodiniaceae 1. *Margalefidinium polykrikoides* (Margalef) F. Gómez, Richlen &D.M. Anderson, 2017	Mediterranean, Eastern Atlantic	Shipping (ballast waters)	2001 North-Eastern Black Sea, 2002 Varna Bay	Established	Terenko (2003), Vershinin and Orlova (2008)
Fam.Gymnodiniaceae *Akashiwosanguinea* (K.Hirasaka) Gert Hansen & Moestrup, 2000 (Syn. *Gymnodinium aeruginosum* Stein and *Gymnodinium blax* Harris)	Mediterranean	Shipping (ballast waters)	2000 Coastal waters of Crimea	Established	Senicheva (2002)
Fam. Amphidiniaceae *Amphidinium acutissimum* J. Schiller, 1932	Northeast Atlantic	Shipping (ballast waters)	1999 Odessa Bay	Established	Terenko and Terenko (2000)
Amphidinium lanceolatum Schröder, 1911	Marine species	Shipping?	Georgian waters	Established	https://www.algaebase.org
Fam.Ceratiaceae 1. *Tripos eugrammus* (Ehrenberg) F. Gómez, 2013	Marine species	Shipping?	Georgian waters	Established	https://www.algaebase.org
Fam. Polykrikaceae *Polykrikos geminatus* (Schütt) D. Qiu & S. Lin, 2013	Marine species	Shipping?	Georgian waters	Established	https://www.algaebase.org
Fam.Dinophysaceae 1. *Dinophysis odiosa* (Pavillard) Tai & Scogsberg 1934	Mediterranean	Shipping (ballast waters)	2001 Coastal waters off Sevastopol	Established	Senicheva (2002)
Phylum Chlorophyta **Class Prasinophyceae** **Order Mamiellales** Fam.Mamiellaceae *Mantoniella squamata* (Manton & Parke) Desikachary 1972	North Atlantic	Shipping (ballast waters)	Beginning of 1980s Rumanian coastal waters	Established	https://www.algaebase.org

Phylum Cyanobacteria **Class Cyanophyceae** **Order Synechococcales** Fam. Prochlorococcaceae *Prochlorococcus marinus* Chisholm, Frankel, Goericke, Olson, Palenik, Waterbury, West-Johnsrud &Zettler, 1992	Mediterranean	Shipping (ballast waters)	1990 Ukraine cost waters 2002 Burgas estuary	Established nov. gen. nov. sp.: an oxyphototrophic marine prokaryote containing divinyl chlorophyll a	Petrova-Karadjova (1990) Moncheva et al. (1995) Zaitsev and Ozturk (2001)
Phylum Haptophyta **Class Coccolithophyceae** **Order Phaeocystales** Fam Phaeocystaceae *Phaeocystis pouchetii* Lagerheim, 1893	Aegean or Marmara seas	Shipping	1989 г. Bulgarian coastal and open waters	Established	Petrova-Karadjova (1990)
Phylum Heterokontophyta **Class Bacillariophyceae** **Order Rhizosoleniales** Fam. Rhizosoleniaceae *Pseudosolenia calcar-avis* (M. Schultze) Sentstrom 1986 (=*Rhizosolenia calcar-avis* (*M. Schultze)*	Atlantic Ocean	Shipping	1925–26 North-western part of the Black Sea	Established	Velikova et al. (1999)
Fam.Hemiaulaceae *Cerataulina bergonii* Ostenfeld, 1903	North Atlantic	Shipping	The 1930s Coastal waters of Sevastopol	Established	Morozova-Vodyanitskaya (1948)
Fam. Leptocylindraceae *Leptocylindrus danicus* Cleve 1889	North Atlantic	Shipping	The 1930s Coastal waters off Sevastopol	Established	Krakhmalny et al. (2012)
Class Bacillariophyceae **Order Chaetocerotanae** Fam. Chaetocerotaceae *Chaetoceros socialis* Lauder, 1864	North Atlantic	Shipping	1908–1928 Coastal waters off Sevastopol	Established	Morozova-Vodyanitskaya (1948)

(continued)

Table 4.10 (continued)

Species	Donor region	Vector	First record	Status of establishment	References
Chaetoceros *tortissimus* Gran	North Atlantic	Shipping	1939 Coastal waters off Sevastopol	Established	Senicheva (2002)
Chaetoceros radians Schutt 1985 (uncertain)	North Atlantic	Shipping	1939 Coastal waters off Sevastopol	Established	Morozova-Vodyanitskaya (1948)
Chaetoceros debilis Cleve 1894 emend Xu, Y.Li & Lundholm in Xu et al. 2020 *Chaetoceros diversus* Cleve, 1873	Northwest Atlantic	Shipping	1999 Coastal waters off Sevastopol	Established	Senicheva (2002)
Fam. Chaetocerotaceae *Bacteriastrum delicatulum* Cleve 1897	Mediterranean Sea	Shipping	2000 Odessa Bay	Established	Terenko and Terenko (2000)
Class Bacillariophyceae **OrderThalassionematales** Fam Talassionemataceae *Lioloma pacificum* (Cupp) Hasle	Mediterranean Sea	Shipping	2001–2003 Coastal waters of Odessa Bay	Established	Senicheva (2002) Alexandrov (2004)
Thalassiosira nordenskjoldi Cleve 1873	North Atlantic, Arctic	Shipping	1986–1988 Near Bosphorus Strait	Established	Senicheva (2002)
Thalassiosira nana Lohmann, 1908	The North Atlantic and the Arctic	Shipping	1930s Coastal waters off Sevastopol	Established	Morozova-Vodyanitskaya (1948)
Fam. Bacillariaceae *Pseudo-nitzschia inflatula* Hasle (=Nitzschia infantula Hasle)	Mediterranean Sea	Shipping	2002 Coastal waters off Sevastopol 1999–2001 Odessa region	Established	Senicheva (2002)

Fam. Thalassionemataceae *Thalassiothrix longissima* Cleve & Grunow, 1880 (uncertain)	Mediterranean Sea	Shipping	2002–2003 Odessa region	Established	Alexandrov (2004)
Thalassiothrix mediterranea Pavill.	Mediterranean Sea	Shipping	Georgian waters	2001 rare species in the Black Sea	Terenko and Terenko (2001)
Fam. Skeletonemaceae *Sceletonema subsalsum* (Cleve-Euler) Bethge, 1928 (uncertain)	Mediterranean Sea	Shipping	1993 Coastal waters off Romania	Established	Alexandrov (2004)
Fam. Protoraphidiaceae *Asterionellepsis glacialis* (Castracane) Round, 1990 (uncertain)	Mediterranean Sea	Shipping	1968 Coastal waters off Sevastopol	Established	Senicheva (1971)
Fam. Fragilariaceae *Diatoma tenue var. elongatum* Lyngbye, 1819	Marine	Shipping	Georgian waters	Established	(2009). Catalogue of Diatom names.
Class Dictyochophyceae **Order Pedinellales** Fam. Dictyochaceae *Apedinella radians* (Lohmann) P.H. Campbell, 1973	Mediterranean Sea, Atlantic Ocean, Pacific Ocean	Shipping? (ballast waters)	1999 Coastal waters off Odessa Bay	Established?	Terenko and Terenko (2000)
Fam. Actinomonadaceae *Octactis speculum* (Ehrenberg) F.H. Chang, J.M. Grieve & J.E. Sutherland, 2017	Coastal area of European part of Atlantic Ocean	Shipping? (ballast waters)	1999 Coastal waters waters off Odessa Bay	Established?	Terenko and Terenko (2000)

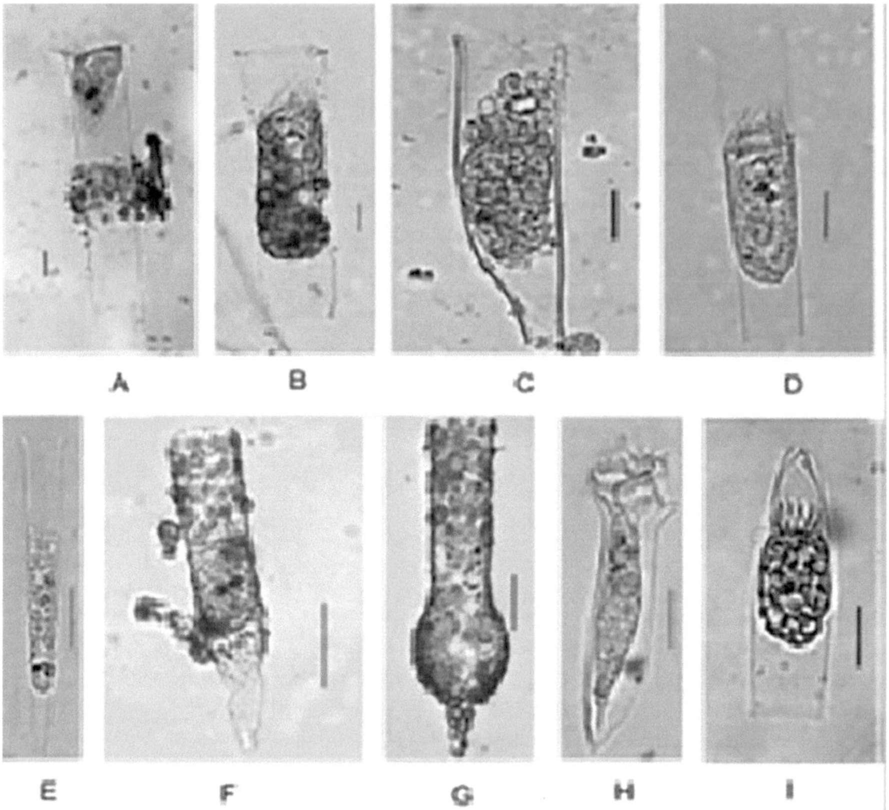

Fig. 4.24 Non-native Tintinnidae species found in the Black Sea in 1997–2014. (**a**) *Eutintinnus lususundae*; (**b**) *E. tubulosus*; (**c**) *E. apertus*; (**d**) *Eutintinnus* sp.; (**e**) *Salpingella decurtata*; (**f**) *Rhizodomus tagatzi*; (**g**) *Tintinnopsis tocantinensis*; (**h**) *Amphorellopsis acuta*; (**i**) *Nolaclusilis* sp. Scale bars: (**a**, **d**, **g**) 20 μm, (**b**, **c**, **i**) 10 μm, (**e**, **h**) 15 μm, (**f**) 50 μm. (After Gavrilova and Dovgal 2016)

Phytoplankton

Among the non-native phytoplankton species that have been established for a long time ago and made a significant contribution to the phytoplankton community, the diatom *Pseudosolenia calcar-avis* (=*Rhizosolenia calcar-avis*) stands out. It was first observed in the northwestern part of the Black Sea in 1924 (Velikova et al. 1999). Currently, *P. calcar-avis* is widespread, mainly in the northwestern part of the Black Sea, with occasional outbreaks of development. On the one hand, this species serves as a food for large zooplankton species; on the other hand, it displaces local phytoplankton species during outbreaks of its development. The diatoms *Cerataulina bergonii* (=*Cerataulina pelagica*) and *Chaetocerus socialis*, *Chaetocerus tortissimus*, and *Chaetocerus diversus*, introduced into the Black Sea in the first half of the twentieth century, probably from the North Atlantic, became widespread (Morozova-Vodianitskaya 1948). Outbreaks of development of *Cerataulina pelagica* and *Chaetoceros socialis* occur in spring and autumn throughout the basin (Georgieva 1993; Krakhmalny et al. 2012). A recently arrived warm-water non-native species *Leptocylindrus danicus* appeared and develops only in summer in the northwestern part of the sea (Krakhmalny et al. 2012).

Mediterranean pelagic species have penetrated into the Black Sea with the Lower Bosphorus current. Some of them were able to survive in conditions of low salinity and existed only in the near-Bosphorus area. However, a number of other species were recorded in the 1960s and 1970s not only near the Bosphorus part of the Black Sea, but also off the coast of Crimea (Kuzmenko 1966; Senichkina 1973; Kovalev et al. 1998). Currently, a number of other phytoplankton species earlier found only near the Bosphorus have been recently registered in the coastal waters off Sevastopol (Senicheva 2002) (Table 4.10) and in the northeastern part of the sea near Utrish (Vershinin and Orlova 2008).

Not all of non-native algae, discovered in the Black Sea for the first time should be considered as newly arrived non-natives or even established species from the Mediterranean Sea or other areas. The identification of new species may also need clarifying of the taxonomic affiliation of individual representatives of phytoplankton. Thus, in the coastal zone of northwestern Crimea as a result of continuous (1968–2002) observations of the succession of the phytoplankton community, new species of flora were identified. They include (i) diatoms—*Asterionellapsis glacilis* (Castracane) F.E. Round, *Chaetoceros tortissimus* Gran, *Thallassiosira nordenskioeldii* Cleve, *Lioloma pacificus* (Cupp) Hasle, *Pseudonitzschi inflatula* (Hasle) and two species of the genus *Chaetoceros*, and (ii) dinophytes—*Dinophysis odiosa* (Pavillard) Tai & Scogsberg (Senicheva 2002).

A significant number of phytoplankton species were found in the port areas of the northwestern part of the sea (Table 4.10) (Alexandrov 2004, 2007; Moncheva et al. 1995, 2000; Terenko and Terenko 2000; Terenko 2003). Fifteen non-native species were found in the port of Odessa alone, with most of the new species (8) belonging to dinophytes (Alexandrov 2004, 2007). In addition, several species were found as cysts in bottom sediments, which we did not include in the list of established species. Many of these species were recorded also in the Georgian area of the Black Sea (Table 4.10).

In May–June 2005 in the northeastern Black Sea, on its shelf, slopes, and in the open sea, intensive (up to 1.92×10^5 cells/l) growth of non-native species, *Chaetoceros throndsenii* (Marino, Montresor & Zingone) was observed for the first time (Pautova et al. 2012). In the following years it was recorded sporadically—in 2006, 2010, and 2015 (Silkin, pers. comm.). In 2011 *Chaetoceros throndsenii* was recorded in the northwestern Black Sea (Terenko 2017).

Toxic algae species, capable of forming cysts, are among the most dangerous invaders arriving with ballast waters. The presence of cysts gives them the opportunity to survive adverse conditions during transportation and existence in new environment. It is known that the most suitable substrate for the accumulation and maturation of settled cysts is silty bottom sediments. When studying the silty sediments near the port of Odessa, quantitative indicators of dinophyte cysts measured for the first time in the Black Sea, ranged from 1.6 to 105.6 million cl. m^{-2} in the upper 5 mm of sediment layer. The most common were cysts of the dinophyte genera Gonyaulax, Scrippsiella, Diplopsalis, Oblea, Protoperidinium, and Alexandrium, which have been successfully germinated in the laboratory. Despite the fact that *Alexandrium tamarense*, *A. affine*, and *A. acatenella* were not found in the plankton, they are among the potential invaders, since their cysts were found in the port areas (Alexandrov et al. 2017; Terenko 2003).

In 2002, massive blooms, caused by the abnormal growth of green algae *Prochlorococcus marinus* was first observed in the estuaries of the northwestern Black Sea (Nesterova 2003). Most of the new species are found only locally in small numbers, but their populations has been increasing in recent years. Some of them spread to wider areas, and demonstrate outbreaks of development (Terenko 2003; Alexandrov 2004, 2017). In the northeastern part of the sea, in the coastal area of Cape Bolshoy Utrish, species, new to the Black Sea but inhabiting the Mediterranean Sea (i.e., *Chaetocerus tortissimus, Cochlodinium polykrikoides, Alexandrum* sp.) were discovered in 2001. *Cochlodinium polykrikoides* formed the bulk of biomass (517 micrograms l^{-1}), whereas the biomass of the remaining algae did not exceed 170 micrograms l^{-1} (Vershinin and Orlova 2008). This species was also discovered in the Odessa Bay in 2002 (Terenko 2003).

Besides the penetration of euryhaline marine species of the Mediterranean flora, representatives of freshwater phytoplankton also are continuously arriving, especially into the northwestern part of the Black Sea influenced by large rivers—the Danube, Dnieper, and Dniester (Ivanov 1965; Nesterova and Terenko 1987). Some of these species occur regularly, but an extent of their penetration depends on the river plume size. These species are not classified as alien. A number of brackish-water and freshwater phytoplankton species have established in brackish- and fresh-water bays and estuaries of the western part of the Black Sea (Moncheva et al. 1995).

On total, the following number of established non-native phytoplankton species have been registered: diatoms species—15, dinoflagellates species (of which *Alexandrium monilatum* and *Mantoniella squamata* are potentially toxic species (Moncheva et al. 1995; Shiganova 2009)—18, green algae—*Mantoniella squamata*—1, blue-green *Prochlorococcus marinus* that is a genus of very small (0.6 μm) marine cyanobacteria with an unusual pigmentation (chlorophyll a^2 and b^2) -1, golden algae *Phaeocystis pouchetii*—1, unicellular phytoflagellate *Apedinella radians* and *Octactis speculum*—2 (Table 4.10).

Currently, the process of penetration of non-native species of phytoplankton continues and intensifying, both as a result of the currents advection through the Bosphorus and with ballast waters, mainly into the north-western and western parts of the sea. An establishment of already invaded species also continues. Such species may even temporarily occupy the place of subdominants, but, as a rule, they occur in small numbers or intensively develop only in particular years (Moncheva et al. 1995; Terenko and Terenko 2000; Senicheva 2002; Alexandrov 2004; Alexandrov et al. 2017), which indicates the high resistance of the Black Sea phytoplankton community to the non-native species (Polikarpov et al. 2003). It is important to emphasize the changing of the origin regions of newly arriving species. In 1960s and 1980s, and even before 2000 most of phytoplankton species penetrated from the North Atlantic (10 established species and 4 questionable), and were boreal or temperate waters' origin. However, in the recent decades more subtropical and even tropical species of Mediterranean origin have been discovered. Of these, 15 species can be assumed as established according to their description, and 17 are questionable (Table 4.10).

With an increase of the surface layer temperature, more Mediterranean phytoplankton species arrive in the Black Sea and found in different areas, with significant increase of their numbers in recent years. First of all, a large number of phytoplankton species previously unknown in the Black Sea, but permanently occurring in the Mediterranean Sea, are recorded the near Bosporus area (Table 4.11) (Georgieva 1993).

New for the Black Sea diatoms *Fragillaria striatula* Lyngb, *Thalassiothrix frauenfeldii*, coccoliphorids *Calyptrosphaera incise* Schill and dinoflagellates *Ceratium macroceros* were found near the Bosphorus area by Scolka and Bodyanu (1963). These species were observed at salinity 34‰ and temperature 14 °C which significantly differ from the usual Black Sea water conditions. Thus, they are unable to establish in current environment.

Kuz'menko (1966) found a number of species, previously unknown in the Black Sea but typical for the Mediterranean Sea, e.g., *Dynophysis schuttii* Murr. et Whitt. and *Podolampas spinifer* Okatumura; they were collected off the southern coast of the Crimea at salinity 18–18.5‰.

In the beginning of 1990s, new for the Black Sea species, like *Katodinium rotundatum* (Lohm) Fott, *Achradina sulcata* Lohm, and *Pronoctiluca* sp., were found by L.G. Senichkina in the shallow-water area off Yalta. She also recorded *Distephanus octonarius* var. *Polyactis* (Jorg) Gleser and *D. speculum* var. *Septenarius* Jorg, which were previously unreported in the Black Sea (Senichkina 1993). Many of these species were found not only in the Mediterranean origin waters with a high salinity, but also in typical Black Sea waters. Their presence in the upper layers of the Black Sea may be explained by penetration of the waters from the Sea of Marmara, followed by their subsequent mixing. As a result, 33 representatives of the Mediterranean phytoplankton were registered in the subsurface layers of the near-Bosporus region, among them are *Syracosphaera cornifera, Ceratium furca var. eugrammum, Pyrocystis hamulus, Pronoctiluca acuta*, and some others (Table 4.11) (Georgieva 1993).

In the north-eastern part of the Black Sea, 93 species belonging to 7 classes were found. Thirteen species from this list are known as toxic according to data obtained in other seas. Two species of diatoms, namely, *Pseudo-nitzschia pseudodelicatissima* and *P. pungens* were identified. They may cause amnesic poisoning of shellfish; the strength of their toxic effect depends on the environmental conditions. In summer of 2001 a dinoflagellate, belonging to the *Alexandrium* genus was encountered for the first time. Many species of this genus can be the source of the lethally dangerous paralytic shellfish poisoning. In February of 2002, hatching of cysts and a population outburst of *Alexandrium* spp. were observed.

Table 4.11 Newly arrived phytoplankton species found in the near-Bosporus area of the Black Sea (Georgieva 1993)

Species	T, °C	S, ‰
Biddulphia alternans (Bail.) V. H.	7.70	19.34
Eucampia cornuta (Cl) Grun	8.40	19.78
Rhizosolenia styliformis Brightw	17.13	17.59
Thalassiothrix mediterranus Pavill	7.89	19.13
Amphidinium conradi (Conrad) Schill.	7.44	18.38
A. vigrense Wolosz	10.79	18.18
A. mannanini Herd.	10.79	18.18
Ceratium hexacanthum f.contortum (Lemm.) Jorg.	7.80	18.39
C.massiliense (Gourret) Jorg	7.81	18.39
C.furca var. eugrammum (Eht.) Jorg.	18.67–24.97	16.44–18.39
C.fusus var.seta (Eht.) Jorg.	7.80	18.39
C.teres Kof.	n/a	n/a
C. trichoceros (Eht.) Kof.	10.79	18.18
C.tripos var. atlanticum Ostf.	7.81	18.39
C.hexacantum f. aestuarium (Schrod.) Schill.	7.81	18.39
Cochlodinium citron Kof.et Sw.	7.39	18.50
Gymnodinium paradoxum Schill.	n/a	n/a
G. pygmaeum Leb.	10.79	18.18
Oxytoxum parvum Schill.	7.41	18.33
O. variabile Schill	23.98	15.73
O. viride Schill.	8.88	20.32
Peridinium sinaicum Matz	18.82	18.36
Pronoctiluca acuta (Lohm.) Schill.	9.14–9.85	18.57–18.96
P. pelagica Pavill	7.64	n/a
Pyrocystis hamulus Cl.	13.20	17.20
P. fusiformis (W.Th.) Mur.	9.51	21.42
P. pseudonoctiluca (W.Th.) Schill.	6.52	18.40
Coccolithus pelagicus (Walich.) Schill.	7.55	18.20
Rhabdosphaera stylifera Lohm.	8.74	20.14
Syracosphaera coronata Lohm.	7.44	17.98
S. cornifera Schill. (Helladosphaera)	16.10	17.98
S .quadricornu (Anthosphaera) Schill.	8.74	29.14
S. spinosa Lohm.	7.31	18.52
Total 33 species		

Phytoplankton species of the *Dinophysis* genus, namely, *D. rotundata*, *D. caudata, D. acuminata, D. hastata, D. fortii, D. norvegica, D. tripos*, and *D. sacculus* are known as producers of toxins, causing diarrhetic poisoning of shellfish. Two of them, including *Alexandrium monilatum* Balech, 1995 (Fig. 4.25), and *Dinophysis norvegica* Claparede Lachmann, 1859, which have been associated with the fish fatal poisoning (May et al. 2010) and may produce harmful algal blooms.

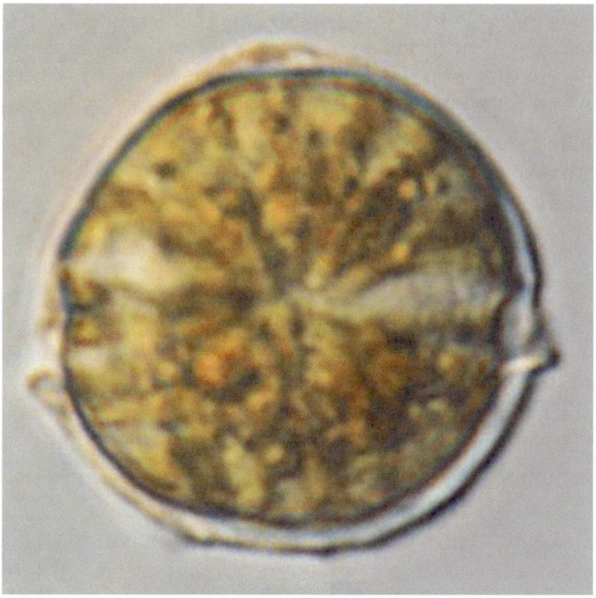

Fig. 4.25 Toxic dinoflagellate *Alexandrium monilatum*

An effect of the toxic dinoflagellate *Alexandrium monilatum* upon the shellfishes is almost unknown. Toxic strains produce endotoxins with hemolytic and neurotoxic action, which may cause a fatal outcome for fish and invertebrates. Survival, grazing and behavioral responses of three shellfish species to *A. monilatum* impact were studied experimentally. Grazing was investigated for two dimensional groups of *Crassostrea virginica*, *Mercenaria mercenaria*, and *Perna viridis*. These species inhabit areas coincided with the *A. monilatum* blooms regions. Clearance rates of each species were depressed when exposed to toxic *A. monilatum* alone or together with the nontoxic *Pavlova* sp. compare to control animals fed only with non-toxic algae. Exposure to toxic A. *monilatum* also caused shellfish to decrease shell valve gape. *A. monilatum* cells remnants were found even in shellfish feces, but the cells were unable to re-establish viable populations following the gut passage. Survival of *M. mercenaria* and *C. virginica* larvae was also tested when exposed to *A. monilatum* cells. Survival rate was significantly lower for larvae, exposed to sonicated *A. monilatum*, in comparison to control larvae tested with non-toxic *A. tamarense*. Overall, the data indicate that blooms of *A. monilatum* can adversely affect some shellfish species by reducing valve gape and clearance rate, and by inducing larval mortality (May et al. 2010).

The epiphytic dinophlagellate *Prorocentrum lima* was intermittently presented in the samples collected in May–November (the maximum abundance was 400 cells/l in September 2001). It was introduced into a laboratory culture and produced diarrhetic toxins, namely, okadaic acid and dinophysis-toxin. *P. lima* is dangerous because it grows on macroalgae fouling mussels and oysters, including cultivated in mariculture. All samples of mussel tissues taken in the summer of 2001 contained diarrhetic toxins that most likely originated from *P. lima*. The toxin concentrations in the mussels did not exceed 0.2 μg/g of tissue and were lower than the maximum permissible concentration accepted in the European Community and Canada. The mixotrophic ichthyotoxic dinophlagellate *Cochlodinium polykrikoides*, which was first encountered off the north-east coast, featured a bloom in the area off Utrish in August of 2001 (the maximum abundance was up to 70 × 103 cells/l; the biomass was up to 620 μg/l) and should be considered as a risk factor for the canned fish (Vershinin et al. 2005, 2008).

Ferns and Macrophytes

Macrophytes are the main phototrophic link of coastal ecosystems; therefore, any changes in their composition entail the restructuring of these ecosystems and, as a result, the transformation of the main biocenoses (Minicheva 2002).

However, first comprehensive monograph on phytobenthos in the Black Sea was published only in 1975 by A.A. Kalugina-Gutnik. She was the first one who mentioned non-native species *Eugomontia sacculata* from class Chlorophyceae (Table 4.12).

Later, cold-waters species of boreal origin were recorded in the Black Sea. In 1990, the cold-water euryhaline coastal species of brown algae *Desmaretia viridis* was discovered in the Odessa Bay for the first time (Minicheva and Eremenko 1993) (Table 4.12). In the winter season of 1994–95, *Desmaretia viridis* became a widespread species in the coastal area of the bay. In recent years, especially in cold ones, *Desmaretia viridis* has been widespreaded in the northwestern part of the Black Sea including Zernov's Phyllophora Field in 2017 and 2019 (Minicheva 2006). It is also known as stringy acid kelp and is the most acidic of the acid kelps with a vacuolar pH of about 1. It is best known for releasing sulfuric acid when damaged, usually destroying itself and other nearby marine plant life. *D. viridis* usually found in colder, shallow, intertidal zones all around the world.

Halosiphon tomentosus (Lyngbye) Jaasund, 1957 is the sole representative of the order Tilopteridales in the Black Sea, which first discovered in the area of the Odessa coast in March 2015.

In 2002 Milchakova for the first time described 38 new for the Black Sea species of macrophytes. Among them are green algae Chlorophyceae (10 species), brown algae Fucophyceae (5 species) and red algae Rhodophyceae (12 species). It was shown that the highest species diversity was recorded at northwestern shelf, The highest number of new flora species was recorded in the Anatolian shelf (Milchakova 2002).

Off the coasts of Romania and Bulgaria, six new species of Cladophora were recorded together with other green alga species; they are euryhaline and eurybiont species and serve as indicators of eutrophication. These species were probably brought with ballast waters (Milchakova 2002). The most significant change of the macrophyte flora in the Black Sea is manifested in almost twofold increase of the number of species of the *Cladophora*, *Ulva*, *Ceramium*, *Polysiphonia*, *Cystoseira* and *Sargassum* genera; many of them play a key role in the bottom communities of the near-shore ecosystems of the Mediterranean Sea (Table 4.12). Most of the new

Table 4.12 Non-native species of ferns and macrophytes in the Black Sea

Species	Donor region	Vector	First record	Status of establishment	Author
Class Polypodiopsida **Order Salviniales** Fam. Azollaceae *Azolla caroliniana* *Azolla filiculoides*	Rivers and lakes of Southeast Asia	Shipping (ballast waters)	1980s Danube Delta	Established	Zaitsev and Ozturk (1989)
Phylum Chlorophyta **Class Chlorophyceae** Order **Ulotrichales** Fam. Chaetophoraceae *Eugomontia sacculata* Kormman 1960	Eastern Atlantic, Mediterranean Sea (?)	Shipping	1975 Coastal waters off Bulgaria	Established	Kalugina-Gutnik (1993)
Class Ulvales **Order Ulvales** Fam. Ulvaceae *Enteromorpha kylinii* Bliding	Mediterranean Sea	Shipping, currents advection	1990s Coastal waters off Turkey	Established	Bavaru et al. (1991)
Ulva kylinii (Bliding) H.S. Hayden, Blomster, Maggs, P.C.Silva, Stanhope & Waaland, 2003	Mediterranean Sea	Shipping	1990s Coastal waters off Turkey	Established	Bavaru et al. (1991), Milchkova (2002), Ausel et al. (1995), algaebase.org
Ulva curvata Kutz. De Toni	Mediterranean Sea	Shipping, currents advection	1990s Coastal waters off Turkey	Established	Aysel and Erdugan (1995)
Ulva fasciata Delile	Mediterranean Sea	Shipping, currents advection	1990s Coastal waters off Turkey	Established	Aysel and Erdugan (1995)
Class Ulvophyceae **Order Cladophorales** Fam. Cladophoraceae (O.F.Muller) Kulzing	Mediterranean Sea	Shipping, currents advection	1990s Coastal waters off Turkey	Established	Aysel and Erdugan (1995)

(continued)

Table 4.12 (continued)

Species	Donor region	Vector	First record	Status of establishment	Author
Cladophora fracta (O.F. Muller ex Vahl) Kutz 1843	Mediterranean Sea	Shipping, currents advection	1990 Coastal waters off Romania	Established	Bavaru et al. (1991)
Cladophora glomerata (L) Kützing, 1843	Mediterranean, Adriatic seas	Shipping	1990 Coastal waters off Romania, Turkey	Established	Bavaru et al. (1991)
Cladophora hutchinsiae Dillw.) Kutz.	Mediterranean, North Atlantic	Shipping	1990 Coast of Romania; 2003–2004 Caucasian coast	Established	Bavaru et al. (1991), Lisovskaya and Nikitina (2007)
Cladophora lehmanniana (Lindenb.)	Mediterranean	Shipping, currents advection	1990s Coastal waters off Turkey	Established	Aysel and Erdugan (1995)
Cladophora pellucida (Huds.) Kutz.	Mediterranean	Shipping, currents advection	1990s Coastal waters off Turkey	Established	Aysel and Erdugan (1995)
Cladophora prolifera (Roth.) Kutz.	Mediterranean	Shipping, currents advection	1990s Coastal waters off Romania	Established	Aysel and Erdugan (1995)
Cladophora ruchingeri (C.Ag.) Kutz	North Atlantic, Adriatic (?)	Shipping	1990 Coastal waters off Romania	Established	Bavaru et al. (1991)
Cladophora rupestris (Linnaeus) Kützing, 1843	Cladophora is one of the largest filamentous green-algal genus and has a widespread distribution in Caspian Sea Coast.	Shipping	2002 Cladophora rupestris (cladophorales, chlorophyta): a potential source of antibacterial compounds.	Established	Milchakova (2002)
Phylum Ochrophyta Class Phaeophyceae	Cool-water	Shipping		Established	Bavari et al. (1991), Gallardo et al.

Fam. Chordariaceae *Leathesia marina* Lyngbye, 1819.			1991., 1993 Coastal waters of Roma-nia, Turkey		(1993), algaebase.org
Fam. Ectocarpaceae *Ectocarpus caspicus* A. Henckel, 1909	Cool-water	Established	2002 Coastal waters of Romania	Established	Milchakova (2002) algaebase.org
Phylum Ochrophyta **Class Phaeophyceae** Fam. Chordariaceae *Sphaerotri chiadisessa* (Setchell & N.L.Gardner) A.D. Zinova, 1958 *Leathesia* S.F. Gray, 1821	Mediterranean Sea (?), European coast of Atlantic (?)	Shipping	1990 Coastal waters off Romania	Established	Bavaru et al. (1991)
Fam. Chordariaceae *Protectocarpus speciosus* (Boerg.) Komm. (not registrated)	Mediterranean Sea (?), European waters of Atlan-tic (?)	Shipping	1990 Coastal waters off Romania	Established	Bavaru et al. (1991)
Order Desmarestiales Fam.Desmarestiaceae *Desmarestia viridis* (Lamouroux, 1813)	Northern Atlantic coastal waters	Shipping	1992 Odessa Bay	Established	Minicheva and Eremenko (1993)
Class Phaeophyceae **Order Tilopteridales** Fam. Cutleriaceae *Cutleria adspersa* (Mertens ex Roth) De Notaris, 1842	Mediterranean Sea, Northern Atlantic	Shipping (ballast waters or fouling)	1992 Coastal waters of Ukraine	Established	Minicheva and Eremenko (1993)
Class Phaeophyceae **Order Tilopteridales** Fam. Halosiphonaceae *Halosiphon tomentosus* (Lyngbye) Jaasund, 1957	Mediterranean Sea,	Shipping	2015 Odessa Bay	Established	Minicheva (2015a, b)

(continued)

Table 4.12 (continued)

Species	Donor region	Vector	First record	Status of establishment	Author
Class Phaeophyceae **Order Fucales** Fam. Sargassaceae *Cystoseira compressa* (Esper) Gerloff & Nizamuddin, 1975	Mediterranean Sea	Shipping (ballast waters or fouling)	Beginning of 1990s Coastal waters off Turkey	Established	Aysel and Erdugan (1995)
Fam.Sargassaceae *Cystoseira corniculata* (Turner) Zanardini, 1841	Mediterranean Sea	Shipping (ballast waters or fouling); currents advection	Beginning of 1990s Coastal waters off Turkey .	Established	Aysel and Erdugan (1995)
Fam. Ceramiaceae *Cystoseira foeniculacea f. latiramosa* (Ercegovic) A.Gómez Garreta, M.C. Barceló, M.A.Ribera & J. R.Lluch, 2001	Mediterranean Sea	Shipping (ballast waters or fouling); currents advection	Beginning of 1990s Coastal waters off Turkey .	Established	Aysel and Erdugan (1995)
Fam. Sargassaceae Kutz. *Sargassum acinarium* (L) C.Ag.=S.Ag.f. linifolium (Turn.)C.Ag. *S. hornschuchii* C.Ag.	Mediterranean Sea	Shipping (ballast waters or fouling); currents advection	Beginning of 1990s Coastal waters off Turkey .	Established	Aysel and Erdugan (1995)
Phylum Rhodophycea **Class Florideophyceae** **Order Ceramiales** Fam. Ceramiaceae *Antithamnion heterocladum* Funk 1955	Mediterranean Sea	Shipping (ballast waters or fouling); currents advection	Beginning of 1990s Coastal waters off Turkey	Established	Aysel and Erdugan (1995)

Fam. Ceramiaceae *Ceramium cimbricum* H.E. Petersen, 1924	Mediterranean Sea (?)	Shipping (ballast waters or fouling); currents advection	Beginning of 1990s Coastal waters off Turkey	Established	Bavaru et al. (1991)
Fam. Ceramiaceae *Gayliella flaccida* (Harvey ex Kützing) T.Cho & LMcIvor, 2008	Mediterranean Sea (?)	Shipping (ballast waters or fouling); currents advection	1990 Coastal waters off Bulgaria	Established	Aysel and Erdugan (1995)
Fam. Ceramiaceae *Pseudoceramium tenerrimum* (G.Martens) Okamura Barros-Barreto & Maggs	Mediterranean Sea (?)	Shipping (ballast waters or fouling); currents advection	Beginning of 1990s Coastal waters off Turkey	Established	Aysel and Erdugan (1995)
Fam. Ceramiaceae *Compsothamnion thuioides* (Smith) F. Schmitz, 1889	Mediterranean Sea (?)	Shipping (ballast waters or fouling); currents advection	Beginning of 1990s Coastal waters off Turkey	Established	Aysel and Erdugan (1995)
Fam. Delesseriaceae *Dasya ocellata* (Gratel.) Harv. 1833	Mediterranean Sea	Shipping (ballast waters or fouling); currents advection	Beginning of 1990s Coastal waters off Turkey	Established	Aysel and Erdugan (1995)
Class Florideophyceae **Order Ceramiales** Fam. Rhodomelaceae *Carradoriella elongella* (Harvey) Savoie & G.W. Saunders, 2019	Mediterranean Sea	Shipping (ballast waters or fouling); currents advection	Beginning of 1990s Coastal waters off Turkey	Established	Aysel and Erdugan (1995)

(continued)

Table 4.12 (continued)

Species	Donor region	Vector	First record	Status of establishment	Author
Fam.Rhodomelaceae *Polysiphonia* Greville, 1823	Mediterranean Sea	Shipping (ballast waters or fouling); currents advection	Beginning of 1990s Coastal waters off Turkey	Established	Aysel and Erdugan (1995)
Fam. Ceramiaceae *Pseudoceramium tenerrimum* (G. Martens) Barros-Barreto & Maggs, 2023	Mediterranean Sea	Shipping (ballast waters or fouling); currents advection	Beginning of 1990s Coastal waters off Turkey	Established	Aysel and Erdugan (1995)
Fam. Ceramiaceae *Compsothamnion thuyoides* (Smith.) F. Schmitz, 1889	Mediterranean Sea	Shipping (ballast waters or fouling); currents advection	Beginning of 1990s Coastal waters off Turkey	Established	Aysel and Erdugan (1995)
Class Florideophyceae **Order Ceramiales** Fam. Delesseriaceae *Dasya ocellata* (Gratel.) Harvey, 1833	Mediterranean Sea	Shipping (ballast waters or fouling); currents advection	Beginning of 1990s Coastal waters off Turkey	Established	Aysel and Erdugan (1995)
Fam. Rhodomelaceae *Polysiphonia elongella* Harvey 1833	Mediterranean Sea, Adriatic Sea (?)	Shipping (ballast waters or fouling); currents advection (?)	Beginning of 1990s Coastal waters off Turkey	Established	Aysel and Erdugan (1995)
Polysiphonia Greville, 1823	Mediterranean Sea	Shipping (ballast waters or fouling); currents advection (?)	Beginning of 1990s Coastal waters of Turkey	Established	Aysel and Erdugan (1995)

Polysiphonia paniculata Montagne 1842	Mediterranean Sea, North and South Atlantic	Shipping (ballast waters or fouling); currents advection (?)	Beginning of 1990s Coastal waters off Turkey	Established	Aysel and Erdugan (1995)
Polysiphonia stricta (Mertens ex Dillwyn) Greville, 1824	Mediterranean Sea (?)	Shipping (ballast waters or fouling); currents advection (?)	Beginning of 1990s Coastal waters off Turkey	Established (?)	Aysel and Erdugan (1995)
Polysiphonia subulata (Ducl.) P.Crouan et H. Crouan	Mediterranean Sea	Currents advection	Beginning of 1990s Coastal waters off Turkey	Established	Aysel and Erdugan (1995), Dimitrova-Konaklieva (1981), Garreta et al. (2001)
Polysiphonia tenerrima Kützing, 1843	Mediterranean Sea	Currents advection	Early 1990s Coastal waters off Turkey	Established	Aysel and Erdugan (1995)
Vertebrata fucoides (Hudson) Kuntze, 1891	Mediterranean Sea	Currents advection	Early 1990s Coastal waters off Turkey	Established	Aysel and Erdugan (1995), Karacuha et al. (2007)
Class Florideophyceae **Order Bonnemaisoniales** 1. Fam. Bonnemaisoniaceae *Bonnemaisonia hamifera* Hariot 1891	Northeastern Atlantic Ocean	Shipping	2015 the Caucasian coast and Crimea	Established	Simakova and Smirnov (2019)
Fam. Hydrocharitaceae *Elodea canadensis* Michx	Fresh water (The water plague)	Currents advection (?)	From European fresh waters?	North America, it is listed in the British waters of Europe, It has spread widely	Aphanas'ev (2006)

species are thermophilic organisms and inhabit the transition zone between the boreal and tropical zones (Milchakova 2002).

Cladophora is one of the largest filamentous green-algal genuses and is widely distributed. It is a genus of reticulated filamentous green algae in the class Ulvophyceae. They may be referred as reticulated algae, branching algae, or blanket weed. The genus has a worldwide distribution and is harvested for the food and medical purposes. Recent studies have shown that seaweeds are a good source of structurally novel secondary metabolites with antibacterial, antialgal, antifungal and antitumoral properties. Therefore, natural products from marine algae appear as a promising alternative valuable source of new compounds for drug development. It is also noteworthy that seaweeds often produce salt-stable antimicrobial compounds, useful in the control of either fish or shellfish pathogens in aquaculture where the high-salinity conditions might reduce antibiotic effectivity. In this investigation the antimicrobial activity of Cladophora rupestris—(Linnaeus) Kützing, 1843, producing a lipidic extract, have been considered.

Later, during studies of macrophytobenthos of the coastal zone in the Tuapse and Greater Sochi region in 2003–2004, 59 species of macrophyte algae were found. Of these, 23 species belong to the Class Chlorophyta, 6 species—Phaeophyta, 30 species—Rhodophyta. 31 species of algae were discovered in this area for the first time; one species is new to the macrophyte flora of the Black Sea. When compared with data obtained by other authors in the 1960s and 1970s, the transformation of the area's flora from oligosaprobic to mesosaprobic, an increase in the species diversity of green algae, and a local shift of the flora from subtropical to tropical (Lisovskaya and Nikitina 2007) were revealed.

The similar tendency appears for the non-native flora of macrophytes after 2000s—an increase in the number of new species of subtropical and even tropical origin introduced from the Mediterranean Sea in recent decades. In the recent years, more warm-water species of Mediterranean origin were found off Turkish coast (Table 4.12).

In 2016 and 2017, off the Northern Caucasus coast a new non-native species of macroalga, *Bonnemaisonia hamifera* Hariot 1891, was discovered (Fig. 4.26). It probably has been brought from the Atlantic where it is also non-native, being originally of the Pacific origin (Simakova and Smirnov 2019; Shiganova et al. 2012).

This algal species exists in two phases. The gametangial phase is always epiphytic, growing on *Cystoseira* and other algae. It forms erect, brownish-red, feathery fronds and grows to a length of up to 35 cm. The tetrasporangial phase forms much-branched, brownish-red tufts of fine filaments growing in small clumps with a width of 2.5 cm. They are epiphytic on coralline algae or occasionally grow direct on rock or other hard substrate. Currently only tetrasporophytic thallomas of the *B. hamifera* life cycle were found in the studied region. They reproduce mainly vegetatively and form tetraspores in autumn (by December). Hydrobotanical research in 2015–2021 found that the expansion of the *Bonnemaisonia hamifera* Hariot filamentous sporophyte stage had reached the southern coast of Crimea by 2017. Massively developing in the epiphyton of native species, it inhibits their growth, and in some areas becomes a dominant. It alters the appearance, structure,

Fig. 4.26 Non-native species *Bonnemaisonia hamifera*

and production indicators of algal communities (up to their degradation in some areas). The *B. hamifera* invasion threatens the biological diversity of the entire Azov-Black Sea area and may produce ecological catastrophe because of isolation of those basins (Sergey et al. 2023).

B. hamifera distribution in the Black Sea covers the entire fetal zone from 0.5 to 40 m. In the shallow water it occurs only as an epiphyte on Cystoseira, Ellisolandia, and Cladostephus. Deeper than 8 m, it becomes one of the dominating vegetation, reaching 100% cover and coat thickness up to several cm (Sergey et al. 2023).

Among macrophytes, the most of Mediterranean species penetrated with the currents, succeeded in establishment in the near-shore waters off the Anatolian coast (27 species). They comprise about 26% of the total number of macrophyte species registered here (Table 4.13).

In addition to marine macrophyte species, two established species of freshwater ferns have been found in the Danube Delta (Table 4.12) (Zaitsev and Prokopenko 1989).

Mesozooplankton

Reported results indicate importance of two target non-native copepods, i.e., *Acartia tonsa* Dana, 1849 and *Oithona davisae* Ferrari F.D. & Orsi, 1984 established in zooplankton community of the Black Sea (Table 4.14 and Fig. 4.27).

Among them, the most significant non-native species is *Acartia tonsa* Dana, 1849, a new species, discovered in 1994 (Benmonte et al. 1994) in the Black Sea. It is a calanoid copepod that inhabit in estuaries and upwelling zones with high food concentration. Like most plankton species common to estuarine ecosystems, *Acartia tonsa* can exist in a wide range of temperature (1–32 °C) and salinity (0.3–36.9‰) (Svetlichny and Hubareva 2014). Wide tolerance of *A. tonsa* explains the fact that it

Table 4.13 Non-native species of Mediterranean macrophytes in the Black Sea (after Shiganova et al. 2012, with additions)

Species	Tolerance to T °C	Year of penetration, place of finding	References to finding
Class Chlorophyceae Fam. Chaetophoraceae *Eugomontia sacculata* Kommann, 1960	Cool-water	1975 Coastal waters off Bulgaria	Kalugina-Gutnik (1975, 1979)
Fam. Ulvaceae *Ulva kylinii* (Bliding) Hayden, Blomster, Maggs, P.C. Silva, M.J. Stanhope & J.R. Waaland, 2003 =*Enteromorpha kylinii* Bliding, 1948	Subtropical	1990 Coastal waters off Romania, Bulgaria; Turkey, early	Zinova and Dimitrova- Konaklieva (1975), Bavari et al. (1991), Aysel et al. (1995), Mil'chakova (2002), algaebase.org
Ulva curvata (Kützing) De Toni, 1889	Subtropical	Early 1990 Coastal waters off Turkey	Aysel et al. (1995), algaebase.org
Ulva fasciata Delile, 1813	Subtropical	Early 1990 Coastal waters off Turkey	Aysel et al. (1995), Taskin et al. (2008), algaebase.org
Fam. Cladophoraceae *Cladophora flexuosa* (O.F. Muller) Kützing, 1843	Subtropical	Early 1990 Coastal waters off Turkey	Aysel et al. (1995), Taskin et al. (2008), algaebase.org
Cladophora fracta (O.F. Muller ex Vahl) Kützing, 1843	Subtropical	Early 1990 Coastal waters off Turkey	Bavari et al. (1991), Gallardo et al. (1993), Aysel et al. (1995), algaebase.org
Cladophora glomerata (Linnaeus) Kützing, 1843	Eurythermal	Early 1990 Coastal waters off Turkey	Gallardo et al. (1993), Aysel et al. (1995), algaebase.org
Cladophora ruchingeri (C. Agardh) Kützing, 1845	Subtropical	Early 1990 Coastal waters off Turkey	Bavari et al. (1991), Gallardo et al. (1993), Aysel et al. (1995), algaebase.org
Cladophora hutchinsiae (Dillwyn.) Kützing, 1865	Subtropical	Early 1990 Coastal waters off Turkey	Bavari et al. (1991), Gallardo et al. (1993), Aysel et al. (1995), algaebase.org
Cladophora lehmanniana (Lindenberg) Kützing var. effusa Kützing, 1849	Subtropical	Early 1990 Coastal waters off Turkey	Gallardo et al. (1993), Aysel et al. (1995), Guiry (2001), algaebase.org
Cladophora pellucida (Hudson) Kützing, 1843	Subtropical	Early 1990 Coastal waters off Turkey	Gallardo et al. (1993), Aysel et al. (1995), Guiry (2001)
Cladophora prolifera (Roth) Kützing, 1843	Subtropical	Early 1990 Coastal waters off Turkey	Gallardo et al. (1993), Aysel et al. (1995), Guiry (2001), algaebase.org

Cladophora ruchingeri (C. Agardh) Kützing, 1845	Eurythermal	Early 1990 Coastal waters off Turkey	Bavari et al. (1991), algaebase.org
Cladophora rupestris var. *mediterranea* Kützing, 1849		Early 1990s Romanian waters,	Bavari et al. (1991), Caraus (2002), algaebase.org
Class Fucophyceae Fam. Corynophlaeace *Leathesia mucosa* Feldmann 1935.	Cool-water	Early 1990s Coastal waters of Romania, Turkey	Bavari et al. (1991), Gallardo et al. (1993), algaebase.org
Fam. Ectocarpaceae *Ectocarpus caspicus* A. Henckel, 1909	Cool-water	2002 Coastal waters of Romania	Milchakova (2002), algaebase.org
Fam. Chaetophoraceae *Cutleria adspersa* (Mert.) De Notaris, 1842	Subtropical	Early 1990s Coastal waters of Ukraine, Turkey	Ribera et al. (1992), Minicheva and Eremenko (1993), algaebase.org
Cutleria chilosa (Falkenberg) P.C. Silva 1957	Subtropical	Early 1990s Coastal waters off Turkey	Bavari et al. (1991), Ribera et al. (1992), algaebase.org
Fam. Sargassaceae *Cystoseira compressa* (Esper) Gerloff & Nizamuddin 1975	Subtropical	Early 1990s Coastal waters off Turkey	Aysel et al. (1995), Guiry (2001), algaebase.org
Cystoseira corniculata (Wulf.) Zanard	Subtropical	Early 1990s Coastal waters off Turkey	Aysel et al. (1995), algaebase.org
Cystoseira ruchingeri (C. Ag.) Kütz.	Subtropical	Early 1990s Coastal waters off Turkey	Aysel et al. (1995), Taskin et al. (2008), algaebase.org
Cystoseira schiffneri G. Hamel, 1939	Subtropical	Early 1990s Coastal waters off Turkey	Ribera et al. (1992), Aysel et al. (1995), algaebase.org
Fam. Sargassaceae *Sargassum acinarium* (L.) C. Ag. Weyler 185	Subtropical	Early 1990s Coastal waters off Turkey	Aysel et al. (1995), algaebase.org
Sargassum hornschuchii C. Agardh, 1820	Subtropical	Early 1990s Coastal waters off Turkey	Aysel et al. (1995), algaebase.org
Class Rhodophyceae Fam. Ceratiaceae *Ceratium flaccidum* (Kütz.) Ardiss	Subtropical	Early 1990s Coastal waters off Turkey	Aysel et al. (1995), algaebase.org

(continued)

Table 4.13 (continued)

Species	Tolerance to T °C	Year of penetration, place of finding	References to finding
Ceramium cimbricum H.E. Petersen f. cimbricum = *C. fastigiatum* Harv. Var. fastigiatum	Subtropical	Early 1990s Coastal waters of Bulgaria,	Bavari et al. (1991), algaebase.org
Ceratium tenerrimum (G. Martens) Okamura var tenerrimum	Subtropical	Early 1990s Coastal waters off Turkey	Aysel et al. (1995), algaebase.org
Compsothamnion thuyoides (Sm.) Nageli	Cool-water	Early 1990s Coastal waters off Turkey	Aysel et al. (1995), algaebase.org
Fam. Dasyaceae *Dasya ocellata* (Gratel.) Harv.	Subtropical	Early 1990s Coastal waters off Turkey	Aysel et al. (1995), algaebase.org
Fam. Rhodomeiaceae *Neosiphonia elongella* (Harv.) M.S. Kim et I.K. Lee =*Pofysiphonia elongella* (Harv.)	Subtropical	Early 1990s Coastal waters off Turkey	Aysel et al. (1995), Garreta et al. (2001), algaebase.org
Polysiphonia deusta (Roth) Spreng	Subtropical	Early 1990s Coastal waters of Turkey,	Aysel et al. (1995), Garreta et al. (2001), algaebase.org
Polysiphonia paniculata Montagne	Subtropical	Early 1990s Coastal waters off Turkey	Aysel et al. (1995), Garreta et al. (2001), Taskin et al. (2008), algaebase.org
Polysiphonia stricta (Dillw.) Grev. =*P. urceolata* (Lightf. Ex Dillw.) Grev.	Subtropical	Coastal waters of Turkey, early 1990s	Aysel et al. (1995), algaebase.org
Polysiphonia subulata (Ducluzeau) Kützing, 1863	Subtropical	Early 1990s Coastal waters of Romania	Bavari et al. (1991), algaebase.org
Polysiphonia tenerrima Kütz.	Subtropical	Early 1990s Coastal waters off Turkey	Aysel et al. (1995), algaebase.org
Polysiphonia tripinnata J. Ag.	Subtropical	Early 1990s Coastal waters off Turkey	Aysel et al. (1995), Garreta et al. (2001), algaebase.org

Table 4.14 Non-native species of mesozooplankton in the Black Sea

Species	Donor region	Vector	First record	Establish	Author
Class Copepoda Order Calanoida Fam. Acartiidae *Acartia tonsa* Dana, 1849	Estuaries and upwelling zones	Shipping (ballast water)	1994 Sevastopol Bay	Established	Benmonte et al. (1994) Gubanova (1997)
Class Copepoda Order Cyclopoida Fam. Oithonidae Oithona *davisae* Ferrari F.D. & Orsi, 1984	Coastal waters of Japan	Shipping (ballast water)	2001 Initially identified as *O. brevicornis*	Established	Zagorodnyaya 2002 as *Oithona brevicornis* Shiganova et al. 2015 as *Oithona davisae*
Class Copepoda Order Calanoida Fam. Pseudodiaptomidae *Pseudodiaptomus Marinus* Sato, 1913	Northwestern Mediterranean, initially waters off Northern Japan	Shipping (ballast water)	2016 coastal waters of Crimea	Established	Garbazey et al. (2016)

Fig. 4.27 Copepoda *Acartia tonsa* (**a**) and *Oithona davisae* (**b**)

easily may be transported with ballast waters into various regions. *A. tonsa* nauplii and adults feed on phytoplankton, planktonic ciliates and rotifers. It acts as a suspension feeder, when feeding on phytoplankton.

Currently, the range of *Acartia tonsa* is expanding. *Acartia tonsa* is classified as a cosmopolitan species distributed along the coasts and usually inhabits estuaries and coastal areas of the Indian, Pacific, Atlantic Oceans, as well as the Mediterranean, Black and the Baltic seas (Figueroa et al. 2020). It is usually the dominant copepod species in bays, estuaries, fjords, straits, and inland canal systems, acting as an important link between primary producers and higher trophic levels (Jørgensen et al. 2019; Aguilera and Bednaršek 2022). It is important that along with summer eggs, *A. tonsa* lays latent eggs. When the temperature decreases and the daylight hour decreases, *A. tonsa* laying latent eggs, which can remain in the silt until favorable conditions occur, and then develops. The population of *A. tonsa* does not occur in the plankton of the Black Sea during the cold season, renewing annually, but latent eggs can persist for several years, and then develop to normal adults. Species like *A. tonsa* increase their abundance and biomass in the summer due to temperature rise and displace more stenothermic native species *Acartia clausi*.

Our preliminary genetic analysis indicate that the Black Sea *A.tonsa* is similar to the Baltic and Caspian seas one, and differ from the species inhabiting northwestern Atlantic coast.

Acartia tonsa is widely distributed in the Black and Azov seas but never has been recorded in Gelendzhik area of the Black Sea (Shiganova, pers. obs.), although it was found near Kerch Strait and in the coastal waters of Novorossiysk (Selifonova 2009).

Another non-native species, cyclopoid copepod of the genus *Oithona* was first recorded in the Black Sea in 2001 and initially identified as *Oithona brevicornis* Giesbrecht, 1891 (Zagorodnyaya 2002; Altukhov and Gubanova 2006; Gubanova and Altukhov 2007). Other authors followed this identification in subsequent studies (e.g., Selifonova et al. 2008; Shiganova et al. 2012; Mihneva and Stefanova 2013; Timofte and Tabarcea 2012).

In 2011 this species from the Black Sea was defined as the subspecies *Oithona brevicornis brevicornis* Giesbrecht, 1891 in the World Registry of Marine Species (WoRMS, cited in Shiganova et al. 2012). Identifications by all previous authors were based on the examination of a few morphological features, such as the shape of the head (rounded anteriorly) and rostrum (sharply pointed) and the numbers of spines on segments 1–3 of legs 1–4 in the female individuals (Zagorodnyaya 2002; Gubanova and Altukhov 2007). However, recent detailed examination of morphological features of this *Oithona* species by Temnykh and Nishida (2012) demonstrated that the cyclopoid species, recently recorded as new to the Black Sea is not *O. brevicornis*, but *O. davisae* Ferrari F.D. and Orsi, 1984, or include one or more species that are similar morphologically to *O. brevicornis* (Temnykh and Nishida 2012). Since this non-native species was morphologically defined as *O. davisae* and our identification was supported by genetic analyses, we used this name throughout this study. *Oithona davisae* spread widely in the Black Sea and is accepted in WoRMS. This species population initially developed primarily in the coastal waters, but now it may be found in deep areas (up to 1000 m) in the northeastern, northwestern and western parts of the Black Sea. Maximum abundance and biomass occurs in August–December (Temnykh et al. 2012; Selifonova 2009, 2011; Shiganova et al. 2012; Michneva and Stephanova 2013). *O. davisae* often dominates in the mesozooplankton community, comprising 80–85% of the total abundance in late autumn (Selifonova 2011; Shiganova et al. 2012). As *Oithona davisae* seasonally develops in the north-eastern area since late spring, reaching the highest abundance in July–August, and continue intensive development with high abundance in autumn-eaarly winter. Its abundance increased during 2016–2019. *O. davisae* became the most abundant in the coastal regions, but it was recorded also in the deep open sea waters. Thus, *O. davisae* occupies its own niche in the Black Sea, replacing extinct *Oithona nana*, and became edible zooplankton useful for planktivorous species feeding. Based on previously cited observations, and on our morphological and genetic findings, we can conclude that the non-native species *Oithona* in the study area is not *O. brevicornis*. Clearly, a comprehensive evaluation of the taxonomic status, including additional molecular genetic analysis, of this species is needed. It should include consideration of specimens from different parts of the Black Sea, northwestern Mediterranean, and the Pacific Ocean, i.e., areas where occurrence of this species (both native and non-native) has been reported.

Another alien species, new for the Black Sea (calanoid copepod *Pseudodiaptomus marinus* Sato, 1913) was reported in the coastal waters of Crimea in 2016 (Seregin and Popova 2016). So, *P. marinus* has been identified as a new species of non-native Copepoda in the Black Sea (Fig. 4.28). This species is native to the coasts of Eastern Asia. In 2019–2020, individuals of this species appeared in July, at temperature from 25.8 to 26.0 °C. It has been observed in samples until the end of November or December at sea surface temperature 13.1 °C or 8.6 °C, respectively.

The maximal abundance of *P. marinus* was observed in the inner part of the Sevastopol Bay. *P. marinus* is original for the northwestern Pacific and was also mentioned in the coastal waters of northern Japan (Walter 1986). Since the species has a high invasive potential, it is widely spread in the World Ocean, e.g., in the San Francisco estuary it has been noted as an invasive species since 1986. Comparing its

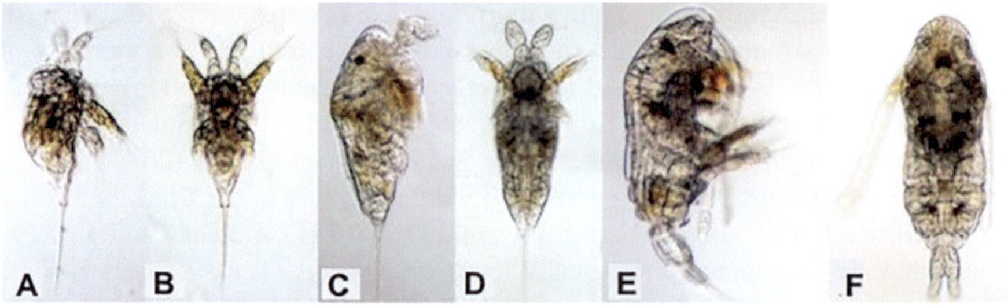

Fig. 4.28 *Pseudodiaptomus marinus* Sato, 1913 development stages. (After Seregin and Popova 2016)

genotype with that of species from waters off Japan, Korea, and China, it was concluded that this species arrived from several local coastal areas of Japan (Ohtsuka et al. 2018). This species was recently recorded at the Atlantic coast of Europe and in the Mediterranean basin: for the first time in 2007 in the Adriatic Sea (De Olazabal and Tirelli 2011); in October 2008 in Sicily (Sabia et al. 2012); in 2010 in the North Sea (Brylinski et al. 2012); in 2016 in the Sevastopol Bay of the Black Sea (Garbazey et al. 2016).

Thus, there are a few species of non-native Copepods that arrived and established in the Black Sea and they became additional valuble food items. The only negative conseqvence of their introduction is the displacement of native species, what was most clearly manifested in the case of *Acartia tonsa* as opportunistic species. Still, it is unknown yet if species arrived from the Mediterranean Sea (Table 4.15) estabilshed in the Black Sea. This process may be intensified because of temperature rise, associated with global warming, which provide for Mediterranean species more opportunities to establish.

Macrozooplankton

Gelatinous plankton is an essential component determining the health of ecosystems. According to their life cycle, ecological aspects and adaptability to various environmental conditions, these organisms may be considered as indicators of climate change, having particular advantages over the other taxa.

Among the hydrozoa species the most pronounced is invasive brackish-water *Blackfordia virginica,* which inhabits worldwide (Fig. 4.29a). It invaded and established in the low-saline parts of the Black and Azov seas, the Mediterranean, Adriatic and Aegean seas, waters of Atlantic coast of southern Europe, in the North and Baltic seas and coastal waters of northern Africa. After the Volga-Don Canal starts to operate, the species penetrated also into the Caspian Sea. However, Mills and Sommer (1995) mistakenly attributed *B. virginica* to the Black Sea native species. This statement was then repeated several times in other publications. *B. virginica* prefers waters with a temperature of 15–25 °C and a salinity of 0.5–3.5%. Its polyps inhabit in the upper sublithoral, attaching themselves to rocks, pieces of wood, and various floating objects, and being a component of the periphyton. Its medusae live in the near-surface layers and are a component of

Table 4.15 List of Mediterranean Copepoda, found in the Black Sea (after Shiganova et al. 2012)

	Reference					
Species	1	2	3	4	5	6
Calanus tenuicornis Dana					+	
Calanus gracilis Dana		+				
Calanus minor Claus				+		
Eucalanus sp.					+	
Mecynocera clausi Thompson				+	+	
Paracalanus nanus Sars				+	+	
Paracalanus aculiatus Giebr.					+	
Clausocalanus arcuicornis (Dana)	+	+	+			
Clausocalanus paululus Farr.		+	+	+		
Clausocalanus (Brady)		+		+		+
Clausocalanus pergens Farr.		+		+		+
Clausocalanus parapergens Frost, Flem		+				
Clausocalanus mastigophorus (Claus)				+		
Calocalanus pavo Dana	+	+	+	+	+	
Calocalanus plumulosus Claus					+	
Calocalanus pavoninus Farr.	+		+			
Calocalanus plumatus Shmel.		+		+		
Calocalanus(tenuis?) Farr.			+			
Microcalanus pusillus O. Sars						N
Ctenocalanus vanus Giesbr					+	+
Aetideus armatus Boeck						N
Euchaeta marina Prestandrea						N
Phaenna spinifera Claus					+	
Scolecithrix danae Lubb		+				
Temora stylifera Dana				+		
Metridia lucens Boeck						N
Pleuromamma abdominalis Lubb				+		
Pleuromamma gracilis Claus				+		+
Pleuromamma sp.		+				
Centropages typicus Kroyri				+		+
Lucicutia flavicornis Claus			+			
Lucicutia gemina Farr				+		
Euterpina acutifrons Claus	+	+	+	+		
Candacia athiopica Dana		+				
Micsetelia gracilis Dana	+	+	+	+		+
Macrocetelia gracilis Dana		+				
Paroithona parvula Farr.					+	
Oithona sp.		+	+			+
Oncaea obscura Farr.						N
Oncaea minuta Gieshr.	+	+	+	+		+
Oncaea dentipes Giesbr.	+	+	+	+		+

(continued)

Table 4.15 (continued)

Species	Reference					
	1	2	3	4	5	6
Oncaea similis Sars		+	+	+		
Oncaea media Giesbr.		+				
Oncaea subtilis Giesbr.		+				
Oncaea curva Sars		+				
Oncaea conifera Giesbr.	+		+	+	+	
Oncaea mediterranea Claus	+		+	+	+	
Oncaea subtilis Giesbr.					+	
Oncaea venusta Philippi					+	
Corycaeus furcifer Claus	+	+	+		+	
Corycaeus latus Dana		+		+		
Corycaeus typicus Kroger	+		+	+		
Corycaeus flaccus Giesbr.	+		+			
Corycaeus clausi F. Dahl	+		+			
Corycaeus limbatus Brady				+		
Corycaeus sp.	+		+	+		
Corycella gracilis Dana		+				
Corycella rostrata Claus				+		
Corycella sp.	+		+			
59 species						

1-Pavlova (1964, 1965), Pavlova and Baldina (1969), **2**-Kovalev et al. (1976), **3**-Kovalev (1971), **4**-Kovalev et al. (1987), **5**-Porumb (1980), **6**-Kovalev et al. (1998), **N**- new species recently found

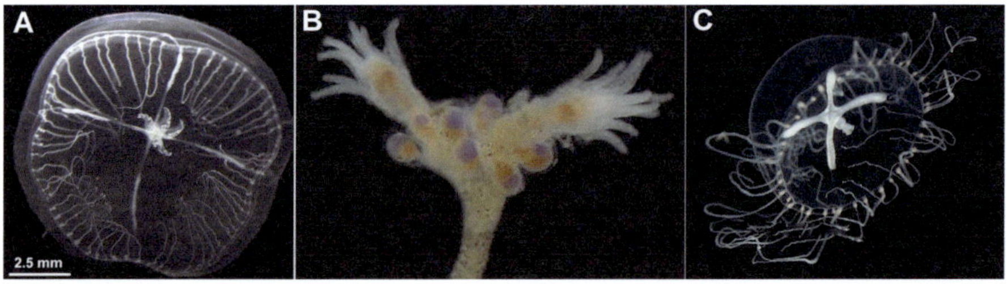

Fig. 4.29 Hydrozoa *Blackfordia virginica* (**a**), *Calyptospadix cerulea* (**b**) and *Odessia maeotica* (**c**)

zooplankton (Valkanov 1936; Mills and Sommer 1995; Shiganova 2009; Jaspers et al. 2018; Faasse and Melchers 2014).

Another hydrozoan is *Calyptospadix cerulea* Clarke, 1882 (Fig. 4.29b). This is quite a mysterious species. It is uncertain where the species originates from or how it arrived in the Black Sea. Commonly known as the Rope-Grass hydroid, lacks a planktonic medusa stage. It was first described from Fort Wool, off Norfolk, Virginia, (Clarke 1882), It was later collected and described from San Francisco Bay, California (Clarke 1882). It is possible that this species has always been present here, but was never observed before. Consequently, this hydroid is a cryptogenic species, which means that it is uncertain if the species is native or not.

Other hydrozoa species recorded in the coastal waters of Crimea are listed in Table 4.16 (Grishicheva and Shadrin 1999).

Hydrozoan species *Odessia maeotica* (Ostroumoff, 1896) was recorded in the Bulgarian coastal waters (Naumov 1968), but later it was defined as Ponto-Caspian origin species (Fig. 4.29c). However, we keep it in the list of invaders because of its extensive penetration from the Black Sea.

We also would like to focus on the effects of the small lacunae jellyfish *Odessia maeotica* in a species-poor habitat, Mediterranean salt marshes in the wetlands of Empordà. The presence of *O. maeotica* changed the plankton composition through top-down effects. Changes were strong in zooplankton, because *O. maeotica* can suppress almost the entire trophic level of large zooplankton (>50 μm). Weak indirect effects on phytoplankton composition were observed as well. When *O. maeotica* was present, changes in the relative abundance of the phytoplankton species were found, but there was no net increase in phytoplankton biomass. These weak indirect effects may be the result of trophic cascade effects coupled with the oligotrophic conditions of these salt marshes. Thus, trophic cascade effects lead to an increase in ciliate biomass, and these ciliates would feed on small algae (jellyfish–copepods–ciliates–small algae), while oligotrophic conditions would prevent increases in algal biomass (Compte et al. 2010).

The increase of the numbers of species, abundances and areas of distribution of gelatinous plankton both native and invaders are the main concern for the Mediterranean, Sea of Marmara and Black Sea ecosystems (Shiganova and Ozturk 2010).

In the Black Sea, *Pleurobrachia pileus* (O.F. Müller, 1776) is the only native ctenophore species. During the last three decades, two North American ctenophores species were unintentionally introduced into the Black Sea. The zooplanktivorous *Mnemiopsis leidyi* (Agassiz, 1865) arrived first in 1988 (Vinogradov et al. 1989), affecting all trophic webs of the Black Sea ecosystem. Then, 10 years later, its predator *Beroe ovata* Bruger 1789 arrived from the same area and established in the Black Sea in 1997 (Konsulov and Kamburska 1998). After that, the Black Sea ecosistem began to recover. These two species play a crucial role, being the drivers of the whole Black Sea ecosystem (Shiganova 2009) and will be discussed in details in the following sections, specially dedicated to them.

Recently, several Mediterranean jellyfishes penetrated the Sea of Marmara for the first time. Among them were scyphomedusae *Chrysaora hysoscella* (Linnaeus, 1766), *Cassiopea andromeda* (Forskål, 1775), *Paraphyllina ransoni* Russell, 1956, trachymedusa *Liriope tetraphylla* Chamisso and Eysenhardt, 1821 and three hydromedusae *Aglaura hemistoma* Péron and Le Sueur, 1810, *Neoturris pileata* (Forskål, 1775), *Solmundella bitentaculata* (Quoy and Gaimard, 1833) (Ozgur and Ozturk 2008; Öztürk and Topaloglu 2009; Isinbillir et al. 2010). At that, *C. hysoscella* already was noticed in the Black Sea near Bosphorus in 2009, and *Bolinopsis vitrea* was recorded near the Bosphorus and in the Bulgarian waters (Ozturk al al. 2011). Still, both were not established.

Table 4.16 Non-native macrozooplankton species in the Black Sea

Species	Donor region	Vector	First record	Establish	Author
Phylum Cnidaria (Coelenterata) Class Hydrozoa Order Leptothecata Fam. Blackfordiidae *Blackfordia virginica* Mayer, 1910	Chesapeake Bay Atlantic coast of North America	Shipping	1925 Bulgarian coast	Established	Valkanov (1936)
Fam. Campanulinidae *Campanulina pumila* (Clark, 1875)		Shipping	1990	Established	Grishicheva and Shadrin (1999)
Order Anthoathecata Fam. Bougainvilliidae Calyptospadix cerulea Clarke, 1882 (syn. Bougainvillia megas) (Kinne, 1956)	Chesapeake Bay Atlantic coast of North America	Shipping	1935 Bulgarian and Danube Coast	Established	Volcanov (1935), Morduhai-Boltovasky (1968)
Order Anthoathecata Fam. Eudenriidae *Eudentrium annulatum* Norman, 1864	Atlantic coast of North Europe	Shipping	1990 Coastal waters of Crimea	Established	Grishicheva and Shadrin (1999)
Order Anthoathecata Fam. Corynidae *Coryne pusilla* (Gaertner, 1774)	Atlantic coast of North Europe	Shipping	1990 Coastal Crimea	Established	Grishicheva and Shadrin (1999)
Fam. Corynidae *Stauridia producta* Wright.1858	Mediterranean Sea	Shipping	1990 Coastal Crimea	Established	Grishicheva and Shadrin (1999)
Order Anthoathecata Fam. Moerisiidae *Odess Odessia maeotica* (Ostroumoff, 1896) Syn.: *Moerisia maeotica* (Ostroumov, 1896)	Ponto Caspian species	Ponto-caspian species, became invasive	1968 Bulgarian coast	Established	Naumov (1968)

(continued)

Table 4.16 (continued)

Species	Donor region	Vector	First record	Establish	Author
Class Scyphozoa Order Semaeostomeae Fam.Pelagiidae *Chrysaora hysoscella* (Linne, 1766)	Mediterranean Sea	Penetrated	2009 near Bosporus	Not established	Öztürk and Topaloglu (2009)
Phylum Ctenophora ClassTentaculata Order Lobata Esch Fam. Bolinopsiidae *Mnemiopsis leidyi* A.Agassiz, 1865	Northen America	Shipping	1982 Coastal Crimea	1988 Established	Pereladov (1988) Vinogradov et al. (1989)
Bolinopsis vitrea (L. Agassiz, 1860)	Mediterranean Sea	Shipping Penetration		Not established	Bayram et al. (2011)
Order Beroida Eschscholtz, 1829 Class Nuda Fam. Beroidae *Beroe ovata* Bruguière, 1789	Northen America	Shipping	1997 Varna Bay	1997 Established	Konsulov and Kambursky (1998)

Benthos

Polychaeta Harmfull invasive species belonging to polychaetes of fouling communities are listed in Table 4.17. Among them is *Ficopomatus enigmaticus* (Fauvel, 1923) (formerly known as *Mercierella enigmatica* Fauvel, 1923), which forms extensive settlements on the solid substrates. It is probably native to the brackish coastal lakes of India. In 1923, it was discovered at the mouth of the Seine River in France, later in the Adriatic Sea. From the coastal waters of Europe, or from the Adriatic, it entered the Black Sea as a component of the ships' fouling where spread widely and penetrated also into the Azov and Caspian seas (Naumov 1968). Being a fouling species, it damages ships hulls and hydraulic equipment; however its planktonic larvae is useful being a feeding items for fish (Zaitsev and Mamaev 1997). This species build calcareous tubes, which in a high aggregation may forms a kind of underwater biogenic reef. Thus, in the Varna-Burgas and Danube regions, this non-native species has formed new fouling communities, in which its biomass reaches 7–8 kg.m^{-2} (Marinov 1957; Tsvetkov and Grancharova 1976; Marinov 1977). The ability to change the biotope by building reefs is a key feature that allows *F. enigmaticus* to create a new type of community. The fouling community with a predominance of alien polychaetes differs in species composition from the

Table 4.17 Non-native benthic species in the Black Sea

Species	Donor region	Vector	First record	Establishment status	Author
Class Entoprocta (Kamptozoa) Fam. Urnatellidae *Urnatella gracilis* (Leidyi, 1851)	Coastal waters of North America, then coastal waters of Belgium	Shipping (hull fouling)	1954–1962 in lower reaches of the Danube, Dniester and Don	1969 Established	Mordukhai-Boltovskoi (1969a, b, c)
Phylum Annelida **Class Oligochaeta** **Order Tubificida** Fam. Naididae 1. *Tubificoides benedii* (d'Udekem, 1855)	North Atlantic	Shipping (hull fouling)	2006 North-western and Danube area	2006 Established	Shurova (2006)
Class Polychaeta **Order Serpulimorpha** Fam. Serpulidae *Ficopomatus enigmaticos* Fauvel 1923 *(Mercierella enigmatica)*	English Channel, North Sea, coastal waters of France	Shipping (hull fouling)	1929 near Poti	Established	Annenkova (1930)
Class Polychaeta **Order Phyllodocida** Fam. Microphthalmidae *Hesionides arenarius* (Friedrich, 1936)	Atlantic or Pacific oceans, Mediterranean Sea	Shipping (hull fouling)	1970s Northwest shelf	Established	Marinov (1977)
Class Polychaeta **Order Spionida** Fam. Spionidae *Polydora cornuta* (Bosk, 1802)	Atlantic coast of Northern America	Shipping (hull fouling)	2013	Established	Radashevsky and Selifonova (2013)
Class Polychaeta **Order Spionida** Fam. Spionidae *Polydora websteri* Hartman in Loosanoff & Engle, 1943	Atlantic part of Northern America	Shipping (hull fouling)	2005 Romanian waters	Established	Surugiu (2005)

Fam.Spionidae *Streblospio gynobranchiata* Rise & Levin, 1998	From Florida and Gulf of Mexico	Shipping (hull fouling)	2007 The north Black Sea	Established	Boltacheva et al. (2020)
Fam. Scalibregmatidae *Hyboscolex* cf. *pacificus* (Moore, 1909)	Eastern Mediterranean	Shipping (hull fouling)?	2017 The north Black Sea	Established	Lisitskaya et al. (2019)
Class Polychaeta **Order Phyllodocida** Fam. Pilargiidae *Sigambra tentaculata* (Treadwell, 1941)	North Atlantic	Shipping (hull fouling)	1950s Mediterra-nean Sea 1960s Black Sea	Established	Kiselyova (1964)
Order Phyllodocida Fam. Syllidae *Strebtocyllis varians* Webster et Benedict, 1884	North Atlantic	Shipping (hull fouling)	1950–1960 North-western part of the Black Sea	Established	Marinov (1966)
Order Phyllodocida Fam. Nephthydidae *Nephthis ciliata* (Muller, 1776)	North Atlantic, North Sea, English Channel	Shipping (hull fouling)	1972 Western part of the Black Sea, Bulgarian coast	Established	Mordukhai-Boltovskoi (1972)
Order Spiomorpha Fam. Cirratulidae *Streblospio shrubsolii* (Buchanan, 1890)	Europien waters	Shipping (hull fouling)	1957 Bulgarian coast	Established	Marinov (1957)
Order Drilomorpha Fam. Capitellidae *Capitellethus dispar* (Ehlers, 1908)	Mediterranean North Atlantic, Indowest Pasific	Shipping (hull fouling)	Before 1972 Bosphorus area	Established	Mordukhai-Boltovskoi (1972)
Order Phyllodocida Fam. Glyceridae *Glycera capitata* Orsted 1842	North Atlantic, North Sea, English Channel	Shipping (hull fouling)	Before 1972 Bosphorus area	Established	Mordukhai-Boltovskoi (1972)

(continued)

Table 4.17 (continued)

Species	Donor region	Vector	First record	Establishment status	Author
Class Polychaeta **Order Spionida** Fam. Spionidae *Dipolydora quadrilobata* Jacobi, 1883	North Atlantic Europien waters, White, Japanise Seas	accidental	2003 Noth-eastern Black Sea Later in Romanian coastal waters	Established	Todorova and Panayotova (2006), quoted by Surugiu (2009)
Class Polychaeta **Order Spionida** Fam. Spionidae *Marenzelleria neglecta* Sikorski & Bick, 2004	North Atlantic Europien waters, Baltic Sea	Shipping (hull fouling)	2015	Established	Syomin et al. (2017)
Phylum Mollusca **Class Gastropoda** **Order Neogastropoda** Fam. Muricidae *Rapana venosa* (Valenciennes, 1846)	Sea of Japan	Shipping (the egg clutch was attached to the bottom)	1947 Novorossiysk area	Established	Drabkin (1953)
Class Gastropoda **Order Littorinimorpha** Fam. Tateidae *Potamopyrgus antipodarum* (J. E. Gray, 1843)	Estuaries of Europe	Shipping	1952 Dnieper lagoons	established	Gomoiu and Skolska (1996)
Class Gastropoda **Order Littorinimorpha** Fam.Aporrhaidae *Aporrhais pespelecani* Linnaeus, 1758	North Atlantic	Shipping (hull fouling)	1960 coastal waters of Georgia, occurs rare	Established	Alexandrov (2004)
Class Gastropoda **Order Nudibranchia** Fam. Corambidae *Corambe obscura* (A. E. Verrill, 1870)	Atlantic coastal waters of North America	Shipping (ballast waters, hull fouling)	1980 Bulgarian coast, Varna Bay	Established	Roginskaya and Grintsov (1990)

Order Nudibranchia Fam.Limapontiidae *Ercolania viridis* (A. Costa, 1866)	North Atlantic	Shipping (ballast waters)	2001 Odessa port	Established	Alexandrov (2004)
Class Gastropoda **Order Neogastropoda** Fam. Buccinidae *Neptunea arthritica* (Valenciennes, 1858)	Pacific Ocean	Shipping (ballast waters)	2001 Crimea peninsula near Sevastopol	Established	Shadrin et al. (2002)
Class Bivalvia **Order Myida** Fam. Teredinidae *Teredo navalis* (Linne, 1758)	North Atlantic	Shipping (hull fouling)	1990	Established	Zaika et al. (1992)
Class Bivalvia **Order Myida** Fam. Myidae *Mya arenaria* Linne., 1758	The Atlantic coasts of Europe or North America	Shipping (hull fouling)	1966 Odessa area	Established	Beshevli and Kolyagin (1967)
Class Bivalvia **Order Mytilida** Fam**.** Mytilidae *Mytilus edulis* Linne, 1758	East Atlantic, Mediterranean	Shipping	2001 Odessa area	Established	Alexandrov (2004)
Mytilus trossulus Gould, 1850	Northwest and Northeast Pacific Coast	Shipping	2001 Odessa area	Established	Alexandrov (2004)
Mytilus galloprovincialis Lamarck, 1819	Mediterranean	Shipping	2001	Established	Alexandrov (2004)
Arcuatula senhousia (W. H. Benson, 1842)	Mediterranean	Shipping	2017, Azov, Black Sea 2021–2022 North-eastern Black Sea	Established	Kovalev et al. (2017) Kolyuchkina et al. (2022)

(continued)

Table 4.17 (continued)

Species	Donor region	Vector	First record	Establishment status	Author
Class Bivalvia **Order Arcida** Fam. Arcidae *Anadara kagoshimensis* (Tokunaga, 1 906)	The northern part of the Adriatic Sea, there from the Pacific Ocean	Shipping (hull fouling)	1968 Caucasian shelf 1992 entire Black Sea	Established	Scarlato and Starobogatov (1972), Kiseleva (1992)
Anadara inaequivalvis (Bruguiere, 1789)	Mediterranean Sea	Natural expansion	1998 Bosphorus area	Established	Öztürk (1998) Turkey Country report
Anadara cornea (Reeve, 1844)	Mediterranean Sea	Natural expansion	1998 Bosphorus area	Established	Öztürk (1998)
Class Bivalvia **Order Ostreida** Fam. Ostreidae *Magallana gigas* (Thunberg, 1793) (*Ostrea gigas* Thunberg, 1793)	The northern part of the Adriatic Sea (got there from the Sea of Japan)	Shipping (ballast waters) (?) Aquaculture	1980 Caucasian shelf	Established	Zolotarev (1996)
Class Bivalvia **Order Gastrochaenida** Fam.Gastrochaenidae *Rocellaria dubia* (Pennant, 1777)	Adriatic Sea (got there from the Pacific Ocean, Sea of Japan)	Shipping	1980 Adriatic Sea	Established	Scarlato and Starobogatov (1972)
Phylum Arthropoda **Class Malacostra** **Order Amphipoda** Microprotopus maculatus Norman, 1867 M. cf. maculatus	Mediterranean Sea	Shipping (hull fouling) (?)	Before 1996 2022	Established	Grese (1966), Grintsov (2022)

Class Thecostraca **Order Balanomorpha** Fam. Balanidae *Amphibalanus improvisus* (Darwin, 1854)	Atlantic coast of North America	Shipping (hull fouling)	1990 Romanian coastal waters	Established	Gomoiu and Skolska (1996)
Amphibalanus eburneus (Gould, 1841)	Atlantic coast of North America	Shipping (hull fouling)	1990	Established	Gomoiu and Skolska (1996)
Class**Thecostraca** Order **Balanomorpha** Fam. Balanidae *Amphibalanus amphitrite Amphitrite* (Darwin, 1854)	Eastern Atlantic-Mediterranean	Shipping (hull fouling)	2001 Odessa Bay	Established	Alexandrov (2004)
Class**Thecostraca** Order **Balanomorpha** *Chthamalus montagui Southward,* 1976	North Atlantic-Mediterranean	Shipping	1995 Crimea coast	Established (?)	Zevina and Poltarukha (1999)
Phylum Arthropoda **Class Malacostraca** **Order Decapoda** Fam. Panopeidae *Rhithropanopeus harrisii* Gould, 1841	Zuiderzee Bay (North Sea), previously from the Atlantic waters of North America	Shipping (hull fouling) (?)	1938 Dniester-Bug estuary	Established	Kobyakova and Dolgopolskaya (1969)
Class Malacostraca **Order Decapoda** Fam. Portunidae *Callinectis sapidus* (Rathbun, 1896)	Atlantic coast of North America	Shipping (hull fouling) (?)	1967 Bulgarian coast	Established	Bulgurkov (1968)
Class Malacostraca **Order Decapoda** Fam. Pirimelidae *Sirpus zariquieyi* Gordon 1953	Mediterranean Sea	Shipping (?)	1980, 2000 Single individuals were found off the eastern Turkish coast	Established in near Bosporus area	Gordon (1953) Zaitsev and Ozturk (2001)

(continued)

Table 4.17 (continued)

Species	Donor region	Vector	First record	Establishment status	Author
Class Malacostraca **Order Decapoda** Fam. Grapsidae *Eriocheir sinensis* Milne-Edwards, 1853	Introduced to Europe (North Sea) in 1912 from the estuaries of East Asia and the coastal waters of the South China and Yellow Seas	Shipping (ballast waters, hull fouling)	1997 female with eggs was collected near the Danube Delta	Established	Gomoiu and Skolka (1998)
Class Malacostraca **Order Decapoda** *Palaemon macrodactylus* Rathbun, 1902	Native to Japan, Korea, and Northern China	Shipping (ballast waters)	2002 Romanian waters 2009 Bulgarian waters 2022 Don delta	Established	Micu and Niţă (2009) Stankevich (2022)
Class Malacostraca **Order Decapoda** Fam. Cambaridae *Orconectes limosus* Rufnesque	The east coast of North America, has also been introduced to Europe. Widespread in Europe	Natural expansion, probably via Danube	2004 Romanian waters	Established	Pâvlovjc et al. (2009)
Class Malacostraca **Order Decapoda** Fam. Panopeidae *Dyspanopeus sayi* (S.I. Smith, 1869)	Introduced to Europe,spread widely, possibly from there entered the Black Sea with ships	Shipping	2009 Romanian waters	Established	Micu et al. (2009)
Class Malacostraca **Order Decapoda** Fam. Varunidae *Hemigrapsus sanguineus* (De Haan, 1835)	Introduced to the coastal waters of Germany	Shipping (hull fouling)	2008 Romanian waters	not established	Micu and Niţă (2009)

Class Malacostraca **Order Decapoda** Fam. Penaeidae *Pandalus latirostris* Rathbun, 1902	Sea of Japan	Intentional introduction	1954–1962 lower reaches of the Danube, Dniester and Don	Failed introduction	Salsky (1964)
Class Malacostraca **Order Decapoda** Fam. Penaeidae *Penaeus japonicus* Spence Bate, 1888	Sea of Japan	Intentional introduction	1954–1962 lower reaches of the Danube, Dniester and Don	Failed introduction	Salsky (1964)
Phylum Echinodermata **Class Echinoidea** **Order Arbacioida** *Arbacia lixura* (Linnaeus, 1758)	Mediterranean Sea	Natural expansion	2007 The Turkish coast of the Black Sea	(?)	Öztürk (2006)
Class Asteroidea **Order Forcipulatida** Fam. Asteriidae *Asterias rubens* (Linnaeus, 1758)	Mediterranean Sea	Natural expansion shipping (?)	1996, 2003 Bosphorus area	Established	Albayrak (1996), Karhan et al. (2007)
Phymut Echenodermata **Class Ophiuroidea** **Order Amphilepidida** Fam. Amphiuridae *Amphiura filiformis* O.F. Müller, 1776	Mediterranean Sea	Natural expansion, shipping (?)	2009 Bosphorus area	Established (?)	Sergin and Kideys (2009)
Phymut Echenodermata **Class Steroidea** **Order Valvatida** Fam.Asterinidae *Asterias gibbosa* Pennant, 1777	Mediterranean Sea	Natural expansion, shipping (?)	Bosphorus area	Established (?)	Sergin and Kideys (2009)
Class Malacostraca **Order Decapoda** Fam. Palaemonidae *Paleamon longirostris* H. Milne-Edwards 1837	Mediterranean Sea	Natural expansion	2005 Bosphorus area	Established (?)	Sezgin et al. (2007)

(continued)

Table 4.17 (continued)

Species	Donor region	Vector	First record	Establishment status	Author
Class Malacostraca **Order Decapoda** Fam. Pirimelidae *Sirpus zariquieyi* Gordon 1953	Mediterranean Sea	Shipping (?)	2000 Bosphorus area	Established	Gordon (1953), Zaitsev and Ozturk (2001)
Order Decapoda Fam. Panopeidae *Dyspanopeus sayi* (S.I. Smith, 1869)	along the entire Atlantic coast of North America	Shipping	2009 Romanian waters	Established	Micu et al. (2009)
Phylum Chordata **Class Ascidiacea** Fam. Molgulidae *Molgula manhattensis* (De Kay, 1843)	Mediterranean Sea, North Atlantic	Shipping	1971 Romanian waters	Established	Băcescu (1971), Micu and Micu (2004)

neighboring community of the mollusk *Mytilus galloprovincialis*. After its discovery in the Black Sea (Annenkova 1929), the range of the species habitat was limited for several decades to three regions: Varna-Burgas, Danube and Caucasian. By the end of the twentieth century Polychaete *F. enigmaticus* was discovered off the Crimean coast and in the north-western part of the Black Sea. It is assumed that the colder water temperature in these areas was a limiting factor (Shurova et al. 2020). An increase of the surface water temperature in the last few decades allowed this species to settle in the northern regions of the sea. Dense clusters of polychaete *F. enigmaticus* do not observed in these two areas. Nevertheless, in some hydraulic structures of the Sevastopol Bay, the number of this non-native species in the autumn-winter reached 2001–2002 ind.m^{-2}, and the biomass was about 0.7–3.0 g. m^{-2} (Grintsov and Murina 2002). The invasive polychaete *F. enigmaticus* was also noted as facilitator of parasite transmission (Kopry et al. 2022).

In September 2019, the polychaete *F. enigmaticus* was found in mass colonial settlements in the coastal area of the Kerch Strait near the Arshintsevskaya Spit. This species is currently known to be widely distributed along the coasts of the Atlantic, Pacific, Indian oceans and North, Mediterranean, and Caspian seas.

Apart from *F. enigmaticus*, other 14 non-native polychaete species have spread to several areas of the Black Sea shelf (Table 4.17): *Dipolydora quadrilobata* (Jacobi, 1883), *Glycera capitata* (Oersted, 1843), *Hesionides arenaria* (Friedrich, 1937), *Magelona mirabilis* (Johnston, 1845), *Marenzelleria neglecta* Sikorski & Bick, 2004, *Polydora cornuta* (Bosk, 1802), *Polydora websteri* Hartman in Loosanoff & Engle, 1943, *Prionospio pulchra* Imajima, 1990, *Sigambra tentaculata* (Treadwell, 1941), and *Streblospio gynobranchiata* Rice & Levin, 1998. A detailed analysis of the settlement with an assessment of the biological parameters of these non-native species may be found in a number of review papers (Kiseleva 2004; Surugiu 2005; Boltacheva et al. 2020, etc.). After a century of invasion, the non-native oligochaete *Tubificoides benedii* (Udekem, 1855) mastered the coastal zone of the sea and estuaries in the north-western and Danube regions. The species habits on various soils at the depths up to 26 m. Its concentration may reach 700 ind. m^{-2} (Shurova 2006; Shalovenkov 2020).

The process of polychaete establishment in the Black Sea is going further. At the beginning of the twenty-first century, seven new species which can be classified as invaders (mainly from the Mediterranean Sea) have been recorded in the northern part of the sea. The main reasons for the spreading of the new species are the intensification of shipping and aquaculture. Climate warming also may be an affecting factor. Once in favorable conditions, the species begin to reproduce, forms stable populations of high density and become a permanent component of the fauna of the area (Kiselyova 2004; Boltacheva 2020). Some of this species, like *Hydroides dianthus* (Verrill, 1873) settles on the shells of juvenile mollusks and forms the tubes which, as they grow, prevent the opening of the shell flaps, e.g., at oyster farm in Sevastopol Bay, the density of this species at the surface of single shell counting up to 30 individuals was registered. The *H. dianthus* is classified as a dangerous invasive species that is currently actively spreading to various areas (Boltacheva et al. 2020).

Bivalvia (Fam. Teredinidae) *Teredo navalis* (Linne, 1758) (Table 4.17) This is a drilling mollusk ("shipworm"), which penetrated into the Black Sea as early as the 7th–5th century BC (Gomoiu and Skolka 1996). The first studies of the *T. navalis* were carried out in the middle of the nineteenth century (Markusen, 1867). The following studies were aimed to assess the negative impact of this mollusk on wooden hydraulic structures in the Black Sea (Soldatova 1961). Two main abiotic factors influence an existence of *T. navalis,* i.e., temperature and salinity. Mass reproduction of *T. navalis* usually occurs in the summer, although larvae persist in the plankton during the cold season (Braiko 1958). High fertility and the possibility to preserve the larval stage for a long time allow *T. navalis* to survive during searching for a suitable woody substrate in the sea. The settling density of the larvae of the shipworm can reach 500,000–1,000,000 ind. m^{-2} in wooden substrat. A wooden beam can lose up to 70% of its original weight in three months. An average lifespan of the woodworm mollusk is about 2 years (Braiko 1958).

Thecostraca (Fam. Balanida) Next harmfull benthic invaders were the fouling marine balanuses, *Amphybalanus improvisus* (Fig. 4.30) and *Amphybalanus eburneus,* which arrived and established in the Black Sea in the nineteenth century (Marinov 1990). Both species were introduced from the coastal Atlantic waters of North America. *A. improvisus* also was introduced to the Baltic Sea (Leppakovski and Olenin 2000) (Table 4.17). Currently, both species are widespread in the coastal waters of the Black Sea with the first one being more common. The negative impact of these species is appeared in the coating of bottoms of ships, pipelines, piers, and dams by their colonies. On the other hand, their pelagic larvae constitute a significant part of the coastal feeding meroplankton. Amphybalanuses are hermaphrodites, and *Amphybalanus improvisus* is capable of self–fertilization, when cross-fertilization is impossible (Furman and Yule 1990). In addition, the eggs of all *Amphybalanuses* are fertilized and develop in the mantle cavity to the stage of the nauplius, which provides additional protection for the offsprings. The larvae settle in clusters to allow for cross-fertilization (Zevina 1994).

Fig. 4.30 *Amphybalanus improvisus*

The high growth rate of *Amphybalanuses* provides its ability to reproduce already at an age of one month (Zevina 1994). Its biomass may constitute 3% to 29% of the total biomass of fouling communitie (Zaitsev and Oztürk 2001). The maximum concentration was recorded in the hydraulic structures of the Odessa port, where it reached 81,360 ind.m^{-2} with the biomass equal to 5884.0 g.m^{-2} (Vinogradov et al. 2012). In addition to fouling communities living on the rocky substrate, barnacles are also found in benthic communities on soft soils, settling mainly on the shells of living and dead mollusks at a depth up to 50 meters (Kiseleva 1992, 2002). Their maximum abundance were recorded on the shelf in the transformation zone of riverine Danubian waters, where it reached 41,378 ind.m^{-2} (Gomoiu 1986). The largest biomass of *A. improvisus* was found in the area of the Varnensky and Burgassky bays, where it reached 7 kg.m^{-2} (Marinov 1990). In the Kerch Strait area, bottom community consists only of *A. improvisus* (Ivanov and Sinegub 2008).

A. improvisus is a cosmopolitan species occuring in temperate and tropical zones of the Atlantic Ocean, the Arctic Ocean and in the Baltic, the North, the Mediterranean and the Black seas. The species natural origin location is unknown, but presumably it came from the North America. It colonised wast areas of the World Ocean, including Indo-Pacific and Australasia as a biofouling agent on the hulls of ships. It was one of the first recorded invaders in the Baltic Sea, being found in Sweden and Lithuania in 1844, in the Elbe estuary in 1854 and off Great Britain coast in 1880s. It is tolerant to both high and low salinity and is often found in estuaries and low-salinity bays. As an invasive species, it competes with native organisms and it is an unwanted coloniser of the shells of cultivated oysters and mussels and aquaculture cages.

In 2001, species of the family Balanidae, *Amphybalanus amphitrite*, introduced with the ballast water, presumably from the Mediterranean Sea, was discovered in the Odessa Bay (Alexandrov 2004).

Barnacle crayfishes are considered as harmful crustaceans, as they can grow by the tons at the bottoms of vessels, weighing them down and reducing shipping speed, which leads to additional costs for cleaning the bottoms and financial losses.

In 1995, another representative of Thecostraca—*Chthamalus montagui*, was found for the first time off the coast of Crimea and included in the list of barnacles of the Black Sea (Zevina and Poltarukha 1999).

Bivalvia (Fam. Myidae and Mytilidae) *Mya arenaria* Linne, 1758, *Mytilus edulis* Linne, 1758, *Mytilus galloprovincialis* Lamarck, 1819, *Mytilus trossulus* Gould also are non native species of the edible mussels (Fig. 4.31). Now they are widespreaded in the Black Sea and also cultivated as an object of aquaculture.

M. galloprovincialis is one of the three principal closely related species of blue mussels complex, which collectively are widely distributed at the temperate to subarctic coasts of both the Northern and Southern Hemispheres, and often are dominant inhabitants on hard substrates of the intertidal and nearshore habitats. *M. galloprovincialis* often hybridize with its sister taxa, the closely related *M. edulis* and *M. trossulus*, when they are together in the same location. *M. galloprovincialis* is

Fig. 4.31 *Mya arenaria* (**a**), *Mytilus edulis* (**b**), *Mytilus galloprovincialis* (**c**), *Mytilus trossulus* (**d**)

considered as the most tolerant to warm water species of the three, and occupy the southernmost locations in Europe and North America (Skurikhina et al. 2001; Terentev 2024).

Bivalvia (Fam. Arcidae) Mollusk *Anadara kagoshimensis* (Tokunaga, 1906) is the most abundant non-native species of Family Arcidae (Fig. 4.32). After its first discovery in 1968 (Kiseleva 1992), it settled during 40 years almost all over the Black Sea shelf. The maximum values of *A. kagoshimensis* biomass were recorded in the 1980s at the coastal shelf of the Burgas Bay, where the biomass of the alien species exceeded 4000 g.m^{-2}, with a maximum abundance of 400 ind. m^{-2}. In subsequent years, the biomass of mollusks in this area decreased and now does not exceed 1100 g.m^{-2} (Marinov 1990). In the last 10–15 years, high rates of biomass and abundance of mollusks have also been recorded on the shelf of the Caucasus—1180 g.m^{-2} and 2590 ind.m^{-2} (Chikina 2009) and in the estuarine areas of the shelf of the Danube and Dniester rivers 2700 g.m^{-2}, 1160 ind.m^{-2}, respectively. In these two areas, as well as in the Kerch Strait, high rates of *A. kagoshimensis* population development were recorded. It resulted in its dominance in the zoobenthos, in terms of biomass and abundance. The *A. kagoshimensis* settlement leads to the formation of new bottom communities in the Black Sea. In the Black Sea, it lives on sandy and rocky ground, silt, and sand with thickets of seagrasses at depths up to 20 m. The average size of its shells may reach 70–75 mm. Breeding occurs in June–September, since the species needs high temperatures for spawning and larval development (Chikina and Kucheruk 2005; Chikina 2009). Mollusk is also widespreaded in the Azov, Mediterranean and Adriatic seas.

Fig. 4.32 *Anadara kagoshimensis*

Clam shells are an affordable natural material promising for use as a biosorbent. It demonstrated high efficiency in removing heavy metal ions (e.g., copper and lead) from polluted water.

Two more species of genus *Anadara*—*Anadara inaequivalvis* (Bruguiere, 1789) and *Anadara cornea* (Reeve, 1844) were recorded in the Bosporus area (Ozturk 1998) (Table 4.17).

Bivalvia (Fam. Ostreidae) Pacific oyster *Magallana gigas* (Thunberg, 1793) *(Crassostrea gigas)*, is a non-native species for the Black Sea ecosystem (Fig. 4.33). This species with the high commercial value and the highest production rate is widely exploited in the aquaculture. It has been cultivated near the Crimean and Romanian coasts since 1980s. *Magallana gigas* and the mussel *Mytilis galloprovincialis* Lamarck, 1819 are the objects of aquaculture in the Sevastopol region, namely in the Kazachya Bay.

Recently *Magallana gigas* has created breeding population off the Turkish coasts becoming the dominant species on some hard substrate, including rocky bottoms and large rocks used for creating bulk land in the sea. As a result, the population density rised to 94.58 ind.m^{-2} and an average shell length to 3.33 + 0.724 cm. It indicates that *M. gigas* has adapted to the Turkish coasts' conditions and become able to create breeding populations (Aydin and Gul 2021).

The meat of this edible shellfish has a high nutritional value due to a balanced high-grade protein composition, glycogen, essential PUFA, a rich composition of mineral elements, and biologically active substances. The body of the shellfish is enclosed in a shell, consisting of organic matter. Also, it is important that its life cycle allow species to form wast accumulations in a relatively short time which greatly facilitates their catch. Almost all global catches provided by oysters, scallops, and mussels. The clam group includes various types of bivalves that partially or completely burrow into soft soils, significantly facilitating their fishing. In the Black Sea, the species of genus *Anadara* do not used as a human food, but in the Southeast Asia *Anadara granosa and A. subcrenata* are cultivated as edible product.

Fig. 4.33 *Magallana gigas*

Gastropoda. *Rapana venosa* (Valenciennes) In 1947 this mollusk was discovered in the Novorossiysk area as a part of mussel Biocenosis (Fig. 4.34). It is a predator that consumes oysters, mussels, and other bivalves (Drabkin 1953). However, rapana is a valuable delicacy food product. Rapana inhabits the Japanese, Yellow, and East China Seas, both the areas with high salinity (25–32%) (Chung et al. 1993; Golikov and Scarlato 1967) and desalinated waters (Razin 1934).

Rapana was brought to the Black Sea from the Sea of Japan with a vessel, on the bottom of which a clutch of her eggs was attached (Drabkin 1953). Rapana possesses high ecological plasticity, tolerance to low salinity, oxygen deficiency and pollution. It is highly fertile, rapidly grows and reaches puberty at 1–3 years old, having size of 50–70 mm (Harding and Mann 2001). The long-lasting planktonic phase (14–80 days) provides an opportunity for transporting rapana larvae with ballast water (Mann and Harding 2000). Rapana successfully reproduced in the Black Sea, especially off the Caucasian coast, and in the 1950s almost destroyed the oyster community on the Gudauta bank, followed by the mussel *M. galloprovincialis* and the scallop *Chlamys glabra*, which live together with the oyster. Later, Rapana began to destroy mussel settlements off the southern coast of Crimea, and then off the coast of Bulgaria, where it also reached high abundance. By the beginning of 1970, rapana had spread almost throughout the whole Black Sea and is currently absent only in the most desalinated areas of its northwestern and western parts. Initially, only the Rapana shell was in demand for the handicraft production. In the 1980s, intensive commercial catches of rapana as a valuable food item began, at the coast of Turkey and then off the coast of Bulgaria. Later, Rapana was mined on the Caucasian shelf and in the Kerch Strait areas, where its total biomass was estimated

Fig. 4.34 *Rapana venosa*

at 2800 and 6000 tons (Zaitsev and Mamaev 1997). Later, due to overfishing, Rapana population significantly derease in the northeastern part of the Black Sea. Rapana food objects (small bivalves) significantly dropped due to the development of the ctenophore *Mnemiopsis leidyi*, which consumes their pelagic larvae.

Rapa whelk is the most important component of the demersal fauna in the southern part of the Black Sea. Its commercial catches began in 1986. About 67% of the landings were exploited by Turkey. This species became an important income source for the individual fishermen due to its export value. The Rapa whelk stocks are now overexploited and badly managed. The stocks must be monitored continuously, since in the absence of its natural predator (e.g., sea stars) and limited food resourses, Rapana harms endemic species, predominantly the bivalve species. To reduce the impact of fishing on the ecosystem, Rapa whelk stocks should be managed with ecosystem-oriented approach. Recently, all Black Sea countries fishing the Rapana and the total annual catch reached 10,000–15,000 tons. In Bulgaria, this fishery has been commenced in the 1990s and rapa whelk was initially caught by scuba divers (Daskalov and Rätz 2010), with catches around 90% of the total shellfsh catch (FAO 2000). Since 2012, the catch of *R. venosa* with beam trawl has been allowed in Bulgaria (Fishery and Aquacultures Act 2012); with catches of 4000–5000 tons/year, this species catchs take the first place in the country's marine fisheries.

Gastropoda (Fam. Tateidae) ***Potamopyrgus antipodarum*** **(J. E. Gray, 1843)** It is another non-native species which arrived from the New Zealand (Grossu 1951). However, *P. antipodarum* first appeared and settled off the Atlantic coasts of

Europe, then penetrated into the Mediterranean and Black Seas. It is found in brackish estuaries with a salinity of 4–6‰, from which it gradually penetrates rivers and spreads in fresh waters. *P. antipodarum* was first discovered in Great Britain in 1883 in the Thames estuary (Smith 1889). Later, it was discovered in the North Sea and coastal waters of the Germany, the Netherlands, in the English Channel on the northern coast of France, as well as along the shores of the Baltic Sea (Bondensen and Keiser 1949). It was found in the Black Sea off the coast of Romania (Grosu 1951) and now *P. antipodarum* is existed in the estuaries of the northwestern part of the Black Sea (Chukhchin 1984). This species was accepted as harmful since it consums fish's food and disrupt water supply systems (Orlova and Feneva 2018).

Gastropoda (Fam. Buccinidae) ***Neptunea arthritica*** **(Bernardi, 1857**) in 2001 was discovered in Kamyshevaya Bay (Sevastopol, Crimea). *N. arthritica* is a predator, arrived from the Far East that tolerates desalination; establishment of this species can cause serious changes in the bottom communities of the Black Sea (Shadrin et al. 2002).

Decapoda (Fam. Panopeidae) Crab ***Rhithropanopeus harrisii*** **(Gould, 1841**) is present in the communities based both on the hard substrates (rocks, stones) and on the loose soils (Fig. 4.35a). This non-native species does not have a high biomass and abundance. Its size is relatively small: the width of the carapace does not exceed 24 mm (Zolota 2017). Crab features, such as high fertility, planktonic larva, a wide range of nutrition and temperature tolerance contribute to the widespreaded settlement of this species (Reznichenko 1967; Zolota 2017). Off the Crimean coast, the average crab biomass did not exceed 4 g.m^{-2} (maximum 32.066 g.m^{-2}), and the average abundance is 5 ind.m^{-2} (maximum—100 ind. m^{-2}). Higher parameters were recorded in the Danube region, where maximum biomass reached 37.348–64.009 g.m^{-2}, and the population reached 311–533 ind. m^{-2} (Surugiu and Zamfirescu 2004). High rates of development of the *R. harrisii* population were also recorded in the saline lakes near Varna and Beloslav, where the maximum biomass of crab population reached 640 g.m^{-2}, with the abundance of 260 ind.m^{-2} (Todorova and Konsulova 2008). Its diet is extensive: *R. harrisii* is able to eat both vegetable and animal food, and often feeds on carrion.

It may compet with the native crabs. It was the way how the Sea of Azov *Brachynotus sexdentus* was displaced (Zolota 2017). On the other hand, crab may be the fish food (Forsstrom et al. 2015).

Decapoda (Fam. Portunidae) The blue crab ***Callinectes sapidus*** **Rathbun, 1896** is rapidly expanding in the northwestern Mediterranean coast of Morocco (Fig. 4.35b). The blue crab *C. sapidus*, one of the largest representatives of its order, was discovered in the Black Sea off the Bulgarian coast in 1967 (Bulgurkov 1968). Then, individual specimens were found in the Kerch Strait in 1975 and later in the Varna Bay and near the Bosphorus. Still, these cases were isolated findings (Zaitsev 1998). The blue crab is native to the Atlantic coastal waters of North America and was introduced to the Black and Marmara seas with ballast water. It is not known exactly where the blue crab was imported from. The blue crab was found for the first time in the Mediterranean Sea in the 1940s and in the Adriatic Sea

Fig. 4.35 *Rhithropanopeus harrisii* (**a**) and *Callinectes sapidus* (**b**)

in 1959. Thus, it could have been introduced into the Black Sea from the Mediterranean (Zaitsev 1998). Although, it cannot be ruled out that it was introduced directly from the Atlantic coast of America, or from the Adriatic Sea, where conditions are closer to the Black Sea. In the recent years, the blue crab has been found off the coast of Sevastopol, and its population has been increasing (Revkov et al. 2003).

The native geographical areal of the Atlantic blue crab includes the Antilles, the Gulf of Mexico and Bermuda, and extends from Nova Scotia in Canada, Maine and northern Massachusetts to northern Argentina (Williams 1974; Nehring 2011; Castriota et al. 2012). Recent climate changes appear to favour *C. sapidus'* metabolic response and proliferation around the world (Marchessaux et al. 2022). In the Mediterranean Sea, *C. sapidus* is recognized as an invasive species (Mancinelli et al. 2016) and categorized amongst the top 100 worst problematic invasive species due to the large number of records (Streftaris and Zenetos 2006). The first recorded occurrence of *C. sapidus* in North Africa in Morocco was discovered near the Marchica Lagoon at Nador in 2017 (Chartosia et al. 2018). Recently, this species was reported in the ports of Tangier Med (Chairi and González-Ortegón 2022). Additionally, it has been observed in other nearby southern Mediterranean African countries, including Algeria (Kara and Chaoui 2021) and Tunisia (Mills et al. 2020).

Decapoda (Fam. Grapsidae) Crab ***Eriocheir sinensis*** **Milne-Edwards, 1853.** Female individual of the Chinese hairy-armed crab *E. sinensis* with eggs was caught for the first time in the Bay of Garbage near the Danube Delta in 1997 (Gomoiu and Skolka 1998). Later, single specimens were found in the northwestern desalinated areas of the Black Sea and in the Sea of Azov. It is a very widespread catadromic species in European waters. It was introduced in the North Sea in 1912 from the estuaries of East Asia and the coastal waters of the South China and Yellow Seas (Herborg et al. 2003).

Decapoda (Fam. Pirimelidae) Crab *Sirpus zariquieyi* Gordon, 1953. This subtropical species was found and established in the area near Bosporus (Gordon 1953). *S. zariquieyi* is a newcomwer from Cadaqués on the northeast coast of Spain.

Decapoda (Fam. Palaemonidae) ***Palaemon macrodactylus*** **H. Milne-Edwards 1837** This shrimp is native to Japan, Korea, and Northern China. In the second half of the twentieth–early twenty-first century, the Asian shrimp *P. macrodactylus* was discovered in number locations world-wide. The first cases of capture of the oriental shrimp *P. macrodactylus* are described in the Don delta and the eastern part of the Taganrog Bay.

Currently, it is known that the East Asian alien species has successfully established in the Black Sea and in the Taganrog Bay of the Sea of Azov. Mature individuals with riped eggs are found in high quantity from August to September. The species succesfully lives and reproduces in both salt and fresh water (Ctankevich 2022).

Decapoda (Fam. Panopeidae) Crab ***Dyspanopeus sayi*** **(S.I. Smith, 1869).** It was discovered in the Black Sea off the coast of Romania in 2009 (Micu et al. 2009) and expanded its distribution area over the Azov and the Black Sea basins in the past 10 years. In August 2022, a mud crab *D. sayi* was discovered for the first time in the coastal zone of the Kazantip peninsula (Crimea), on the northern coast of the Kerch Peninsula and in the area of the Arshintsevskaya Spit.

The native habitat of this species locates along the Atlantic coast of North America, while the invasion area covers the waters of the Atlantic coast of Europe and the northern part of the Mediterranean Sea.

Ascidiacea (Fam. Molgulida) **Molgula manhattensis (De Kay, 1843)** commonly known as "sea grapes," is a species of ascidian found along the East Coast and Gulf Coast region of the United States. It was introduced in coastal waters off Romania in 1971 (Băcescu 1971). Its larva has only the chord, the spine and other bones do not develop, and the chord resolves with age. An adult individual is clad in a hairy tunic, a hard protective shell with a complex structure. It resides on soft ground just below the littoral being disguised by sand, and filters food from seawater using siphons (Table 4.18).

Asterias rubens (Fam. Asteriidae) (the common starfish or common sea star or sugar starfish) is the most common and familiar starfish in the northeastern Atlantic. It has five arms and usually grows up to 10–30 cm. The common starfish is found on rocky and gravelly substrates where it feeds on mollusks and other benthic invertebrates.

Asterina gibbosa (Fam. Asteriidae) (Cushion star) lives at the sea bottom. This sea star possesses an entirely benthic, lecithotrophic development.

Amphiura filiformis (Fam. Amphiuridae) (Brittle star). It is found on the seabed in the northeastern Atlantic at a depth up to 200 m.

Clausinella fasciata Bivalvia (Fam.Veneridae) is a marine bivalve mollusc.

Hiatella rugosa (Fam. Hiatellidae) exists as a shellfish fouling.

Paleamon longirostris (Fam. Palaemonidae). It is a shrim, considered as invasive species in the Black Sea.

Sirpus zariquieyi (Fam. Pirimelidae) is considered as invasive species in the Black Sea.

Table 4.18 Non-native benthic species of the Mediterranean origin, discovered in recent decades in the Black Sea (after Shiganova et al. 2012, with additions)

Species	Tolerance to *T* °C	First record	References
Class Asteroidea Fam. Asteriidae ***Asterias rubens*** (Linneus, 1758)	Subtropical	1996, 2003 The Bosporus region	Albayrak (1996), Karhan et al. (2007)
Fam. Asterinidae ***Asterina gibbosa*** Pennant, 1777	Subtropical	2010 The Bosporus region, along the Turkish coast	Sergin and Kideys (2010)
Fam. Amphiuridae ***Amphiura filiformis*** (O.F. Müller, 1776)	Subtropical	2010 The Bosporus region, along the Turkish coast	Sergin and Kideys (2010)
Class Bivalvia, Fam.Mytilidae ***Mytilus edulis*** Linne, 1758	Subtropical	2001 Coastal waters Odessa	Alexandrov (2004)
Fam. Veneridae ***Clausinella fasciata*** (Da Costa, 1778)	Subtropical	1990 Bosporus area, Coast of Crimea	Revkov (2003)
Class Bivalvia Fam. Hiatellidae ***Hiatella rugosa*** (Linnaeus, 1767)	Subtropical	2001 Bosporus area, Coast of Crimea Coastal waters off Turkey, Crimea, Sevastopol Bay	Revkov (2003)
Fam. Cardiidae ***Acanthocardia tuberculata*** (Linnaeus, 1758)	Subtropical	1998 Costal water Turkey, Kerch area	Terentyev (1998)
Fam. Anadaridae ***Anadara demiri*** (Piani, 1981)	Subtropical	1998 Bosporus area, coastal waters Turkey, north-western Black Sea	Öztürk (1998), Alexandrov 2004
Anadara cornea (Reeve, 1844)	Subtropical	1998 Bosporus area	Öztürk (1998)
Class Polychaeta Fam. Capitellidae ***Capitellethus dispar*** (Ehlers, 1907)	Subtropical	1972 Bosporus area	Öztürk (1998)
Fam. Spionidae ***Streblospio gynobranchiata*** Rice et Levin, 1998	Subtropical	2007 Sevastopol Bay	Boltacheva (2008)
Fam. Spionida ***Polydora cornuta*** Bocs 1802	Subtropical	2007 Balaklava Bay	Murina (2008)
Order Decapoda **Class Malacostraca** Fam *Palaemonidae* ***Paleamon longirostris*** H.Milne-Edwards	Subtropical	2007 Bosporus area	Sergin et al. (2007)

(continued)

Table 4.18 (continued)

Species	Tolerance to T °C	First record	References
Order Decapoda **Class Malacostraca** Fam. Pirimelidae ***Sirpus zariquieyi*** Gordon 1953	Subtropical	2000 Bosporus area	Zaitsev and Ozturk (2001)
OrderAmphipoda **Class Malacostraca** Fam. Oedicerotidae ***Synchelidium maculatum*** Stebbing, 1906	Subtropical	1980 Bosporus area western shores of Crimea, eastern shores of the Caucasus	Grese and Fedorina (1979)

Synchelidium maculatum (Fam. Oedicerotidae) is Amphipoda, which probably also established in the Black Sea.

Thus, the list of discovered species from the Mediterranean Sea in near Bosporus includes three species of starfish, which will not established in the Black Sea, two species of Bivalvia, which already established in the Black Sea and Amphipoda, which also could be establish in the Black Sea.

Pisces

Intentional Introductions

First non-native fish species entered the Black Sea as a result of intended introductions for increasing the fishing stock. Of these intentionally introduced fish (22 species), only a very small number of species established and become commercial, despite the large-scale efforts for increasing the fishery potential of the basin (Plotnikov 2000; Stroganova et al. 2001).

Planiliza haematocheilus (Temminck & Schlegel, 1845)**.** The most significant and successful case was the introduction of a large Far Eastern mullet, *Planiliza haematocheilus* (Temminck & Schlegel, 1845) (=*Mugil soiuy* (Basilewsky 1855) (Table 4.19), which become a aaaaaaaaaavaluable fishing object in the Black and Azov Seas (Fig. 4.36).

Fry and juveniles of *P. haematocheilus* were imported from the estuaries of the Sea of Japan in 1972–1980, and released into the estuaries of the Black and Azov seas and directly into the northwestern part of the Black Sea and the Sea of Azov. Since the 1980s, *P. haematocheilus* has become a widespread commercial fish in the the Black and Azov Seas, and some individuals have been caught on the Turkish shelf, in the Marmara and even the Mediterranean seas (Pryakhin and Volovik 1997). *P. haematocheilus* was supposed to feed on detritus. However, the food spectrum of pilengas has expanded. In addition to detritus, it actively consumes small benthic organisms, competes for food with the local mullet species, which stocks have decreased lately. The growth rate of *P. haematocheilus* in the Black Sea changed, and it may reach a much larger size and weight (Tsarin et al. 1999). There was not only a 1.5–3–fold increase of its growth rate (the average weight in the Sea of Japan is 3 kg and size 65 cm, while in the Sea of Azov it may reach 4.5 kg and 71 cm), but also an acceleration of puberty for 1 year. In 1992 *P. haematocheilus* was included in the register of commercial fish of the Azov-Black Sea basin (Stroganova and Zadorenko 2000). In 1993, fishing regulations were established and fishing was allowed. *P. haematocheilus* fishing in the Azov and Black Seas is conducted by Russia, Ukraine, and since 1999 by Georgia.

***Salmo gairdneri* Richardson, 1836** Of the intentionally introduced salmonids, only the rainbow trout *Salmo gairdneri* occurs as singular individuals. Rainbow trout is one of the breeding facilities of the Adler Industrial and Experimental Salmon Hatchery and the Adler trout Breeding Plant, located on the Mzymta River. From these plants, juveniles were purposefully released into the rivers of the Black Sea coast—Mzymta and Shakhe. The conditions in these areas are suitable for habitat of rainbow trout and the cases of capture of this species by amateur fisherman have been reported (Tuniev 1999; Drogan 2002; Luzhnyak 2003). However, no natural reproduction of rainbow trout was observed: seven specimens caught in 2004 were immature (stage II), and the gonads of the eighth were in the stage of resorption. Probably, natural spawning of rainbow trout does not occur in the Mzymta River, and its abundance in the river is replenished by fish of factory origin (Pashkov 2005).

***Oncorincus keta* (Walbaum)** Keta is anadromous species, which was intentionaly introduced in the western part of the Black Sea in 1977–1979; however, only single individuals did occur (Zaitsev and Mamaev 1977).

Two species of labraxes—the Japanese labrax *Lateolabrax japonicus* and the common labrax *Dicentrarchus labrax* (Table 4.19) were imported from the coastal low-salinity waters of the Sea of Japan. Both species occur now as singular individuals in the northwestern part of the Black Sea. The common labrax *D. labrax* is regularly caught in the Black Sea, although in small quantities. Thus, in October 1992, one 40 cm-long male was caught in the Tendrovsky Bay (Tkachenko 1994). In the autumn of 1995, two specimens were caught in the area of Betta Bay, at a depth of 2–5 m among the rocks (Pashkov and Plotnikov 1996; Pashkov 2001). In autumn in the area of Fiolent Cape, directly off the coast at a depth of about 2 m, one specimen of a the spawning female labrax was found with the total length of 87 cm and weight of 9750 g. In January 2002 three more specimens of labraxes were caught off the Fiolent River, and another one was caught off the western coast of Sevastopol near the mouth of the Kacha River (Boltachev 2003).

***Plecoglossus altivelis* (Ayu sweetfish)** This delicious fish was brought from the Sea of Japan in 1963 to replenish fish resources in the western part of the Black Sea, and individual specimens are currently found there (Zaitsev and Mamaev 1977).

White ***Hypophthalmichthys molytrix*** and the mottley ***Hypophthalmichthys nobilis*** (species of silver carp) were used for commercial fish farming. First of

Table 4.19 Non-native species of fish in the Black Sea

Species	Donor region	Vector	First record	Establishment status	Author
Phylum Chordata **Class Teleostei** **Order Mugiliformes** Fam. Mugillidae *Planiliza haematocheilus* (Temminck & Schlegel, 1845)(*Mugil soiuy* (Basilewsky)	Estuaries of northern part of the Sea of Japan	Intentional introduction	1972–1980 Shelf and estuaries of the Black Sea, the Sea of Azov	Established	Pryakhin and Volovik (1997)
Class Teleostei **Order Cypriniformes** Fam. Cyprinidae *Pseudorasbora parva* (Temminck & Schlegel, 1846)	Far Eastern rivers	Incidentally introduced with *Hypophthalmichthys molytrix*	Early 1960s lower reaches of the Dniester	Established	Kozlov (1974)
Class Teleostei **Order Cypriniformes** Fam. Cyprinidae *Hypophthalmichthys molytrix* (Valenciennes, 1844) *Hypophthalmichthys nobilis*	Amur River Central and Southern China	Intentional introduction Intentional introduction	1960 Deltas of Dniester, Danube, Dnieper	Supported by artificial reproduction Supported by artificial reproduction	Kozlov (1974)
Order Cypriniformes Fam. Cyprinidae *Mylopharyngodon piceus* (Richardson)	Far Eastern rivers	Intentional introduction	1960s found among algae in the coastal waters of Crimea	Supported by artificial reproduction	Russ (1993)

Class Teleostei **Order Eupercaria *incertae sedis*** Fam. Moronidae *Morone saxatilis* (Walbaum, 1792)	The East Coast of North America	Intentional introduction	1970s North-western part of the Black Sea, southern coast of Crimea	Established, but few in number	Zaitsev and Mamaev (1977)
Class Teleostei **Order Osmeriformes** Fam. Plecoglossidae *Plecoglossus altivellis* Temminck et Schlegel 1846	Sea of Japan	Intentional introduction	1963 single individuals in the western part of the Black Sea	Single individuals	Zaitsev and Mamaev (1977)
Class Teleostei **Order Perciformes** Fam. Percichthyidae *Lateolabrax japonicus* (Cuvier, 1828)	Sea of Japan	Intentional introduction	1970s Western part of the Black Sea	Single individuals	Zaitsev and Mamaev (1977)
Class Teleostei **Order Eupercaria *incertae sedis*** Fam. Moronidae *Dicentrarchus labrax* (Linnaeus)	Sea of Japan	Intentional introduction	1977 Western part of the Black Sea	Single individuals	Zaitsev and Mamaev (1977)
Order Salmoniformes Fam. Salmonidae *Oncorincus keta* (Walbaum)	Mediterranean Sea	Intentional introduction	1977–1979 Western part of the Black Sea	Single individuals	Zaitsev and Mamaev (1977)
Order Salmoniformes Fam. Salmonidae *Oncorhynchus mykiss* (Walbaum, 1792)	Atlantic coast of North America	Intentional introduction	1960s The North-western Black Sea	Supported by artificial reproduction	Zaitsev and Mamaev (1977)

(continued)

Table 4.19 (continued)

Species	Donor region	Vector	First record	Establishment status	Author
Order Salmoniformes Fam. Salmonidae *Oncorhynchus gorbusha* (Walbaum)	Pacific Ocean	Intentional introduction	1970s estuaries of the Black Sea rivers, currently undiscovered	Not established	Zaitsev and Mamaev (1977)
Order Salmoniformes Fam. Salmonidae *Salmo gairdneri* Richard-son, 1836	Adler trout Breeding Plant released juveniles	Intentional introduction	1999–2002 the rivers of the Black Sea coast Mzymta and Shakhe.	Supported by artificial reproduction	Pashkov (2001)
Class Teleostei **Order Centrarchiformes** Fam. Centrarchidae *Lepomis gibbosus* (Linne, 1758)	Aquarium species, got from Europe via Danube River	Penetration via Danube River	1949 found in the Danube Delta	Established	Svetovidov (1964)
Class Teleostei **Order Cypriniformes** Fam. Poeciliidae *Gambusia holbrooki* (Girard 1859)	The Adriatic Sea, there from America	Acclimated in aquarium.	1925 Colchis swampy waters Delivered with ships for mos-quito control	Established	Lindbergh (1933)
Class Teleostei **Order Beloniformes** Fam. Adrianichthyidae *Oryzias latipes* (Temimninck et Schlegel)	Estuaries of Southeast Asia (?)	Delivered by aquarists	1970s desalinated areas of the north-western part of the Black Sea	Established	Zaitsev and Mamaev (1977)
Fam. Cyprinidae *Carassius auratus* Lin-naeus, 1758	Goldfish are immensely popular ornamental cyprinid fish	Delivered by aquar-ists traded in more than 100 countries.	2000 Desalinated waters of the Black Sea	Established	Abramenko (2000)

Order Perciformes Fam. *Chaetodontidae* *Heniochus acuminatus* Linnaeus, 1758	Indo-Pacific	Shipping (ballast waters) (?)	2003	Single individuals	Boltachev (2006)
Order Perciformes Fam. Centracanthidae *Centracanthus cirrus* Raffinesque	Mediterranean Sea	Natural expansion	1982	Increase in number	Tsokur (1988), Abasa et al. (2006), Shiganova and Ozturk (2009)
Phylum Chordata **Class Teleostei** **Order Gadiformes** Fam. Gadidae *Micromecisthis poutassou* (Ricco)	Mediterranean Sea	Natural expansion	1999 North-western the Black Sea; Single individuals regularly observed in Turkish waters	Single individuals	Boltachev (2006)
Class Teleostei **Order Carangaria *incertae sedis*** Fam. Sphyraenidae *Sphyraena obtusata* Cuvier	Indo-Pacifica, entered the Mediterranean as Lessepsian migrants, then the Black Sea	Lessepsian migrants, entered from Mediterranean Sea	2004 Turkish waters 2014 Crimea	Single individuals	Fishbase, Öztürk (2006)
Class Teleostei **Order Carangaria *incertae sedis*** Fam. Sphyraenidae *Sphyraena pinguis* Gunter, 1874	Pacifica, entered Mediterranean Sea as Lessepsian migrants, then the Black Sea	Lessepsian migrants, entered from Mediterranean Sea	2009 in the waters of Crimea	Single individuals	Boltachev (2009), Golani et al. (2021)
Class Teleostei **Order Clupeiformes** Fam.Clupeidae *Sardinella aurita* Valenciennes	Mediterranean Sea, Sea of Marmara	Natural expansion	1997 Near Sevastopol, Turkish waters	Migration, increasing numbers in the Black Sea	Boltachev (2006), Fishbase, Özturk (2006)

(continued)

Table 4.19 (continued)

Species	Donor region	Vector	First record	Establishment status	Author
Class Teleostei **Order Mugiliformes** Fam.Mugilidae *Chenon labrosus* (Risso, 1827)	Mediterranean Sea	Natural expansion	1999 near Sevastopol, southern part of the Black Sea	Migration	Boltachev (2006) Fishbase
Class Teleostei **Order Mugiliformes** **Fam.Mugilidae** *Chelon ramada* (Risso, 1827)	Mediterranean Sea	Natural expansion	1950, 2007 North-western Black Sea, Crimea	Migration	Boltachev (2006) Fishbase
Order Eupercaria ***incertae sedis*** Fam. Labridae *Thalassoma pavo* (L)	Mediterranean Sea, Sea of Marmara	Natural expansion	2006 Turkish waters	Migration	Fishbase, Özturk (2006)
Order Eupercaria ***incertae sedis*** Fam.Sparidae *Sarpa sarpa* (L)	Mediterranean Sea, Sea of Marmara	Natural expansion	2001 reappeared in the Southern part of the Black Sea, increases the number	Migration	Fishbase, FAO Özturk (2006)
Order Eupercaria ***incertae sedis*** Fam.Sparidae *Sparus aurata* (L)	Mediterranean Sea, Sea of Marmara	Natural expansion	2006 The coastal waters of Crimea; increasing population in recent years	Migration	Fishbase, FAO, Boltachev (2006), Ozturk (2006)
Order Eupercaria ***incertae sedis*** Fam.Sparidae *Boops boops* (L)	Mediterranean Sea, Sea of Marmara	Natural expansion	1989 Southern part of Black Sea	Migration	Fishbase, FAO; Ozturk (2006)
Order Eupercaria ***incertae sedis*** Fam.Sciaenidae *Umbrina cirrhosa* (L)	Mediterranean Sea	Natural expansion	1962 occurs in small quantities in the northern part of the Black and Azov Seas; its number has increased in recent years	Migration	Pashkov (2001), Fishbase

Order Eupercaria *incertae sedis* Fam. Sparidae *Spicara maena* (L)	Mediterranean Sea	Natural expansion	1964–89 Increase the number in the Black Sea	Migration	Svetovidov (1964), Salekhova (1979), Salekhova et al. (1989)
Order Gobiiformes Fam. Gobiidae *Gobius auratus* Risso 1810	Mediterranean Sea	Natural expansion (?)	1967 coastal area of North-western Black Sea	Established	Gordina (1974), Shiganova and Ozturk (2009)
Order Gobiiformes Fam. Gobiidae *Gobius cruentatus*	Mediterranean Sea	Shipping (?) Migrations (?)	2006 Crimea Southern Black Sea	Established	Boltachev (2006), Ozturk (2006), Fishbase
Order Gobiiformes Fam. Gobiidae *Gobius xanthocephalus*	Mediterranean Sea, Sea of Marmara	Shipping (?) Migrations (?)	2006, 2009 Crimea Southern Black Sea	Established	Boltachev, (2009), Ozturk (2006)
Order Gobiiformes Fam. Gobiidae *Tridentiger irigonocephalus* (Gill, 1859)	Pacific Ocean	Shipping (ballast waters) (?)	Sevastopol Bay	Established	Boltachev (2009)
Order Blenniiformes Fam.Blenniidae *Parablennius incognitus* Bath, 1968	Mediterranean Sea, Sea of Marmara	Shipping (ballast waters), migrations (?)	Crimea Southern Black Sea	Established	Bogorodsky (2006), Fishbase
Class Teleostei Fam. Sparidae *Centracanthus cirrus* Rafinesque, 1810	Mediterranean	Shipping	2004 Romanian waters	Established	Tsokur (1988), Abaza et al. (2006)
Order Syngnathiformes Fam. Syngnathidae *Syngnathus acus* L. 1758	Atlantic Ocean (?)	Shipping (?)	Crimea	Established	Rass (1993a, b)
Monogenoidea *Ligophorus kaohsianghsieni* (Gusev, 1962) Gusev, 1985	Estuaries of northern part of the Sea of Japan	Incidentally introduced with pilengas during his naturalization	1994 1995 1995	Established Established Established	Dmitrieva (1996), Maltsev (1957)

Fig. 4.36 *Planiliza haematocheilus*

Fig. 4.37 *Pseudorasbora parva*

them was brought from the Amur River and another one from the Southern China to the ponds and estuaries of the Azov-Black Sea basin. They now become important commercial freshwater fish, spreading into estuaries and rivers up to deltas, but they do not reproduce in natural conditions (AZNIIRH data).

The black ***Mylopharyngodon piceus*** and the white ***Ctenopharyngodon idella*** are two species of Amurs. Being brought from the Far-eastern rivers, now are established in the Dniester River and found in the coastal waters of Crimea (Rass 1993a, b).

Gambusia holbrooki The gambusia (**mosquito-eating**) was intentionally introduced from low-saline areas of the Adriatic Sea in 1925 to reduce the number of malarial mosquitoes in the area of Colchis. Gambusia successfully established in freshwater marshy reservoirs, and then penetrated and widely spreaded in the Black and Azov seas, turning out to be an euryhaline species, able to occurs in the wide salinity range 0–17‰ (Zaitsev and Mamaev 1977).

Weedy Fish Imported with the Commercially Introduced Fishes

Pseudorasbora parva (Amur chebachok) (Fig. 4.37) along with the carp and white cupid was introduced from Far Eastern rivers to fresh water bodies in Romania (Kozlov 1974).

From there, it spread through canal systems into the rivers of the Black Sea basin, then into the rivers of the Azov basin, and then penetrated into the rivers of Europe. Currently, it is a widespread type of weed fish. The species is extremely eurybiont, and can live in various freshwater bodies—rivers, ponds, reservoirs, discharge channels, and highly eutrophicated waters. Its food spectrum is similar to that of juvenile commercial fish species. Planktonic crustaceans form the base of the diet of its larvae and fry, and the proportion of benthic organisms in the diet increases with their growth. In this regard, the Amur chebachok competes with the juveniles of most fish, both planktophages and benthophages. In addition, chebachok can eat the eggs of valuable commercial fish (Frolova 2002). *P. parva* carriers the parasite *Sphaerothecum destruens*, which is harmful to other fishes, but not to himself.

Unintended Fish Species Penetration Due to Aquarists' Activity
Carassius auratus Linnaeus, 1758 (Far Eastern silver carp) (Fig. 4.38a), which was also introduced along with introduced commercial fish. It was recently discovered in the basin and low-saline waters of the Black Sea (Abramenko 2000). The silver carp has replaced the endemic golden carp and is currently a real scourge for valuable fish species in the Taganrog Bay, as well as in inland waters (Boltachev 2008).

Lepomis gibbosus (aquarium sunfish) (Fig. 4.38b) inhabiting in the North American freshwater basins was introduced by aquarists to Europe in the eighteenth century (Svetovidov 1964; Russ 1993). From the ponds in which it was bred, the sunfish penetrated in the large rivers Rhine, Oder, Danube and the associated inland reservoirs. It was first recorded in the Danube Delta in 1949, from where, in subsequent years, it gradually settled along the adjacent shores of the Black Sea to the south into the coastal lakes of Romania and Bulgaria, and to the northeast into separate reservoirs of the Danube-Dnieper interfluve (Svetovidov 1964). Currently, the sunfish lives in the basins of the Tiss and Bug rivers, in the lower reaches of the Danube, Dniester and Dnieper, in the Odessa Bay, Berezansky Estuary, and is rarely found in the low-saline areas of the Black Sea. It is very common in numerous

Fig. 4.38 *Carassius auratus* (**a**), *Lepomis gibbosus* (**b**), *Oryzias latipes* (**c**), and *Heniochus acuminatus* (**d**)

channels of the Dnieper, in the Kakhovsky reservoir, the Dnieper-Bug estuary, from where it penetrated into the lower reaches of the Southern Bug. At the end of September 2002, more than 200 specimens of sunfish were caught in a fish pond located in the Northern Crimea, where they probably entered from the Kakhovsky reservoir through the North Crimean Canal with Dnieper waters. The canal runs along the northeastern part of Crimea from its isthmus to the Kerch Peninsula (Boltachev et al. 2003). The fish turned out to be euryhaline, spread widely in the coastal waters, occurs at salinity up to 14–15‰ and is eurythermal; it tolerates both high summer temperatures and can wintering under the ice, even in a very small bodies of water (Svetovidov 1964). Juvenile sunfish feed on crustaceans and adults—on insects and small fish. In natural and artificial water bodies, it damages fisheries by eating eggs, larvae, and fry of valuable fish species and competing with some of them for food (Svetovidov 1964). This species according to the fish farming classification belongs to weedy fish and has no commercial value (Boltachev et al. 2003). The process of spreading of sunfish into the inland waters of Ukraine and Russia occurs both as a result of spontaneous expansion of the range and as a result of accidental movement from fish hatcheries with young bred fish species. The analysis of distribution trends makes it possible to predict in the coming years its penetration eastward into the basins of the rivers of the Sea of Azov (Don and Kuban), the artificial and natural reservoirs associated with them, and in the future—the expansion of its range to the Volga. Given the high ecological plasticity of sunfish and favorable conditions in the southern Ukraine and Russia, it may be assumed that in the coming years they will exert a negative impact on the ecosystems of inland waters of the region and cause economic damage to fisheries. Thus, in this case, it is another aggressive non-native species, which requires immediate actions to limit its spread and reduce its population (Boltachev et al. 2003).

Oryzias latipes **(Temminck & Schlegel, 1846),** another aquarium fish (Japanese oryzia) (Fig. 4.38c), was also accidentally released by aquarists. It was imported from the estuaries of the Sea of Japan, established and widely spread in the low-saline areas of the northwestern part of the Black Sea.

Heniochus acuminatus **Linnaeus, 1758** (Fig. 4.38d). The white-finned kabuba, or white-finned pennant butterfly, or pennant bristletooth is a species of ray-finned fish from the bristletooth family. The fish has a white, curved and flat body, a typical long white dorsal fin and two black stripes on the sides. The posterior fin and part of the dorsal fin are yellow. The white-capped kabuba reaches a length of 25 cm. It lives alone, in pairs or in small groups and is found in the Indo-Pacific region. It lives near the coast and reefs, feeds on plankton. Due to its bright color, the white-bellied kabuba is often found in aquariums. In the Black Sea was brought probably with the ballast waters (Boltachev et al. 2010).

All four above mentioned aquarium freshwater species are ecologically plastic both in terms of temperature, salinity, oxygen content, and nutrition and reproduction. They compete fiercely with valuable fish species.

Spontaneously Penetrated Commercial Species

Centracanthus cirrus **Rafinesque, 1810 (Curled picarel)** is a species of marine ray-finned fish from the family Sparidae, the only one in the genus of Acanthus. It was recorded in 2004 in Romanian waters, where only 3 specimens were caught until 2013 (Abaza et al. 2006). Later, a significant increase of population provided its inclusion in commercial fish at the Romanian shelf (Shiganova and Özturk 2010). It originally inhabits in the eastern Atlantic, from the Mauritania to Portugal, including Madeira, the Azores and the Canary Islands, as well as in the Mediterranean, Aegean and Marmara seas.

Anguilla anguilla (**the river eel**), which is occasionally found in the Black Sea basin, which can probably be considered as a migrant. Drapkin (1964) mentioned 16 cases of eel capture in the Novorossiysk area since 1946, and in 1958 one case of capture was registered in Anapa. In addition, the eel was found in the Kiziltash estuary system located on the Taman Peninsula (Plotnikov et al. 1990).

Commercial Species Penetrated from Mediterranean Sea

As mentioned above, some Mediterranean fish species make regular feeding and/or spawning migrations to the Black Sea. These are, first of all, valuable large predatory fish—the Mediterranean-Atlantic horse mackerel ***Trachurus trachurus trachurus*** (Limnarus), the Atlantic pelamid ***Sarda sarda*** (Bloch), the moonfish ***Pomatomus saltatrix*** (Limnaeus), the Atlantic mackerel ***Scomber scombrus*** (Limnaeus), the Mediterranean mackerel ***S. japonicus colias*** Gmelin. Swordfish ***Xiphias gladius*** Linnaeus (and even its spawning (Gordina and Bagnyukova 1992)), common tuna ***Thynnus thynnus thynnus*** (Linnaeus), Mediterranean smear ***Spicara moena*** (L), ***Sardina pilchardus*** (Walbaum) (Svetovidov 1964) have sometimes been observed in the northwestern and western parts of the sea. In the 1970s and 1980s, the number of migrating fish decreased significantly, and most of the species almost stopped entering the Black Sea. However, in recent years, feeding conditions have improved due to an increase in stocks of small pelagic fish, associated with the decrease of fishing pressure and the introduction of the ctenophore *Beroe ovata*, which has reduced the population of *M. leidyi*. As a result, some of the Mediterranean species reappeared both in the western part of the sea (the Mediterranean-Atlantic horse mackerel, ***Trachurus trachurus trachurus***, the Atlantic pelamid ***Sarda sarda***, the moonflowers ***Pomatomus saltatrix*** (Abaza, 2006)), and in the northwestern and northeastern part (Mediterranean smear ***Spicara moena*** (L), ***Sardina pilchardus*** (Walbaum), petropsaro ***Labrus viridis*** (Linnaeus), black-headed tropper ***Tripterygion tripteronotus*** (Risso)). These fish are seasonal migrants from the Mediterranean Sea and are not considered as non-natives, although most of them play, or at least have played, a significant role in the trophodynamics of the ecosystem. In addition, the feeding and spawning areas of some of them in the Black Sea are expanding (Pashkov 2001; Nadolinsky 2004; Boltachev 2006).

Since 1999, three Mediterranean fish species have appeared and are actively breeding in the coastal waters of the Crimea: Dorado ***Sparus aurata*** Linnaeus, salpa ***Sarpa salpa*** (Linnaeus) and sponge ***Chelon*** (=Mugil) ***labrosus*** (Risso) (Boltachev 2006). If Dorado was previously recorded in the northwestern part of the Black Sea (Svetovidov 1964), then there is no sponge or thick-lipped mullet. The Chelon labrosus grouper was first caught in October 1981 in Donuzlav Bay. In October 1983, schools of 10–15 fish were already observed in the Sevastopol area during underwater observations (Salekhova et al. 1987). Since 1999 *C. labrosus* began to appear regularly in the coastal area of Sevastopol. It approaches to the Crimean shores by May, and in early summer the species can be found from Cape Aya to Cape Chersonesos both singly and in flocks of 40–50 individuals. In the second half of the summer, it is found in mixed schools with Black Sea mullet species. By November, the number of the species is decreasing, and in the winter, it does not appear at the Crimea coast. Those species are observed regularly in the southern part of the Black Sea. Apparently, there is a gradual development of the Black Sea by those species for feeding, with a return in the cold season to the Mediterranean for spawning (Boltachev 2003).

Mediterranean Species Established in the Black Sea

Sarpa salpa (Salpa) has also not been previously observed in the coastal waters of Crimea (Boltachev 2006). It was only noted as single specimens off the coasts of Turkey, Georgia, Bulgaria, and Romania (Georgiev 1954; Svetovidov 1964). One individual was first caught in September 1999 by a fixed seine in Balaklava Bay. It was caught alive and passed to the Sevastopol Demonstration Aquarium, where stayed for almost 11 months. During the period of its maintenance, its standard length increased from 11 to 20.9 cm (Boltachev 2003). Currently, the salpa population is also growing rapidly in the coastal waters of Crimea. In 2005 during scuba diving, large salpa flocks about 100–150 individuals were recorded. In February 2006, a 23.5 cm long salpa was caught in the open part of the Balaclava.

Sparus aurata (Dorado) now can be often found in small flocks and as individuals in Balaklava Bay and adjacent coastal waters. In summer, Dorado spawning behavior can be observed by scuba divers. Fish is actively developing in the coastal waters of Crimea. In January 2004, Dorado was first caught in the Dniester-Bug estuary. It is likely that in the warm winters, Dorado and Salpa can hibernate in the coastal waters off Crimea (Boltachev 2006).

Umbrina cirrosa, (the shi drum), is a species of marine fish from the warmer waters of the eastern Atlantic Ocean, Mediterranean Sea and now the Black Sea.

Thus, three above mentioned fish species had not been previously observed in the Black Sea but now they established there. They are commercially valuable species which is trawled for and farmed in aquaculture.

Micromecisthis poutassou (northern whiting)—in January 1999 an individual of 15.7 cm long was caught above a depth of 60 m in the vicinity of Balaklava. This typical Atlantic boreal species, widespreaded in the Mediterranean Sea, probably penetrated from there. The whiting makes long-distance migrations and is known as stenohaline and eurythermal, occurring at an oceanic salinity of at least 33‰ and was first discovered at a salinity of 18‰ (Boltachev 1999, 2006). There are two possible explanations of the appearance of these two species in the Black Sea. First, the fish could have migrated to the Black Sea from the Marmara or Mediterranean seas, or, which looks more likely, given their small size, they could have been introduced with ballast water (Boltachev 2006).

A number of Mediterranean species have penetrated in recent years and can probably be classified as established in the Black Sea.

Parablennius incognitus, (the green dog), was observed as individuals, including a male guarding eggs in the nest, for the first time in 2002, but in the following years the species was observed in a large number in open areas of the coast from Sevastopol to Fiolent. In recent years, it becomes obvious that the ichthyofauna of the coastal zone of Crimea has been replenished with several more Mediterranean-Atlantic species. *Parablennius incognitus* began to be recorded in Turkish waters and appeared in the waters of Abkhazia (Ozturk 2006; Bogorodsky 2006).

***Gobius auratus* Risso** (golden goby). It is the Mediterranean species, which was first discovered in thickets of coastal bottom algae in the Crimea region in the early 1970s (Gordina 1974), and in the 2000s it was found in a large number in the northeastern part of the sea (Nadolinsky 2004).

Gobius cruentatus (red-mouthed goby) since the year 2008 has been observed in Martynova Bay in the area where mussel collectors are installed, which was also discovered last year in the area of Tolstoy and Sevastopol Bay (Chesalina) (Boltachev 2008). The goby was recently described off the coast of Turkey (Ozturk 2006). This species of goby native to the Eastern Atlantic Ocean from southwestern Ireland to the coasts of Morocco and Senegal, and also in the Mediterranean Sea where it occurs in inshore waters at depths of from 15 to 40 meters (49–131 ft) in areas with rocky or sandy substrates or in meadows of sea-grass. This species can reach a length of 18 centimeters.

Gobius xanthocephalus (golden goby) in 2006, 5 specimens of were caught in Kazachya Bay, and more than 20 were observed visually. Specimens of the golden goby were first caught in the Black Sea off the coast of Abkhazia around the same time (Vasilyeva and Bogorodsky 2004; Bogorodsky 2006).

Tridentiger trigonocephalus (Chinese striped three-toothed goby) is endemic of the Far Eastern seas. For the first time one specimen was caught in 2006 in the estuary of the Kutovaya part of Sevastopol Bay, but in subsequent catches, conducted in this site twice a month all the year, this species was absent and was classified as an accidentally occuring species. However, at the end of July 2008, in Starosevernaya Bay, during the mussel's collection from of the pier, another specimen was found in one of the shells, which, together with the mussel flaps, was delivered alive to INBYUM. The analysis showed that this is a fairly large male Chinese three-toothed bull with the total length 85 mm. The inner surfaces of both mussel flaps were 80–85% covered with a single-layer layer of eggs in the amount of about 8 thousand pieces, and after moving to the aquarium, the male began active care about the eggs. After 4 days, a massive hatching of the larvae happened. In August, 17 individuals of the species were caught in Starosevernaya Bay and in the estuary of Sevastopol Bay, and several dozen were observed visually. The striped bullhead was found mainly on piles covered with mussels, and in case of danger they hide in their beds. Thus, it has been proved that this species was established in the Sevastopol bays. The entry of this species may be associated with the discharge of ballast waters or with the unauthorized release of several specimens by aquarists. Thus, the 6 introduced fish species are the bottom species. Representatives of the

Gobiidae and Bleniidae are small demersal fish living near the bottom, so they were most likely introduced with ballast waters.

Syngnathus acus (common sea needle). Two females and a male were found in Sevastopol Bay (Boltachev 2008). Its origin has not been precisely clarified. This species was reported in the Black Sea by FNAM (1986) and Rass (1993a, b), but without a location indication. If it is *Syngnathus acus*, then it is likely that this species was introduced from the northeastern Atlantic Ocean. If it is *Syngnathus acus rubescens*, then it is a Mediterranean species and could have been imported from there.

Centrocanthus cirrus. Curled picarel, which also probably penetrated from the Mediterranean Sea, has been repeatedly observed in Romanian waters. *C. cirrus* has already significantly increased its abundance in the littoral zone of Romania (Abaza et al. 2006) and the Turkish part of the Black Sea (Ozturk 2006). For the first time, developing eggs of the baleen lycanthus was found in the central part of the sea in June 1982 (Tsokur 1988). Thus, of the imported fish, only *Planiliza haematocheilus* has become a widespreaded and valuable commercial species. Cupids and carp also have commercial value, but their reproduction is supported artificially. Six more species are found as individuals.

Along with the cupids and carp, the weed fish Amur chebachok *Pseudorasbora parva* was introduced. Three species of fish parasites were introduced along with *Planiliza haematocheilus*. In addition to commercial species, the intentionally introduced *Gambusia holbrooki* has successfully esteblished and expanded its range. Two species of freshwater aquarium fish became established due to the negligence of aquarists. Of these, the sunfish *Lepomis gibbosus* expanded its range and caused significant damage to the ecosystems of the estuaries of the Black Sea. The expansion of the habitats and increasing of the duration of stay in the Black Sea of Mediterranean migrants become a new phenomenon. The authors have already classified nine Mediterranean fish as established in the Black Sea. However, further field and taxonomic studies of these species are necessary for complete certainty. In recent years, there has been a new trend—the appearance of Indo-Pacific fish species that migrated from the Mediterranean Sea, where they came as Lessepsian migrants, or brought into the Black Sea with ballast water.

Sphyraena pinguis (Red barracuda) (Fig. 4.39). Among others, two specimens of red barracuda were caught by bottom trawling in the Balaklava Bay in August 1999. This is a typical Indo-Pacific species that entered the Aegean Sea through the Suez Canal as a Lessepian migrant, and then, probably, migrated into the Black Sea (Boltachev 2006, 2009). This Lessepian migrant, which appeared in the waters near Palestine in 1931, is now spreading futher in the eastern Mediterranean Sea, including the Aegean Sea (Golani et al. 2021). Specimens of *Sphyraena obtusata* have been recorded in the southern part of the Black Sea, but it is possible that these are the same species, since the specimens found were initially identified by Boltachev as ***Sphyraena obtusata*** and only after detailed analysis were redefined as ***Sphyraena pinguis*** (Boltachev 2009).

Parasite ***Monogenoidea*** is representated by *Ligophorus kaohsianghsieni* (Gusev 1962), which was incidentally introduced with fish *Planiliza haematocheilus* in the

Fig. 4.39 *Sphyraena pinguis*

Black Sea during its naturalization. It is included in the class of parasitic flatworms, or monogeneans, or monogenetic flukes, belonging to the phylum Platyhelminthes. Monogeneans are extremely harmful for fisheries. They feed on mucus, epithelium or (less often) blood of the fish host, sometimes causing mass death of fish.

4.6 *Mnemiopsis leidyi* A. Agassiz 1865 in the Black Sea

The lobate ctenophore *Mnemiopsis leidyi* is indigenous to the estuaries and bays found in temperate and subtropical zones along the Atlantic coasts of both North and South America. It thrives across a broad spectrum of temperature and salinity levels (Harbison et al. 1978; Kremer 1994; Purcell et al. 2001; Costello et al. 2012; Mianzan 1999; Oliveira et al. 2016).

In 1982, the species *M. leidyi* was introduced into the Black Sea via ship ballast water and had spread throughout the sea by 1988 (Pereladov 1988; Vinogradov et al. 1989; Shiganova et al. 2003, 2004a, b, c; Finenko et al. 2003).

Subsequent genetic analysis definitively established the species' precise identification and geographic origin. Figure 4.40 illustrates the donor and recipient regions for *M. leidyi*. The diagram displays allele distribution (inner circle for internal transcribed spacer (ITS)) and haplotype distribution (outer donut for Cytochrome c oxidase subunit I (COI)). Each color corresponds to a distinct allele/haplotype. Unique alleles/haplotypes are emphasized in grey.

***Mnemiopsis leidyi* Taxonomy and Morphology**
Phylum Ctenophora Esch
Class Tentaculata Chun
Order Lobata Esch
Fam. Bolinopsidae Bigelow, 1912
Genus Mnemiopsis L.Agassiz, 1860
Mnemiopsis leidyi (A.Agassiz) 1865

M. leidyi (Fig. 4.41) is distinguished by its two mobile lateral (oral) lobes, which develop from the ctenophore spherosome (body). These oral lobes begin at the level of the aboral organ or above, differing from members of the genus *Bolinopsis*, whose oral lobes start roughly midway between the mouth and the aboral organ (Mayer 1912). Four smaller auricles are positioned beneath the two oral lobes. When closed, the oral lobes fully enclose the mouth region and completely cover the auricles

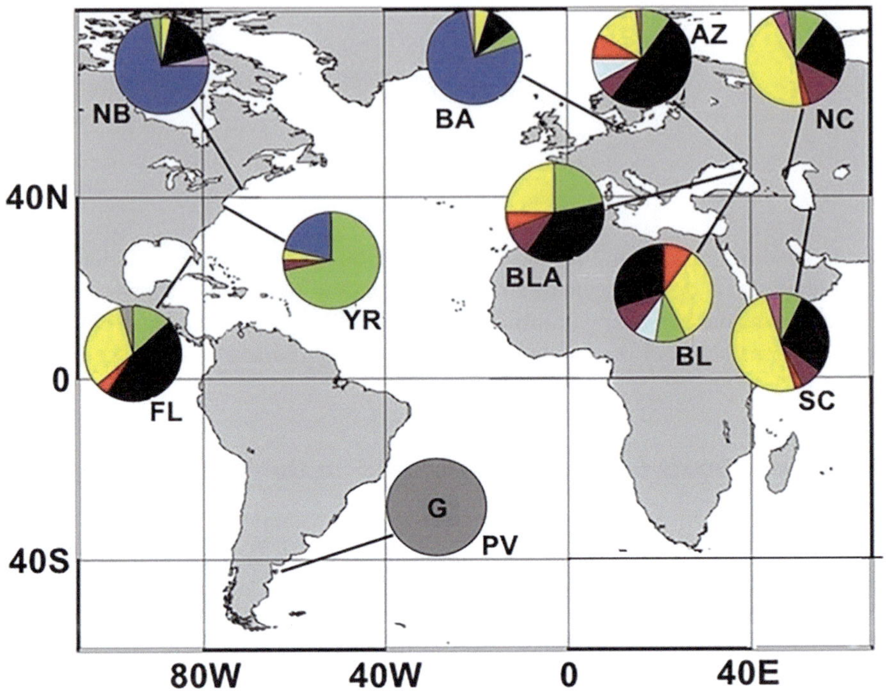

Fig. 4.40 Genetic research has confirmed that *M. leidyi* invaded the Black Sea (BS) from Florida (FL) in the Gulf of Mexico, and subsequently spread to the Azov (AZ) and South Caspian (SC) seas, as well as from Narragansett Bay (NB) into the Baltic Sea (BA). No invasions were detected along the Peninsula Valdes coast in Argentina (PV) (Ghabooli et al. 2011). Sample origins: AZ—Sea of Azov, BL—Black Sea, BLA—Black Sea (Gelendzhik Bay), NC—North Caspian Sea (Makhachkala coast), SC—South Caspian Sea, BA—Baltic Sea (Kiel Fiord), FL—Florida (Tampa Bay), NB—Narragansett Bay, YR—York River (Virginia), PV—Peninsula Valdes coast (Argentina)

(Mayer 1912). The testicular apparatus is found at the oral edge of the body. Its central portion lies above the lips of the slit-like mouth. Each side of the elongated slit-like mouth is lined with a series of short, simple tentacles, which connect continuously with those in the deep lateral furrows.

M. leidyi is a hermaphroditic organism possessing two sets of gonads—ovaries and testicles—within its gastrodermis. These reproductive organs are situated along the eight meridional canals of the gastrovascular system, occupying the spaces between the ctenes, or comb flappers. The ovarian rows are oriented toward the main symmetry planes: the sagittal and testicular planes. Opposite each meridional canal, rows of testicles are positioned. Consequently, in neighboring canals, the testicular rows are mirrored—a distinctive feature of the Lobata order. Gonad development occurs in the central sections of the canals, extending from the statocyst level nearly to the outermost ctenes near the oral edge of the body.

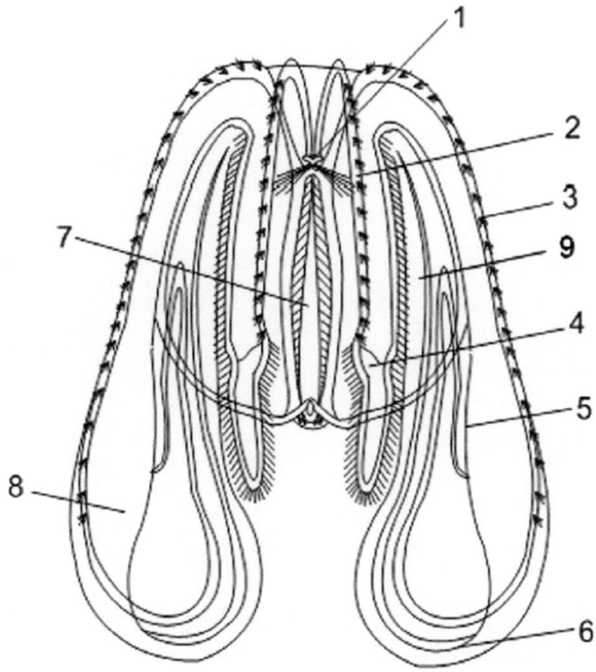

Fig. 4.41 Anatomical overview of *M. leidyi.* (1) aboral organ; (2) subtentacular comb row; (3) subsagittal comb row; (4) auriculus; (5) subsagittal canal; (6) translobal tube; (7) tentacular canal; (8) lobe; (9) spherostome. (Drawing T. Shiganova)

Impact on the Black Sea Ecosystem

When *M. leidyi* first invaded, the Black Sea had no established predators for this species. This absence of natural enemies enabled *M. leidyi* to successfully establish itself and achieve high population densities. Furthermore, favorable environmental conditions, including suitable temperatures and abundant zooplankton prey, supported this expansion. Adult *M. leidyi* primarily consume edible zooplankton, fish eggs, and small larvae (Tsikhon-Lukanina and Reznichenko 1991; Shiganova et al. 2004a, b, c, 2014a, b), whereas their larvae feed on microzooplankton (Sullivan and Gifford 2004). Consequently, in the years after the invasion, significant grazing on mesozooplankton led to cascading effects across multiple levels of the Black Sea ecosystem (Fig. 4.42).

Consequently, a major reduction in zooplanktivorous fish stocks occurred, accompanied by a loss of species diversity. Bottom-up effects involved the collapse of planktivorous fish populations, the depletion of large pelagic fish species composition, a drop in their numbers, and a corresponding decline in dolphin populations. Top-down effects consisted of (i) a depletion of edible zooplankton to a maximum annual biomass of 0.5 mg C m^{-3}, nearly two orders of magnitude lower than in earlier periods, (ii) an increase in phytoplankton biomass due to reduced grazing pressure from zooplankton, and (iii) a decrease in species diversity. Furthermore, bacterioplankton production rose as a result of increased mucus release and degradation products from *M. leidyi*. Subsequently, heterotrophic flagellates and ciliates,

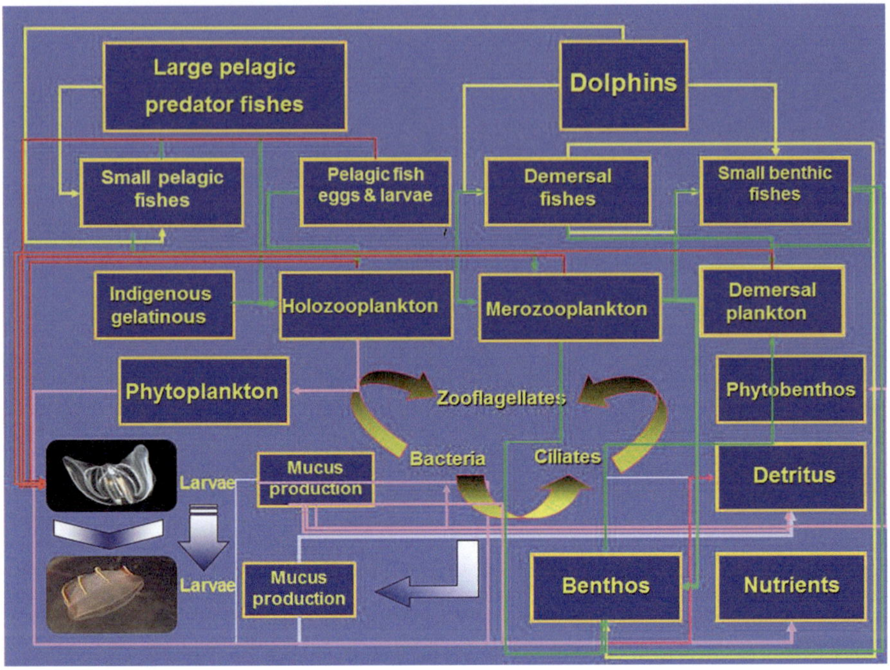

Fig. 4.42 The framework illustrating the influence of *M. leidyi* on the Black Sea's ecological system

which consume bacteria, experienced a significant biomass increase and grew in abundance (Shiganova et al. 2004a, b, c, 2019a, b, 2020). By the late 1980s, the pelagic ecosystem had shifted to dominance by gelatinous plankton, with *M. leidyi* making up the majority of the biomass (Shiganova et al. 2003).

Due to a decline in edible zooplankton biomass, the populations of small pelagic planktivorous fish have also decreased (Fig. 4.43).

The Black Sea Functions as Both a Recipient and a Door Area of the Comb Jelly *M. leidyi* Across Eurasian Seas

The Black Sea served as the donor for *M. leidyi*'s spread across Eurasian seas, where it became established. Its expansion was aided by the global growth of shipping routes and the construction of canals, which enhanced connectivity between marine basins.

The establishment and subsequent spread of *M. leidyi* were accelerated by rising climate temperatures and the degradation of marine ecosystems, especially in coastal zones. The altered conditions within the enclosed and semi-enclosed seas of Eurasia created an ideal environment for the successful colonization of newly arriving *M. leidyi* (Shiganova et al. 2019a, b). Research into the species' genetics has further elucidated the invasion chronology of *M. leidyi* across Eurasia.

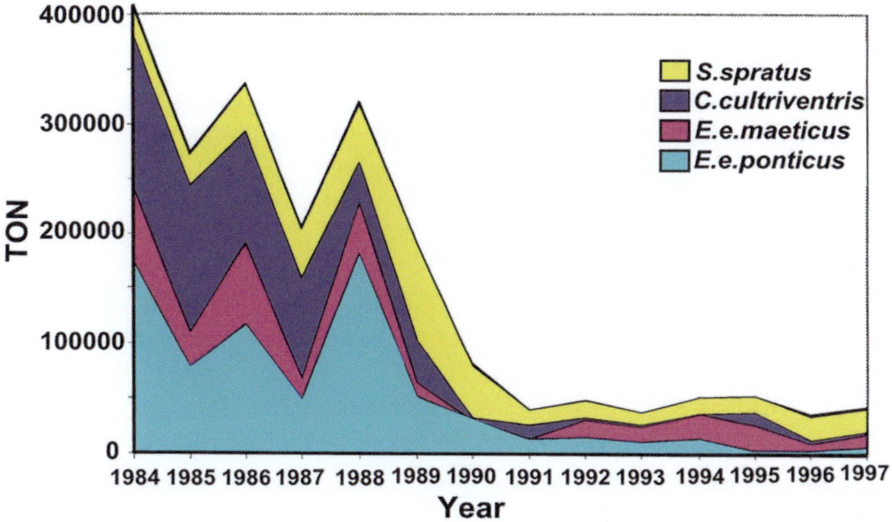

Fig. 4.43 Populations of small pelagic fish species (*Sprattus sprattus* (Linnaeus, 1758), *Clupeonella cultriventris*, *Engraulis engrasiolus ponticus*, *Engraulis engrasiolus maioticus*) decreased following the introduction of *M. leidyi.* (After Shiganova and Bulgakova 2000)

Despite some inconsistencies, earlier research and our own investigations have verified the occurrence of multiple invasions: *M. leidyi* was introduced into the Black Sea from the Gulf of Mexico (Ghabooli et al. 2011; Bayha et al. 2015; Reusch et al. 2010) and subsequently dispersed across the Southern seas, encompassing the Ponto-Caspian basin, the Mediterranean Sea, and nearby oceanic waters. The coastal region of North America (Narragansett Bay) serves as the source population for the North and Baltic seas. In both instances, *M. leidyi* was introduced from coastal or estuarine zones, with no indication of an invasion originating from the open Atlantic waters (Ghabooli et al. 2011; Bayha et al. 2015). Presently, the geographic distribution of *M. leidyi* is extensively widening throughout the Eurasian seas, highlighting its remarkable capacity to establish itself in new environments (Fig. 4.44).

Present distribution patterns of *M. leidyi* involve (1) continuous presence in source regions characterized by high yearly productivity, strong retention, and consistent annual persistence, with movement into refuge habitats during adverse environmental conditions; (2) localized, current-driven spread into sink areas where reproduction and population expansion occur once conditions become favorable (Costello et al. 2012; Shiganova et al. 2019a, b).

Morphological Adaptations of *Mnemiopsis leidyi* Across Diverse Habitats
M. leidyi is a holoplanktonic species that lacks a benthic phase and does not produce resting eggs, cysts, or specialized overwintering forms. It is capable of self-fertilization and exhibits high reproductive output; normal development rates of up to 100% have been consistently observed in self-fertilized eggs from isolated

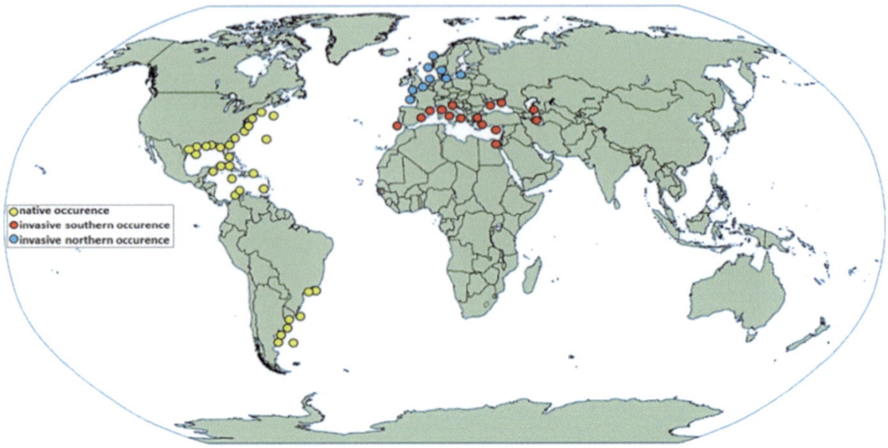

Fig. 4.44 The natural range of *M. leidyi* along the Atlantic coasts of North and South America (indicated by yellow dots) and its subsequent expansion into Eurasian recipient areas (blue and red dots). (Based on GESAMP 1997; Costello et al. 2012; Shiganova et al. 2019a, b; compilation and drawing by authors))

individuals. Although the role of outbreeding in ctenophores remains debated, self-fertility may be viewed as a pre-adaptive trait that facilitates rapid colonization (Pianka 1974; Kremer 1976). Furthermore, *M. leidyi* can regenerate from body fragments as small as one quarter of the original organism (Coonfield 1937). As a generalist carnivorous feeder (Tzikhon-Lukanina et al. 1991; Javidpour et al. 2009a, b; Costello et al. 2012), it can endure extended periods without food under adverse conditions by lowering its metabolic rate and reducing body size through consumption of its own mesoglea (Anninsky et al. 2005).

Furthermore, *M. leidyi* populations from various marine environments display notable adaptive morphological diversity. For the first time, our research investigated how environmental factors influence the morphological characteristics of *M. leidyi* during its varied phenotypic development (Shiganova 2020). We analyzed *M. leidyi* specimens originating from the Black Sea, which served as the initial source region, and later spread through the Sea of Marmara to the Aegean Sea, as well as to the Sea of Azov and the Caspian Sea (Fig. 4.45a). Despite sharing a common origin, these populations exhibit differences in morphological traits that, as previously noted, were historically used to classify them as three distinct species (Fig. 4.45b) (Mayer 1912; Agassiz 1860, 1865).

In the low salinity environment of the Sea of Azov and the Caspian Sea, *M. leidyi* morphologically fits well into the description of *M. gardeni*, provided by Agassiz L. (Fig. 4.45a, b). In the Sea of Azov, they are usually small (up to 20–72 mm at mature stage). In the Caspian Sea mature individuals measure 15–65 mm in total length. The oral lobes are small, roundish and comprising one-fifth to one-sixth of the body length. The auricles may be about one-third of the total length or longer (Fig. 4.45a).

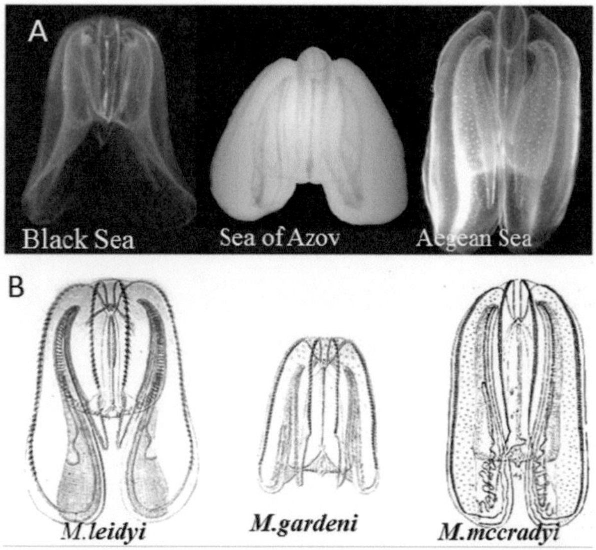

Fig. 4.45 (**a**) *M. leidyi* specimens from the Black Sea. (Photo by T. Shiganova), the Sea of Azov. (Photo courtesy of S. Dudkin), and the Aegean Sea. (Photo courtesy of E. Christou); (**b**) *M. leidyi*, *M. gardeni*, and *M. mccradyi*. (Adapted from Mayer 1912)

The subsagittal canals are straight, lacking any curves. Individuals from the Sea of Azov frequently exhibit warts on their bodies, a feature absent in those from the Caspian Sea (Fig. 4.46). These distinctions between the Azov and Caspian populations are likely a consequence of their unique environments, which are shaped by the limnological characteristics of each basin. *M. leidyi* found in the Caspian Sea displays the smallest body size, reduced rigidity, and a transparent appearance (Fig. 4.46a). The most notable variations between the Caspian and Azov *M. leidyi* are attributable to the lower salinity levels in the Caspian, even though the Azov population is annually replenished by inflows of Black Sea water.

Consequently, phenotypic distinctions grow more evident over time, as the Caspian specimens experienced a longer period for evolutionary separation. Nevertheless, *M. leidyi* from hypersaline Egyptian lakes (Fig. 4.46c) are also of a small size and resemble the Caspian individuals (Fig. 4.46a).

Consequently, phenotypic distinctions grow more evident over time, as the Caspian specimens experienced a longer period for evolutionary separation. Nevertheless, *M. leidyi* from hypersaline Egyptian lakes (Fig. 4.46c) are also of a small size and resemble the Caspian individuals (Fig. 4.46a) (Fig. 4.47).

The characteristics previously outlined for the Black and North seas are also representative of specimens from the Aegean and Mediterranean seas (Fig. 4.48).

Specimens of *M. leidyi*, which live in the Aegean and Mediterranean seas and are acclimated to their elevated salinity, correspond to *M. mccradyi* (Fig. 4.48b), as documented by Mayer (1912). These organisms exhibit notable firmness and frequently display a whitish or pinkish hue (Fig. 4.48a), featuring prominent body warts and elongated central cilia suspended beneath the mouth (Fig. 4.48a, b). Their structural modifications are probably due to the influence of high salinity, elevated

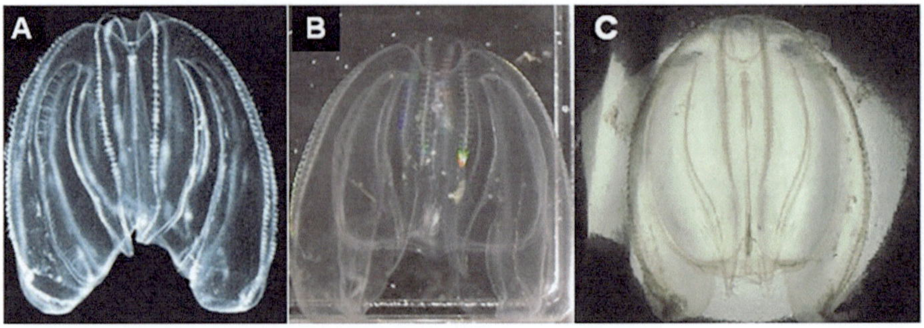

Fig. 4.46 (**a**) *M. leidyi* from the Caspian Sea: length 38 mm. (Photo courtesy of A. Kamakin); (**b**) *M. leidyi* from the Baltic Sea, length 30 mm. (Photo courtesy of J. Javidpour), (**c**) *M. leidyi* from the Egyptian Lake. (Photo courtesy of G. El-Shabrawy)

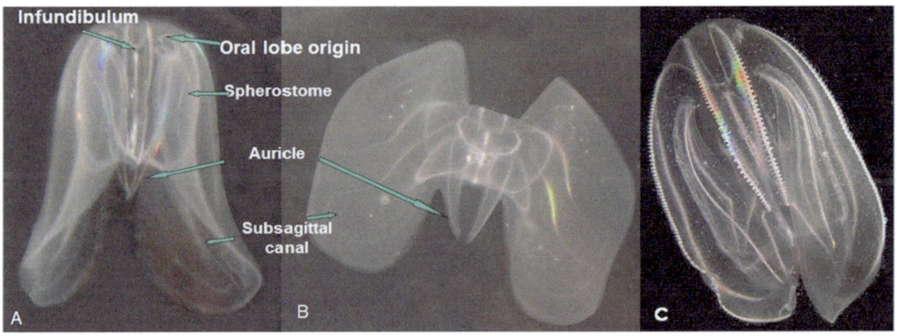

Fig. 4.47 (**a**, **b**) *M. leidyi* collected in the Black Sea, with a total length of 135 mm. (Photo by T. Shiganova), (**c**) *M. leidyi* from the North Sea. (Southern Norway, photo courtesy of H. Ringvold)

temperatures, and oligotrophic environments. A comparison of specimen sizes from various regions is illustrated in Fig. 4.49.

Shorter adult specimens were observed in low-salinity environments, such as the Sea of Azov, the Caspian Sea, and the Baltic Sea, with lengths spanning 20–45 mm and reaching a maximum of 60–165 mm. However, when salinity levels surpass 45‰, the length of *M. leidyi* declines once more, as seen in locations like the Mar Menor Lagoon and the Fayum Lakes in Egypt.

Key morphological characteristics of *M. leidyi*, once employed to classify them into distinct species, include their phenotypic variability, which is linked to environmental factors. In newly colonized habitats, *M. leidyi* is pre-adapted for fast colonization because of its elevated metabolism and reproductive speed. This has resulted in the formation of large populations exhibiting diverse annual cycles and distribution patterns, significantly affecting local ecosystems (Shiganova 2020).

Life Cycle of *Mnemiopsis leidyi*

The expansion of *M. leidyi* populations is governed by both temperature and food availability, making its development patterns vary by location and time of year. In

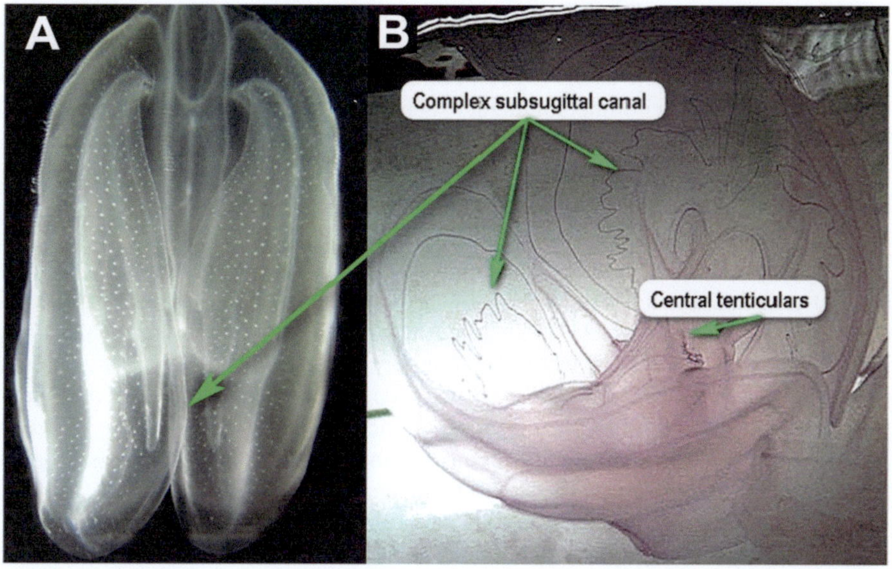

Fig. 4.48 A specimen of *M. leidyi*, collected in the Aegean Sea, with a total length of 140 mm. (Photo courtesy of E. Christou)

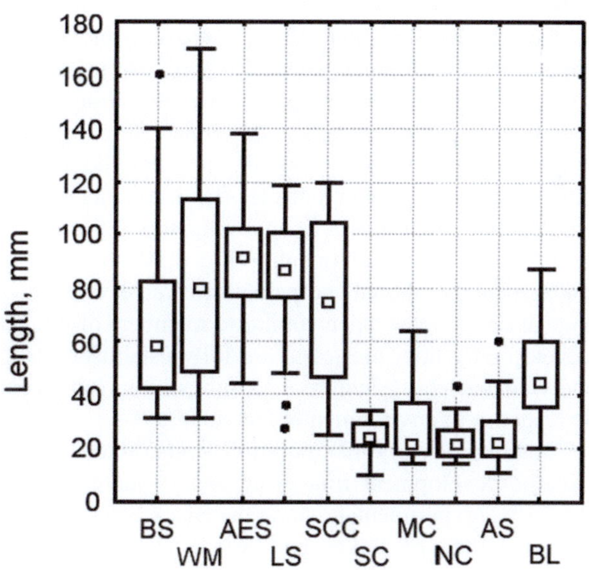

Fig. 4.49 Total length of adult *M. leidyi* with lobes (BS—Black Sea; WM—Western Mediterranean (collaboration with M. Lilley, unpublished)); AES—Aegean Sea (T. Shiganova); SCC—South Catalonian coast (Marambio et al. 2013, pers. comm.); SC—Southern Caspian Sea; MC—Middle Caspian Sea; NC—Northern Caspian Sea; AS—Sea of Azov) (T. Shiganova); BL—Berre Lagoon (Delpy et al. 2016). Black boxes—Wicker plots with mean and size range

temperate regions during the spring, juvenile *M. leidyi* commence growth, form reproductive organs, and initiate breeding once rising temperatures and adequate food supplies enable metabolic development.

Reproduction through larval or juvenile stages, known as paedogenesis, has also been observed (Pianka 1974). Reproductive activity intensifies as temperatures rise, peaking at the same time as the highest population levels of *M. leidyi*. Fecundity is also influenced by environmental factors such as temperature and food availability. Individuals spawn repeatedly during their reproductive phase, leading to a wide range of sizes among reproducing individuals—from newly spawned generations to fully mature adults. These observations point to a lifespan of approximately 1 year, though it may be shorter (Pianka 1994; Shiganova 2009).

Size is not always a reliable measure of age, as individual growth rates within a population vary significantly across different habitats, and shrinkage can happen during starvation (Greve 1979). The size at which individuals reach maturity is influenced by environmental factors, particularly salinity. In the low-salinity waters of the Caspian and Azov seas, *M. leidyi* starts reproducing at a total length of 15 mm (Shiganova et al. 2001a, b), while in the Baltic Sea, individuals mature between 13 and 35 mm (Javidpour et al. 2009a, b). In the Black Sea, *M. leidyi* begins reproduction at a total length of 30 mm (Shiganova et al. 2001a, b).

M. leidyi eggs possess a spherical shape, measuring 0.3–0.4 mm in diameter, and are enclosed by a thin, non-structured membrane. Each egg has an average dry mass of approximately 0.0005 mg, with organic carbon accounting for 2% of its composition (Reeve et al. 1989). The larva is fully developed within the egg capsule once it reaches a length of about 0.3 mm. As the embryo gains mobility, the capsule loses its elasticity and its form becomes irregular. Prior to hatching, the embryo is equipped with double rows of small tentacles, enabling movement and feeding in the plankton. Embryonic development spans 16–24 h, varying with temperature (Fig. 4.50).

Upon hatching, the cydippid larva, measuring less than 0.5 mm, closely resembles a larva from the genus *Pleurobrachia* (Fig. 4.50a, b). This spherical larva possesses eight rows of meridional canals equipped with comb flappers and is characterized by double tentacles (tentilla), a hallmark of the cydippid form.

Once the larva grows to approximately 3.5 mm, oral lobes begin to emerge on both sides of the mouth opening (Fig. 4.50c, d). Subsequently, the lower sections of the vertical meridional and paragastric canals extend, causing the *Mnemiopsis* larva to take on an appearance similar to members of the genus *Bolinopsis*. The auricles develop only after the lobes are fully established (Fig. 4.50f). A fully developed individual reaches a length of 15 mm, with its lobes continuing to grow and elongate in response to environmental factors.

Duration of early maturation depends on temperature and food availability. The average development time from egg to maturity is 19 ± 5 days at temperature 23–25 °C (experimental data in the Black Sea (Shiganova 2009)). This is consistent with the data of Baker and Reeve (1974), who reported duration of growth 13–23 days. Size of individuals at the stage of early maturity and beginning of reproduction depends on environmental conditions, first of all on salinity. The number of eggs, produced by individuals of any size and stage of maturity is highly variable, depending on temperature and food. During the period of favorable conditions *M. leidyi* may reproduce almost every day. In some basins it penetrates further in a sink area, where it can live temporally and reproduces during period of

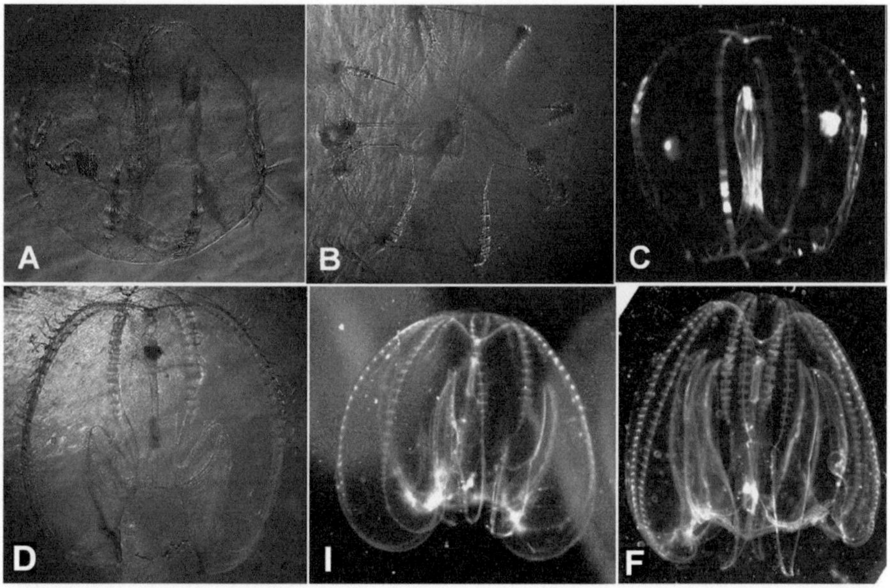

Fig. 4.50 (**a**, **b**) Cydippid larvae of *M. leidyi*; (**c**, **d**) initial stage of lobes development; (**i**) transitional stage of lobes' development; (**f**) transitional stage with lobes and auricles' development. (Photo by T. Shiganova)

sufficient conditions (Costello et al. 2006, 2012). Adults continue to grow to maximal size and finally die when favorable conditions cease. In the sink areas the whole populations eliminate, while in the source areas a reduced population, comprising of new generation of individuals, migrates deeper and continues feeding. They may stop feeding when the temperature is unfavorable or/and food concentration is low (Shiganova et al. 2001a, b). In this case they stay in refuge near the bottom without feeding, using their ability to survive by utilizing energy, stored in own mesoglea (Reeve et al. 1989). In spring, with warming and zooplankton development, *M. leidyi* again raises to the upper layer. Such seasonal pattern of *M. leidyi* is specific for the source areas in temperate basins, where they may live throughout the year.

Fecundity

A defining characteristic contributing to the success of *M. leidyi* is its elevated reproductive capacity, which fluctuates across habitats in response to temperature, salinity, and prey density. Experimental data from the Black Sea determined the average fecundity (eggs laid per individual per day) to be 2534 $\pm$ 1818 eggs ind^{-1} day^{-1} (mean $\pm$ std. dev.) under conditions of 17‰ salinity and a temperature of 25 °C.

Larger specimens exhibited greater egg production rates, achieving a peak fecundity of 6200 eggs ind^{-1} day^{-1} from a 92 mm individual (Shiganova et al. 2004a). *M. leidyi* elevates egg output with rising temperatures, yet ceases reproduction at 26.6 °C in these experiments (Fig. 4.51).

Fig. 4.51 The daily egg output per individual of *M. leidyi* in the Black Sea is influenced by temperature ($R = 0.71$, $p < 0.05$; $N = 45$)

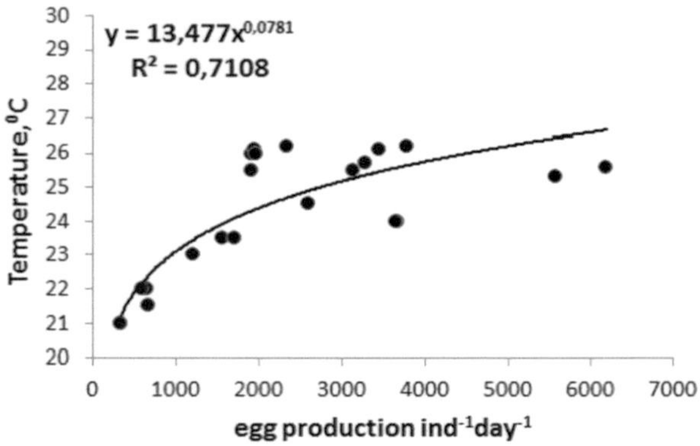

In studies conducted in the Sea of Azov, average fecundity was reduced, reaching 897 ± 228 eggs ind^{-1} day^{-1} at a salinity of 13‰ (Fig. 4.52a), and 600 ± 455 eggs ind^{-1} day^{-1} at a salinity of 11‰ (Shiganova, unpublished). In Taganrog Bay, *M. leidyi* reproduces actively at an average salinity of 6.5–9‰ and elevated zooplankton densities, but reproduction ceases when salinity drops below 6‰ (Mirsoyan et al. 2006).

During our research in the Southern Caspian Sea, peak reproductive output was recorded in newly gathered specimens measuring 20–30 mm in length. At 25 °C, daily egg output per individual varied between 21 and 1650 eggs, with an average fecundity calculated at 867 ± 517 eggs per individual per day (Finenko et al. 2006) (Fig. 4.52a).

During experiments in the northern Aegean Sea, the reproductive output of freshly gathered *M. leidyi* was found to be low, varying from 24 to 1172 eggs per individual daily, and averaging 278 ± 224 eggs ind^{-1} day^{-1}. This limited fecundity is likely attributable to oligotrophic environments characterized by sparse zooplankton densities (Fig. 4.52a) (Shiganova et al. 2004c).

In the northern Adriatic Sea, recorded fecundity levels ranged from 136 to 13,512 eggs per individual daily (mean ± std.: 4320 ± 3980 eggs ind^{-1} day^{-1}, $N = 10$), with the peak output coming from the biggest and heaviest specimen (total length 9.8 cm, wet mass 65 g) (Malej et al. 2017). This represents the highest fecundity documented to date in Mediterranean waters.

In certain regions of the Mediterranean Sea where reproductive patterns have not been thoroughly examined, existing evidence suggests that *M. leidyi* has successfully established itself and formed self-sustaining populations across all coastal zones where it has been observed in significant numbers.

M. leidyi is particularly abundant in bays, lagoons, and coastal zones with lower salinity and terrigenous nutrient inputs from river discharge (Shiganova et al. 2019a, b). Consequently, among temperate southern seas, the highest per capita egg production was recorded in the Black Sea, followed by lower values in the Sea of Azov and the Caspian Sea (Fig. 4.52a). In the high-salinity (35–39‰) Mediterranean Sea, egg production increased only in productive regions such as the northern

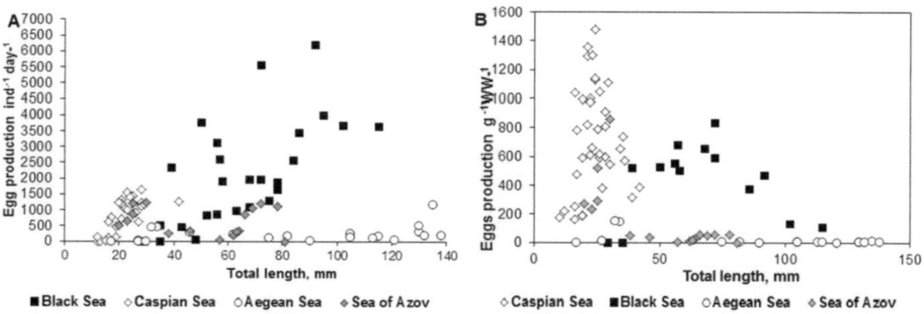

Fig. 4.52 The total egg production per individual *M. leidyi* (eggs ind^{-1} day^{-1}) (**a**) and the weight-specific reproductive output (eggs g WW^{-1} day^{-1}) (**b**) measured at 25° C in the Black, Caspian, Azov, and Aegean seas

Adriatic, while it remained lower in the oligotrophic northern Aegean Sea (Fig. 4.52a). However, the highest weight-specific fecundity (eggs released per gram of ctenophore wet weight) was noted in the Caspian Sea, with lower levels in the Sea of Azov and the Black Sea, and the lowest in the Aegean Sea, where individuals are larger and thus heavier (Fig. 4.52b) (Shiganova et al. 2004a, b, c).

In the Northern seas, the quantity of eggs deposited also diminishes with lower salinity levels (Jaspers et al. 2013), although wet weight-specific fecundity increases significantly under low salinity conditions, as demonstrated by studies in the less saline Kiel Bight (estimated by Shiganova using data from Javidpour et al. 2009a, b) (Fig. 4.52a, b). This indicates that population fecundity (eggs g^{-1}WW day^{-1}) is greater for smaller adult individuals in environments with reduced salinity.

A key element influencing reproductive success is the density of available prey. Typically, periods of intense reproduction and population peaks align with times of high zooplankton concentration. Experimental studies have demonstrated, however, that *M. leidyi* can sustain reproduction at a diminished rate following 2–4 days without food. Remarkably, certain individuals managed to reproduce even after 11–12 days of starvation (Jaspers et al. 2015).

Observational summaries from various recipient areas show that adequate salinity levels for egg production range from 10–38‰ in the Southern seas (Fig. 4.49a) and from 10–20‰ (or higher) in the Baltic Sea, although egg output is reduced at 10‰ salinity (Shiganova, this study; Lehtiniemi et al. 2011). In the Caspian and Azov seas, *M. leidyi* is capable of reproducing at a minimum salinity of 6.0–6.5, with typical hatching and larval development occurring under these conditions (Shiganova et al. 2003).

Seasonal Dynamics, Time, and Duration of Reproduction

The seasonal and year-to-year variability in the persistent and temporary presence of *M. leidyi* across the Eurasian recipient seas, including its source and sink regions, has been thoroughly examined. Research indicates that *M. leidyi*, due to its strong physiological tolerance and adaptive abilities, exhibits varied temporal distribution patterns and phenology in response to differing environmental conditions (Table 4.20).

Table 4.20 Periods of *M. leidyi* reproduction (grey stripes) and peaks of abundance and reproduction (black stripes) and corresponding favorable environmental conditions: temperature (T), temperature of reproduction (T, R), salinity (S), peak of zooplankton (Z) in the recipient habitats (Shiganova et al. 2004a, b, c; Shiganova et al. 2018, Mirsoyan et al. 2006; Shiganova 2011; Isinibilir 2012; Shiganova 1993; Siapatis 2014; Gülşahin 2013; Galil pers.comm.; Malej et al. 2017; Delpy et al. 2016; Javidpour et al. 2009a; Riisgård et al. 2015–17; Jaspers et al. 2011; 2013; Haraldsson et al. 2013; Van Walraven et al. 2013)

AREA/MONTH	J	F	M	A	M	J	J	A	S	O	N	D	T °C	T °C, R	S	Z	References
Black Sea (before 2006)													5-25	21-25	12-22	+	Shiganova et al., 2004
Black Sea (after 2006)													8-30	18-26	12-22	+	Shiganova et al., 2018
Sea of Azov													-0.3-30	18-26	6-14	+	Mirzoyan et al., 2006
Southern Caspian													10-30	10-30.5	12.6-13	+	Shiganova, 2011
Middle Caspian													5.8-26	18-26	12.1-12.6	+	Shiganova, 2011
Northern Caspian													3.8-28	23-27	6.1-10	+	Shiganova, 2011
Sea of Marmara													8-29	21-26	22-29	+	Isinibilir, 2012 Shiganova, 1993
Northern Aegean									ND	ND	ND		12-27	21-25	33-39	+	Siapatis, 2014
Southern Aegean													14.5-29	18-26.5	37.3-39.6	+	Gulsahin, 2013
Levantine Sea				ND	ND	ND							17.5-31	19.5-23.5	39.3-40.0	ND	Galil, pers. comm.
Northern Adriatic													6-29	17-28	11-38	ND	Malej et al., 2017
Ligurian Sea			ND										1-27	17-21.9	37.9-39.2	+	Lilley, Lombard, pers. comm.
Catalan coast													10-30	10-25	34-38	ND	Marambio, pers. comm.
Berre Lagoon													3.4-28.2	20-25.2	15.9-26.2	+	Delpy et al., 2016
Bages-Sigean													9.6-27.1	18-27	18.9-34.3	+	Delpy et al., 2016
Portugal coast					ND	ND	ND	ND	ND	ND	ND		17.2-22.2	17.4-20	34.9-35.3	+	Crus et al., 2018
Kiel Bight													3.5-22.3	13.1-22.2	14-20	+	Javidpour et al., 2009
Danish Fjords													-0.5-24	12-24	19-34	+	Riisgård et al., 2015–17
Central Baltic													2-20	8.4-10.7	7.8±0.3	+	Jaspers et al., 2011, 2013
Kattegat													0-21	11-21	25±0.3	-	Jaspers et al., 2011 Haraldsson et al., 2013
Skagerrak													6-20	9-20	25-29	-	Haraldsson et al., 2013
Wadden Sea													1.7-22	18-21	34-35	ND	Van Walraven et al., 2013

M. leidyi alters its phenology to optimize the length of its seasonal growth cycle and to broaden its distribution zones across various Eurasian seas, encompassing both source and sink regions. Furthermore, physiological and environmental constraints, along with the corresponding reproductive period of *M. leidyi*, vary among the Eurasian seas. The timing and length of the reproductive season are influenced by temperature, salinity, and the availability of food for larvae, as well as mesozooplankton for juveniles and adults (Fig. 4.53).

Within temperate climate zones (such as the Black, Azov, Caspian, Baltic, and North seas), distinct source and sink regions arise as a result of intense winter cooling. The Sea of Azov, along with the Middle and Northern Caspian, serve as sink areas where *M. leidyi* proliferates and reproduces during warmer months and is eliminated by the cooling waters of late autumn. In the Southern Caspian, reproduction has been documented in recent years even during the winter season. Nonetheless, a peak in reproduction is consistently observed in August–September across all sub-regions of the Caspian Sea. It is important to note, however, that reproduction rates also decline when water temperatures approach the higher threshold of 27° C. In these seas, the peak reproductive period occurs in the summer to early autumn, influenced by both temperature and zooplankton abundance peaks (Table 4.20).

Over the past 10 years, reproductive activity has shifted to begin sooner, typically in May–June, as a result of rising spring and early summer temperatures and the associated earlier emergence of warm-water zooplankton in temperate marine environments (Shiganova et al. 2014a, b, 2019a, b).

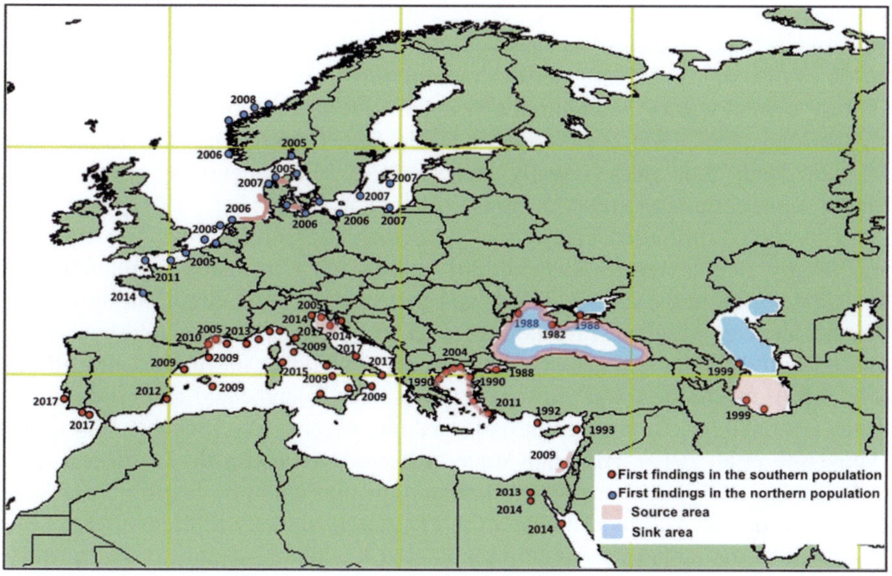

Fig. 4.53 Chronology of *M. leidyi* introduction and spread across Eurasian marine environments. The years shown correspond to the initial detection in each region (references: Pereladov, 1988; Vinogradov et al. 1989; Mutlu 1999; Studenikina et al. 1991; Mirsoyan et al. 2006; Shiganova et al. 2001a; Shiganova 1993; Shiganova and Malej 2009; Galil et al. 2009; Boero et al. 2009; Lilley et al. 2014; Fuentes et al. 2010; Marambio et al. 2013; Javidpour et al. 2006; Boersma et al. 2007; Faasse and Bayha 2006; Hansson 2006; Oliveira 2007; Tendal et al. 2007; Van Ginderdeuren et al. 2012; Antajan et al. 2014; Hosia and Falkenhaug 2013; Delpy et al. 2012; Cruz et al. 2018)

Within the subtropical, nutrient-poor Mediterranean climate, which features hot summers, *M. leidyi* exhibits increased seasonal abundance and reproduces during spring and autumn, occasionally in winter. This reproductive timing aligns with favorable temperatures and the seasonal development of zooplankton in these regions, such as the Aegean Sea, Levantine Sea, Ligurian coast of France and Italy, and Catalan coastal waters (Fig. 4.53). Throughout the hot summer months, *M. leidyi* vanishes from the upper water layers, and a smaller, new generation occupies deeper refuge zones (Shiganova et al. 2004c; Lombard, Lilley, Shiganova, not published). In the more productive northern Adriatic Sea, *M. leidyi* achieves high population densities and elevated reproduction rates in late summer and autumn (Malej et al. 2017). The open Mediterranean Sea serves as a sink area where *M. leidyi* can be distributed by local currents. A portion of these dispersed individuals may successfully reproduce, potentially establishing populations in coastal zones, lagoons, estuaries, and bays. Another segment undergoes sterile dispersal, meaning they can survive temporarily without reproducing and are ultimately eliminated.

The seasonal patterns observed in the Northern European seas resemble those in the temperate Southern seas (Table 4.20), although the reproductive period is more condensed because of a briefer warm season. Spawning takes place during the hottest months and frequently aligns with the highest levels of zooplankton. In the

Wadden Sea, reproduction was documented in two phases (May–June and mid-August) and continued at elevated rates until October (Van Walraven et al. 2013) (Table 4.20). The impacted zones of the North Sea (such as Dutch and German coastal waters, including the Wadden Sea) have become significant source regions that supply the south-western Baltic Sea. Consequently, in Limfjorden, *M. leidyi* likely re-invades annually from the North Sea, functioning as a sink area for the south-eastern coastal populations of the North Sea.

In contrast, Limfjorden likely serves as a source region for other Danish fjords and nearby marine areas (Riisgård 2017). The most intensive reproduction of *M. leidyi* occurs in the Danish Fjords and the eutrophic Kiel Bight during the late summer and early autumn (Javidpour et al. 2009a, b; Riisgård and Goldstein 2014). Subsequently, these regions act as source populations for the Central Baltic (Fig. 4.53), where *M. leidyi* reproduction has been documented and where it can reside seasonally and reproduce provided salinity exceeds 7‰ (Schaber et al. 2011; Haraldsson et al. 2013). Although some modeling studies for the North Sea (e.g., Collingridge et al. 2014) suggested that reproduction could be feasible for a large portion of the year (up to 212 days in 2011), this is unlikely in the Northern seas. Nonetheless, the authors accurately pinpointed the source regions. Consequently, observational data indicate that as water temperatures rise, the reproductive period for *M. leidyi* starts earlier and concludes later in the temperate Southern seas. A similar pattern is presumed to occur in the Northern seas. However, in the subtropical Mediterranean, the duration of reproduction is anticipated to shorten with rising water temperatures, as the temperature is already excessively high across most of the basin during the summer.

The native habitat of *M. leidyi* includes the subtropical estuaries of Biscayne Bay (Florida), where temperatures vary between 18 and 32 °C and salinity fluctuates from mesohaline (below 20‰) to hypersaline (exceeding 40‰), influenced by seasonal and regional factors. *M. leidyi* ceases reproduction during the summer months when temperatures are elevated (Table 4.21). In the central region of

Table 4.21 Periods of *M. leidyi* reproduction (grey stripes) and peaks of abundance and reproduction (black stripes) and corresponding favorable environmental conditions: temperature (T), temperature of reproduction (T, R) salinity (S), peak of zooplankton (Z) in the native habitats (Costello et al. 2012; Lonsdale 1981; Olson 1987; Purcell et al. 2001; Baker 1973; Buskey 1993; Mianzan et al. 1996; Sorarrain 1998; Mianzan and Sabatini 1985; Mianzan 1986; Mianzan et al. 1996, 2010)

AREA/MONTH	J	F	M	A	M	J	J	A	S	O	N	D	T °C	T °C, R	S	Z	References
Narragansett Bay, RI								■	■				1-25	10-23	21-32	+	Costello et al., 2012
Mid Chesapeake Bay, MD				░	░	■	■	■	■				2-30	12-29	5-16	+	Lonsdale, 1981; Olson, 1987; Purcell et al., 2001
Biscayne Bay, FL									■	■	■		18-32	18-28	14-45	+	Baker, 1973
Nueces Estuary, TX						■	■	■					10-30	10-30	20-38	+	Buskey, 1993
Rio de la Plata estuary, ARG			■	■	■								7.5-25	10-25	9-24	+	Mianzan et al., 1996; Sorarrain, 1998
Blanca Bay, ARG			■	■	■				■	■	■		5-24	10-24	24-38	+	Mianzan & Sabatini, 1985; Mianzan, 1986
Nord Patagonic Tidal front, ARG						■	■	■					10-16	10-16	33	+	Mianzan et al., 1996, 2010; Mianzan, pers. comm.

Biscayne Bay, the abundance of *M. leidyi* declines in summer and reaches its highest levels in late autumn and early winter. Along the Gulf of Mexico's coastal waters, such as the subtropical zones of St. Andrei Bay (Florida) and the Nueces Estuary (Texas), the yearly temperature span is 10–30 °C, with salinity levels ranging from 20–33‰.

In these areas, *M. leidyi* is present throughout the year, with population peaks occurring in late summer when zooplankton biomass is highest (Kremer 1994). In colder environments such as Narragansett Bay, which is near the northern limit of *M. leidyi*'s range, annual temperatures vary between 1 and 25 °C, with readings above 20 °C from June through September (Table 4.21). Limited egg release has been observed at temperatures as low as 6 °C, but 10 °C serves as a practical threshold for egg production, and rates of production rise as temperatures increase. The full temperature range for reproduction spans 10–23 °C, with some reproductive activity also noted at 25 °C (Costello et al. 2012).

In the Chesapeake Bay, the reproductive period for *M. leidyi* extends from mid-April through early June, with peak activity occurring between June and September (Table 4.21). This reproduction takes place within a temperature range of 12–29 °C and a salinity range of 5–16‰. Notably, in this region, *M. leidyi* can reproduce at a minimum salinity level of 5‰ (Purcell et al. 2001).

Our analysis has compared reproductive timing and rates in the recipient Eurasian seas with those in the species' original habitats (Reusch et al. 2010; Ghabooli et al. 2011, 2013). In the high-salinity (up to 45‰) and warm (18–32 °C) waters off southern Florida, *M. leidyi* population peaks and reproductive activity were documented in spring, autumn, and early winter (Purcell et al. 2001). A comparable seasonal pattern was noted in the Mediterranean Sea.

In the Black, Caspian, Marmara, and Azov seas, the peak reproductive period occurs from August to September (Table 4.20). In contrast, Narragansett Bay—the origin of *M. leidyi* introduced into the Baltic and North seas—has salinity (21–32‰) and temperature (1–25 °C) conditions comparable to those of the receiving Northern European Seas. Reproduction begins there in April at 14 °C, with the peak arriving earlier, in July–August, than in the Baltic Sea (Table 4.21). Still, the temperature at which reproduction initiates is similar, and the range remains alike (10–23 °C), with some reproductive activity also observed at 25 °C.

Following an assessment of environmental factors in both donor and recipient regions, we determined the following limitations on the presence, reproduction, and population expansion of *M. leidyi:*

Occurrence conditions Self-sustaining populations of *M. leidyi* exist in specific regions where they can reproduce during certain seasons. Additionally, there are areas where M. *leidyi* can temporarily spread and survive, even without reproduction, under unfavorable environmental conditions. Ideal environmental circumstances for the presence of *M. leidyi* are detailed in Fig. 4.54a, c. The appropriate water temperature for *M. leidyi* is flexible, spanning from 3–30 °C in southern seas to 1–24 °C (peak temperatures) in northern recipient areas (Fig. 4.54a). The comb jelly

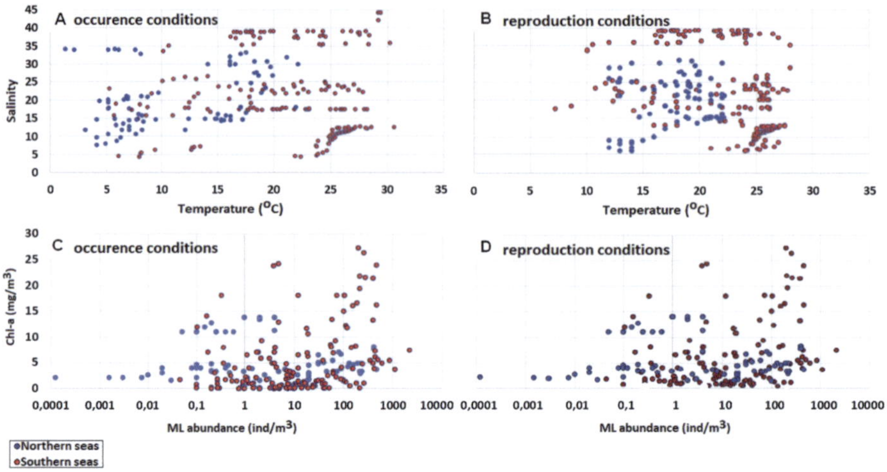

Fig. 4.54 Optimal ranges for sea surface temperature (SST), salinity (SSS), and chlorophyll-a (Chl-a) concentration that support the presence of *M. leidyi* (**a**, **c**) and its reproduction and population expansion (**b**, **d**). (After Shiganova et al. 2019a, b)

M. leidyi thrives in salinity levels from 3.5‰ to 45‰ in southern marine environments and between 4.5‰ and 35‰ in northern waters (Fig. 4.54a) (Costello et al. 2012; Shiganova et al. 2019a, b).

Reproduction conditions The environmental factors that support the reproduction and expansion of a self-sustaining *M. leidyi* population, covering both source and sink regions, are detailed in Fig. 4.54b, d. The optimal water temperature range for *M. leidyi* reproduction and population expansion in southern seas is 12–27 °C. Active reproduction ceases when temperatures fall below 11.8 °C. Nevertheless, in highly productive zones of the Southern Caspian and Northern Adriatic, reproduction has been observed at elevated temperatures of 29 °C and 28 °C, respectively (Fig. 4.54b). In northern seas, the temperature range conducive to reproduction is 10–24 °C (Fig. 4.54b). Limited reproduction has also been documented at 6 °C in the Northern Caspian, Narragansett Bay, and Chesapeake Bay (Costello et al. 2012; Purcell et al. 2001). The reproductive and population growth conditions for M. *leidyi* occur within a water salinity range of 6–40‰, as illustrated in Fig. 4.54b. Environmental conditions for food productivity indicate that Chlorophyll a concentration serves as a reliable proxy for food availability. Reproduction in *M. leidyi* commences once Chl-a levels surpass 0.5 mg. m^{-3} (Fig. 4.54d). Information regarding the occurrence, reproduction, and population expansion of *M. leidyi*, gathered in this research, was further utilized to construct a model. This model effectively replicated both historical and present trends while also forecasting future invasive dynamics under climate change scenarios (Shiganova et al. 2019a, b).

4.7 *Beroe ovata* Bruguière, 1789 in the Black Sea and Its Impact on *Mnemiopsis leidyi*

The main reason for the development of high abundance and biomass of *Mnemiopsis leidyi* in the Black Sea compare to its native habitats (near-shore waters of the North and South Atlantic) was the absence of its predator, i.e., *Beroe ovata,* which is able to control *M. leidyi* (Purcell et al. 2001).

When *M. leidyi* appeared in the Black Sea, an idea to introduce its predator, another ctenophore *B. ovata* to control *M. leidyi* population often aroused. In the native habitat *B. ovata* feeds exclusively on *M.leidyi* and forms a predator-prey pair (Seravin 1994; GESAMP (Joint Group of UN Experts on Scientific Aspects of Marine Protection)). However, those proposals were not implemented because of the concerns that the appearance of new gelatinous species might further amplify the deterioration of the Black Sea ecosystem, affected by the pressure of *M. leidyi.*

Nevertheless, in 1997 by a chance (with the ships' ballast waters) a new non-native ctenophore, *Beroe ovata* appeared in the Black Sea (Konsulov and Kamburska 1998). In August 1999, in the northeastern part of the sea, the first outburst of the development of the new species was registered (Shiganova et al. 2000a, b; Vinogradov et al. 2000) followed by the same event in the northwestern Black Sea (Finenko et al. 2000).

***Beroe ovata* Taxonomy and Morphology**
B. ovata is an important ecosystem trophic link with the planktivorous ctenophores of the Lobata order; meanwhile, information about its ecology and even taxonomy before an appearance in the Black Sea is insufficient.

After an appearance of *B. ovata* in the Black Sea, arose the question about its species statute. Konsulov and Kamburska (1997) defined it as *Beroe ovata.* In August–September 1997, the ctenophore appeared in the northeastern part of the Black Sea and was also identified as *B. ovata.* It was suggested that it could invade from the Mediterranean Sea (Shiganova et al. 2000a; Vinogradov et al. 2000). However, after performing of simple morphological measurements, contradictory conclusions have been obtained (Shiganova et al. 2000b). We carried out detailed studies of the new invader's morphology and taxonomy under the supervision of Prof. L.N. Seravin and concluded that it belongs to the *Beroe ovata* species, which dwells in the same areas as *M. leidyi* does, i.e., in the near-shore waters of North America (Seravin et al. 2002). After genetic analysis, performed by Bayha with co-authors (2004), later confirmed by our own genetic analyses (Shiganova and Abuzova 2021), identification of this species was defined as follow:

Type Ctenophora Eschscholtz, 1829
Order Beroida Eschscholtz, 1829
Family Beroidae Eschscholtz, 1825
Genus *Beroe* Browne, 1756
Beroe ovata Bruguière, 1789

Synonyms:
Beroe ovata Eschscholtz, 1829 (Chun, 1880).
Beroe ovata Bosc, 1802 (Harbison et al., 1978).
Beroe ovata Chamisso & Eysenhardt, 1821 (Mayer, 1912) (see for complete synonyms until 1912).
Beroe ovata Moser, 1903, 1910.
Beroe clarkii Mayer, 1900.
Beroe 'ovata' Eschscholtz, 1829 (Mills et al., 1996).
Beroe ovata Mayer 1912 (Seravin et al., 2002).
Beroe ovata sensu Mayer 1912 (Bayha et al., 2004).

B. ovata has a wide, mitre-shaped flattened body, extended laterally in the voluminous stomodaeum. The lateral compression of the *B. ovata* body is remarkable, being more than three times flattened in the paragastral plane (Fig. 4.55).

Younger individuals are wider both in the oral and aboral parts of the body. Meanwhile, under certain conditions, the body shape of the ctenophore is variable to some extent; e.g., the aboral end may be significantly stretched (Fig. 4.56a). The almost rectilinear oral part, which is a distinguishing species feature of *B. ovata,* may sometimes be slightly oval (Fig. 4.56a, b). Usually, ctenophores feature a pink coloring (Figs. 4.55 and 4.56); the largest individuals are more intensively colored with a brownish tint. The size of large adult individuals in the Black Sea ranges from 81 to a maximum of 162 mm at an average value of 40–70 mm.

Individuals of *B. ovata* have a wide-opening mouth (Fig. 4.57), which provides a possibility to capture preys without hunting by tenticulars or lobes as the ctenophore *M. leidyi* does. The mouth continues with a vast stomodeum, sometimes referred to as pharynx, though actually it represents a stomach.

Because of the vast stomodeum, the ctenophore body looks bag-shaped; it occupies 4/5 of the animal body width and extends up to the flattened wall of the

Fig. 4.55 *Beroe ovata* Bruguière, 1789. (Photo by T. Shiganova)

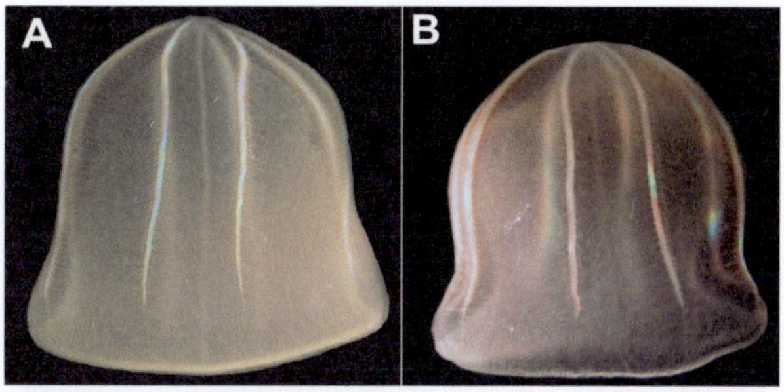

Fig. 4.56 *Beroe ovata* from the Black Sea (length: (**a**) 76 mm, (**b**) 72 mm). (Photo by T. Shiganova)

Fig. 4.57 Hunting *Beroe ovata*. (Photo by T. Shiganova)

aboral end of the body, where a relatively small infundibulum is located. If one cuts out the lips of the mouth, it can be seen under binocular that at a certain distance from the lip end of the mouth, the inner surface of the frontal edge of the stomodeum is covered with large ciliate structures (microcline). Microcline rims the frontal part of the stomodeum (behind the corners of the mouth) in the form of a ring; they are used as teeth that can help to bite off parts of large preys, if it cannot swallow it as a whole. After swallowing the food, the ends of microcline in the mouth area join together, closing the mouth for the time when the prey is in the stomodeum (Fig. 4.58).

The aboral end of the body is supplied with the aboral organ. It is located between two small rounded polar fields, which are rimmed by branched papillae at their outer edge (Fig. 4.59). The aboral organ is sometimes referred to as statocite, which seems incorrect. At present, it is established that this organ represents a sensitive polyfunctional system that not only serves as a receptor of gravity and also a receptor of the hydrostatic pressure.

Fig. 4.58 *Beroe ovata* after swallowing a large *M.leidyi.* (Photo by T. Shiganova)

Fig. 4.59 Aboral part of the *B. ovata* body. (Photo by T. Shiganova)

All these features are specific for *B. ovata,* which dwells in the same coastal Atlantic waters of North and South America as *M. leidyi* does (Shiganova and Abusova 2021).

The ctenophore *B. ovata* are capable to luminarize. The centers of luminescence are enclosed in rows of locomotive plates; the latter are so sensitive that even at a minor irritation one can see the light coming from the eight luminescent ambulacras. The light is produced by cells in the meridional canals; they are arranged asymmetrically with respect to the canal axes and look normal to the surface of the comb flapper. Observations of a ctenophore in the water allow to see green lights, running along the meridional canals.

Ecological and Physiological Features of *Beroe ovata*

The relation of the wet weights of *B. ovata* individuals to body lengths, obtained in our experiments, is expressed by the following equation: $W = 0.0036\ L^{2.02}$ ($n = 36$, $p < 0.01$; $L = 8$–162 mm) (Fig. 4.60).

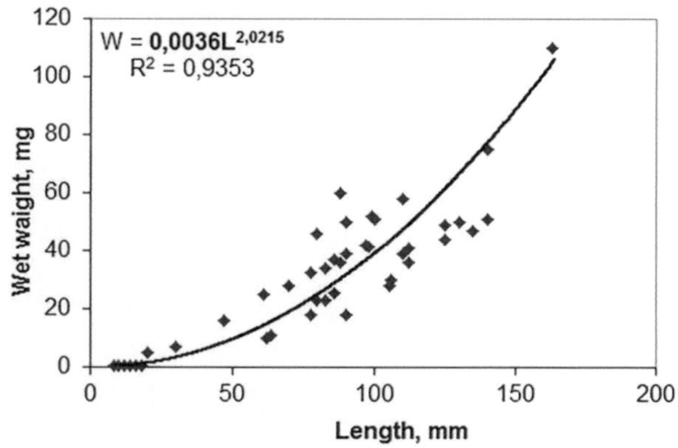

Fig. 4.60 Correlation between the wet weights of *B. ovata* individuals and body lengths (Shiganova et al. 2001a, b)

The correlation between the dry weight (W) and body length (L) is expressed by the following equation: $W = 0.0016\ L^{1.36}$ ($R^2 = 0.93$, $n = 31$, $p < 0.01$). The proportion of dry matter in the *B. ovata* body decreases with the enlarging of individual. On average, for 31 examined individuals with body lengths from 40 to 112 mm, the proportion of dry matter comprised 2.38% of the wet body weight (Shiganova et al. 2001a, b).

Structure of the Reproductive System and Reproduction of *Beroe ovata*
Beroe ovata is similar to the most of the ctenophore species, being a hermaphrodite, capable for self-impregnation. Intermediate and large individuals of *B. ovata* feature the ribbon-shaped gonads (testicles and ovaries), extending along the meridional channels and containing mature sexual products. The testicles are located in peripheral parts of the canals, while ovaries occupy their central parts. Often, ovaries may also be encountered in diverticulae (Fig. 4.61).

Spawning is implemented via the breaks of the temporary gonatopores; after spawning, the gonatopore closes with no damages for the ctenophore (Oliveira and Migotto 2006). A possibility exists for internal impregnation and even for embryonic development up to the appearance of a larva inside the diverticula. One or several spermatozoa penetrate the *B. ovata* egg and the impregnation happens in less than 1 min (Oliveira and Migotto 2006).

The bulk of eggs and sperm are usually thrown outside and the impregnation occurs beyond the ctenophore body. Carre and Sardet suggest that *Beroe ovata* features only cross-impregnation. Meanwhile, in our experiments on the fertility, single individuals were kept separately; nevertheless, we regularly observed impregnated and developing eggs and got larvae, though the percentage of non-impregnated eggs was very high.

Fecundity of *Beroe ovata*
Experiments on the reproduction of *B. ovata* in the Black Sea were performed in September 2002 with the *B. ovata* individuals of 30–89 mm lengths and a wet weight of 7.0–28.48 g. Freshly caught ctenophores were placed individually into aquariums in order to study their fertility and the percentage of individuals, developed after self-

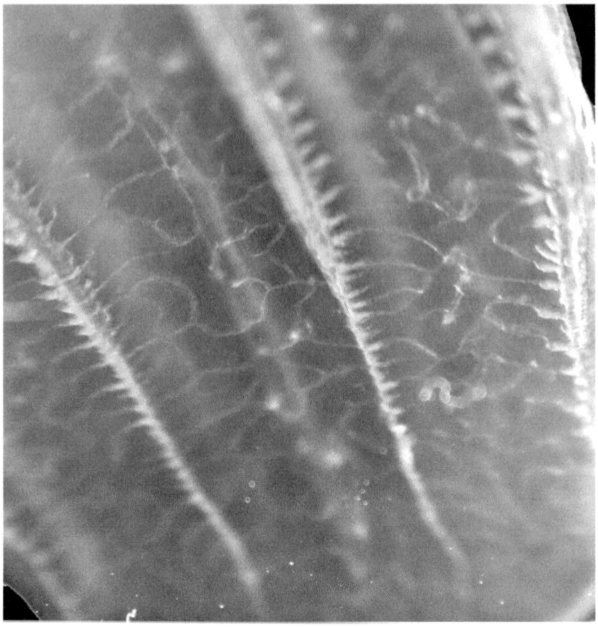

Fig. 4.61 Structure of the *B. ovata* meridional canals with the diverticulae. (Photo by T. Shiganova)

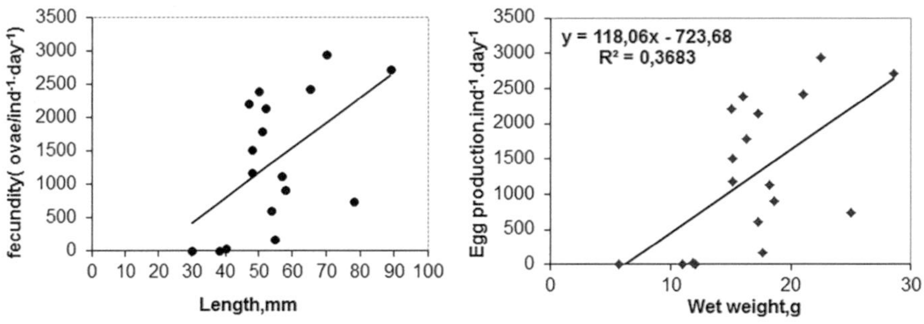

Fig. 4.62 Fecundity of *B. ovata* in the Black Sea depending of the individuals' body length and wet weight

impregnation. *B. ovata* were fed with *M. leidyi* during 2 days. Two days later, the aquariums were inspected to evaluate the appearance of eggs. Individuals with the lengths less than 40 mm showed no spawning. All other individuals of larger sizes, especially those longer than 50 mm and heavier than 15 g actively reproduced (Fig. 4.62). The number of spawned eggs varied from 24 (40 mm long individual) to 2945 (70 mm long individual); the average number of spawned eggs amounted to 1989 $\pm$ 886 eggs/ind.$^{-1}$ day^{-1}.

Our experiment on tolerance of *B. ovata* to salinity indicated that at salinity of 17‰, fertility was at maximum. A decrease of salinity to 11–12‰ resulted in fertility decrease. When salinity fell to 10‰, only a few eggs were obtained

(T.A. Shiganova, unpublished). Under natural conditions, the fertility depends also both on the individual sizes and on the concentration of the prey (*M. leidyi*).

Embryonic and Postembryonic Development of *Beroe ovata*
The hatched eggs of *B. ovata* are rounded in shape (Fig. 4.63), sometimes slightly oval, with a diameter of eggs from 800 to 1000 μm and an inner diameter of 300 μm. Immediately after hatching the surface of the egg looks vitreous since the newly hatched egg is mantled with a reflecting network, disappearing during the first hours after hatching. The gelatinous layer around the egg also gradually disappears with the embryo growth.

Under natural conditions, if several individuals of adult *B. ovata* are located close to eggs, the latter are rapidly impregnated immediately after spawning. The embryogenesis is quick and stereotyped and is free from metamorphosis (Martindale and Henry 1997). It proceed during 28–36 h at temperature of 24–25 °C. The hatched larvae resembls in shape an adult individuals of Beroida, though lacking any tentacles characteristic of a cydippida larva (Fig. 4.64). The larva is large in size and has alredy formed meridional canals. It does not have diverticules and its oral part is more rounded and stretched than that of an adult *B. ovata*. Thus, it resembles asult individuals of *Beroe cucumis*, which, probably, is a more ancient species (Fig. 4.64).

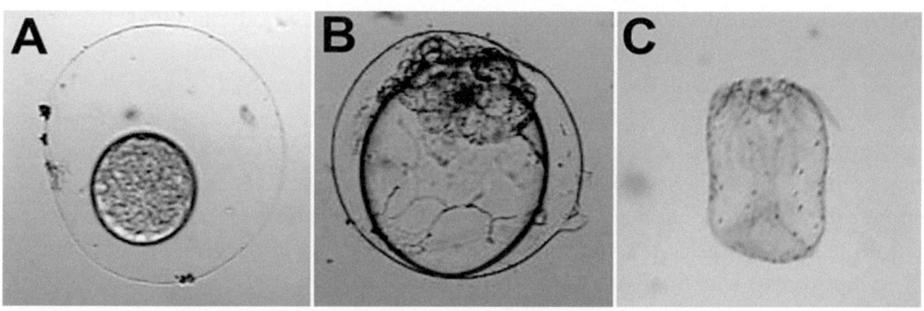

Fig. 4.63 Stages of the of *Beroe ovata* larva development ((**a**) newly hatched egg; (**b**) embryo; (**c**) larva)

Fig. 4.64 Larva of *Beroe ovata*

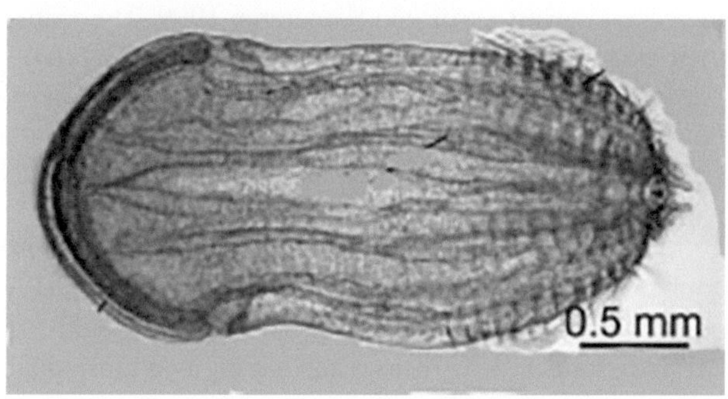

Growth of *Beroe ovata*

The growth of *B. ovata* is an exponential and its specific rate doesn't depend on the size and is rather related to the specific ration, which, in turn, depends considerably on its prey, *M. leidyi*. In our experiments, the specific growth rate of larvae made up 0.01–0.20% day^{-1} at a concentration of *M. leidyi* larvae 10–30 ind. l^{-1}.

In the experiments of other authors, the specific growth rate made up 0.05–0.20 mm when the ration increased from 27 to 150% of the body weight (Finenko et al. 2003). Similar proportions were obtained for the gross growth efficiency *K*, which was negative (about −0.26) when a ration was less than 27% of the predator wet weight. Under these feeding conditions, large *B. ovata* lost their weight, while the efficiency became positive with the ration increase.

Feeding, Digestion, and Ration of *Beroe ovata*

Food Spectrum

Our laboratory experiments on the feeding of *B. ovata* and its food spectrum were carried out in 1999, when *B. ovata* has just spread over the entire Black Sea and probably tried to find appropriate food in a new, biotope swallowing different food objects. We caught individuals of *B. ovata* which just swallowed *M. leidyi*, fish fry (horse mackerels and mullets), or medusa *Aurelia* and placed them in aquariums with 1.5–20 l volume to observe their digestion. Fish fry were not digested by *B. ovata*. Larva of horse mackerel was spited out by *B. ovata* after 15–20 min being already dead and with a slightly macerated integument, while a hatchling of mullet was attached to the outside surface of the *B. ovata* body and was also spited out; however, it was alive and continued to dwell in the aquarium.

Medusa *Aurelia* with the size of about 32 mm, swallowed by *B. ovata,* was kept in the stomodeum of the latter for over 8 h. During this period, it was alive and performed pulsed movements of its bell in the same way it do under usual conditions. Finally, it was also spited out and continued to live in the aquarium. A similar situation was also observed with the Copepoda *Acartia clausi*: *B. ovata* didn't consume it. In case *Acartia clausi* occasionally got into the stomodeum of *B. ovata*, it also was finally spited out.

In addition, we often observed the cases of cannibalism: the swallowing of *B. ovata* by another individual (Fig. 4.65). Still, it was also spited out without any damage, and often the swallowed individual lived longer than the cannibal itself.

Experiments on the clarification of *B. ovata* food spectrum were continued aboard the vessel during a cruise in September 1999. The viable individuals of *B. ovata* with their potential preys (*M. leidyi*, *Pleurobrachia pileus* and medusa *Aurelia*) were placed into a single pool with 0.25 m^3 volume. There, *B. ovata* readily consumed *M. leidyi*: immediately after digesting the previous prey, it was capable of swallowing the next one. The sizes of the preys swallowed by *B. ovata* sometimes were comparable with the sizes of the predator itself.

B. ovata also readily consumed *Pleurobrachia pileus*; a single *B. ovata* individual could swallow up to three *P. pileus* or small *M. leidyi* at a time. *B. ovata* didn't consume medusa *Aurelia*. In the one of experiments, *B. ovata* managed to swallow two individuals of *P. pileus,* each having two individuals of *Calanus euxinus* in their

Fig. 4.65 *B. ovata* swallowed another *B. ovata* individual. (Photo by T. Shiganova)

gastrovascular cavinity. *P. pileus* was digested, while Calanuses were spited out via the mouth.

In the experiments performed by Oliveira, larvae of *B. ovata* from the coastal waters of South America fed on small hydromedusas (Oliveira and Migotto 2006). Adult individuals there also consume *Bolinopsis vitrea* in addition to *M. leidyi*.

The representatives of Beroidae have very low content of the chitinase enzyme required for crustacean zooplankton digesting; therefore, they are unable to consume it (Hoeger and Mommsen 1984).

Feeding Behavior of *Beroe ovata*

Both under natural conditions and in an aquarium, the feeding *B. ovata* having discovered *M. leidyi* nearby begining to make circles around the potential prey and then either swallows it entirely with a widely opened mouth, or bites off parts of it. *B. ovata* can swallow *M. leidyi* individual, which is even more than itself. In aquarium, the swallowing proceeded in two ways: *B. ovata* makes circles around the prey, then opens and rounds the mouth, and pulls the prey inside the gastrovascular interstice owing to an abrupt bulging of its entire body. The alternative way is the biting off parts of its body, mainly lobes. After this, the mouth is tightly closing and *B. ovata* resting while digests the food. Meanwhile, in some cases, when the food still has not been fully digested, *B. ovata* may capture a new prey. The feeding behavior of *B. ovata* is more active than that of *M. leidyi*.

Table 4.22 Rate of digestion of different preys by the *Beroe ovata* (Shiganova et al. 2000a)

Beroe weight, g	Weight of preys, g		Duration of digedstion, h		Temperature, °C	Rate of digestion, % of the wet body weight per hour
	M. leidyi	***P. pileus***	***M. leidyi***	***P. pileus***		
37.4	19.1		5		21.2	10.2
10.9	4.0		5		21.5	7.3
21.8	3.0	1	4	7	24.8	4.6
13.5	7.7		5		25	11.4
26.4		3		8	25	1.4
23.1	21.8		3		26	33.3

Feeding Intensity of *Beroe ovata*

In order to estimate *B. ovata* control over the *M. leidyi* population in the Black Sea, its feeding intensity have been investigated. Obviously, that first of all it depends on the *M. leidyi* concentration. In aquarium conditions, *B. ovata* can swallow either one large *M.leidyi* individual and then passively digest it, a few small *M. leidyi* or a small *M. leidyi* with 2–3 individuals of *Pleurobrachia pileus.* As usual, the intervals between the preys swallowing last for 1–5 h. In other experiments, one *B. ovata* consumed 2–6 *M. leidyi* individuals per day. Herewith, the swallowing of the subsequent prey might occur before the digesting of the previous prey is completed, e.g., once *B. ovata* individual of 55 mm long with 11 g weight sequentially swallowed three *M. leidyi* individuals of 35–45 mm long with 13 g weight (Table 4.22).

According to the observations of Finenko *B. ovata* used to swallow one large prey. The average interval between the full digestion and the subsequent swallowing was 2.3 h (aquarium conditions; Finenko et al. 2001).

Later, the feeding, estimated from the filtration rate (water volume filtered by individual during a specified period per unit of weight) was considered by Finenko et al. (2003). According to their data, the filtration rate (FR) do not depend on the prey (*M. leidyi*) concentration in the interval of 0.8–6.2 ind. l^{-1} (0.47–7.36 g l^{-1}); nevertheless, they found a negative correlation between the filtration rate and the wet weight of a ctenophore (W): $FR = 0.105\ W^{-0.939}$ ($n = 36$, $r = 0.59$, $p < 0.01$).

Process of Food Digestion

After swallowing, the prey of *B. ovata* and goes into the stomodeum. The stomodeum has no ferruginous rolls; therefore, the swallowed prey is macerated, rather than digested here. The cilia of the stomodeum help to transport the small particles formed in the course of the food destruction to the aboral part of the body, where the pre-infundibular set of structures is located. It includes a fissure lying between two contractable pre-infundibular lips. If they are joined together, the fissure is tightly closed; when they spread apart, a gap, leading to the infundibulum (funnel) opens. The pre-infundibular set of structures also includes four cilia galleries located in pairs on either side of the pre-infundibular fissure. Through the open fissure the cilia galleries provide the transport of food particles and fluid in which they are suspended, inside the funnel (infundibulum). Infundibulum is often regarded as a stomach; meanwhile, in the opinion of Prof. L.N. Seravin, this is incorrect since no food digestion occurs in it (Seravin 1998). The walls of the funnel are inlaid with a monolayer epithelium of an endodermal origin. The digested food is supplied to it and immediately, without any delay is transferred to the canals of the radial system. That is, the funnel or infundibulum implements rather the function of distribution of food objects than that of digestion. The inedible or too large particles are thrown away by the gallery cilia. They get to the sides of the stomodeum, where they can be stuck together by mucus and removed outside (on the sides of the mouth opening).

B. ovata does not have infundibular canal; therefore, two minor effluent channels originates immediately from the infundibulum (funnel). Near their distal ends, they are connected with flat tanks lying below the polar fields (see Fig. 4.59). These tanks accumulate undigested particles that occurred in the canal systems. On the opposite sides of the fields, asymmetrically, episodically, excretory cones (one at a time) are turned out. On their tops, effluent pores open and the particles accumulated in the tanks are issued outside in the form of a fluid jet.

In our experiments, the duration of the digestion varied (depending on the size of the predator and the size and number of the preys swallowed) from 2.5 h for digesting of a single 0.60 g prey to 7 h for simultaneous digesting of three preys with a total weight of 1.20 g at a temperature 21–26 °C (Table 4.22).

The digestion of three individuals of *P. pileus* lasted 7 h at a temperature 25 °C. The mean digestion rate comprised 5% of the body weight per hour, though in some cases it reached 24%. The daily food consumption comprised 68–210% of the wet body weight (with 117% on average) (Shiganova et al. 2001b).

Finenko and coauthors defined the duration of digesting of *M. leidyi* individuals by *B. ovata* as 0.5–5.5 h at 21 ± 1 °C (Finenko et al. 2001). They also found out that the duration of digestion (D) depends on the ratio between the weights of the prey and predator (P, range 0.01–2): $D = 4.26\ P^{0.478}$ ($n = 19$, $r = 0.65$).

Ration of *Beroe ovata*

The daily food ration of *B. ovata* was calculated in two ways (Shiganova et al. 2001b). The first method permitted to define the minimal daily requirements of *B. ovata* in food; it was calculated after estimation of the respiration rhythm at a temperature of 24–26 °C and comprised 2.4 cal h^{-1} g^{-1} of dry weight. Correspondingly, the minimum daily ration of *B. ovata* is about 80 cal h^{-1} g^{-1} of dry weight or about 2 cal h^{-1} g^{-1} of wet weight. When *M. leidyi* is consumed (caloric value 10 cal g^{-1} of wet weight; Vinogradov et al. 1989), the minimal daily ration of *B. ovata* should comprise about 20% of its wet weight.

The second method assumes that *B. ovata* feeds continuously on *M. leidyi*, consuming one individual at a time; then, using the calculated average time of digestion, the daily ration of *B. ovata* should be from 80 to 400% of wet weight of its prey *M. leidyi*. This ration is not feasible in real marine conditions. The in situ observations performed in August–September 1999 showed that about 20% of the *B. ovata* individuals had *M. leidyi* in their stomodeums. Therefore, we calculated that the relative (specific) daily ration of *B. ovata* comprised 16–80% of its wet weight (Shiganova et al. 2001b). Meanwhile, it should be noted that concentrations of both *B. ovata* and *M. leidyi* were not high at the moment (Shiganova et al. 2001b).

B. ovata Respiration Intensity

The oxygen consumption by *B. ovata* Q (ml ind.$^{-1}$ h^{-1}) almost linearly correlates with its wet weight (W) as $Q = 0.02\ W^{0.86}$ ($R^2 = 0.78$, $n = 21$, $p < 0.001$) (Fig. 4.66). The oxygen consumption in terms of dry weight varies from 0.25 to 0.71 ml ind.$^{-1}$ h^{-1} with an average value 0.54 ± 0.131 ml ind.$^{-1}$ h^{-1}.

Our results converted to the standard temperature (20 °C) using normal Crogue's curve are consistent with the results of Finenko with coauthors (Finenko et al. 2000, 2001), though twice as high compare with the values for *B. ovata* in Narragansett Bay, reported by Kremer et al. (1986) (Table 4.23). The probable reason may be the differences in the salinities of the two regions, which influences the dry weight of

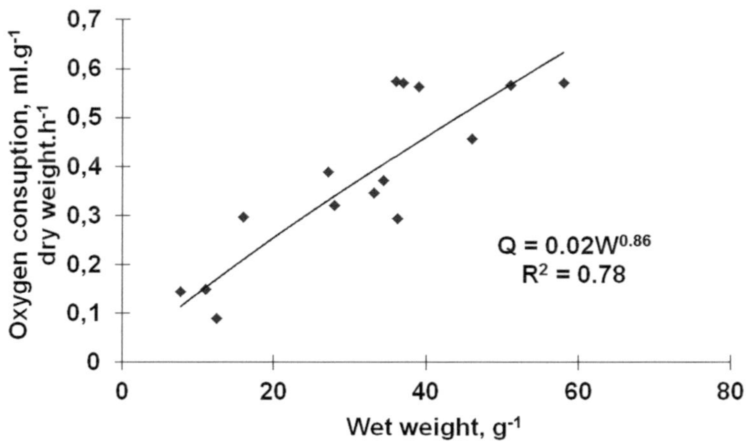

Fig. 4.66 Correlation between *Beroe ovata* respiration rate (ml ind.$^{-1}$ h^{-1}) and its wet weight

Table 4.23 *B. ovata* oxygen consumption (converted to the standard temperature of 20 °C)

Dry weight, g	Duration, h	Temperature, °C	Salinity	Oxygen consumption, ml g^{-1} of DW h	Oxygen consumption, ml g^{-1} of W h at 20 °C	References
0,01-0,56	3–5	25	35	0,27	0,19	Kremer et al. (1986)
0,31–1,04	4	24,6–26	18	0,25–0,71 (0,54 ± 0,131)	0,37	Shiganova et al. (2000a, b)
0,01–0,9	15–18	21	18	0,35–0,44	0,34	Finenko et al. (2000)

B. ovata. Our results, converted into the carbon contents of the ctenophore's body, yield a value of 6.9 ml O_2 g^{-1} of C. This value is close to value, reported by Kremer et al. (1986), which is equal to 7.3 ml O_2 $g^{-}1$ of C.

Although our data limitation does not allow quantitative estimation of the digestion rate dependence the on the temperature, we dare to note that digestion duration tends to decrease with the temperature rise. In addition, we noted a significant difference between the rates of digestion of *M. leidyi* and *Pleurobrachia pileus* individuals: the digestion of the latter lasted much longer, despite its size is smaller by an order of magnitude. Consistent data for *B. ovata* digestion were also obtained by other researchers. Thus, according to Swanberg (1974), at a temperature 24–25 °C *B. ovata* digests *Bolinopsis* during 4–5 h. Kamshilov (1960) reported that in the Barents Sea *Beroe cucumis* digested small *Bolinopsis infundibulum* during 3 h.

To evaluate the influence of the oxygen content on the digestion rate of *M. leidyi* by *B. ovata*, we performed a special experiment. Individual of *B. ovata* with a large *M. leidyi* inside was placed into a closed 1.5 l respirometer. In this case, the duration of digestion comprised 10.5 h, while the oxygen content during this period decreased from 6.39 to 2.38 ml O_2/l.

Studying the intensity of the metabolism, it was noted, that the intensity of the metabolism of ctenophores may change from 3 to 7 times, depending on the activity level. In other experiments, the difference in the respiration rhythms between active and passive *B. ovata* in the Black Sea differed by 4.5 times (Svetlichny et al. 2004). Ability to decrease an intensity of metabolism is a vital feature of *B. ovata,* required for survival of unfavorable environmental conditions.

Seasonal Variability of *Beroe ovata* and *Mnemiopsis leidyi* Populations in the Black Sea

Seasonal dynamics of the considered ctenophores' populations follows the development of lower trophic levels: in spring, mainly due to seasonal warming, intensive development of spring zooplankton species occurs followed by appearance of their consumer, *M. leidyi.* When *M. leidyi* reproduction peaks, a population of *B. ovata,* feeding on *M. leidyi,* appears. Shortly after its appearance, *B. ovata,* in case of sufficient food (i.e., larvae of *M. leidyi*) availability for its larvae, starts reproduction. In 10–14 days, *B. ovata* reaches the peak of reproduction. However, the phenology of both ctenophores differs from year to year as well as quantitative parameters of the peaks of development and reproduction (Fig. 4.67).

After *M. leidyi* is fully eaten out, *B. ovata* disappears from the pelagic zone. Curious observations have been made during the first years of *B. ovata* invasion: in late September, when *M. leidyi* population was already eaten out, hungry *B. ovata* individuals in search for food captured plastic bags. At that time *B. ovata* individuals begin to group in clusters of 5–20 members. In a quiet Gelendzhik Bay, free from strong currents, we observed accumulations of *B. ovata* individuals in the near-bottom layer. Most probably, under unfavorable conditions (e.g., an absence of food), *B. ovata,* like many other ctenophore species, descends to bottom layer and reduces metabolism. *B. ovata* is able to reduce it more than sevenfold. In this way,

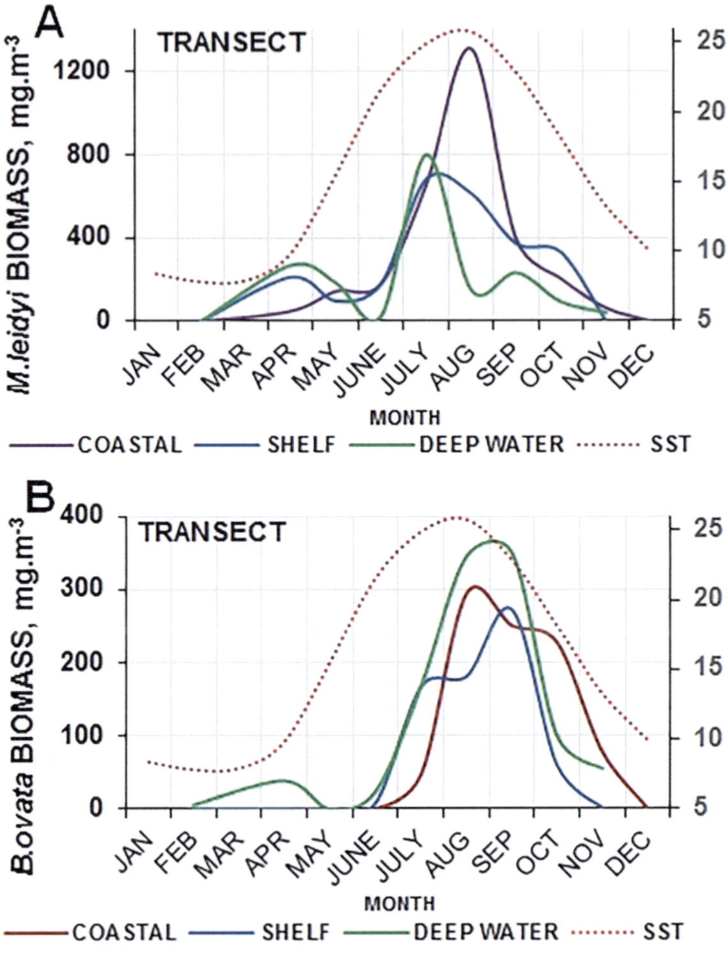

Fig. 4.67 Seasonal variability of *M. leidyi* (**a**) and *B. ovata* (**b**) biomass at transect stations. Dotted lines are seasonal variability of SST

B. ovata overcomes unfavorable conditions, like *M. leidyi* do. The adaptive significance of clustering may be explained by the genetically formed aspiration of individuals to keep closer to each other in order to retain the possibility of cross-impregnation required for the existence of a viable population.

Since 1999, spring warming in the Black Sea takes place earlier. As a result, adult *M. leidyi* also appears earlier (in April–May) and disappears by the end of November–December (Fig. 4.68). Its biomass increases in June, when water temperature rises to about 20 °C, and peaks in July–August, reaching the maximum values in coastal waters at water temperature 23–25 °C. *M. leidyi* biomass is higher in the shelf waters and the lowest in the area, deeper than 500 m. However, despite

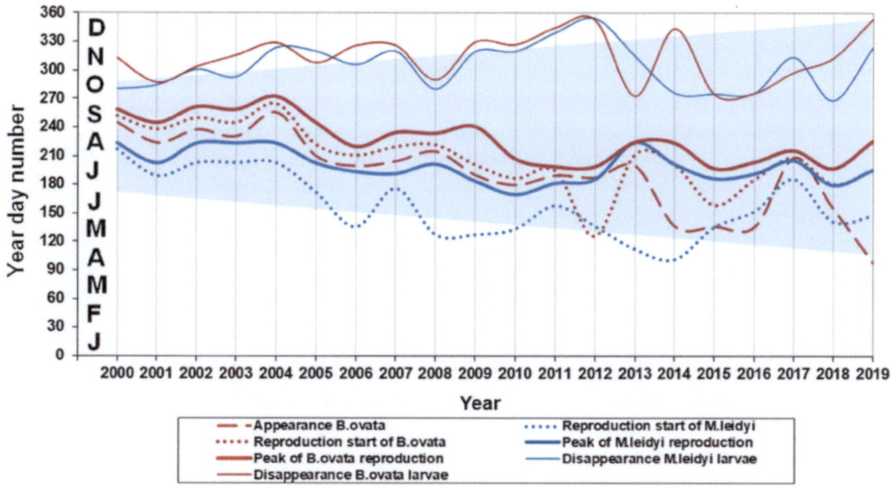

Fig. 4.68 Interannual variability of appearance, beginning of reproduction and disappearance of *M. leidyi* (blue lines) and *B. ovata* (red lines) in the Black Sea. The blue fill illustrates an increase of duration of both ctenophores' annual life cycle

the changes noted in recent years, the sequence of development and reproductive stages of *M. leidyi* and *B. ovata* is the same every year.

The reproductive sequence comprised seven steps (Fig. 4.68): (1) adult *M. leidyi* starts to reproduce; (2) reproduction of *M. leidyi* reaches its annual peak value; (3) adult *B. ovata* appears; (4) adult *B. ovata* starts to reproduce; (5) reproduction of *B. ovata* reaches its annual peak value; (6) *M. leidyi* larvae disappears, and (7) *B. ovata* larvae disappears.

The interweaving of the reproductive sequences of the two species indicates that the timescales of the two sequences were consistent, which allowed the strong interaction between the prey and its predator. The above sequence repeated itself every year from 2000 until present time, although with some temporal variations (Fig. 4.68). *B.ovata* follows *M.leidyi* and disappears depending on availability preys.

Impact of *Beroe ovata* on the *Mnemiopsis leidyi* Population

After invasion of *B. ovata*, the ecosystem of the Black Sea obtained a predator, which is able to control harmful invader *M. leidyi*, and the predator-prey pair, specific for native areas, was formed. Thus, we have an example of successful occupation of an empty ecological niche by non-native species. Spontaneous appearance of this non-native species was a lucky positive event for the Black Sea ecosystem preventing ecosystem degradation at all the trophic levels (Shiganova et al. 2004a, b, c).

So, to what extent *B. ovata* affects *M. leidyi* population? As an example, we estimated the grazing of zooplankton by *M.leidyi* and grazing of *M. leidyi* by *B. ovata*, using the data obtained in cruises aboard the R/V "Ashamba" in the layer 0–100 m from 10 to 550 m depth. Observations were carried out in the period,

Table 4.24 Grazing of zooplankton by *M. leidyi* and grazing of *M. leidyi* by *B. ovata*

St. No	Depth	Catch depth	Zooplankton biomass, mg m^{-2}	Zooplankton biomass, cal m^{-2}	Biomass of *M. leidyi*, g m^{-2}	Grazing of zooplankton by *M. leidyi* per day, %	Grazing of *B. ovata* by *M. leidyi* per day, %
1	12	10	266	186,21	18,05	7,2	1,30
2	32	29	764	535,10	9,71	1,35	25,52
3	50	45	3324	2327,00	4,57	0,15	191,2
4	105	100	5181	3626,67	6,52	0,13	211,59
5	257	100	13,721	9604,44	25,49	0,20	167.62
6	550	100	9992	6994,17	22,64	0,24	0,0
Mean value			55,41 ± 5336,65	3878,93 ± 3735,66	14,5 ± 8,77	1,54 ± 2,81	99,51 ± 100,59

when *B. ovata* has already significantly reduced the population of *M. leidyi*. Nevertheless, the mean percentage of the grazing of *M. leidyi* by *B. ovata* comprised 100%. Accordingly, the pressure on zooplankton by M. leidyi was ceased (Table 4.24).

B. ovata is able to increase its biomass rapidly when *M. leidyi* concentration is high, and on the contrary, its biomass decreases if no *M. leidyi* available.

After the appearance of *B. ovata,* the phenology of *M. leidyi* somewhat changed. Now, only single *M. leidyi* individuals may be encountered in the spring. Meanwhile, starting from 2001, *M. leidyi* began to reproduce earlier with the spring temperature growth and the first larvae may appear as early as May. The active development starts at the end of June and the peak of the reproduction occurs in middle–late July; only in some years, it is observed in early August (Fig. 4.68). In the period of the seasonal absence of *B. ovata,* under the conditions of high zooplankton concentration and high temperature, *M. leidyi* can gain high biomass and affect zooplankton, fish eggs, and fish larvae. Meanwhile, the duration of its impact has reduced to 1–2 months instead of 8–9 months before the appearance of *B. ovata.*

In the last years adult *M. leidyi* appears in the surface layer earlier (in April–May) (Fig. 4.68a) due to increasing spring temperature and start to develop spring zooplankton. Its biomass increases in June, when water temperature rises to about 20 °C, and peaks in July–August, reaching the maximum values in coastal waters at water temperature 23–25 °C. *M. leidyi* biomass is higher in the shelf waters and the lowest in the area, deeper than 500 m.

Interannual Variability of *M. leidyi* and *B. ovata* Biomass in 2016–2023

The interannual variability of the *M. leidyi* biomass depends on changes in the surface water temperature, food concentration (i.e., edible micro- and mesozooplankton), and also by *B. ovata* control (Shiganova et al. 2014a, b, 2018). As *B. ovata* used to suppress *M. leidyi* in the recent years, we observed decline of *M. leidyi* population and rise of *B. ovata* in this period (Fig. 4.69a, b). The same tendency is evident also in the Blue and in Gelendzhik bays (Fig. 4.70).

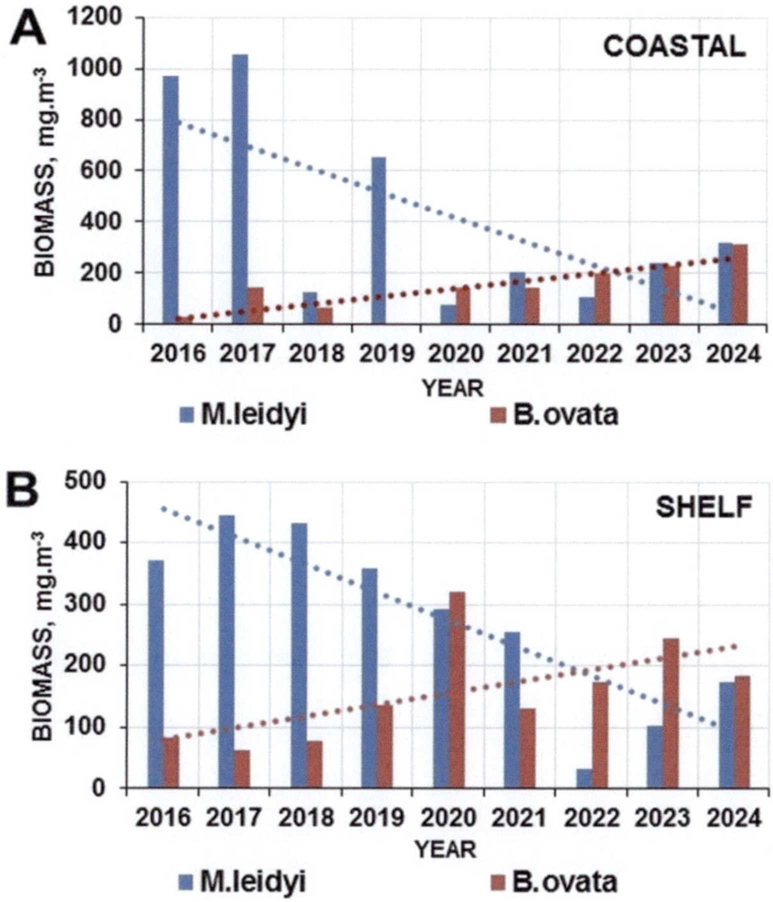

Fig. 4.69 Interannual variability of *M. leidyi* and *B. ovata* biomass at coastal (**a**) and shelf (**b**) along transect stations. Dotted lines—linear trends. Cross-correlations between *M. leidyi* and *B. ovata* biomass is negative with the correlation coefficients $R = -0.69$ for coastal and $R = -0.87$ for shelf locations

Zooplankton Population Recovery After Invasion of *Beroe ovata*

After predator of *M. leidyi* (*B. ovata*) was introduced in the Black Sea, zooplankton stocks and its diversity began to recover, second seasonal zooplankton peak reappeared, and the ecosystem state began to restore.

As expected, the main feature of the total zooplankton biomass seasonal cycle is the spring peak (observed in May, except the deep-water station, where it occurs in April), associated with the water warming and intensive growth of phytoplankton.

In summer, after decline of harmful dinoflagellate *N. scintillans* population, total mesozooplankton biomass increases and reaches maximum (up to 160 mg. m^{-3}) at marine stations. Its biomass is slightly higher in deep waters, and could reach 230 mg. m^{-3} in the bays in July–August, which coincides with the maximum of SST and development of *B. ovata*, preying on *M. leidyi* (Fig. 4.71a). At this time,

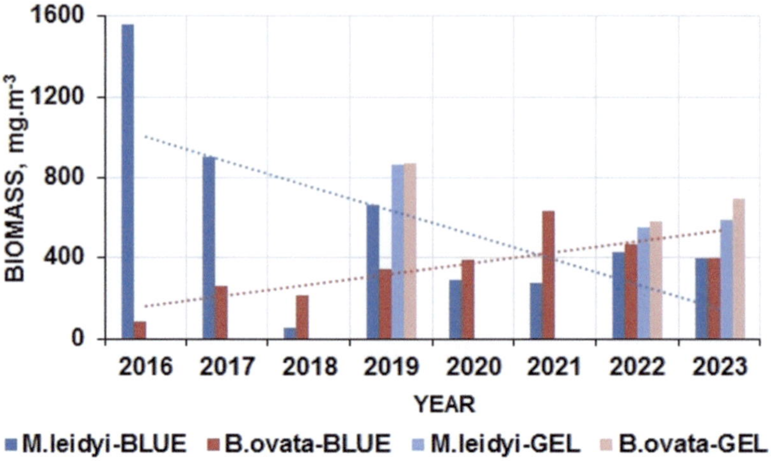

Fig. 4.70 Interannual variability of *M. leidyi* and *B. ovata* biomass in the Blue and in the Gelendzhik Bays (only for 2019, 2022 and 2023). Dotted lines—linear trends of *M. leidyi* and *B. ovata* biomass variability in the Blue Bay. Cross-correlation between *M. leidyi* and *B. ovata* biomass in the Blue Bay is negative with the correlation coefficient $R = -0.9$

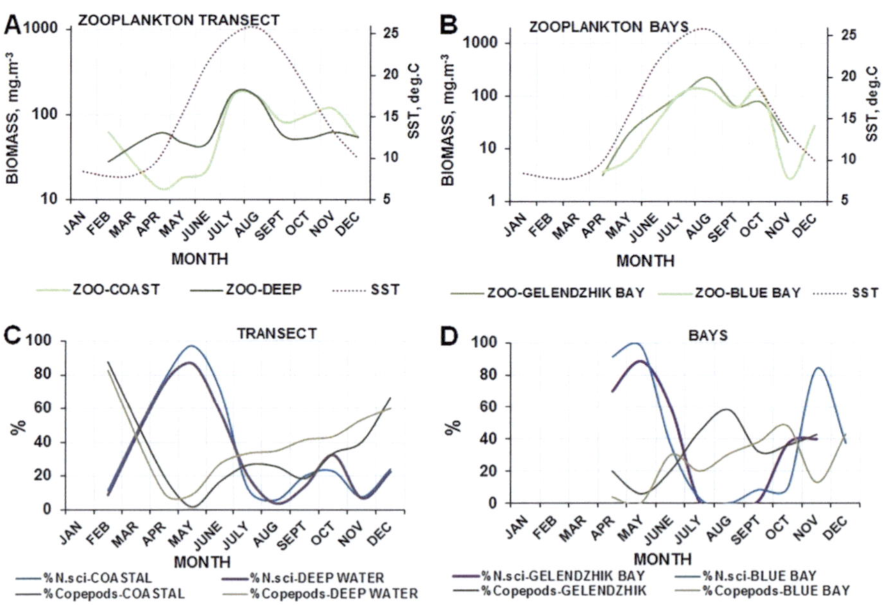

Fig. 4.71 (**a**, **b**) seasonal variability of total zooplankton biomass (mg.m^{-3}); (**c**, **d**) contribution (%) of *Noctiluca scintillans* and copepods to total mesozooplankton biomass

edible zooplankton biomass is almost equal to total biomass due to practically complete absence of *N. scintillans* and intensive reproduction of warm-water Copepoda, Diplostraca and meroplankton (Fig. 4.71c). This second zooplankton peak reappeared again after invasion of *M. leidyi* predator *B. ovata.* Then, second, less intensive autumn maximum of *N. scintillans* biomass occurs and total zooplankton biomass declines again. This is especially pronounced in the Blue Bay, where, in fact, sharp minimum of edible zooplankton biomass is coinciding with the secondary peak of *N. scintillans* (Fig. 4.71b, d).

Seasonal variability of mesozooplankton demonstrates significant increase of its biomass during the peak of seasonal development of *B. ovata* in August–September, when *M. leidyi* population is suppressed by *B. ovata* preying. At that time, warm-water copepods *Centropages ponticus*, eurythermic *Paracalanus parvus*, *Acartia clausi* and *Oithona similis*, non-native *Oithona davisae* and Chaetognata *Parasagitta setosa* are abundant, whereas most of these species almost disappeared during *M. leidyi* blooms (Shiganova 2009).

In April 2017, hyponeustonal Copepoda *Pontella mediterranea,* which also completely disappeared after *M. leidyi* arrival, was recorded again in the northeastern Black Sea, with the maximum concentrations at the shelf. At the deeper locations, the most abundant remained Copepoda *Calanus euxinus.* However, *Oithona nana* completely disappeared in the Black Sea now.

Among the Diplostraca, the most abundant is bacteriophage *Penilia avirostris,* which occurred even during *M. leidyi* blooms due to increase of mucus and dead gelatinous remains, released by *M. leidyi*, which are suitable matter for bacteria, consumed by *P. avirostris.*

The biomass of *P. setosa,* which also almost completely disappeared during *M. leidyi* outbreak, considerably increased after *M. leidyi* decrease. However, *P. setosa* biomass is variable and depends on the location (Fig. 4.72). Its spring peak is observed only at deep-water stations, where *N. scintillans* occur in low abundance or absent. Its highest biomass reaches 66 mg.m^{-3}. The second maximum with about half of the intensity (up to 35 mg.m^{-3}) is observed in July–August at all transect stations and in the Gelendzhik Bay, which also coincides with the active

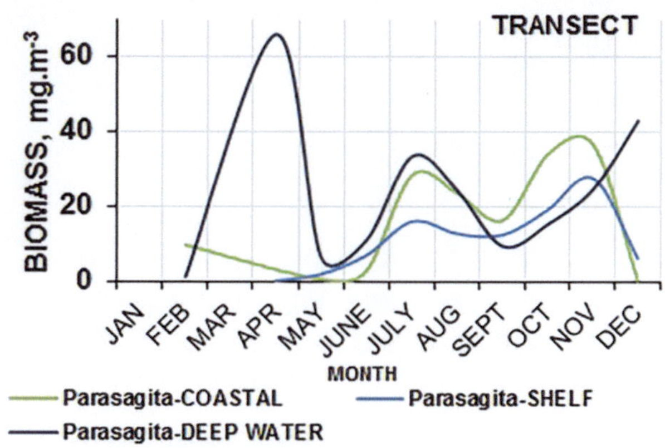

Fig. 4.72 Seasonal variability of *Parasagita setosa* biomass

phase of *B. ovata* development. Finally, in the coastal waters *P. setosa* demonstrates relatively intensive (up to 40 mg.m^{-3}) autumn maximum in October–November. In deep waters its biomass also increases in late fall, when *M. leidyi* population has already vanished (Fig. 4.72).

Interannual Variations of Zooplankton Species Composition After Arrival of *Beroe ovata*

Our observation in 2016–2023 indicates general improvement of the zooplankton state after *B. ovata* invasion. However, it was not observed in all locations, since in some areas negative impact of the native Black Sea gelatinous species, including *N. scintillans,* continued. It was especially noticeable in the coastal waters (Fig. 4.73). The biomass and diversity of Copepoda species was diminished in the coastal waters. Thus, in 2016–2017 only *Acartia clausi* and *Paracalanus parvus* dominated, with other species being represented only by single individuals. In 2018, biomass of Copepoda increased. Biomass of *Acartia clausi* and species, recoded as the single individuals in previous years (*Oithona similis*, *Pseudocalanus elongates*, *Paracalanus parvus*, *Centropages ponticus*), also increased. However, in the following years and especially in 2022 (an outbreak of *N. scintillans*), biomass and species diversity dropped again. It worth to be mentioned a presence of non-native *Oithona davisae* in the coastal area. Then, in 2023 species diversity and their biomass restored. Species, completely disappeared during *M. leidyi* outbreak, were recorded again, indicating the probable restoration of the Copepoda species composition (Fig. 4.74).

Overall, at all stations (both coastal and deep-water), Copepoda total biomass and species diversity were higher. *Calanus euxinus* prevailed at most of the stations; its individuals were found beginning from 2016 with the higher biomass in 2019. *Pseudocalanus elongates*, *Centropages ponticus* and *Acartia clausi* also significantly added to Copepoda total biomass. The other species were also recorded, but in the lesser amount. In 2024, significant increase of biomass (first of all *C. euxinus*)

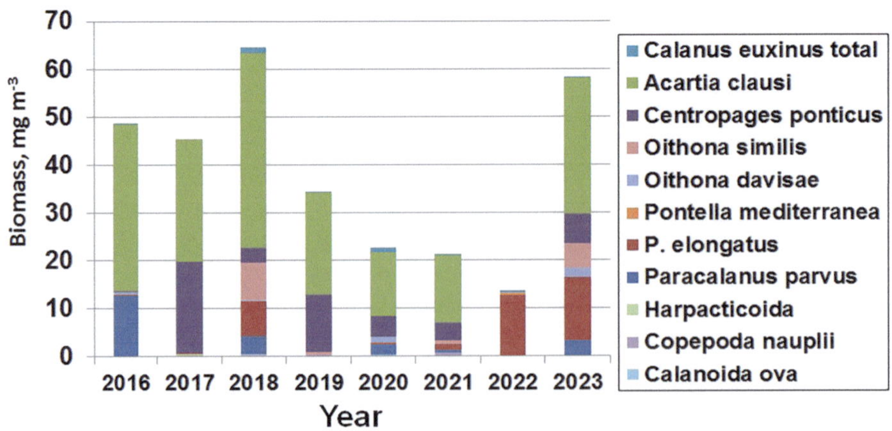

Fig. 4.73 Interannual variability of year-averaged Copepoda species biomass in the coastal waters

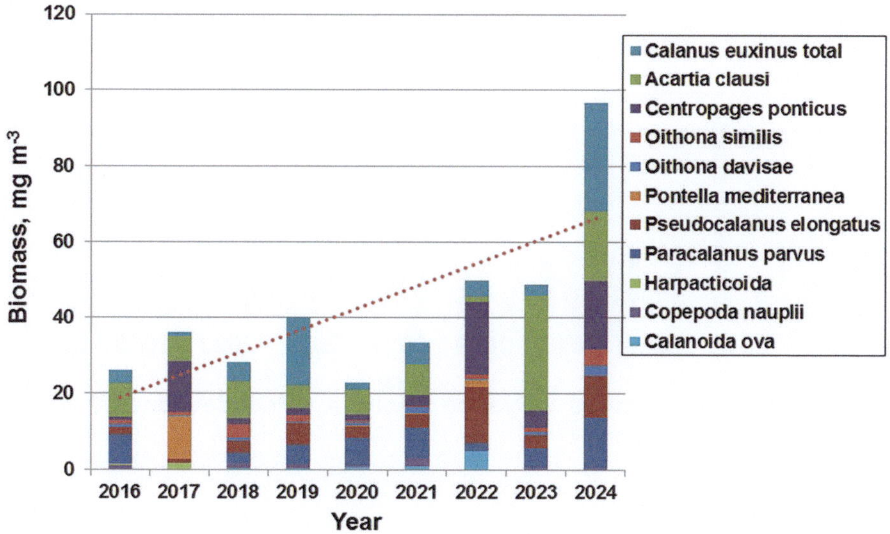

Fig. 4.74 Interannual variability of year-averaged Copepoda species biomass for all stations along the transect. Dotted line—linear trend (approximation reliability $R^2 = 0.52$)

was recorded (Fig. 4.74). In general, we found positive linear trend of Copepoda species biomass during 2016–2024 with sufficiently high approximation reliability ($R^2 = 0.52$), which may additionally confirm an improvement of environment state.

Species of Diplostraca (former Cladocera) were represented by four warm-water species: *Penilia avirostris, Pleopis polyphemoides, Evadne spinifera, Pseudevadne tergestina.* Among them, the most abundant was *Penilia avirostris.* It is reasonable, since the *P. avirostris* is a widely distributed species in coastal areas around the world. Among the other species, in order of decreasing numbers *were Pleopis polyphemoides*, *Pseudevadne tergestina* and *Evadne spinifera. M*inimal biomass of all species was recorded in 2017–2018 and in 2022 during outbreak of *N. scintilans.* In 2023 only two species *(Pleopis polyphemoides* and *Penilia avirostris)* were registered, being slightly more abundant compare to 2022 (Fig. 4.75). In 2024, in addition to previously mentioned two species, *Pseudevadne tergestina* with sufficiently high biomass had appeared (Fig. 4.75).

Mesozooplankton Biodiversity and Ecological Status of Environment After Arrival of *Beroe ovata*

While estimating mesozooplankton variations under the impact of gelatinous species, we considered seasonal and interannual variability of Shannon biodiversity index. The main feature of Shannon index seasonal pattern is a sharp decrease of its value from 1.5 to 2 to less than 0.5 in April–May, both at transect and in the bays (Fig. 4.76a, b), associated with the spring bloom of *N. scintillans*. In this period, *N. scintillans* contributes up to 99% in total biomass, suppresses the rest of mesozooplankton species and, consequently, contributes to the decreasing of

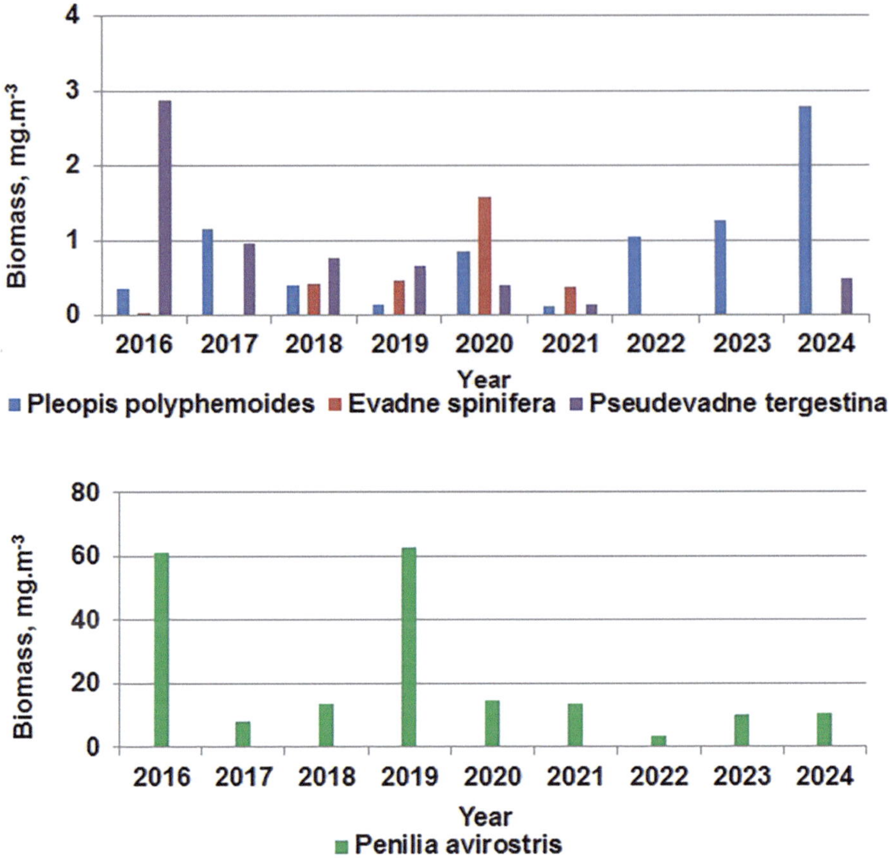

Fig. 4.75 Interannual variability of year-averaged Diplostraca species biomass

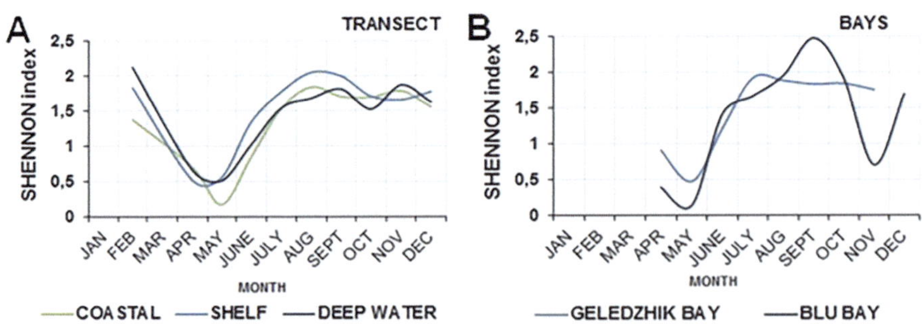

Fig. 4.76 Seasonal dynamics of Shannon diversity: (**a**)—along the transect and in the bays (**b**)

biodiversity. After that, within two months (late June–July), along with a decrease of *N. scintillans*, the Shannon index is restored to its normal values (1.5–2.5). This also was facilitated by the decrease of *M. leidyi* population, consumed by *B. ovata*. This pattern is observed at all locations except the Blue Bay, where low *N. scintillans*

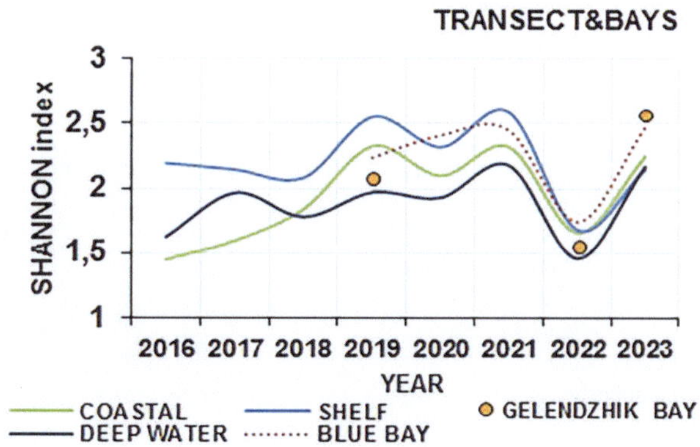

Fig. 4.77 Interannual variability of Shannon index along transect and in bays

abundance observed until October. The second intensive peak of *N. scintillans* bloom occurred in November in Blue Bay, contributes to sharp decrease of zooplankton biodiversity to less than 0.7 (Fig. 4.76b).

In the considered time span, interannual variability of mesozooplankton biodiversity demonstrated a pronounced positive trend (Shannon index grows from 1.5–2 to 2.2–2.6) (Fig. 4.77), interrupted only in 2022 due to extreme *N. scintillans* bloom. However, already in 2023, the index recovered to its previous values, which may indicate that after the disturbances, associated with the invasion of *M. leidyi*, the ecosystem restored an ability to self-maintain its normal status. It is worth to note that in 2017 *N. scintillans* bloom, which was not as strong as in 2022, was not reflected in the Shannon index (Fig. 4.77). This may indicate that the ecosystem is already sufficiently balanced and able to overcome modest negative impacts.

Interannual variability of integrated index of ecological status (IES, Shiganova and Kazmin 2024) has been assessed for the harmful gelatinous species *M. leidyi* and *Aurelia* sp. (Fig. 4.78).

The results indicate that IES, estimated for *M. leidyi* (Fig. 4.78a) was "Good" in the whole observational period, during which *B. ovata*, presented (arrived in the 1990), and was already active.

Regarding the Black Sea native *Aurelia* sp., the result is not so definite. In each year, both positive and negative values of IES have been observed, depending on location (Fig. 4.78b). In most cases, values were strongly positive (+1.0), pointing on relatively "good" ecological status. Still, moderately negative IES (about −0.4; except 2018, when IES dropped to −1.0) also have been observed (Fig. 4.78b). Thus, situation with the impact of *Aurelia* sp. is rather controversial and needs deeper study. However, similar estimation was not received for *Rhisostoma pulmo*.

Recovering of the Small Pelagic Fish After Arrival of *Beroe ovata*

Since 2000, small pelagic species stocks and catches gradually recovering in the Black Sea. The highest catches provided by Azov anchovy and sprat (Fig. 4.79), followed by Mediterranean horse mackerel and red mullet.

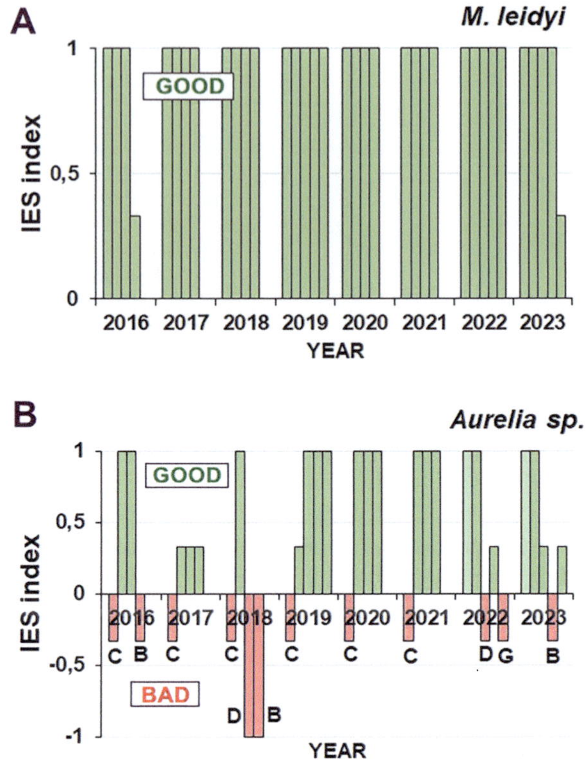

Fig. 4.78 Interannual variability of integrated index of ecological status (IES) (dimensionless) for *M. leidyi* (**a**) and *Aurelia* sp. (**b**). Green color—good ecological status, red color—bad ecological status. C, B, G, and D stand for coastal water, the Blue Bay, the Gelendzhik Bay, and deep water consequently

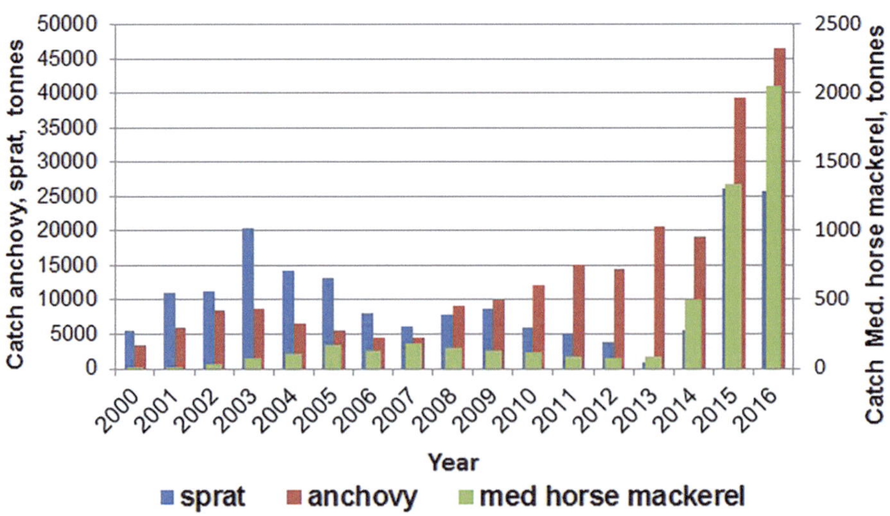

Fig. 4.79 Recover of small pelagic fish catches following the arrival of *B. ovata* in the Black Sea in 1999

Investigation of genus *Beroe* species is crucial issue as they have proven to be able to control populations of harmful zooplanktivorous ctenophores like widely spreading around the World Ocean invasive *Mnemiopsis leidyi* (Shiganova et al. 2024), being the most effective and harmless to ecosystems.

4.8 The Marmara and Aegean Seas as Nearest Neighbors of the Black Sea: Biota Composition, Non-native Species, Ecosystem State

Water masses exchange The Sea of Marmara and the Aegean Sea are the closest to the Black Sea marine basins, through which it is connected to the Mediterranean Sea and further to the Atlantic Ocean. The Black Sea is connected to the Sea of Marmara via Bosphorus Strait, and the Sea of Marmara is connected to the Aegean Sea via Dardanelles Strait (Fig. 4.80).

The Bosphorus and the Dardanelles control water and matter exchange between the Black Sea and the Mediterranean Sea and provides a link between the ecosystems of these two neighboring basins.

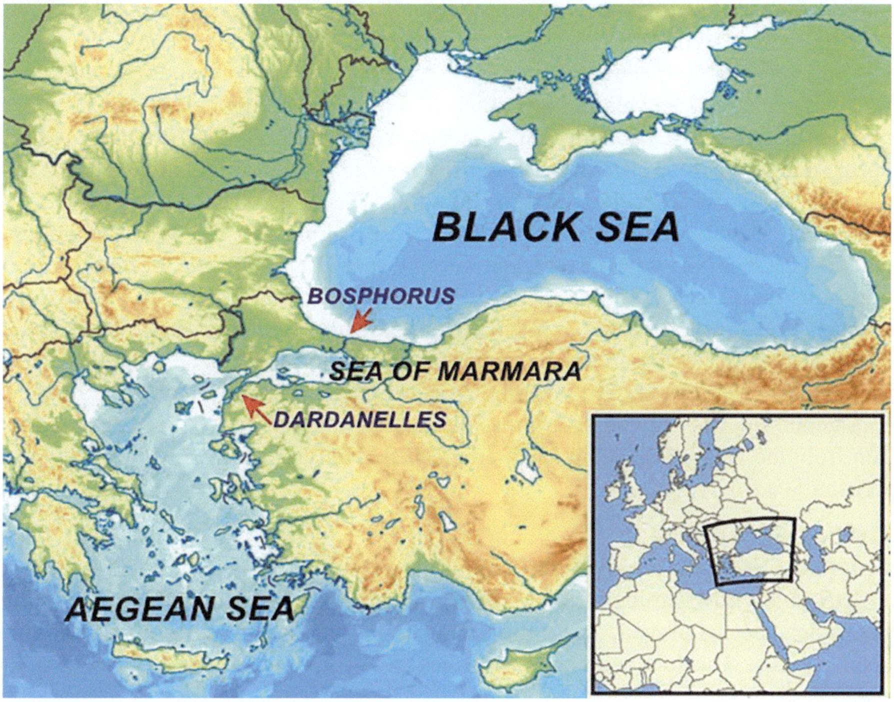

Fig. 4.80 Geographical location of the Black Sea, the Sea of Marmara and the Aegean Sea and connecting straits (Bosphorus and Dardanelles). Insert—general view

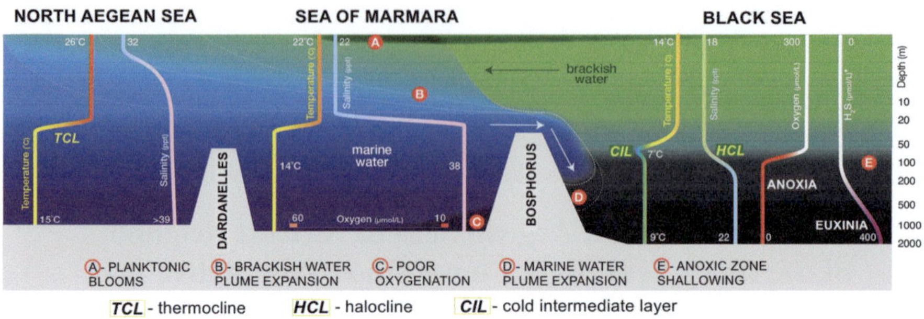

Fig. 4.81 Schematic cross-section of the water masses and flows structure between the North Aegean Sea, Sea of Marmara, and Black Sea basins. (After Krijgsman et al. 2022; modified)

The Bosphorus has a length of about 30 km, a mean depth less than 100 m and width ranging in between 0.7 and 3.5 km. Bosphorus Strait has a two-layer flow system, with the upper-layer Black Sea brackish waters ($S = 18‰$) and the lower-layer Mediterranean marine waters ($S = 38‰$) flowing in opposite directions (Fig. 4.81). A sharp pycnocline occurs between the two layers at 50 m at the Black Sea entrance, and at 20–25 m at the Sea of Marmara entrance to the Bosphorus channel. The upper- and lower-layer current velocities in the Bosphorus vary in the range 1.5–2 m/s, and 0.9–1 m/s, respectively. The present average annual fluxes of the Black Sea and Mediterranean Sea waters through the Bosphorus are about 350 and 170 km^3, respectively.

The Dardanelles has a length of about 61 km, a mean depth of 55 m and width ranging in between 1.2 and 7.0 km. Water exchange through the Dardanelles Strait occurs due to the difference in water density between the Marmara and Aegean seas and also has a two-layer flow system (Fig. 4.81). Surface current carries freshened, less dense transformed water of the Black Sea origin from the Sea of Marmara (salinity 25.5–29‰) with an average current speed 0.8–1.3 m/s. Bottom current transports dense, saline (up to 38.5‰) Mediterranean water in the Sea of Marmara with an average speed about 0.3 m/c. In the upper layer, 350 km^3/year of desalinated Black Sea water flows into the Aegean Sea through the Dardanelles Strait while in the bottom layer, 180 km^3/year of highly saline Mediterranean water flows out.

Satellite images provide an expressive illustration of the water masses propagation from the Black Sea to the Mediterranean Sea through the Sea of Marmara and Aegean Sea (Fig. 4.82).

The Sea of Marmara
Ecosystem State Before the Non-native Species Invasion

Due to the flows structure (Fig. 4.81), in the upper layer of the Marmara Sea the Black Sea species prevail, while the deeper layers are inhabited by the Mediterranean species. The total number of species in taxonomic groups is presented in Table 4.25.

Until 1970s, the Turkish canals system had the rich fish stocks. Among the commercial fish, the most numerous were the moonfish *Pomatomus saltatrix*

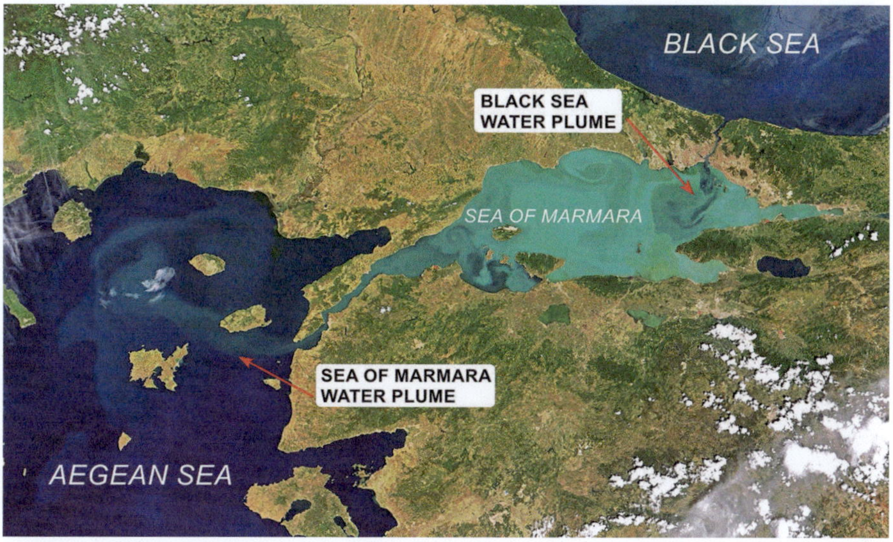

Fig. 4.82 Satellite optical image of the North Aegean Sea, Sea of Marmara, and Black Sea basins. (Captured by the Moderate Resolution Imaging Spectroradiometer (MODIS) on NASA's Terra satellite (https://visibleearth.nasa.gov/images/116502/sea-of-marmara-turkey/116504l; modified). Dynamic features are visualized due to phytoplankton spring bloom)

(Linnaeus, 1766), horse mackerel *Trachurus trachurus* (Linnaeus, 1758), mull barbatus *Mullus barbatus* Linnaeus, 1758, *Mullus surmuletus* Linnaeus, 1758, Lichia amia *Lichia amia* (Linnaeus, 1758), *Gaidropsarus mediterraneus* (Linnaeus, 1758), *Mugil cephalus* Linnaeus, 1758, kalkan *Scophthalmus maximus* (Linnaeus, 1758), mackerel *Scomber japonicus* Houttuyn, 1782, swordfish *Xiphias gladius* Linnaeus, 1758, tuna *Thunnus thynnus* (Linnaeus, 1758), sturgeon *Acipenser sturio* Linnaeus, 1758, *Atherina boyeri* Risso, 1810, *Gaidropsarus mediterraneus* (Linnaeus, 1758) and a number of other fish species.

Later, the share of *Acipencer sturio*, *S. scombrus, Psetta maxima, Xiphias gladias*, *Pomatomus saltator, Lichia amia, Atherina boyeri*, and *Gaidroparus mediterraneus* the in catches in the Marmara Sea significantly decreased (Ozturk and Ozturk 1996). The fall in the fish stocks was associated not only with the overfishing, but also with the illegal fishing, pollution of various origins, variations of the hydro-physical conditions and other anthropogenic impacts.

Non-native Species Introduction in the Sea of Marmara

The Sea of Marmara is located in between the Black and Mediterranean Seas and the species from the both seas can enter it through the straits. Species compositions by types are presented in Tables 4.26, 4.27, 4.28, and 4.29.

The water temperature increase in the Mediterranean area due to the global warming resulted in penetration of increased number of thermophilic species from the Mediterranean in the Black Sea. It may provoke destructive changes in the Black Sea ecosystem.

Table 4.25 Number of species in taxonomic groups in the Sea of Marmara

Taxonomic group	Number of species
Sponges	28
Anthozoa	15
Scypohoza	4
Hydozoa	38
Ctenophora	3
Plathelminthes	10
Nemertinea	6
Polychaeta	118
Briozoa	47
İsopoda	17
Amphipoda	41
Cirripedia	4
Mollusca	428
Cephalopoda	15
Scaphopoda	6
Polyplacophora	6
Caudofoveata	2
Crustacea	109
Natantia	34
Reptantia	8
Anomura	21
Brachyura	46
Echinodermata	51
Ascidia	14
Piesces	200
Elasmobranches	33
Cyanobacteria	53
Rhodophyta	296
Heterokontophyta	107
Chlorophyta marine plants	104 4
Total	1868

Mediterranean species enter the deep layers of the Sea of Marmara with the bottom current (Fig. 4.81). Many of these species penetrate further into the southern part of the Black Sea and stay there, temporarily or permanently. Lots of ships pass through the Sea of Marmara to and from the Black Sea and this traffic are increases every year. Thus, the shipping (both ballast waters discharge and fouling) significantly contribute to the non-native species arrival in the Sea of Marmara.

Phytoplankton

Among the non-native species of phytoplankton, *Pseudosolenia calcar-avis* algae (Table 4.26) were discovered first in the Sea of Marmara and it is a widespread species there (Kosatas et al. 1993). It probably arrived in the Sea of Marmara with

Table 4.26 Non-native species of phytoplankton and macrophytes in the Sea of Marmara

Species	Donor region	Vector	First record	Status of establishment	References
Class Bacillariophyta **Order** Rhizosoleniales **Fam.** Rhizosoleniaceae *Pseudosolenia calcar-avis* (M. Schultze) Sentstrom 1986 (=*Rhizosolenia calcaravis (M. Schultze)*	North Atlantic	With current through the Bosphorus or with ballast water	1993	Established	Kocatas et al. (1993)
Class Dinophyceae **Order** Gonyaulacales **Fam.** Pyrocystaceae *Alexandrium monilatum* (Howell) (F.J.R. Taylor 1979)	Atlantic coast of North America	Shipping (ballast waters)	?	Established	Zaitsev and Ozturk (2001)
Class Coccolithophyceae **Order** **Phaeocystales** **Fam.** Phaeocystaceae *Phaeocystis pouchetii* (Hariot) Lagerheim, 1893	Atlantic Ocean	Penetration? Shipping?	2001	Established	Zaitsev and Ozturk (2001)
Class Chlorophyceae **Fam.** Bryopsidaceae *Codium flagile* Suringar 1867	Atlantic Ocean	Penetration, shipping	1940s	Established	Zaitsev and Ozturk (2001)
Class Chlorophyceae *Bryopsis pennata* Lamouroux, 1809	Atlantic Ocean	Penetration, shipping	1986	Established	Cinar et al. (2005)
Class Rhodomelaceae *Acrochaetium codicolum* Borgesen, 1927	Indian, Pacific, Atlantic oceans ?	Penetration, Shipping ?	1986	Established	Zeybek et al. (1986)
Class Rhodomelaceae *Acrochaetium armata* Harvey, 1855	Indian, Pacific, Atlantic oceans ?	Penetration, Shipping ?	1986	Established	Zeybek et al. (1986)

(continued)

Table 4.26 (continued)

Species	Donor region	Vector	First record	Status of establishment	References
Bonnemaisonia hamifera Hariot, 1891	Pacific, Atlantic oceans ?	Shipping ?	1986	Established	Zeybek et al. (1986)
Chondria collinsiana Howe, 1920	Pacific, Atlantic oceans ?	Shipping ?	1986	Established	Zeybek et al. (1986)
Chondrophycus papillosus (C. Agardh) Garbary & Harper, 1998	Red Sea	Lessepsian immigrant ?	1957	Established	Cinar et al. (2005)
Gracilaria arcuata Zanardini 1858	Red Sea, Indo-Pacific	Lessepsian immigrant	1986	Established	Zeybek et al. (1986)
Griffithsia corallinoides L.	Atlantic Ocean	Penetration through Gibraltar	1993	Established	Cinar et al. (2005)
Hypnea variabilis Okamura, 1909	Pacific Ocean	Shipping	1986	Established	Zeybek et al. (1986)
Radicilingua thysanorhizans (Holmes) Papenfuss, 1956	Atlantic Ocean	Shipping Penetration?	1986	Established	Zeybek et al. (1986)
Rhodophysema georgii Batters, 1900	Atlantic, Pacific oceans?	Shipping Penetration?	1986	Established	Zeybek et al. (1986)
Class Fucophyceae *Chorda filum* (L.) Stackhouse 1797	Atlantic, Pacific oceans?	Shipping?	1986	Established	Zeybek et al. (1986)
Ectocarpus siliculosus (Dillwyn) Lyngbye, 1819	Atlantic Ocean	Shipping Penetration?	1899	Established	Cinar et al. (2005)
Halothrix lumbricalis (Kutzing) Reinke, 1888	Atlantic, Pacific oceans?	Shipping Penetration?	1993	Established	Aysel et al. (1993)
Pilayella littoralis (L.) Kjellman, 1872	Atlantic Ocean	Shipping Penetration?	1993	Established	Aysel et al. (1993)
Class Chordariaceae *Protectocarpus speciosus* (Boerg.) Komm.	Atlantic coastal waters of Northern Europe	Penetration	1993	Established	Cinar et al. (2005)
Sargassum latifolium (Turner) C/Agardh, 1820	Red Sea	Lessepsian immigrant?	1986	Established	Zeybek et al. (1986)

(continued)

Table 4.26 (continued)

Species	Donor region	Vector	First record	Status of establishment	References
Sphaerotrichia divaricata (Agardt) Kylin, 1940	Atlantic Ocean	Shipping Penetration?	1986	Established	Zeybek et al. (1986)
Posidonia oceanica	Mediterranean Sea	Penetration	2004	Established	Yüksek and Okus (2004)

the surface currents from the Black Sea or with the ballast waters (Zaitsev and Osturk 2001). Presumably it was a secondary invasion, occurred earlier than 1928, since the *P. calcar-avis* at first was introduced from the North Atlantic to the Black Sea, where it also became a widespread species (see Chap. 3 above).

Another non-native species in the Marmara Sea is dinoflagellate *Alexandrium monilatum*. This alga was introduced into the Mediterranean Sea from the coastal waters of North America in 1985 and into the Black Sea (Gulf of Varna) in 1991 (Moncheva et al. 1995), from where it probably penetrated into the Sea of Marmara (Zaitsev and Ozturk 2001). *A. monilatum* produces toxins which in high concentrations ("red tides") may harm the fish and reduce growth rates of shellfish.

The golden alga *Phaeocystis pouchetii* was introduced in the Black Sea in 1989 (Petrova-Karadjova 1990) and the Aegean Sea (Zaitsev and Ozturk 2001), and most likely penetrated into the Sea of Marmara.

Macrophytes

Since the 1990s, due to the water warming in the Marmara Sea, a number of macrophyte species, including green algae (2 species: *Codium flagile*, *Bryopsis pennata*), brown algae (6 species: *Chorda filum*, *Ectocarpus siliculosus*, *Halothrix lumbricalis*, *Pilayella littoralis*, *Protectocarpus speciosus*), *Sargassum latifolium, Sphaerotrichia divaricata*, red algae (7 species: *Acrochaetium codicolum, Acrochaetium armata, Chondrophycus papillosus, Gracilaria arcuata, Hypnea variabilis, Radicilingua thysanorhizans, Rhodophysema georgii*) probably penetrated it with a bottom current from the Aegean Sea. These species are non-native also for the Aegean Sea (Table 4.26). Some of them at first penetrated with the currents or with ships from the Red Sea or from the Indo-Pacific region to the eastern Mediterranean and several were imported from the Atlantic and Pacific Oceans. The total number of non-native macrophytes species is 19 (Table 4.26). In 2004, a Mediterranean species of sea grass, *Posidonia oceanica*, was discovered, which is widespread in the southern part of the Sea of Marmara (Yüksek and Okus 2004).

Zooplankton

The number of Black Sea zooplankton species in the surface layer depends on the season, e.g., in summer 16 Black Sea zooplankton species were discovered in the Bosphorus Strait area. Among them are *Calanus euxinus, Parasagitta setosa,*

Table 4.27 Non-native species of zooplanktonic invertebrates in the Sea of Marmara

Species	Donor region	Vector	First record	Status of establishment	References
Class Crustacea **Order** Copepoda **Fam.** Acartiidae *Acartia tonsa* (Dana, 1849)	Atlantic coast of North America (?)	Shipping (ballast waters), adults or resting eggs transported	1998	Established	Unal et al. (2000)
Centropages furcatus (Dana, 1846)	Red Sea, Indo-Pacific	Lessepsian immigrant	2000	Established	Unal et al. (2000)
Pavocalanus latus Andropov, 1972	Red Sea, Indo-Pacific	Lessepsian immigrant	2000	Established	Unal et al. (2000)
Pavocalanus elegans Andropov, 1972	Red Sea, Indo-Pacific	Lessepsian immigrant	2000	Established	Unal et al. (2000)
Trachimedusa *Liriope tetraphylla*	Mediterranean Sea	Penetration	2005	Established	Yilmaz and Yuksek (2009)
Cnidaria, Scyphozoa Rhisostomea *Cassiopea andromeda*	Mediterranean, Aegean seas	Penetration	2008	Established	Ozgur and Ozturk (2008)
Cnidaria Coelenterata Scyphozoa *Chrysaora hysoscella* (Linne, 1766)			2000	Established	Inanmaz et al. (2003)
Class Ctenophora **Order** Lobata Esch **Fam.**. Bolinopsiidae **Genus** *Mnemiopsis* L.Agassiz *Mnemiopsis leidyi* A. Agassiz, 1865	Atlantic coast of North America	Penetrated with currents through the Bosphorus	1990	Established	Shiganova (1993)

(continued)

Table 4.27 (continued)

Species	Donor region	Vector	First record	Status of establishment	References
Order Beroida Eschscholtz, 1829 **Fam.** Beroidae **Genus** *Beroe* Browne, 1756 *Beroe ovata* Bruguière, 1789	Atlantic coast of North America	Penetrated with currents through the Bosphorus	1999	Established	Isinibilir et al. (2004), Shiganova and Abusova (2021)
Bolinopsis vitrea (L. Agassiz, 1860)	Mediterranean Sea	Penetrated with currents	2009	?	Ozturk et al. (2011)

Table 4.28 Non-native benthic species in the Sea of Marmara

Species	Donor region	Vector	First record	Status of establishment	References
Class Polychaeta **Order** Sabellida **Fam.** Serpulidae *Ficopomatus enigmaticos* Fauvel 1923 (former Mercierella enigmatica)	English Channel, North Sea, French coastal waters	Shipping	1952	Established	Demir (1954)
Phylum Mollusca **Class** Gastropoda **Fam.** Muricidae *Rapana venosa* (Valenciennes, 1846)	Sea of Japan	Shipping and penetration through the Bosphorus from the Black Sea	Before 1980	Established	Ozturk and Ozturk (1996)
Phylum Mollusca **Class** Bivalvia **Order** Venerida **Fam.** Myidae *Mya arenaria* Linne, 1758	North Sea? Atlantic coasts of Europe or North America?	Shipping (ballast waters) or larva penetration with current	Before 1999	Established	Zaitsev and Ozturk (2001)

(continued)

Table 4.28 (continued)

Species	Donor region	Vector	First record	Status of establishment	References
Fam. Teredinidae *Teredo navalis* (Linne, 1758)	Boreal Atlantic?	Fouling or larva transfer with ballast waters or penetration through the Bosphorus	Before 1954	Established	Demir (1954)
Order Mytilida **Fam.** Anadaridae *Anadara inaequivalvis* (Bruguiere, 1789)	North Adriatic Sea, there from Pacific Ocean	Shipping	1970s?	Established	Zaitsev and Ozturk (2001)
Class Crustacea **Order** Decapoda **Fam.** Portunidae *Callinectis sapidus* (Rathbun, 1896)	Atlantic coastal waters of North America?	Shipping (ballast waters)?	Before 1986	Established	Muller (1986)
Class Malacostraca **Order** Decapoda **Fam.** Penaeidae *Penaeus (Marsupenaeus) japonicus* Spence Bate, 1888	Indo-Pacific, may have penetrated into the Mediterranean from the Red Sea (Lessepsian immigrant)	Commercially valuable aquaculture item	1960s	Established	Zaitsev and Ozturk (2001)
Class Echinoidea **Order** Arbacioida *Arbacia lixura* Linnaeus, 1758	Mediterranean and Aegean seas	Penetration?		Established	Öztürk (2006)
Class Asteroidea **Order** Forcipulatida *Asterias rubens* Linnaeus, 1758	Mediterranean Sea	Shipping	1996	Established	Albayrak (1996), Karhan et al. (2007)
Class Malacostraca **Order** Stomatopoda *Erugosquilla massavensis* (Kossmann, 1880)	Indian Ocean	Through the Suez Canal	2002	Established	Katagan et al. (2004)

Pseudocalanus elongates, Acartia clausi, Oithona similis, Paracalanus parvus, and others (Tarkan et al. 2005).

Copepoda *Acartia tonsa* was found in the Marmara Sea in autumn 1996 (Kovalev et al. 1998). *A. tonsa* was re-discovered in the Marmara Sea in 1998 (Unal et al. 2000). In 2000, three species of Copepoda (inhabitants of the Indo-Pacific and the Red Sea, which had previously entered the Levantine Sea as Lessepsian migrants, and then probably penetrated with the currents into the Sea of Marmara) were discovered (Table 4.27).

Among the other representatives of zooplankton, only non-native species of gelatinous macroplankton were observed. The most prominent is ctenophore *Mnemiopsis leidyi*, introduced from the Black Sea, probably in 1988 (Shiganova 1993). Later, in 1999–2000, its predator, the ctenophore *Beroe ovata* was introduced (Isinibilir et al. 2004). The *M. leidyi* arrival resulted in decrease of the abundance, biomass, and species diversity of edible zooplankton (Shiganova et al. 1995). The introduction of *B. ovata*, despite its low abundance in the Sea of Marmara, resulted in a noticeable reduction of the *M. leidyi* population (Isinibilir et al. 2004). Both species spread with the surface current from the Black Sea through the Bosphorus and in the Sea of Marmara. In August–September 2000, another non-native representative of the gelatinous plankton, *Chrysaora hysoscella*, was found in the coastal zone of Erdek Bay in the southern part of the Marmara Sea (Inanmaz et al. 2003). This jellyfish is common in the waters around the British Islands (Russell 1970), in the Mediterranean Sea, the Adriatic Sea (Mills et al. 1996), and off the coast of South Africa (Fearon et al. 1992). It was probably introduced from the Mediterranean Sea to the Sea of Marmara (Inanmaz et al. 2003). *Chrysaora hysoscella* is a predatory jellyfish that feeds on zooplankton, and fish eggs and larvae. This jellyfish also has the stinging cells that are dangerous for humans when contacted. An introduction of such a species is harmful to the ecosystem. In the recent years, the penetration of gelatinous species into the Marmara Sea has intensified. Thus, in 2008, the large jellyfish *Cassiopea andromeda* was discovered, which entered the Mediterranean Sea as a Lessepcian migrant (Ozgur and Ozturk 2008). In 2008, the trachymedusa *Liriope tetraphylla* was discovered, which caused an outbreak of development in the Marmara Sea (Yilmaz and Yuksek 2009). In 2009, the Mediterranean ctenophore *Bolinopsis vitrea* was discovered (Ozturk et al. 2011).

Summarizing, we can conclude that in the 1990s, the species penetrated the Sea of Marmara mainly from the Black Sea. However, in the recent years a new trend has emerged: the penetration of species, first of all gelatinous plankton, from the Mediterranean Sea, including both own Mediterranean species and Lessepcian migrants. Moreover, some of these species have already penetrated further in the near-Bosphorus part of the Black Sea. An example of such species are the predatory scyphomedusa *Chrysaora hysoscella* and the ctenophore *Bolinopsis vitrea*. *Chrysaora hysoscella* was recorded only near the Bosphorus, whereas *Bolinopsis vitrea* was discovered twice in the Bosphorus waters and off Bulgaria (Ozturk et al. 2011).

Benthos

The earliest benthic invader to the Sea of Marmara was probably the *Teredo navalis* shipworm (Demir 1954) (Table 4.28), possibly arrived from the Black Sea, where it was introduced in the Attic period (750–500 BC) (Gomoiu and Skolka 1996) by the Greek conquerors, who appeared in the Mediterranean Sea before 1800 BC (Streftaris et al. 2005). However, their occurrence in the Bosphorus and the Prince Islands (Zaitsev and Ozturk 2001), located near the Bosphorus, may indicate the penetration of pelagic larvae of these worms directly with currents from the Black Sea into the Bosphorus or with the ships possessing wood constructions.

In 1952, the polychaete fouling worm *Ficopomatus enigmaticos* (former *Mercierella enigmatica* (Demir 1954)) was discovered, which had previously been introduced into the Black Sea (Chap. 3, above).

The next benthic non-native species discovered in the Sea of Marmara was the bivalve mollusks *Anadara inaequivalvis*, introduced probably in the 1970s and also noted in the Bosporus and around the Prince Islands (Zaitsev and Ozturk 2001). Anadara was introduced to the Adriatic Sea from the coastal waters of the Philippines (the Pacific Ocean), then to the Black Sea in 1968 (Scarlato and Starobogatov 1972). *A. inaequivalvis* was probably also introduced to the Sea of Marmara from the Black Sea.

Another bivalve mollusk, *Mya arenaria*, was probably introduced into the Marmara Sea at larval stages with the ballast water. Zaitsev and Ostürk believe that *Mya arenaria* was introduced from the North Sea or from the coastal waters of North America (Zaitsev and Ozturk 2001). However, in 1966, *Mya arenaria* was discovered in the Black Sea in the coastal waters off Odessa (Beshevli and Kolyagin 1967). *M. arenaria* has spread widely in the Black Sea (see Chap. 3 above). In the Sea of Marmara, *M. arenaria* is widespread in the area of the Bosphorus at its entrance to the Black Sea and in the Bosphorus itself, where it reached a population of 1300 individuals.m^{-2}. *M. arenaria* was also found in the Prince Islands, where its biomass was very high. In 1999, *M. arenaria* settled mainly on sandy and clay soils in the coastal zone of the Sea of Marmara at depths of 2–15 m. According to the distribution of *M. arenaria* in the Marmara Sea, it is most likely that it was imported from the Black Sea. This was possible both with vessels and by direct penetration with the Black Sea currents at the larval stages. In the Sea of Marmara, young mussels are consumed by kalkan, mullets, and gobies. Adult individuals of *M. arenaria* are consumed by rapana.

The Gastropoda *Rapana venosa* was discovered in the Black Sea in 1946. In 1973, *R. venosa* was discovered in the Adriatic Sea and in 1974 in the Mediterranean Sea. The exact date of rapana's appearance in the Sea of Marmara is unknown. Zaitsev and Ozturk wrote about rapana's arrival from the Black Sea (Zaitsev and Ozturk 2001). Since the 1980s, the *R. venosa* has multiplied intensively in the Bosporus waters, around the islands and in the coastal zones of the Sea of Marmara and in the Dardanelles. In the early 1990s, *R. venosa* was found in large numbers in all Turkish channels, where it practically destroyed all local mollusks (Ozturk and Ozturk 1996). At the same time, *R. venosa* has become an important aquaculture item for Turkey. Since the 1990s, *R. venosa* has been exported by Turkey to Japan and Korea. The annual export volume is estimated at USD 2 million. Annual *R. venosa* meat production was over 4000 tons in 1997. There are 9 temporary and 2 permanent *R. venosa* export facilities on the Turkish coast of the Black Sea (Zaitsev and Ozturk 2001).

The blue crab *Callinectis sapidus* has been observed in the Sea of Marmara by several researchers (Muller 1986; Froglia et al. 1998; EEA 1999). It was discovered in the Black Sea off the Bulgarian coast in 1967 (Bulgurkov 1968), then individual specimens were found in the Kerch Strait in 1975, later in the Gulf of Varna, and in the Bosphorus. The blue crab is native to the Atlantic coastal waters of North America and was introduced to the Black and Marmara Seas with the ballast waters. However, the first findings of the blue crab in the Mediterranean Sea were made in 1951 and in the Adriatic Sea in 1959, so it is possible that the blue crab was introduced to the Sea of Marmara from the Mediterranean Sea (Froglia et al. 1998).

Another non-native representative of Decapoda is the shrimp *Penaeus (Marsupenaeus) japonicus*. In the 1960s, it was imported from the Sea of Japan (Izkenderan Bay) and cultivated as a mariculture. *P. japonicus* is a valuable commercial species, reaching large sizes: the male may be up to 19 cm, and the female up to 23 cm. Currently, *P. japonicus* is found only in the Sea of Marmara (Zaitsev and Ozturk 2001). An attempt to cultivate this shrimp was made in 1954–1962 in the Black Sea in the lower reaches of the Danube, Dniester and Don rivers (Salsky 1964). However, the salinity in the north-western part of the Black Sea turned out to be too low for this species: the shrimp did not take root.

Fishes

The fish *Gambusia holbrooki* (mosquito-eater) was introduced in the Lake Amik (Turkey) and the marshy reservoirs of the Sea of Marmara coast to control mosquitoes. Later, gambusia was acclimatized in other wetlands of the Anatolian coast of Turkey (Zaitsev and Ozturk 2001).

Pilengas *Planiliza haematocheilus* (Temminck and Schlegel, 1845) (former *Mugil soiuy*) was introduced, established, multiplied, and spread widely in the Azov and Black Seas (see Chap. 3 above). In 1989, two adult specimens were for the first time discovered off the Turkish coast of the Black Sea. Currently, *P. haematocheilus* migrates westward, passing through the Bosphorus to the Sea of Marmara and through the Dardanelles to the coastal waters of the Aegean Sea (Zaitsev and Ozturk 2001).

Summarizing available data on the non-native species in the Sea of Marmara, it can be concluded that until the late 1990s, mostly species from the Black Sea penetrated and established into the Sea of Marmara (Zaitsev and Ozturk 2001). Then, in the last decade more and more the Sea of Marmara species began to move in the Black Sea. This may be associated with the increase of temperature and salinity in the Black Sea, or by a change of the environmental conditions of the Marmara Sea ecosystem itself. In the recent years, gelatinous species have been increasingly penetrating, and mucus accumulations have been observed in the surface layers of the Marmara Sea (M. Isinibilir, pers. comm.). This is an alarming sign for the Black Sea, since in the last decade, the Sea of Marmara has become an acclimatization

Fig. 4.83 *Sphyraena pinguis*

place for the Mediterranean species, which than penetrate in the southern part of the Black Sea. Currently, 6 species of macrophytes and 3 species of zoobenthos are already known, which penetrated into the Marmara Sea, and then into the southern part of the Black Sea. In 2009, predatory scyphomedusa *Chrysaora hysoscella* and ctenophore *Bolinopsis vitrea*, which had previously penetrated from the Mediterranean Sea to the Marmara Sea, were found in the Bosphorus region.

Majority of fish species in the Sea of Marmara belong to the Eastern-Atlantic or Mediterranean ichthyofauna and migrate to the Black Sea for feeding or spawning. Some species were attributed as a non-native (see Chapler 3 above): *Sardinella aurita*, *Chenon labrosus, Chelon ramada, Thalassoma pavo, Sarpa sarpa, Sparus aurata, Boops boops, Umbrina cirrhosa, Spicara maena, Gobius auratus, Gobius cruentatus, Gobius xanthocephalus, Parablennius incognitus, Syngnathus acus.* They penetrated in the Black Sea from the Sea of Marmara or Mediterranean via the Sea of Marmara (see Chap. 3, Table 4.19). The harmful species of Indo-Pacific origin, (Lessepsian migrants, i.e., red barracuda *Sphyraena pinguis* Günther, 1874 (Fig. 4.83) and tetraodon *Lalocephalus sceleratus* (Fig. 4.84)) were autonomously penetrated to the south-western Crimean shore of the Black Sea in 2014 (Boltachev and Karpova 2014).

Lagocephalus sceleratus, commonly known as the silver-cheeked toadfish, or Sennin-fugu is an extremely poisonous marine bony fish in the family of Tetraodontidae (puffer fishes). The species is common in the tropical waters of the Indian and Pacific oceans. It is a Lessepsian migrant into the eastern Mediterranean Sea, which it reached through the Suez Canal, and it is spreading towards the western Mediterranean. It had been caught off the Israel, at the south coast of Turkey, in the water around the Cyprus, at the south coasts of mainland Greece, Crete, and Rhodes. In 2013 it was reported in the waters off Lampedusa Island in the central Mediterranean.

Non-native Biota Composition in the Sea of Marmara

From the available data (Tables 4.25, 4.26, 4.27, 4.28, and 4.29), we can conclude that a total of 48 non-native species have been recorded in the Sea of Marmara, including 3 species of phytoplankton, 19 species of macrophytes, 1 species of sea

Fig. 4.84 *Lalocephalus sceleratus*

Table 4.29 Non-native fish species in the Sea of Marmara

Species	Donor region	Vector	First record	Status of establishment	References
Phylum Pisces Osteichthyes Cyprinodontiformes Poeciliidae *Gambusia holbrooki* (Girard 1859)	The coast of Italy, The Adriatic Sea, there from America	Acclimatization in aquariums. Delivered with ships for mosquito control		Established	Zaitsev and Ozturk (2001)
Class Osteichthyes **Order** Mugiliformes **Fam.** Mugillidae Planiliza haematocheilus (Temminck & Schlegel, 1845)	Estuaries of northern part of the Sea of Japan	Spread from the Black Sea through the Bosporus	1989 two adults were caught at Turkish coast of the Marmara Sea	Single individuals occurred	Zaitsev and Ozturk (2001)

grass, 4 species of zooplankton, 6 gelatinous species, 10 benthic species, 2 fish. Obviously, that this uncompleted list and there should be many more non-native species in the Marmara Sea, unreported due to the lack of investigations.

From the Mediterranean Sea, 31% species were introduced in the Sea of Marmara (Fig. 4.85a). Most of them are also non-native species to the Mediterranean Sea itself: 11 of them are Lessepsian migrants which first penetrated into the Aegean or northeastern part of the Mediterranean Sea, and then into the Sea of Marmara or brought by ships from the Indo-Pacific region, 10 species of macrophytes and 1 species of phytoplankton were imported from the Atlantic Ocean, one species of

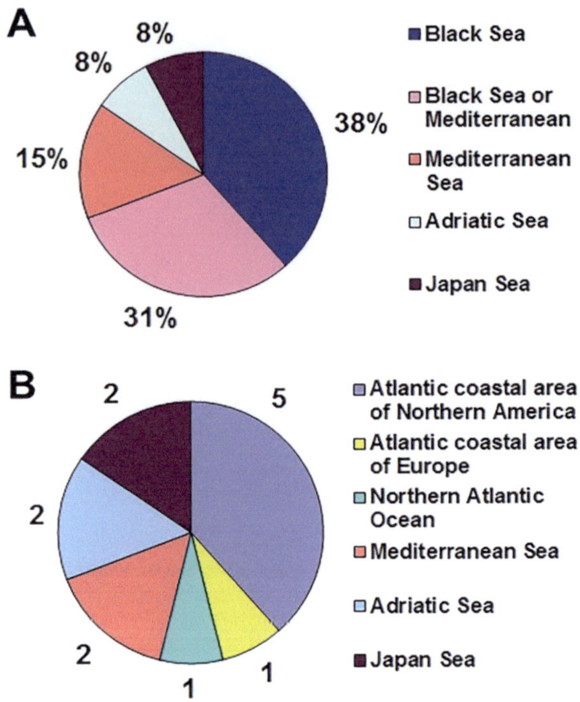

Fig. 4.85 The origin of non-native species in the Sea of Marmara. (**a**) percentage of non-native species in the Sea of Marmara; (**b**) the numbers of species, which penetrated first in the Black Sea and then in the Sea of Marmara. (After Shiganova 2009)

fish *Gambusia holbrooki* was introduced intentionally from the Adriatic Sea basin. Only six native Mediterranean species have entered the Sea of Marmara; they begun to penetrate in the recent years and their number is growing. Thirteen species initially penetrated in the Black Sea and then in the Sea of Marmara (Fig. 4.85b). Among them is important to note ctenophores *Mnemiopsis leidyi* and *Beroe ovata*.

An Introduction of Ctenophores *Mnemiopsis leidyi* and *Beroe ovata* in the Sea of Marmara

M. leidyi has spread from the Black Sea to the Sea of Marmara, where it develops almost to the same sizes as in the Black Sea. The maximum size of the specimen caught in the Sea Marmara was 130 mm. *M. leidyi* established a reproductive population in the Sea of Marmara and in some years (e.g., in 1992) its abundance may be very large (Shiganova 1993). In October 1992, *M. leidyi* was widespread in the Marmara Sea, with an average abundance of 1350 ind.m^{-2}. Its concentration increased toward the Dardanelles, and was especially high in the narrowest part of the Marmara Sea near the Dardanelles. In the Bosphorus, its population was 676–1652 ind.m^{-2}, while in the Dardanelles a population was 2300–2992 ind.m^{-2} abundance. The size and weight of the individuals also increased as one move towards the Dardanelles. The entire population was concentrated in the surface layer above the thermocline. During those observations, small and medium-sized individuals prevailed, especially in the very superficial layer, indicating an intensive reproduction, despite the autumn conditions with the temperature as low as 17–19 C.

Thus, *M. leidyi* individuals penetrated with the currents from the Black Sea and spread throughout the Marmara Sea. While the individuals spread, they grew and multiplied intensively, and as a result, a huge cluster of *M. leidyi* individuals formed near the Dardanelles Strait.

Two specimens of *Beroe ovata* were first discovered in the Marmara Sea in 1999, when it created a first outbreak of abundance in the Black Sea. In the subsequent years, *B. ovata* regularly penetrated through the Bosphorus into the Sea of Marmara and, despite its low abundance, effectively controlled the population of *M. leidyi* (Isinibilir et al. 2004). The abundance of both ctenophores depends on their abundance in the near-Bosphorus region of the Black Sea.

The Aegean Sea

In the northern part of the Aegean Sea, the surface water with a reduced salinity (29‰) is modified Black Sea water supplied via the Dardanelles. This water affects the surface layer (20–30 m) in the northern part of the Aegean Sea (Zervakis et al. 2005). It flows to the south-west (Theocharis et al. 1993; Stergiou et al. 1997), mixing with the Levantine water. Further, the Black Sea water move southward along the eastern coast of the Hellenic Peninsula and after mixing with the saline (38‰) waters enter the southern part of the Aegean Sea.

The northern part of the Aegean Sea has a high bio-productivity, which may be confirmed by the increased concentration of chlorophyll *a* (Fig. 4.86). Oligotrophic conditions increase southward with moving toward the Levantine Sea (Stergiou et al. 1997; Ignatiades et al. 2002; Sempere et al. 2002; Siokou-Frangou et al. 2002). In the northern part of the Aegean Sea, the concentration of chlorophyll *a* range between 29.75 and 50.65 mg.m^{-2}, the primary production varies from 41.48 to 157.14 mg C m^{-2} h^{-1}, while in the southern part of the Aegean Sea, the contents of chlorophyll *a* vary within 20.12 and 30.76 mg m^{-2} and the primary production is within 29.99–42.39 mg C m^{-2} h^{-1} (Ignatiades et al. 2005). It is important to note that the Black Sea waters, supplied to the northern part of the Aegean Sea are enriched in organic nitrogen (Polat and Tugrul 1996); moreover, in the region close to the Dardanelles, an increase in the chlorophyll *a* concentration, primary production, zooplankton biomass, and production of copepods occurs (Siokou-Frangou et al. 2002).

In the surface layer (0–100 m) of the Aegean Sea, the abundance of zooplankton changes from 389 to 1738 ind.m^{-3} in the open waters and from 595 to 3460 ind.m^{-3} in the near-shore area (Siokou-Frangou et al. 2004). Almost all of the zooplankton species in the northern part of the Aegean Sea are of Mediterranean origin.

In the Aegean Sea, one encounters a series of taxonomic groups that are completely absent or rarely occured in the Black Sea such as Doliolids, Salps, Pteropods, Siphonophors, and Euphausiids; therefore, the total biological diversity of the Aegean Sea is higher than that of the Black Sea (Furnestin 1979).

In the Aegean Sea, the stock of commercial fishes is rather high: about 20% of the Turkish catches are those from the Aegean Sea. Selected large pelagic fishes migrate from the Aegean to the Black Sea, among them the bonito *Sarda sarda* (Bloch), the bluefish *Pomatomus saltatrix* (Linne), two mackerel species (*Scomber scombrus*

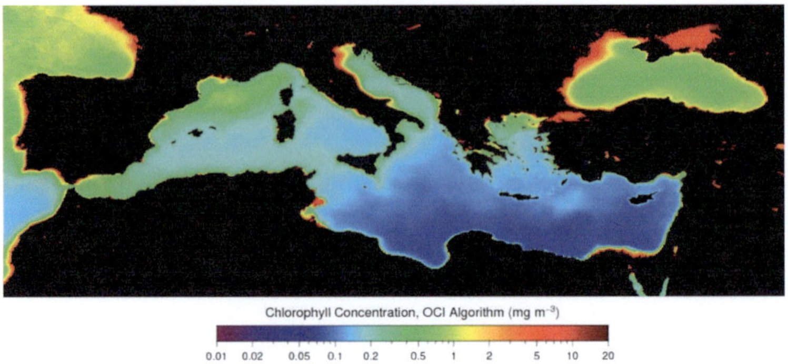

Fig. 4.86 Seasonal climatology of chlorophyll concentration, spring. (NASA Aqua-MODIS data; https://oceandata.sci.gsfc.nasa.gov/l3/)

Linne and *S. japonicus colias* Gmelin), the horse mackerel *Trachurus trachurus* (Linne), and the tuna *Thunnus thunnus* (Linne); in 1954, ten specimens of the tuna *Euthynnus alleteratus alleteratus* (Rafinesque) were recorded even off the coasts of Bulgaria (Svetovidov 1964).

The Mediterranean and Aegean Seas as a Recipient Basin for Non-native Species

The Mediterranean Sea is the first on the list of the seas most receptive to naturalization of the non-native species. At present, the Mediterranean Sea has become the largest recipient basin for non-native species due to its geographical position (Golash 2004). The Mediterranean Sea is connected with the Atlantic Ocean (via the Gibraltar), with the Black Sea (via the Bosporus and Dardanelles), and with the Red Sea, Indian, and Pacific oceans (via the Suez Canal).

After the Suez Canal was opened, the Mediterranean Sea acquired a transport connection with the basins of the Indian and Pacific oceans and the latter oceans connected with the Atlantic Ocean and the Black Sea via the Mediterranean Sea; together with the new navigation routes, new corridors for species invasion appeared.

From the pharaohs' times, canals for navigation have repeatedly been built, though then they passed via the eastern channels of the Nile or via the presently extinct freshwater lagoons in the Nile's delta, which formed barriers that prevented marine fauna from penetrating to the Mediterranean Sea via the canal (Sneh et al. 1975). In contrast to them, the present-day Suez Canal has no obstacle such as the presence of fresh water which stops the further penetration of marine fauna. Meanwhile, features such as the long length of the canal, its shallowness, current's regime and large temperature and salinity increments present no barriers to the invasion of large number of Indian–Pacific origin species from the Red Sea to the Mediterranean Sea. Therefore, organisms, which pass the canal, established in the Mediterranean Sea; first, they form populations along the Levantine coasts and then in other regions of the Mediterranean Sea (www.ciesm.org/atlas/).

In the modern literature, the following terminology is used to attribute the species that penetrate via the Suez Canal, established in it, and propagate farther to the Mediterranean Sea. Some scientists (Calman 1927; Monod 1932; Ben Tuvia 1970, 1983; Galil and Zenetos 2002; etc.) use the term "Erythrean invaders" to indicate the species that migrated via the Suez Canal in the northern direction. Meanwhile, Por (Por 1971) suggested the term "Lessepsian immigrants" for all the Indian–Pacific species that penetrate to the Mediterranean Sea via the Suez Canal in honor of the French engineer Ferdinand de Lesseps, who directed the construction of the Suez Canal (1859–1869).

A series of authors (Galil and Zenetos 2002) apply the term "Lessepsian invaders" to refer to marine organisms that dwell in the Red Sea, penetrate to the Mediterranean Sea via the Suez Canal, established along the Levantine coasts, and are capable of further propagation to the West.

Oliverio discussed the meaning of the term "Lessepsian immigrants" and inferred that it is reasonable to apply this term to indicate the species that penetrated via the Suez Canal, established in it, and then widely spread over various regions of the Mediterranean Sea (Oliverio 1995). In our study, similar to other recent publications (Gofas and Zenetos 2003), we join these two terms into a single term "Lessepsian migrants" or "Lessepsian immigrants" to specify the species that naturally penetrated to the Suez Canal from the Red Sea, are regularly encountered in it, and gradually spread over the eastern part of the Mediterranean Sea.

In addition to the direct propagation of organisms via the Suez Canal, many Indian–Pacific species are brought to the Mediterranean and Black seas and further to various regions of the Atlantic Ocean with ballast waters and with ship fouling communities.

The Mediterranean Sea is also subjected to the invasion of non-native animals and plants species from the Atlantic Ocean via the Gibraltar and from the Black Sea via the Bosporus and Dardanelles.

By December 2005, the Mediterranean Sea has acquired 745 valid established non-native species, though 98 of them are doubtful. This comprises more than 50% of the total number of invaders to the European seas. Of them, 385 species establishes and formed their populations, 262 species are encountered in low abundances or are represented by isolated individuals (Zenetos et al. 2005; Streftaris et al. 2005).

The best-known non-native species that strongly influenced the ecosystem of the Mediterranean Sea is the green bottom alga *Caulerpa taxifolia.* With respect to its origin, it is a decorative aquarium alga, which was brought by aquarium operators to the State Aquarium of the Oceanographic Museum of Monaco and has been cultivated in it since 1982 (Meinesz and Hesse 1991). Then, it penetrated into the near-shore waters and rapidly spread over the western part of the Mediterranean Sea with boat anchors and fish nets. It occupied more than 6000 ha of the near-shore aquatic areas off Herzegovina, France, Monaco, Spain, Italy, and Tunisia, where it crucially altered the condition of the bottom substrate and significantly affected the ecosystems of these regions (Thibaut and Meinesz 2004; Pancucci-Papadopoulou et al. 2005). Probably, *Caulerpa taxifolia* might penetrate into the eastern part of the Mediterranean Sea as well, though the thickets of *Posidonia oceanica* widely spread in the eastern part of the Mediterranean Sea serve as a natural barrier that prevents *Caulerpa taxifolia* from colonizing this region (Occhipinti-Ambrogi and Savini 2003). *Posidonia oceanica*, commonly known as Neptune grass or Mediterranean tape weed, is a sea grass species that is endemic to the Mediterranean Sea. It forms large underwater meadows that are an important part of the ecosystem.

Owing to its geographical position, the eastern part of the Mediterranean Sea and, in particular, the Aegean Sea accepts many invading species. On the one hand, selected Black Sea species and Black Sea invaders arrive here directly with currents via the Bosporus, the Sea of Marmara, and the Dardanelles. On the other hand, large numbers of species have penetrated and continue penetrating from the Red Sea and the Indian and Pacific oceans via the Suez Canal; many of them also propagate to the Aegean Sea. Finally, the Aegean Sea is crossed by numerous navigation routes from the Atlantic Ocean to the Black Sea and from various regions of the World Ocean to the Mediterranean and Black seas and to the Atlantic via the Suez Canal. This provides conditions favorable for non-native species penetration with ballast waters and fouling communities from any region of the World Ocean. Additional sources for the non-native species appearance are aquaculture, aquarium keeping, and tourism, which are widely spread in the near-shore regions of the entire Mediterranean Sea and, in particular, in the Aegean Sea.

Non-native Species in the Aegean Sea and Their Role in the Ecosystem
Composition of non-native species in the Aegean Sea by taxonomic groups is presented in the Tables 4.30, 4.31, 4.32, and 4.33.

Phytoplankton and Macrophytes
To date, representatives of 12 non-native phytoplankton species have been registered in the Aegean Sea. Among them are the diatom *Pseudosolenia calcar-avis* and the dinoflagellates *Alexandrium taylori, A.andersonii, A. catenella, A. minutum, Gymnodinium catenatum, G. Fusus, G. mikitomoi, Gonyaulax grindley, Trichodesmium erythreum, Coolia monotis,* and *Gonyaulax grindley.*

In this study, we focused also on the species that invaded the Black Sea. In the eastern part of the Mediterranean Sea, the diatom alga *Pseudosolenia calcar-avis,* which was first detected in Bodrum Bay of the Levantine Sea, it was also observed in the Aegean Sea (Koray and Kesici 1994). *Pseudosolenia calcar-avis* is widely spread in the Atlantic and Pacific oceans; probably, it arrived in 1928 from the North Atlantic to the Black Sea, where it became a mass species, featuring significant recurring outbursts of abundance (see Chap. 3, above). Later, *P. calcar-avis* was encountered in the Sea of Marmara and then in the eastern part of the Mediterranean Sea. *P. calcar-avis* might be brought to this region both from the Black Sea and from the North Atlantic.

The dinoflagellate *Alexandrium monilatum* was probably brought to the Mediterranean Sea in 1985 from the near-shore Atlantic waters off North America; Turkish scientists suggest that it also dwells in the Aegean Sea (Gotsis-Scretas and Friligos 1990). In 1991, it was encountered in Varna Bay in the Black Sea (Moncheva et al. 1995).

Table 4.30 Non-native species of phytoplankton and macrophytes in the Aegean Sea

Species	Donor region	Vector	First record	Status of establishment	References
Phylum Bacillariophyta **Fam.** Soleniaceae *Pseudosolenia calcar-avis* (M. Schultze) Sentstrom 1986 (=*Rhizosolenia calcar-avis* (*M.Schultze*)	North Atlantic	?	1994	Established	Koray and Kesici (1994)
Fam. Chaetocerotaceae *Chaetoceros coarctatus* Lauder	Atlantic, Pacific oceans ?	Shipping	1987	Established	Cinar et al. (2005)
Class Dinophyceae **Order** Peridiniales **Fam.** Gonioaulaca-ceae *Alexandrium monilatum* (Howell) (F.J.R.Taylor 1979)	Atlantic coast of North America	Shipping (ballast waters)	1985	Established	Gotsis-Scretas and Friligos (1990)
Class Prymnesiophyceae **Order** Prymnesiales Phaeocystaceae *Phaeocystis pouchetii* (Hariot) Lagerheim, 1893	Unknown	Shipping? Cosmopolitan?	1989	Established	Skretas and Friligos (1990)
Class Chlorophyceae **Order** Bryopsidales Fam. Caulerpaceae *Caulerpa racemosa var. lamourouxii* (Turner)	Indo-Pacific, Red Sea	Shipping (through the Suez Canal; Lessepsian immigrant)	1950s	Established	Cirik and Ozturk (1991)

(continued)

Table 4.30 (continued)

Species	Donor region	Vector	First record	Status of establishment	References
Siphonophyceae **Class** Chlorophyceae **Order** Siphonales **Fam**. Bryopsidaceae ***Codium flagile*** Suringar 1867	Atlantic or Pacific oceans	Incidentally	1950s	Established	Zaitsev and Ozturk (2001)
Class Rhodo-phyceae Ceramiales Rhodomelaceae *Acanthophora muscoides* (Linnaeus, 1753) Bory de Saint-Vincent	Indo-Pacific, Red Sea	Through the Suez Canal; Lessepsian immigrant	Before 1977	Established	Aysel (1997)
Class Rhodo-phyceae **Order** Ceramiales Oltm. **Fam**. Rhodomelaceae *Laurencia intermedia* (Yamada 1931)	Indo-Pacific, Red Sea	Through the Suez Canal; Lessepsian immigrant	?	?	Zaitsev and Ozturk (2001)
Class Rhodo-phyceae Ceramiales Phodomelaceae	Red Sea or Indian Ocean	Penetration through the Suez Canal	Before 1987	Established	Athanasiadis (1987)
Chondria collinsiana M.F. Howe *C. polyrhiza* Collins et Hervey	Indian Ocean	Shipping (fouling)	Before 1987	Established	Athanasiadis (1987)
Class Rhodo-phyceae Gigartinales Hypneaceae	Indian Ocean	Shipping (fouling)	Before 1987	Established	Athanasiadis (1987)

Hypnea spinella (C. Agardh) Kutzing *H. valentiae* (Turner) Montagne	Indian Ocean	Shipping (fouling)	Before 1987	Established	Athanasiadis (1987)
Class Rhodo-phyceae **Order** Bonnemaisoniales **Fam**. Bonnemaiso-niaceae *Asparagopsis armata* Harvey	Atlantic Ocean ?	Shipping (fouling)	Before 1971 In the Mediterranean in the 1920s	Established	Guven and Oztig (1971), Schnetter and Schnetter (1981)
Class Rhodo-phyceae Ceramiales Rhodomelaceae *Womersleyella setacea* (Hollenberg) R.E. Norris	From other areas of the Mediterranean Sea	Shipping (fouling)	Before 1997	Established	Athanasiadis (1987)
Class Rhodo-phyceae *Hypnea esperi auctorum*	Red Sea	Penetration through the Suez Canal	1972	Established	Lipkin (1972)
Class Rhodo-phyceae *Lophocladia lallemandii* (Mont) F. Schmitz	Red Sea	Penetration through the Suez Canal	1918	Established	Petersen (1918)
Class Phaeophyceae *Cladosiphon zosterae* (J. Agardh) Kylin	Atlantic Ocean	With mariculture	1988	Established	Verlague (2001)
Class Phaeophyceae *Stypopodium schimperi* (Kutz) Verlague	Red Sea	Penetration through the Suez Canal	1982	Established	Mayhoub (1989)
Class Spermatophyta *Halophila stipulacea* (Forssk.) Ascherson	Red Sea	Penetration through the Suez Canal	1894	Established	Fritsch (1895)

Table 4.31 Non-native species of planktonic invertebrates in the Aegean Sea

Species	Donor region	Vector	First record	Status of establishment	References
Class Coelenterata **Order** Riscomedusae **Fam.** *Rhopilema nomadica* Galil 1990	Red Sea	Through the Suez Canal; Lessepsian immigrant	1977	Established	Galil (2000)
Class Coelenterata **Class** Scyphozoa **Order** Rhizostomeae **Fam.** Cassiopeidae *Cassiopea andromeda* (Forsskal 1775)	Pacific Ocean Indian Ocean	Through the Suez Canal; Lessepsian immigrant	2008	Established	Ozgurk and Ozturk (2008)
Class Ctenophora **Order** Lobata Esch **Fam.** Bolinopsiidae **Genus** *Mnemiopsis* L. Agassiz *Mnemiopsis leidyi* A. Agassiz, 1865	Atlantic coast of North America	Penetrated with currents via Bosphorus and Dardanelles. In bays with ballast water	1990	Appears regularly with Black Sea current	Shiganova et al. (2001a, b)
Phylum Ctenophora Order Beroida Eschscholtz, 1829 **Class Nuda Fam.** Beroidae **Genus** *Beroe* Browne, 1756 Beroe ovata Bruguière, 1789	Atlantic coast of North America	Penetrated with currents via Bosphorus and Dardanelles	1999	Appears regularly with Black Sea current	Shiganova et al. (2007a, b)

(continued)

Table 4.31 (continued)

Species	Donor region	Vector	First record	Status of establishment	References
Phylum Ctenophora Order Beroida Eschscholtz, 1829 **Class Nuda Fam.** Beroidae **Genus** *Beroe* Browne, 1756 *Beroe mitrata* (Moser, 1907)	Mediterranean Sea	Penetrated with currents	?	Established Occurring regular	Shiganova and Abusova, (2021)
Class Crustacea **Order** Copepoda **Fam.** Acartiidae *Acartia tonsa* (Dana, 1849)	Atlantic coast of North America ?	Shipping (ballast waters)	Before 1997	Established	Tarkan et al. (2001)

Table 4.32 Non-native benthic species in the Aegean Sea

Species	Donor region	Vector	First record	Status of establishment	References
Class Crustacea **Order** Decapoda **Fam.** Portunidae *Callinectis sapidus* (Rathbun, 1896)	Atlantic coast of North America ?	Shipping (ballast waters) ?	1940s	Established	Giordani and Soika (1951), Rield (1983)
Class Crustacea **Order** Decapoda **Fam.** Penaeidae *Penaeus semisulcatus* de Haan 1844	Indo-West Pacific	Penetration via Suez Canal	Before 1991	Established	Kocatas et al. (1991)
Class Crustacea **Order** Decapoda **Fam.** Brachyura *Ixa monody* Holthuis & Gottlieb, 1956 (pebble crab)	Red Sea, eastern coastal waters of Africa, Persian Gulf	Through the Suez Canal; Lessepsian immigrant	Before 1956	Established	Holtius and Gottlieb (1956)

(continued)

Table 4.32 (continued)

Species	Donor region	Vector	First record	Status of establishment	References
Class Crustacea **Order** Decapoda **Fam.** Brachyura *Charybdis (Goniohellenus) longicollis* Leene, 1938	Red Sea, Indian and Pacific Oceans	Through the Suez Canal; Lessepsian immigrant	Before 1961	Established	Holtius (1961)
Class Crustacea **Order** Decapoda **Fam.** Brachyura *Talamita poissonii* (Audouin, 1826)	Indo-West Pacific	Penetration via Suez Canal		Established	http://www.ciesm.org/atlas
Class Crustacea **Order** Decapoda **Fam.** Xanthidae *Pilumnopeus vauquelini* (Audouin, 1826)	Red Sea, Persian Gulf	Through the Suez Canal; Lessepsian immigrant		Established	http://www.ciesm.org/atlas
Class Crustacea **Order** Decapoda *Marsupenaeus japonicus*	Indo-Pacific	Through the Suez Canal; Lessepsian immigrant	1970s	Established	http://www.ciesm.org/atlas
Class Crustacea **Order** Stomatopoda *Erugosquilla massavensis* (Kossmann, 1880)	Red Sea, Persian Gulf	Through the Suez Canal; Lessepsian immigrant		Established	http://www.ciesm.org/atlas
Class Polychaeta **Order** Serpulimorpha **Fam.** Serpulidae *Hydroides dianthus* (Verrill, 1873)	Northwest Atlantic	Shipping	Before 1999	Established	Kocak et al. (1999)
Phylum Mollucca **Class** Gastropoda **Order** Prosobranchia **Fam.** Nacellidae	Red Sea, Indian and Pacific Oceans	Through the Suez Canal; Lessepsian immigrant	1989	Established	Fountoulakis and Sabelli (1999)

(continued)

Table 4.32 (continued)

Species	Donor region	Vector	First record	Status of establishment	References
Cellana rota (Gmelin, 1791)					
Fam. Neritidae *Smaragdia souverbiana* (Montrouzier, 1863)	Red Sea	Through the Suez Canal; Lessepsian immigrant	Before 1996	Established	Buzzurro and Greppi (1966)
Nerita sanguinolenta Menke, 1829	Red Sea		1994		Nordsieck (1973)
Fam. Trochidae *Trochus erythraeus* Brocchi, 1821	Red Sea	Through the Suez Canal; Lessepsian immigrant	1994	Established	Cosenza and Fasulo (1997)
Fam. Strombidae *Strombus persicus* Swainson 1821	Persian Gulf	Shipping	1986	Established	Nicolay (1986)
Phylum Mollusca **Class** Gastropoda **Fam.** Calyptraeidae *Crepidula fornicata* (Linnaeus, 1758)	Atlantic Ocean ?	Shipping or introduction with cultivated oysters and mussels	1994	Established	Dellamotte and Vardala-Teodorou (1994)
Phylum Mollusca **Class** Gastropoda **Fam.** Muricidae *Murex forskoehli* Roeding, 1798	Red Sea	Through the Suez Canal; Lessepsian immigrant	1966	Established	Settepassi (1967)
Phylum Mollusca **Class** Gastropoda **Fam.** Muricidae *Rapana rapiformis* (Von Born, 1778)	Indian and Pacific Oceans	Through the Suez Canal; Lessepsian immigrant	1970	Established	Barash and Danin (1988/89)
Phylum Mollusca **Class** Gastropoda **Fam.** Muricidae *Rapana venosa*	Japan Sea	Shipping, possibly spreading from the Black Sea	Before 1990s	Established	Koutsoubas and Voultsiadou-Koukoura (1991)

(continued)

Table 4.32 (continued)

Species	Donor region	Vector	First record	Status of establishment	References
(Valenciennes, 1846)					
Phylum Mollusca **Class** Gastropoda **Order** Opisthobranchia **Fam.** Bullidae *Bulla ampulla* L. 1758	Indian and Pacific Oceans	Through the Suez Canal; Lessepsian immigrant	1999	Established	Vardala-Theodorou (1999)
Phylum Mollusca **Class** Gastropoda **Fam.** Bullidae *Pyrunculus fourierii* (Audouin, 1826)	Through the Suez Canal	Penetration	Before 1990	Established	Tringali and Villa (1990)
Fam. Haminoeidae *Haminoinoea cyanomarginata* Heller &Thompson, 1983	Through the Suez Canal	Penetration Lessepsian immigrant	2001	Established	Gosliner & Mollo (cited after Zenetos et al. 2005)
Fam. Cylichnidae *Acteocina mucronata* (Philippi, 1849)	Through the Suez Canal	Penetration Lessepsian immigrant	1997	Established	Storsberg (1997)
Fam. *Cerithium scabridum*	Through the Suez Canal	Penetration Lessepsian immigrant		Established	http://www.ciesm.org/atlas
Fam. Aplysiidae *Bursatella leachi* De Blainville, 1817	Red Sea, Indian Ocean	Penetration through the Suez Canal	1986	Established	Barash and Danin (1986)
Fam. Polyceridae *Polycerelia emertoni* Verrill, 1881	Atlantic Ocean	Shipping	1995	Established	Koutsoubas et al. (2000)
Fam. Tethydae *Melibe fimbriata=M. viridis* Alder & Hancock1864	Indian and Pacific oceans	Through the Suez Canal; Lessepsian immigrant	1982	Established	Thompson and Crampton (1984)

(continued)

Table 4.32 (continued)

Species	Donor region	Vector	First record	Status of establishment	References
Phylum Mollusca **Class** Gastropoda **Fam.** Retusidae *Cylichnina girardi* (Audouin, 1826)	Indian and Pacific oceans	Through the Suez Canal Eritrean immigrant	1994	Established	Buzzurro and Greppi (1966)
Phylum Mollusca **Class** Bivalvia **Fam.** Arcidae *Anadara demiri* (Piani, 1981)	Chinese seas	Shipping, through the Suez Canal	1993	Established	Zenetos (1994)
Fam. Mytilidae *Brachidontes pharaonis* (Fischer P., 1870)	Red Sea, Indian Ocean	Through the Suez Canal Eritrean immigrant	1975	Established	Koroneos (1979)
Fam. Ostreidae *Crassostrea gigas* (Thunberg, 1793)	Indian and Pacific oceans	Aquaculture	1989	Established	Dimitrakis (1989)
Fam. Pteridae *Pinctada radiata* (Leach, 1814)	Red Sea, Indian Ocean	Through the Suez Canal Eritrean immigrant; aquaculture	1963	Established	Serbetis (1963)
Fam. Malleidae *Malvufundus regulus* (Forskal, 1775)	Red Sea, Indian Ocean	Through the Suez Canal Eritrean immigrant	2001	Established	Giannuzzi-Savelli et al. (2001)
Fam. Chamidae *Pseudochama corbieri* (Jonas, 1846)	Red Sea	Through the Suez Canal; Lessepsian immigrant	1946	Established	Ralli-Tzelepi (1946)
Fam. Cardiidae *Fulvia fragilis* (Forsskai in Niebuhr, 1775)	Red Sea, Indian and Pacific oceans	Through the Suez Canal; Lessepsian immigrant	1999	Established	Vardala-Theodorou (1999)
Fam. Petricolidae *Petricola pholadiformis* Lamarck, 1818	West Atlantic	Shipping	1994	Established	Delamotte and Vardala-Theodorou (1994)

(continued)

Table 4.32 (continued)

Species	Donor region	Vector	First record	Status of establishment	References
Fam. Myiidae *Mya arenaria* L.1758	East Atlantic, Black Sea?	Shipping	1984	Established	Zenetos et al. (2005)
Fam. Gastrochaenidae *Gastrochaena cymbium* (Spengler, 1783)	Indo-Pacific	Through the Suez Canal Eritrean immigrant	1974	Established	Tenekides (1989)

Golden alga *Phaeocystis pouchetii* probably originates from the North Atlantic. Its blooming was first registered in the Aegean Sea in February–March 1989 (Gotsis-Scretas and Friligos 1990). Simultaneously, in 1989 *Phaeocystis pouchetii* was also found in the western part of the Black Sea (Petrova-Karadjova 1990). However, at present, this species is removed from the list of the Mediterranean invaders because it has been recognized as a cosmopolitan species (Afgiro Zenetos, pers. comm.).

Among the non-native species of macrophytes in the Aegean Sea there is no any species of the Black Sea origin. The green macrophytes algae are presented by *Caulerpa racemosa var. lamourouxii*. Originally described in Suez (the Red Sea) by Montagne (1856, p. 454) as *Herpochaeta requienii*, later transferred by Agardh (1873) to the genus *Caulerpa* and then relegated by Weber-van Bosse (1898) to a form of *C. racemosa* under *var. lamourouxii*. It differs from the type form by simple or irregularly divided uprights rather than dichotomously divided ones and is not known to produce vesicular side branches. Doubt attaches to the status and molecular identity of form *requienii* from the Red Sea (Olivier De Clerck and Thomas Sauvage, pers. comm.). Ongoing molecular sequencing of the Dubai material and comparison to Red Sea Caulerpa species should shed further light on its taxonomic identity. Besides *C. requienii,* there are two other forms of *C. racemosa* in the Mediterranean of which one is considered by Verlaque et al. (2000, 2003) to be "invasive." Morphological and molecular-genetic studies (see Kleinand Verlaque 2008, Verlaque et al. 2003, Renoncourt and Meinesz 2002) on this invasive form have shown it to be *C. racemosa var. cylindracea* (Sonder) Verlaque, Huisman & Boudouresque, an endemic from south-western Australia. In 1950, another green alga *Codium flagile* was encountered in the Mediterranean Sea. Its primary donor region was the northwestern part of the Pacific Ocean. From there, it was brought to the North Sea in 1900, to the Baltic Sea in 1932, and to the North Atlantic in 1946 (Streftaris et al. 2005), where from it probably migrated to the Mediterranean Sea.

The red alga *Asparagopsis armata* is a cosmopolitan species probably arrived from the Atlantic Ocean. First encountered in the 1920s in the near-shore waters off Algeria (Sauvageau 1925), it is now widely spread in the Mediterranean Sea. In the Aegean Sea, it is recorded in the near-shore waters off Turkey and Greece throughout the year (Guven and Oztig 1971; Schnetter and Schnetter 1981).

Table 4.33 Non-native fish species in the Aegean Sea

Species	Donor region	Vector	First record	Status of establishment	References
Phylum Pisces **Class** Osteichthyes **Order** Perciformes **Fam.** Carangidae *Alepes djeddaba* (Forsskal, 1775)	Indo-Pacific	Through the Suez Canal; Lessepsian immigrant	Before 1960	Established	Bini (1960)
Order Perciformes **Fam.** Callionimidae *Callionymus filamentosus* Valenciennes 1837	Indo-Pacific	Through the Suez Canal; Lessepsian immigrant	1994	Established	Corsini et al. (2005), Pancucci et al. (2005)
Fam. Apogonidae *Apogon nigripinnis* Cuvier, 1828	Indo-Pacific	Through the Suez Canal; Lessepsian immigrant	1987	Established	Golani et al. (2002)
Class *Osteichthyes* ***Order*** *Atheriniformes* ***Fam.*** *Atherinidae* *Atherinamorus lacunosus* (Froster, in Bloch and Schneider, 1801)	Indo-Pacific	Through the Suez Canal; Lessepsian immigrant	Before 1986	Established	Whitehead et al. (1986)
Class Osteichthyes **Order** Beloniformes **Fam.** Hemiramphidae *Hemiramphus far* (Forsskal, 1755)	Indo-Pacific	Through the Suez Canal; Lessepsian immigrant	Before 1966	Established	Ben-Tuvia (1966)
Class Osteichthyes **Order** Beloniformes **Fam.** Exocoetidae *Parexocoetus mento* (Valenciennes, 1846)	Indo-Pacific	Through the Suez Canal; Lessepsian immigrant	Before 1966	Established	Ben-Tuvia (1966)

(continued)

Table 4.33 (continued)

Species	Donor region	Vector	First record	Status of establishment	References
Class Osteichthyes **Order** Tetraodontiformes **Fam.** Tetraodontidae *Lagocephalus spadiceus* (Richardson, 1844)	Indo-Pacific	Through the Suez Canal; Lessepsian immigrant	1950	Established	Whitehead et al. (1986)
Class Osteichthyes **Order** Perciformes **Fam.** Leiognathidae *Leiognathus klunzingeri* (Steindachner, 1898)	Indo-Pacific	Through the Suez Canal; Lessepsian immigrant	1980	Established	Whitehead et al. (1986)
Class Osteichthyes **Order** Perciformes **Fam.** Mugilidae *Liza carinata* (Valenciennes, 1836)	Indo-Pacific	Through the Suez Canal; Lessepsian immigrant	1956	Established	Whitehead et al. (1986)
Class Osteichthyes **Order** Mugiliformes **Fam.** Mugillidae *Planiliza haematocheilus* (Temminck & Schlegel, 1845) (=*Mugil soiuy* (Basilewsky 1855)	Estuaries of northern part of the Sea of Japan	From the Black Sea through the Bosporus	1997	Migration	Kaya et al. (1998)
Class Osteichthyes **Order** Beryciformes **Fam.** Holocentridae *Sargocentron rubrum* (Forsskal, 1775)	Indo-Pacific	Through the Suez Canal; Lessepsian immigrant	Before 1966	Established	Ben-Tuvia (1966)

Class Osteichthyes **Order** Aulopiformes **Fam.** Synodontidae *Saurida undosquamis* (Richard-son, 1948)	Indo-Pacific	Through the Suez Canal; Lessepsian immigrant	Before 1971	Established	Ben-Tuvia (1971)
Class Osteichthyes **Order** Perciformes **Fam.** Pempheridae *Pempheris vanicolensis* Cuvier, 1831	Indo-Pacific	Through the Suez Canal; Lessepsian immigrant	Before 1978	Established	Ben-Tuvia (1978)
Class Osteichthyes **Order** Perciformes **Fam.** Scombridae *Scomberomorus commerson* (Lacepede, 1802)	Indo-Pacific	Through the Suez Canal; Lessepsian immigrant	1981	Established	Papaconstantinou (1987)
Class Osteichthyes **Order** Perciformes **Fam.** Siganidae *Siganus luridus* (Ruppell, 1878)	Indo-Pacific	Through the Suez Canal; Lessepsian immigrant	Before 1987	Established	Papaconstantinou (1987)
Class Osteichthyes **Order** Perciformes **Fam.** Siganidae *Siganus rivulatus* (Forsskal, 1775)	Indo-Pacific	Through the Suez Canal; Lessepsian immigrant	Before 1986	Established	Whitehead et al. (1986)
Class Osteichthyes **Order** Perciformes **Fam.** Sphyraenidae *Sphyraena chrysotaenia* (Klunzinger, 1884)	Indo-Pacific	Through the Suez Canal; Lessepsian immigrant	Before 1966	Established	Ben-Tuvia (1966)

(continued)

Table 4.33 (continued)

Species	Donor region	Vector	First record	Status of establishment	References
Class Osteichthyes **Order** Tetraodontiformes **Fam.** Monacanthidae *Stephanolepis diaspros* (Fraser-Brunner, 1940)	Indo-Pacific	Through the Suez Canal; Lessepsian immigrant	Before 1978	Established	Por (1978)
Class Osteichthyes **Order** Perciformes **Fam.** Mullidae *Upeneus moluccensis* (Bleeker, 1855)	Indo-Pacific	Through the Suez Canal; Lessepsian immigrant	Before 1986	Established	Whitehead et al. (1986)
Fam. Fistulariidae *Fistularia commersonii* Ruppell, 1838		Through the Suez Canal from Red Sea	2002	Established	Corsini et al. (2002)
Class Osteichthyes **Order** Cyprino-dontiformes **Fam.** Poeciliidae *Gambusia holbrooki* (Girard 1859)	The coast of Italy, The Adriatic Sea, there from America	Acclimatization in aquariums Delivered with ships for the purpose of mosquito control	2001	Established	Zaitsev and Ozturk (2001)

Also, in the near-shore waters, first off Greece and later off Turkey, the red alga *Chondria collinsiana* was registered; it is a pantropic species that probably penetrated via the Suez Canal (Athanasiadis 1987; Ausel et al. 1993). Another species of this genus, *C. polyrhiza* was also found near the coasts of Greece (Athanasiadis 1987). Isolated findings of two more species of red alga *Acanthophora muscoides* and *Laurencia intermedia* were reported in the eastern part of the Aegean Sea (Aysel 1997). A single individual of *Acanthophora muscoides* was found in Izmir Bay on rocky grounds; it is a Lessepsian immigrant that penetrated via the Suez Canal. An individual of *Laurencia intermedia* was encountered in the Karakale region on the rocky grounds. This alga might either penetrate from the Red Sea via the Suez Canal or be brought by a ship from the Indian-Pacific region (Zaitsev and Ozturk 2001).

The red algae *Hypnea spinella*, *H. Valentiae,* and *Womersleyella setacea* were found in the near-shore waters off Greece. They arrived here from other regions of the Mediterranean Sea being initially brought to the Mediterranean Sea with the ship's fouling (Athanasiadis 1987). The red algae *Hypnea esperi auctorum* and *Lophocladia lallemandii* are Lessepsian migrants from the Red Sea.

In addition to the above-mentioned species, the representatives of Phaeophyceae *Cladosiphon zosterae* and *Stypopodium schimperi* (= *S. tubruqense*, also referred to as *S. zonale*), which also penetrated from the Red Sea, should be noted.

Summarizing the information about non-naive species of macrophytes in the Aegean Sea and their origination, we can conclude that all of them were either brought with ship's fouling from the Atlantic, Indian, or Pacific Oceans or penetrated with currents via the Suez Canal from the Red Sea and the Indian and Pacific oceans as Lessepsian immigrants (Cormaci et al. 2004).

Zooplankton

As it was mentioned above, some Black Sea planktonic species episodically appear in the region of the Aegean Sea near the Dardanelles, which is influenced by the transformed Black Sea water. Among them is the copepod *Pseudocalanus elongatus* Boeck; although some scientists regard this species as an invader (Pancucci-Papadopoulou et al. 2005), it is not. First, regarding its origin, it is a Mediterranean species, penetrated to the Black Sea and second, it appears in the Aegean Sea only episodically and in small abundances. There is no information about its ability to form an independent reproductive population in the Aegean Sea.

Another representative of copepod, *Acartia tonsa* was recorded in the Mediterranean Sea in the mid-1980s (Belmonte et al. 1994). In the Aegean Sea, *Acartia tonsa* was found relatively recently in Izmir Bay and off the Crete coast (Kocak et al. 1999). Later, *Acartia tonsa* was recorded in the northern part of the Levantine Sea. Recently it was found that *Acartia tonsa* was first brought to the Black Sea in the early 1970s (Gubanova 2000). Then it was registered in the Sea of Marmara (Kovalev et al. 1998) and later in other regions of the Mediterranean Sea, including the Adriatic Sea, the western part of the Mediterranean Sea, the near-shore waters off France, and in lagoons with high water salinity, where it could reach high abundances. In the Berre Lagoon, *Acartia tonsa* became a dominant species (Gaudy et al. 2000).

Some zooplankton species like the copepods *Calanopia elliptica* and *Centropages furcatus* and the echinoderm *Ophiactis savignyi* appeared in the eastern part of the Mediterranean Sea as Lessepsian immigrants (Pancucci-Papadopoulou et al. 2005). Recently, this list has been complemented by new non-native representatives of zooplankton, though this information has not been published yet (Argiro Zenos, pers. comm.). Among the other known non-native zooplankton species, the representatives of gelatinous plankton, ctenophore *Mnemiopsis leidyi* should be mentioned. It is regularly arriving from the Black Sea via the Bosphorus and Dardanelles (Shiganova et al. 2001a, b). *M. leidyi* was first registered in the northern part of the Aegean Sea in the spring-summer of 1990 in the Saronikos Bay and then in Elephsis Bay, located nearby; at that time its abundance was the highest. In the subsequent years, small amounts of *M. leidyi* were regularly observed in the region, close to the Dardanelles off the Limnos Islands and in bays and near-shore zone of the northern part of the Aegean Sea (Shiganova et al. 2001a, b). *M. leidyi* was also recorded in Kuzadashi Bay of the Turkish coast of the Aegean Sea (Yusal and Multu 1993; Kideys and Niermann 1994).

In addition, many both large and small individuals of *M. leidyi* were found in Mersin Bay of the Levantine Sea in the late spring of 1992, though no *M. leidyi* were observed in this region in the subsequent years. A few *M. leidyi* individuals were recorded off the Syrian coasts in October 1993 (Shiganova 1997).

Analyzing the appearance and distribution features of *M. leidyi* in the Sea of Marmara and the Aegean Sea, we can suggest that it spreads with the water coming from the Black Sea via the Bosporus and Dardanelles (Shiganova et al. 2001a, b). Also, it could arrive with the ballast waters to Saronikos Bay, where the largest Greek port Piraeus is located. Moreover, *M. leidyi* was recorded near the major ports of Mersin and Latakia (Kideys and Niermann 1994; Shiganova 1997).

The ctenophore *Beroe ovata* was first observed in the Aegean Sea in 2004 in the northern part of Evoikos Bay among the aggregations of large *M. leidyi* individuals (Shiganova et al. 2007a, b).

Another ctenophore, *Beroe mitrata* (Fig. 4.87) was recorded in the Ligurian Sea (Madin et al. 1991) off VilleFranche coast (Tamm and Tamm 1993). Mills also has mentioned *B. mitrada* as an infrequent visitor in the Mediterranean (Mills et al. 1996).

For the first time our observations identified *Beroe mitrata* in Turkish Bays—Gökova, Marmaris, Bodrum and Fethye Bays (Killi et al. 2019; Shiganova and Abyzova 2021). In the Gökova Bay both *M. leidyi* and *B. mitrata* first arrived from the northern Aegean Sea with the southward currents (Gülşahin and Tarkan 2014). Both ctenophores were observed in the Gökova Bay regularly.

Species of Beroe, inhabiting in the Mediterranean Sea and known earlier as *Beroe ovata* Eschscholtz, 1829 (Chun, 1880), have been re-described, based on the morphological and genetic analyses and re-named as *Beroe pseudocucumis* (Fig. 4.88) (Shiganova and Abuzova 2021).

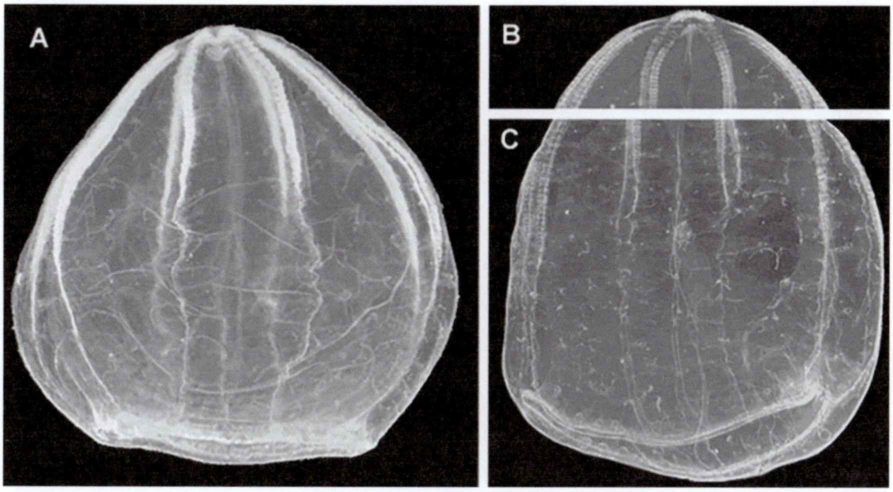

Fig. 4.87 *Beroe mitrata* (Moser, 1907). (Photo by N. Killi)

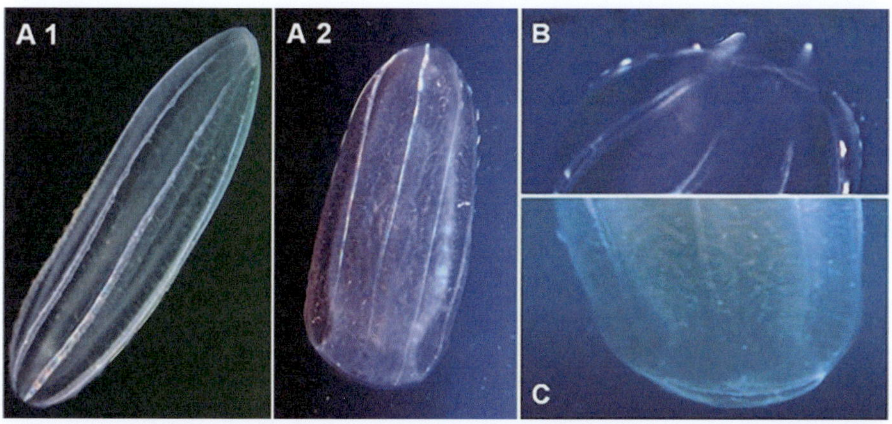

Fig. 4.88 *Beroe pseudocucumis*. (**a**1, **a**2) general view of adult ctenophores with meridional and para gastral canals; (**b**) aboral pole with a row of long branched papillae and subtentacular canal with row of combs at the upper part; (**c**) view from the mouth side with visible oval mouth. (Photographs were taken by Fabien Lombard in the coastal waters of Villefranche, France, Mediterranean Sea (available at https://ecotaxa.obs-vlfr.fr/prj/215))

The scyphomedusa *Rhopilema nomadica* (Fig. 4.89) penetrated to the eastern part of the Mediterranean Sea via the Suez Canal in the middle 1980s (Galil 2000); in 1995, it was observed in the near-shore waters off the southeastern coast of Turkey (Kideys and Gugu 1995). In 1988, one individual was found in Izmir Bay (A. Karatas, pers. comm., cited after Galil 2000). Every summer, large (up to 100 km in diameter) aggregations of these medusae are forming in near the coasts

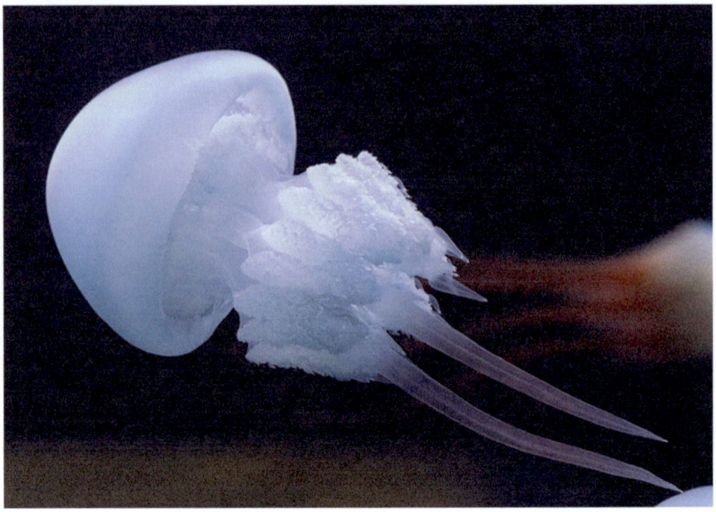

Fig. 4.89 *Rhopilema nomadica*

of the Levantine Sea. Grazing out zooplankton, the medusae strongly affect its abundance in this poor oligotrophic region of the Mediterranean Sea. Large medusae accumulations in the near-shore zone also negatively influence the tourism and recreation in this region. Since the medusae has stinging cells, dangerous for humans, the attractiveness of these areas for recreation is decreasing. Also, in the period of medusa development, fishery trawls and purse nets are jammed with them (Galil and Zenetos 2002).

Thus, the composition of non-native zooplankton species in the Aegean Sea indicate that in the northern part of the Aegean Sea species are mainly arrived from the Black Sea with the currents. These are the copepod *Pseudocalanus elongates*, the ctenophore *Mnemiopsis leidyi,* and the recently recorded ctenophore *Beroe ovata.* The *M. leidyi* individuals were probably brought from the Black Sea with the ballast waters. The vector of invasion of the copepod *Acartia tonsa* has not been reliably identified to date. Other zooplankton species are Lessepsian migrants from the Red Sea or Indian-Pacific region. In the recent years, about 15 new non-native species of zooplankton were registered, though these data are still unpublished. Among them, there are no species originating from the Black Sea (A. Zenetos, pers. comm.).

Benthos

Among the non-native species of benthic crustaceans, encountered in the Black and Aegean seas, the blue crab *Callinectis sapidus* originated from the near-shore Atlantic waters of North America should be mentioned. In the Mediterranean Sea, the blue crab was first registered in the 1940s in the coastal waters off Egypt, Israel, and Italia (Galil and Zenetos 2002). Later, it was also met in the near-shore waters of the northern part of the Aegean Sea, in its lagoons off the Greek coasts, and off the southern coasts of Turkey (Giordani and Soika 1951). The crab has a commercial value. The blue crab was also brought to the western part of the Black Sea, where at present its naturalization seems to be proceeding (Revkov 2003).

Another crab, *Charybdis (Goniohellenus) longicollis* was encountered in the Turkish waters of the Aegean Sea (Holtius 1961). In the Mediterranean Sea, it first appeared in 1924–1925 in the near-shore waters off Israel (Steinnits 1929); then, it was found in the near-shore waters off Egypt, Syria, and Cyprus (Galil 1993) and gradually spread over these regions in the sequence characteristic for a typical Lessepsian migrants (Galil and Zenetos 2002). Two other alien crab species of the Brachyura family (*Talamita poissonii* and *Ixa monody*) were found in the Turkish waters of the Aegean Sea; probably, both of them penetrated via the Suez Canal (http://www.ciesm.org/atlas). The decapods *Pilumnopeus vauquelini* and *Erugosquilla massavensis* penetrated into the Turkish waters of the Aegean Sea via the same route. *Erugosquilla massavensis* became a mass species. The shrimp *Penaeus semisulcatus* penetrated via the Suez Canal; it is of Indian-Pacific origin and at now became a commercial species (Kocatas et al. 1991). Another shrimp, *Marsupenaeus japonicus,* also penetrated via the Suez Canal; it established and formed populations along the Levantine coast from Egypt to Rhodes. It is also successfully cultivated in the central and western parts of the Mediterranean Sea; experiments on its cultivation were also undertaken in Amvrakikos and Vistonikos bays in the Ionian Sea.

In the Aegean Sea, 26 non-native species of mollusks have been registered to date. In the Aegean Sea, 12 non-native species of gastropods and 9 species of bivalves were identified (Table 4.32). The gastropod species *Strombus persicus*, *Bursatella leachi*, and *Melibe fimbriat*, and the bivalve species *Brachidontes pharaonis* and *Pinctada radiata* are widely spread, while species such as *Crepidula fornicata*, *Rapana venosa*, *Bulla ampulla, Anadara demiri,* and *Fulvia fragilis* feature a restricted spreading, although sometimes formed significant populations (Zenetos et al. 2004).

With respect to their origin, non-native mollusks may be classified in the following way. Five Lessepsian migrants (*Cylichna girardi, Bursatella leachi, Brachidontes pharaoni, Malfuvundus regulus,* and *Gastrochaena cymbium*) had established, became abundant in the eastern part of the Mediterranean Sea, and then penetrated in the Aegean Sea. Species that represent another group (*Cellana rota, Trochus erythraeus, Murex forskoehli,* and *Pseudochama corbieri*) are encountered in the Suez Canal; thus, they penetrated via the canal and then spread over the Aegean Sea (Table 4.32). According to the recent terminology, all of them may be referred to Lessepsian immigrants. Three species of alien mollusks in the Aegean Sea (*Neria sanguinolenta, Acteocina mucronata,* and *Haminoea cyanomarginata*) are the Red Sea endemics, four more (*Smaragdia souverbiana, Rapana rapiformis, Bulla ampulla,* and *Melibe fimbriata*) are Indian-Pacific dwellers; presumably, all these species refer to Lessepsian immigrants. Meanwhile, they are absent in the Suez Canal and, having arrived to the Mediterranean Sea, restricted their spreading to the eastern part of the sea (Zenetos et al. 2005).

The Indian-Pacific oyster *Crassostrea gigas* is the species widely spread in temperate and tropical seas; its presence in the Aegean Sea due to cultivation (Zenetos et al. 2005). The oyster was also brought to the Black Sea with the ships; attempts to cultivate the oyster in the Black Sea as an aquaculture were made as well. At present, single oyster individuals are encountered off the Caucasian and Crimean coasts of the Black Sea (see Chap. 3, above).

The gastropod mollusks *Polycerella emertoni* of Atlantic origination was found in the ports and nearby lagoons; therefore, the most probable vector of their invasion is penetration with the ships. The bivalve mollusks *Petricola pholadiformes* of Atlantic origination was probably also brought with the ballast waters.

In some cases, invasion vectors are unknown. Some species, e.g., the bivalve mollusk *Fulvia fragilis* invaded via the Suez Canal (it was found in the Suez Canal) and established in the coastal waters of Israel, Tunisia, and southern Turkey (www.ciesm.org.atlas); meanwhile, recently it was encountered in Saronikos Bay and in its port of Piraeus, where it was most probably brought with the ships.

The Black Sea invaders such as the gastropod mollusk *Rapana venosa* was initially brought from the Black Sea and established in Thermaikos Bay of the Aegean Sea. Supposedly, its larvae were brought with ballast waters or through cultivation of oysters and mussels and then spread farther. Rapana replaced oysters and mussels as it had previously done in the Black Sea. In the Turkish near-shore waters of the Aegean Sea, rapana was also registered (Zaitzev and Ozturk 2001). In the 1970s, rapana independently invaded the Adriatic Sea, to which it was brought from the Black Sea with ships (Emily Chandler, pers. comm. based on the data of molecular analysis, 2006). In 1966, one more invader *Mya arenaria* was found in the Black Sea. At present, it is widely spread in the northwestern part of the Black Sea and penetrated into the Sea of Azov; it invaded the Sea of Marmara and, in 1984, it was encountered in Saronikos Bay of the Aegean Sea (Zenetos et al. 2004). It seems more probable that *Mya arenaria* was brought from the Black Sea rather than from its initial habitat in the North Atlantic.

Fishes

Pilengas *Planiliza haematocheilus* (Temminck & Schlegel, 1845) (Mugil soiuy) is the only Black Sea origin species, in the Aegean Sea. It was introduced in the Black Sea from the Sea of Azov and further entered the Aegean Sea via the Bosphorus, the Sea of Marmara and the Dardanelles (Table 4.33). Pilengas individuals have been caught by fishermen in the Homa Lagoon and in the Izmir Bay (Kaya et al. 1998).

The vast majority of the non-native fish species in the Aegean Sea migrated through the Suez Canal from the Red Sea. Some of the established fish species compete with native species. For example, the Mediterranean species *Mullus barbatus* and *Merluccius merluccius* hake were displaced into deeper and colder waters by their Lessepsian food competitors *Upeneus moluccensis* and *Saurida undosquamis* (Oren, 1957). *Gambusia holbrooki,* mosquito-eating fish, was introduced into the swampy freshwater reservoirs along the Anatolian coast, such as Lake Amik and other lakes for mosquitos' extermination (Zaitsev and Ozturk 2001).

In the recent years, probably due to the global warming, more and more fish have penetrated via the Suez Canal. The total number of new non-native fish species of tropical origin has reached 20, among them are *Callionymus filamentosus*, *Apogon pharaonis*, *Enchelycore anatina*, *Etrumeus teres*, *Fistularia commersonii*, *Iniistius pavo, Lagocephalus sceleratus*, *Lagocephalus suezensis*, *Oxyurichthys petersi*, *Pteragogus pelycus*, *Seriola fasciata*, *Sillago sihama*, *Solea senegalensis*, *Sphoeroides pachygaster*, *Sphyraena flavicauda*, *Tylerius spinosissimus*, *Tylosurus crocodilus*, *Upeneus moluccensis*, *Upeneus pori* (Corsini et al. 2005; Golani et al. 2002).

Many of them are harmful, including:

- *Sphyraena flavicauda*, yellow-tailed barracuda, one of the small species of the barracuda family Sphyraenidae;
- *Fistularia commersonii*, bluespotted cornetfish. It originats from the Indo-Pacific and recently has reached the Mediterranean through the Suez Canal. It is dangerous Lessepsian migrant, expanding their range in the Aegean Sea.
- *Lagocephalus guentheri,* diamondback puffer. It is a species of marine ray-finned fish belonging to the family Tetraodontidae, the puffer fishes. This puffer is native to the western Indian Ocean, including the Red Sea, from where it has colonized the Eastern Mediterranean, likely via the Suez Canal.

Thus, an introduction of fish through the Suez Canal in the Mediterranean Sea and further in Aegean Sea did not contribute to an increase in biodiversity. Many of the introduced species have displaced native species. Also, significant number of new species turned out to be harmful and dangerous to humans.

References

Abaza V, Boicenco L, Moldoveanu M, Timofte F, Bologa AS, Sburlea A, Dumitrache C, Staicu I, Radu G (2006) Evolution of marine biodiversity status at the Romanian Black Sea coast as result of anthropogenic modifications in the last decades. In: 1st biannual Sci. Conf. Black Sea ecosystem and beyond. Commission on the protection of the Black Sea against pollution. Abstract, Istanbul. Turkey, pp 50–51

Abramenko MI (2000) Discovery of the silver crucian carp Carassius auratus gibelio (Bloch) in the Russian section of the Black Sea. Rep Russ Acad Sci 374(3):415–418

Aguilera VM, Bednaršek N (2022) Variations in phenotypic plasticity in a cosmopolitan copepod species across latitudinal hydrographic gradients. Front Ecol Evol 10:925648. https://doi.org/10.3389/fevo.2022.925648

Albayrak S (1996) Echenoderm fauna of the Bosphorus (Turkey) Oebalia. 22:22–32

Alexandrov BG (2004) The problem of the transfer of aquatic organisms by ships and some approaches to assessing the risk of new invasions //. Marine Ecol J 3(1):5–17

Altukhov DA, Gubanova AD (2006) *Oithona brevicornis* Giesbrecht in the Sevastopol Bay in October, 2005. Morskoy Ekologicheskiy Zhurnal (Marine Ecolog J) 5:32

Andrusov NI (1988) An outline of the history of the development of the Caspian Sea and its inhabitants. Reproduction of the original author's publication of the 1889 edition (St. Petersburg Publishing House) - HER Media, 1988

Annenkova NP (1930) Freshwater and brackish-water Polychaeta of the USSR. USSR Academy of Sciences, L, p 47

Anninsky BE (2009) Organic composition and ecological energetics of jellyfish Aurelia aurita L. (Cnidaria, Scyphozoa) under the Black Sea conditions. In: Shulman GE (eds) Trophic relationships and food supply of heterotrophic animals in the pelagic ecosystem of the Black Sea. Black Sea Commission Publications, Istanbul, pp 99–160

Anninsky BE, Finenko GA, Abolmasova GI, Hubareva ES, Svetlichny LS, Bat L et al (2005) Effect of starvation on the biochemical compositions and respiration rates of ctenophores *Mnemiopsis leidyi* and *Beroe ovata* in the Black Sea. J Mar Biolog Assoc UK 85:549–561

Antajan E, Bastian T, Raud T, Brylinski JM, Hoffman S, Breton G (2014) The invasive ctenophore *Mnemiopsis leidyi* A. Agassiz, 1865 along the English Channel and the North Sea French coasts: another introduction pathway in northern European waters? Aquat Invasions 9:167–173

Arkhangelskyi AD, Strakhov NM (1938) The geological structure and history of the development of the Black Sea. USSR Academy of Sciences, Moscow, p 228

Ausel H, Guner H, Dural B (1993) Turkiye Marmara Denizi florasi II. Phaeophyta ve Phodophyta. Ege Universitesi Su Urunleri Fakultesi Dergisi 10:10.115–168

Averkiev FV (1960) Collection of statistical data on fish catches and non-fishing objects in the Azov-Black Sea basin for 1927–1959. Tr AzNIIRKH 1, v.2:94

Aysel V, Erdugan HB (1995) Checklist of the Black Sea sea-weeds, Turkey (1823–1994). Tr J Botany 19:545–534

Baker LD (1973) The Ecology of the Ctenophore *Mnemiopsis mccradyi* Mayer in Biscayne Bay. Florida. M.Sc. Thesis. The Uninversity of Miami, Miami, Florida, USA 131 pp.

Baker ID, Reeve MR (1974) Laboratory culture of the lobate ctenophore *Mnemiopsis mccradyi* with notes on feeding and fecundity. Mar Biol 26:57–62

Barash A, Danin L (1986) Further additions to the knowledge of Indo-Pacific Mollusca in the Mediterranean Sea. Spixiana 9:117–141

Barash A, Danin L (1988/89) Marine mollusca at Rodes. Isr J Zool 35:1–74

Bavari A, Bologa AS, Scolka HV (1991) A checklist of the Benthic marine algae (except the diatoms) along the Romanian Shore of the Black Sea. Rev Roum Biol, Biol Veget 36(1–2):7–22

Bayha KM, Harbison GR, Mcdonald JH, Gaffney PM (2004) Preliminary investigation on the molecular systematis of the invasive ctenophore Beroe ovata. In: Dumont H, Shiganova T, Niermann U (eds) The aquatic invasions in the Black, Caspian and Mediterranean Seas - NATO ASI Series, 2 environment. Kluwer Academic Publishers, pp 167–175

Bayha KM, Chang MH, Mariani CL, Richardson JL, Edwards DL, DeBoer TS et al (2015) Worldwide phylogeography of the invasive ctenophore *Mnemiopsis leidyi* (Ctenophora) based on nuclear and mitochondrial DNA data. Biol Invasions:11–24. https://doi.org/10.1007/s10530-014-0770-6

Bayram Ö, Mihneva V, Shiganova T (2011) Fist records of *Bolinopsis vitrea* (L. Agassiz, 1860): (Ctenophore, Lobata) in the Black Sea. Aquat Invasion 6(3):355–360

Belmonte G, Mazzocchi MG, Prusova IY, Shadrin NV (1994) *Acartia tonsa*: a species new for the Black Sea fauna. Hydrobiologia 292/293: 9-15. In: Ferrari ED, Bradley BP (eds) Ecology and morphology of Copepods. Kluwer Academic Publishers. Printed in Belgium

Ben-Tuvia A (1966) Red Sea fishes recently found in the Mediterranean. Copeia 2:254–275

Ben-Tuvia A (1971) Revised list of the Mediterranean fishes of Israel. Israel J Zool 20:1–30

Ben-Tuvia A (1978) Immigration of fishes through the Suez Canal. Bul Fish 76:249–255

Bini G (1960) Catalogue of the names of fishes of commercial importance in the Mediterranean. FAO, Rome, p 250

Bodeanu N (1993) Microalgal blooms in the Romanian area of the Black Sea and contemporary eutrophication conditions. In: Toxic Phytoplankton blooms in the sea, pp 3–29

Boero F, Putti M, Trainito E, Prontera E, Piraino S, Shiganova T (2009) Recent changes in Western Mediterranean Sea biodiversity: the establishment of *Mnemiopsis leidyi* (Ctenophora) and the arrival of Phyllorhiza punctate (Cnidaria). Aquat Invasions 4:675–680

Boersma M, Malzahn A, Greve W, Javidpour J (2007) The first occurrence of the ctenophore *Mnemiopsis leidyi* in the North Sea. Helgol Mar Res 61:153–155

Bogorodsky SV (2006) Finding of *Parablenius incognitus* (Blennidae) in the Eastern Coast of the Black Sea Northern Abkhazia. Vopr Ikhtiol 46(1):22–28

Boltachev AP (2003) Ichthyofauna of the Black Sea coast of Crimea. Taxonomic diversity. In: Yeremeyev VN, Gaevskaya AV (eds) The current state of biodiversity of the coastal waters of the Crimea (Black Sea sector). EKOSI-Hydrophysics, Sevastopol, pp 363–373

Boltachev AP (2006) The modern state and changes of Ichthyofauna in coastal sea water of the Crimea (Black Sea). In: First biannual Sci. Conf. Black Sea ecosystem 2005 and beyond. Commission on the Protection of the Black Sea against pollution, abstract, Istanbul, Turkey, vol 2, pp 114–116

Boltachev AP (2009) Clarification of the species status of the Sphyraena obtusata group barracuda found in the Black Sea. Quest Ichthyol T.49(1):135–137

Boltachev AR, Karpova EP (2014) Faunal Revision of alien species Fish Species in The Black Sea. RGBI.N 3

Boltacheva NA, Lisitskaya EV, Podzorova DV (2020) The distribution of polychaete invaders in biotopes of the northern part of the Black Sea. Russ J Biol Invasions 4:15–33

Braiko VD (1958) On the biology of the Black Sea woodworm *Teredo navalis* L. Proc Sevastopol Biol Station 10:76–82

Brylinski JM, Antajan E, Raud T, Vincent D (2012) First record of the Asian copepod *Pseudodiaptomus marinus* Sato, 1913 (Copepoda. Calanoida. Pseudodiaptomidae) in the southern bight of the North Sea along the coast of France. Aquat Invasions 7(4):577–584. pp

Bulgurkov KI (1968) *Callinectes sapidus* Rathbun in the Black Sea. Pub NIOPS 9:97–99

Burgasov PN, Pokrovsky VI (1986) Cholera. In: Great medical Encyclopedia, vol 27. Soviet Encyclopedia, Moscow, pp 36–47

Buskey EJ (1993) Annual pattern of micro- and mesozooplankton abundance and biomass in a subtropical estuary. J Plankton Res 15:907–924

Buzzurro G, Greppi E (1966) The lessepsian mollusks of mollusks of Tasucu (South East Turkey). La Conchiglia Supplement to 279:3–22

Caddy JF, Griffiths RC (1990) A perspective on recent fishery-related events in the Black Sea. studies and reviews. General Fisheries Council for the Mediterranean 63:43–71

Chashchin AK (1995) Abundance, distribution and migration of the Black Sea anchovy stocks. J Zool 19. Tubetak:173–180

Chikina MV (2009) Macrozoobenthos of loose soils of the North Caucasian coast of the Black Sea: spatial structure and long-term dynamics. PhD Thesis

Chikina M, Kucheruk N (2005) Long-term changes in the structure of benthic communities in the northeastern part of the Black Sea Influence of alien species. Oceanology 45(1):176–182

Chukhchin VD (1984) Ecology of gastropods of the Black Sea. Naukova Dumka, Kiev, p 176

Cinar ME, Ergen Z, Dagli E, Petersen ME (2005) Alien species of spionid polychaetes (Streblospio gyno-branchiata and Polydora cornuta) in Izmir Bay, eastern Mediterranean. J Mar Biol Assoc U K 85:821–827

Collingridge K, Van der Molen J, Pitois S (2014) Modelling risk areas in the North Sea for blooms of the invasive comb jelly *Mnemiopsis leidyi* A. Agassiz, 1865. Aquat Invasions 9(1):21–36. https://doi.org/10.3391/ai.2014.9.1.02

Compte S, Gascón XD, Quintana D, Boix (2010) Top-predator effects of jellyfish Odessia maeotica in Mediterranean salt marshes. Mar Ecol Prog Ser 402:147–159. https://doi.org/10.3354/meps08453

Coonfield BR (1937) Regeneration in *Mnemiopsis leidyi* Agassiz. Biol Bull 71:421–428

Corsini M, Kondilators G, Economidis PS (2002) Lessepsian migrant *Fistularia commersonii* from the Rhodes marine area. Economidis J Fish Biol Fish Soc Br Isles 61(4):1061–1062

Costello JH, Sullivan BK, Gifford DJ, Van Keuren D, Sullivan LJ (2006) Seasonal refuge, shoreward thermal amplification, and metapopulation dynamics of the ctenophore *Mnemiopsis leidyi* in Narragansett Bay, Rhode Island. Limnol Oceanogr 51:1819–1831

Costello JH, Bayha KM, Mianzan HW, Shiganova TA, Purcell JE (2012) Transitions of *Mnemiopsis leidyi* (Ctenophora: Lobata) from a native to an exotic species: a review. Hydrobiologia 690:21–46

Danilevsky NN, Mayorova AA (1979) Anchovy. In: Raw materials of the Black Sea. Pishevaya prom, M, pp 25–73

Daskalov GM, Grishin AN, Rodionov S, Mihneva V (2007) Trophic cascades triggered by overfishing reveal possible mechanisms of ecosystem regime shifts. PNAS 104(25): 10518–10523. https://www.pnas.org/doi/full/10.1073/pnas.0701100104

De Olazabal A, Tirelli V (2011) First record of the egg-carrying calanoid copepod *Pseudodiaptomus marinus* in the Adriatic Sea. Mar Biodivers Rec 4. https://doi.org/10.1017/S1755267211000935

Dellamotte M, Vardala-Teodorou E (1994) Shells from the Greek Seas. The Goulandris Natural History Museum, Athens, p 320. (in Greek)

Delpy F, Pagano M, Blanchot J, Carlotti F, Thibault-Botha D (2012) Man-induced hydro-logical changes, metazooplankton communities and invasive species in the Berre Lagoon (Mediterranean Sea, France). Mar Pollut Bull 64(9):1921–1932

Delpy F, Albouy-Boyer S, Pagano M, Thibault D, Blanchot J, Guilhaumon F et al (2016) Identifying the drivers of abundance and size of the invasive ctenophore Mnemiopsis leidyi in northwestern Mediterranean lagoons. Mar Environ Res 119:114–125

Demir M (1954) Bogazlar ve Adalar sahillerinin omurgasiz dip hayvanlari. Istanbul Univ. Fen Fakultesi Hidrobiologi Aras. Enst Yayin 3. Istanbul:615. (in Turkish)

Dmitrieva EV (1996) Fauna of monogeneans of the Far Eastern pilengas (Mugil soiuy) in the Black Sea. Vestnik Zool 4-5:95–97

Dobrovolsky AD, Zalogin BS (1982) Seas of the USSR. M., Publishing House of Moscow State University, p 192

Faasse MA, Bayha KM (2006) The ctenophore *Mnemiopsis leidyi* A. Agassiz 1865 in coastal waters of Netherlands: an unrecognized invasion. Aquat Invasions 1:270–277

Faschuk DY, Arkhipov AG, Shlyakhov VA (1995) Concentrations of mass commercial fish of the Black Sea at different stages of ontogenesis and its determining factors. Quest Ichthyol 1:73–92

Fedorov PV (1978) Pleistocene of the Ponto-Caspian Sea. Nauka, Moscow, p 165. (in Russian)

Figueroa NJ, Figueroa DF, Hicks D (2020) Phylogeography of *Acartia tonsa* Dana, 1849 (Calanoida. Copepoda) and phylogenetic reconstruction of the genus *Acartia* Dana, 1846. Mar Biodivers 50. https://doi.org/10.1007/s12526-020-01043-1

Finenko GA, Romanova ZA, Abolmasova GI (2000) A new settler to the Black Sea-the grebe Beroe ovata Brunguiere. Ecol Sea Kiev 50:21–25

Finenko GA, Anninsky BE, Romanova ZA, Abolmasova GI, Kideys AE (2001) Chemical composition, respiration and feeding rates of the new alien ctenophore, *Beroe ovata,* in the Black Sea. In: Purcel JE, Graham WM, Dumont HJ (eds) Hydrobiologia, vol 451. Kluwer Acad. Pub, pp 177–186

Finenko GA, Romanova ZA, Abolmasova GI, Anninsky BE, Svetlichny LS, Hubareva ES et al (2003) Population dynamics, ingestion, growth and reproduct rates of the invader Beroe ovata and its impact on plankton commun Sevastopol Bay, the Black Sea. J Plankton Res 25:539–549

Finenko GA, Romanova ZA, Abolmasova GI, Anninsky BE, Pavlovskaya TV, Bat L, Kideys A (2006) Ctenophores–invaders and their role in the trophic dynamics of the Planktonic Community in the Coastal Regions off the Crimean Coasts of the Black Sea (Sevastopol Bay). Oceanology 46(4):472–482

Fountoulakis E, Sabelli B (1999) Observations on the Malacofauna of the littoral of Saronikos Gulf. In: Hellenic Zoological society (ed) Contributions to the zoogeography and ecology of the Eastern Mediterranean region, pp 465–472

Froglia C, Galil B, Noel P, Turkay M (1998) Alien species in the Mediterranean Sea. Decapoda & Stomopoda, Crustacia. (Working document). 35. Congress CIESM

Fuentes VL, Angel DL, Bayha KM, Atienza D, Edelist D, Bordehore C et al (2010) Blooms of the invasive ctenophore, Mnemiopsis leidyi, span the Mediterranean Sea in 2009. Hydrobiologia 645:23–37

Furman ER, Yule AB (1990) Selffertilisation in *Balanus improvisus*. J Exp Mar B & A Ecol Elsevier 144:235–239

Galil BS, Zenetos A (2002) A sea change—exotics in the eastern Mediterranean. In: Leppakoski E, Gollash S, Olenin S (eds) Invasive aquatic species of Europe. Distribution, impacts and menegment. Kluwer Acad.pub, pp 325–327

Galil BS, Kress N, Shiganova T (2009) First record of *Mnemiopsis leidyi* A. Agassiz, 1865 (ctenophora; lobata; mnemiidae) off the Mediterranean coast of istrael. Aquat Invasions 4(2): 356–362

Garbasey OA, Popova EV, Gubanova AD, Altukhov DA (2016) First record of the occurrence of *Pseudodiaptomus marinus* (Copepoda: Calanoida: Pseudodiaptomidae) in the Black Sea (Sevastopol Bay). Mar Biol J 1(4):78–80. pp

Gavrilova NA (2001) Eutintinnus (Ciliophora, Oligotrichida, Tintinnidae) - a new tintinnid genus for the Black Sea fauna. Ekologiya Morya 58:29–31. (In Russian with English summary)

Gavrilova NA (2005) New for the Black Sea tintinnids species. Ekologiya Morya. 69:5–11. (In Russian with English summary)

Gavrilova NA (2010) Microzooplankton (Tintinnidae). In: Matishov GG, Boltachev AR (eds) The introducers in the biodiversity and productivity of the Sea of Azov and the Black Sea. SSC RAS Publish, Rostov-on-Don, pp 63–69. (In Russian with English summary)

Gavrilova NA, Dovgal IV (2016) Tintinnid ciliates (Spirotrichea, Choreotrichia, Tintinnida) of the Black Sea: recent invasions. Protistology 10, 3:91–96. Article *in* Protistology

Georgieva LV (1993) Species composition and dynamics of the phytocene. In: Kovalev AV, Finenko ZZ (eds) Plankton of the Black Sea. Naukova dumka, Kiev, pp 31–74

Geptner VG, Chapsky KK, Arsenyev VA, Sokolov VE (1976) Mammals of the Soviet Union. Pinnipeds and toothed whales, vol 2, ch.3. M.: Higher School, p 718

GESAMP (IMO/FAO/UNESCO-IOC/WMO/WHO/IAEA/UN/UNEP Joint Group of Experts on the Scientific Aspects of Marine Environmental Protection) (1997) Opportunistic settlers and the problem of the ctenophore Mnemiopsis leidyi invasion in the Black Sea. Rep Stud GESAMP 58:84

Ghabooli S, Shiganova TA, Zhan A, Cristescu M, Eghtesadi-Araghi P, MacIsaac H (2011) Multiple introductions and invasion pathways for the invasive ctenophore Mnemiopsis leidyi in Eurasia. Biol Invasions 13(3):679–690. https://doi.org/10.1007/s10530-010-9859-8

Ghabooli S, Shiganova TA, Briski E, Piraino S, Fuentes V et al (2013) Invasion pathway of the Ctenophore *Mnemiopsis leidyi* in the Mediterranean Sea. PLoS One 8(11):e81067. https://doi.org/10.1371/journal.pone.0081067

Golani D, Orsi-Relini L, Massutf E, Quignard J-P (2002) In: Briand FF (ed) CIESM atlas of exotic species in the Mediterranean, vol 1. CIESM publisher, Monaco, p 256

Golani D, Azzurro E, Dulcic J, Massutf O-RL (2021) In: Briand F (ed) Atlas of exotic fishes in the Mediterranean Sea, 2nd Entirel Revised edn. CIESM, Monako, p 365

Gomoiu T-M (1986) Problems of ecological reconstruction in coastal marine zones at the Romanian Littoral. Ecologie gi protecfia ecosisremelor Conscanfa 5:68–72

Gomoiu M-T, Skolka M (1998) Increasing of biodiversity by immigration of new species for the Romanian Fauna. An. Univ. "Ovidius" Constanta. Seria Biologie-Ecologie 2:181–202. (in Romanian)

Gomoiu MT, Skolska M (1996) Changements recents dans la biodiversite de la Mer Noire dus aux immigrants. In: GEO-ECO-MARINA, RCGGM.v.1. Danube delta –Black Sea system under global changes impact, Bucuresti. Constanta, pp 49–65

Gomoyu MT, Kupriyanov CC (1980) Assessment of abundance and distribution medusa *Aurelia aurita* in the eastern Black Sea. In: Vinogradov ME (ed) Pelagic ecosystems of the Black Sea. "Nauka", pp 191–198

Gordina AD, Bagnyukova TV (1992) On the spawning of swordfish *Xiphias gladius* in the Black Sea. Vopr Ichthyol 32(4):166

Gordon I (1953) On a new crab from Cadaqués on the north east coast of Spain (Sirpus zariquieyi n. g. and sp.). EOS, Revista Española De Entomología, Madrid 27(4):303–314. https://www.marinespecies.org/aphia.php

Gotsis-Scretas O, Friligos N (1990) Contribution to eutrophication and phytoplankton ecology in the Thermaikos Gulf. Thalassographica T13 1:1–12

Grese VN (1979) Zooplankton. In: Fundamentals of biological productivity of the Black Sea. Naukova Dumka, Kiev, pp 143–169

Greze VN, Fedorina AI (1979) Abundance and biomass of zooplankton. In: Fundamentals of biological productivity of the Black Sea. Naukova Dumka, Kiev, pp 157–163

Grintsov VA (2022) New data on the morphology of the amphipod *Microprotopus cf. maculatus* (Microprotopidae, Amphipoda) from the Black Sea (Sevastopol, Crimea). In: Proceedings of the Karadag Scientific Station named after T.I. Vyazemsky - nature Reserve of the Russian Academy of Sciences. No. 3, p 12. https://doi.org/10.21072/eco.2022.23.01

Grishichisheva NP, Shadrin NV (1999) Hydroids as epiphytes of mussels and cystozyres. Aquatoria and shores Sevastopol: ecosystem processes and services to society. Akvavita, Sevastopol, pp 229–237

Gruzov LN, Lumkis PV, Nakadovsky GV (1994) Studies of the spatiotemporal structure of plankton fields of the northern half of the Black Sea in 1992-1992. In: Preceding of tr. Odessa Publishing House, pp 94–127

Gubanova AA (2000) Occurrence of Acartia tonsa Dana in the Black Sea.Was it introduced from the Mediterranean? Mediterranean Marine Sci 1/1:105–110

Gubanova A, Altukhov D (2007) Establishment of *Oithona brevicornis* Giesbrecht, 1892 (Copepoda. Cyclopoida) in the Black Sea. Aquat Invasions 2:407–410. https://doi.org/10.3391/ai.2007.2.4.10

Gucu AC (2002) Can overfishing be responsible for the successful establishment of *Mnemiopsis leidyi* in the Black Sea? Estuar Coast Shelf Sci 54:439–451. https://doi.org/10.1006/ecss.2000.0657

Gülşahin N, Tarkan AN (2014) Seasonal distribution of scyphozoa (Cnidaria) and Ctenophora species in Gökova bay, Mugla Turkey. Rapp Comm Int Mer Médit 40:538

Hansson HG (2006) Ctenophores of the Baltic and adjacent seas—the invader *Mnemiopsis* is here! Aquat Invasions 1:295–298

Haraldsson M, Jaspers C, Tiselius P, Aksnes DL, Andersen T, Titelman J (2013) Environmental constraints of the invasive Mnemiopsis leidyi in Scandinavian waters. Limnol Oceanogr 58:37–48

Harbison GR, Madin LP, Swanberg NR (1978) On the natural history and distribution of oceanic ctenophores. Deep Sea Res 25:233–256

Harrison PJ, Furuya K, Glibert PM, Xu J, Liu HB, Yin K, Lee JHW, Anderson DM, Gowen R, Al-Azri AR, Ho AYT (2011) Geographical distribution of red and green *Noctiluca scintillans*. Chin J Oceanol Limnol 29(4):807–831

HELCOM (2012) Development of a set of core indicators: Interim report of the HELCOM CORESET project. PART B: Descriptions of the indicators. Balt. Sea Environ. Proc. No. 129 B

Holtius LB (1961) Report on a collection of Crustacea Decapoda and Stomatopoda from Turkey and the Balkans. Zoologische Verhandelingen 47:1–67

Holtius LB, Gottlieb E (1956) Two interesting crabd (Crustacea, Decapoda, Brachyura) from Mersin Bay, SE, Turkey. Zoologische Mededelingen 34:287–299

Hosia A, Falkenhaug T (2013) Invasive Ctenophope Mnemiopsis leidyi in Norvey. In: Fourth international Jellyfish Bloom symposium, vol 5–7, Hiroshima, Japan

Inanmaz OE, Bekbolet M, Kideys AE (2003) A new scyphozoan species in the Sea of Marmara: Chryaora hysoscella (Linne, 1766). In: Oceanography of the Eastern Mediterranean and Black Sea. Similarities and differences of two interconnected basins. Symposium, Ankara, pp 857–859

Isinibilir M (2012) The seasonal occurrence and abundance of gelatinous Macrozooplankton in Izmit Bay (the northeastern Marmara Sea). J Black Sea Mediteranian 18(2):155–176
Isinibilir M, Tarkan AN, Kideys AE (2004) Decreased levels of the invasive ctenophore M.*leidyi* in the Sea of Marmara in 2001. In: Dumont H, Shiganova T, Niermann U (eds) The Ctenophore Mnemiopsis leidyi in the Black, Caspian and Mediterranean Seas and other aquatic invasions - NATO ASI Series, 2. Environment. Kluwer Acad. Pub, pp 155–166
Ivanov L, Beverton RJH (1985) The fisheries resources of the Mediterranean. Part two. Black Sea. General Fisheries Council for the Mediterranean, Rome, p 70
Ivanov DA, Sinegub IA (2008) Transformation of biocenoses of the Kerch Strait after the introduction of the predatory mollusk Rapana thomasiana and the bivalves Mua arenaria and Cunearca cornea. In: Modern problems of the Azov-Black Sea region. Checkmate. III international conference on November 10–11, 2007. YugNIRO, Kerch, pp 45–51
Jaspers C, Haraldsson M, Lombard F, Bolte S, Kiørboe T (2013) Seasonal dynamics of early life stages of invasive and native ctenophores give clues to invasion and bloom potential in the Baltic Sea. J Plankton Res 35:582–594
Jaspers C, Huwer B, Weiland-Brauer N, Clemmesen C (2018) First record of the non-indigenous jellyfish *Blackfordia virginica* (Mayer, 1910) in the Baltic Sea. Helgoläander Meeresuntersuchungen 72:13. https://doi.org/10.1186/s10152-018-0513-7
Javidpour J, Sommer U, Shiganova TA (2006) First record of Mnemiopsis leidyi in the Baltic Sea. Aquat Invasions 1:299–302
Javidpour J, Molinero JC, Lehmann A, Hansen T, Sommer U (2009a) Annual assessment of the predation of Mnemiopsis leidyi in a new invaded environment, the Kiel Fjord (Western Baltic Sea): a matter of concern? J Plankton Res 31(7):729–738
Javidpour J, Molinero JC, Peschutter J, Sommer U (2009b) Seasonal changes andpopulation dynamics of the ctenophore Mnemiopsis leidyi after its first year of invasion in the Kiel Fjord, Western Baltic Sea. Biol Invasions 11:873–882
Jazdzewski K, Konopasca A (2002) Invasive ponto-caspian species in waters of the Vistula and Oder basins and the Southern Baltic Sea. In: Leppakoski GS, Olenin S (eds) Invasive aquatic species of Europe. Distribution, impacts and management. Kluwer Acad. Pub, Dordrecht/ Boston/London, pp 384–398
Johnson WM (1998) In: Johnson WM (ed) The Monachus Guardian, vol 1.N1. International Marine Mammal Association (IMMA), Ontario, pp 1–6
Jørgensen TS, Petersen B, Jepsen PM, Greve HS, Hancen L, Benni WH (2019) The genome and mRNA transcriptome of the cosmopolitan calanoid copepod *Acartia tonsa* Dana improve the understanding of copepod genome size evolution. Genome Biol Evol 11(5):1440–1450. https://doi.org/10.1093/gbe/evz067
Kalugina AA (1954) Bottom vegetation of the Black Sea off the coast of the North Caucasus// Stocks of marine plants and their use. M: pp 26–57
Kalugina-Gutnik AA (1975) Phytobentos of the Black Sea. Naukova Dumka, p 248
Kalugina-Gutnik AA (1979) Macrophytobenthos. In: Greze VN (ed) Osnovy biologicheskoi produktivnosti Chernogo moray (principles of biological productivity of the Black Sea). Naukova Dumka, Kiev, pp 123–142
Kaminskaya LI (1968) The type of sponge is Porifera. In: Determinant of the fauna of the Black and Azov seas, vol 1. Sciences, Kyiv, pp 35–55
Karhan SU, Kalkan E, Yokes MV (2007) First record of the Atlantic starfish *Asterias rubens* (Echinodermata, Asteroidea) from the Black Sea. JMBA2. Biodiversity records. Published on line, pp 1–4
Kazmin AS (2016) Persistent thermal fronts in the Black Sea: existence, variability, and response to atmospheric forcing. Oceanology 56(3):336–341. https://doi.org/10.1134/S0001437016030115
Ketelaars HAM (2004) Range extentions of Ponto-Caspian aquatic invertebrates in continental Europe. In: Dumont H, Shiganova T, Nierman U (eds) The ctenophore Mnemiopsis leidyi in the Black, Caspian and Mediterranean Seas and other aquatic invasions - NATO ASI Series, 2. Environment. Kluwer Academic Publishers, pp 209–236

Killi N, Abyzova G, Shiganova T (2019) Occurrence of Beroemitrata (Beroidae) associated with its prey aggressive invaderctenophore Mnemiopsis leidyi A. Agassiz from in Gökova Bay. BioEco2019—International Biodiversity & EcologySciences Symposium, Istanbul
Kiørboe T, Titelman J (1998) Feeding, prey selection and prey encounter mechanisms in the heterotrophic dino-flagellate *Noctiluca scintillans*. J Plankton Res 20:1615–1636. https://doi.org/10.1093/plankt/20.8.1615
Kirac CO, Savas Y (1996) Status of the Monk Seal (Monachus monachus) in the neighbourhood of Eregli, Black Sea coast of Turkey. Zool Middle East 12:5–12
Kirac CO, Savas Y, Guclusoy H, Veryeri NO (1998) Distribution and status of monk seal Monachus monachus (Hermann, 1779) along Turkish coasts. In: 1st world marine mammal science conference. Abstract. 19–20 January 1998. Monako, p 28
Kiseleva MI (1979) Zoobenthos. In: Grese VN (ed) Fundamentals of biological productivity of the Black Sea. Naukova Dumka, Kiev, pp 208–239
Kiseleva MI (1981) Benthos of loose soils of the Black Sea. Naukova Dumka, Kyiv, p 168
Kiseleva MI (1992) Comparative characteristics of bottom communities off the coast of the Caucasus. In: Long-term changes in the zoobenthos of the Black Sea. Naukova dumka, Kiev, pp 84–99
Kiseleva MI, Zaika VE (1972) Chordata. In: The determinant of the fauna of the Black and Azov Seas, vol 3. Sciences. A thought, Kiev, pp 294–304
Kocak F, Ergen A, Cinar ME (1999) Fouling organisms and their development in a polluted and an unpolluted marina in the Aegean Sea (Turkey). Ophelia 50(1):1–20
Kocatas A, Koray T, Kaya M, Kara OF (1993) A review of the fishery resources and their environment in the Sea of Marmara. GFCM. Studies and reviews. N 64. FAO, pp 87–143
Konsulov A, Kamburska L (1997) Sensivity to Antropogenic factors of the Plankton Fauna Adjacent to the Bulgarian Coast of the Black Sea. In: Ozsoy E, Mikaelyan A (eds) Sensitivity of North Sea, Baltic Sea and Black Sea to Antropogenic and climatic changes. Kluwer Academic Publishers, Dordrecht/Boston/London, pp 95–105
Konsulov AS, Kamburska LT (1998) Ecological determination of the new Ctenophora - Beroe ovata invasion in the Black Sea. Trudy Institute of Oceanology, Varna, pp 195–197
Koutsoubas D, Voultsiadou-Koukoura E (1991) The occurrence of Rapana venosa (Valenciennes, 1846) (Gastropoda, Thaididae) in the Aegean Sea. Bollettino Malacologico 26:201–204
Koutsoubas D, Arvanitidis C, Dounas C (2000) Community and dynamics of the molluckan Fauna in the Mediterranean lagoon (Gialova lagoon, SW Greece). Belg J Zool 130 (Supplement):135–142
Kovalev AV, Finenko ZZ (eds) (1993) Plankton of the Black Sea. Naukova Dumka, Kiev, p 280
Kovalev AV, Melnikov VV, Ostrovskaya NA, Prusova IY (1993) Mesoplankton. In: Kovalev AV (ed) Black Sea plankton. Naukova Dumka, Kiev, p 183–204. (in Russian)
Kovalev EA, Zhivoglyadova LA, Revkov NK, Frolenko LN, Afanasyev DF (2017) The first discovery of a Bivalve Mollusc, Arcuatula Senhousia (Benson, 1842). In: the Russian Part of the Azov-Black Sea Transportation Basin. Russ J Biol Invasions 3:24 УДК: 594.1(262.5)
Kozlov V (1974) Amur chebak—Pseudorasbora parva (Cyprinidae)—a new species of ichthyofauna of the Dniester. Bullet Zool 1:77–78
Krakhmalny A, Bryantseva Y, Velikova V, Sergeeva O, Skuratova K, Dereziuk N (2012) Black Sea Dinoflagellata (history of the research and current biodiversity). Turk J Fish Aquat Sci 12:539–546. www.trjfas.org ISSN 1303-2712. https://doi.org/10.4194/1303-2712-v12_2_47
Kremer P (1994) Patterns of abundance for *Mnemiopsis* in U.S. coastal waters: a comparative overview. ICES J Mar Sci 51:347–354
Kremer P, Canino MF, Gilmer RW (1986) Metabolism of epipelagic ctenophores. Mar Biol 90: 403–412
Kreneva KV (2003) A new species of shell ciliates for the fauna of the Sea of Azov. Evolution of marine ecosystems under the influence of intruders and artificial mortality of fauna. Abstracts of reports. Rostov-on-Don, pp 20–21

Krijgsman W, Vasiliev I, Beniest A, Lyons T, Lofi J, Tari G, Slomp CP, Cagatay N, Triantaphyllou M, Flecker R, Palcu D, McHugh C, Arz H, Henry P, Lloyd K, Cifci G, Sipahioglu Ö, Sakellariou D, the BlackGate workshop participants (2022) Mediterranean–Black Sea gateway exchange: scientific drilling workshop on the BlackGate project. Sci Dril 31:93–110. https://doi.org/10.5194/sd-31-93-2022

Kuz'menko LV (1966) Two species of Dinoflagellates, new for the Black Sea, in Novosti sistematiki nizshikh rastenii (bulletin of systematics of low plants). Akad. Nauk USSR, Moscow, pp 51–54

Lehtiniemi M, Lehmann A, Javidpour J, Myrberg K (2011) Spreading and physicobiological reproduction limitations of the invasive American comb jelly *Mnemiopsis leidyi* in the Baltic Sea. Biol Invasions. https://doi.org/10.1007/s10530-011-0066-z

Leppakosky E, Olenin S (2000) Xenodiversity of the European Brackish Water Seas: the North American contribution. In: Pederson J (ed) Proceedings of the first National Conference, January 24–27, 1999. Massachusetts Institute of Technology, pp 107–119

Lilley MKS, Thibault-Botha D, Lombard F (2014) Respiration demands increase significantly with both temperature and mass in the invasive ctenophore *Mnemiopsis leidyi*. J Plankton Res 36: 831–837. https://doi.org/10.1093/plankt/fbu008

Lisitskaya EV, Boltachova NA, Nadolny AA (2019) Hyboscolex sp.: the first find of the family Scalibregmatidae (Annelida) in the Black Sea. Invertebrate Zool 16(3):226–232. https://doi.org/10.15298/invertzool.16.3.03

Lisovskaya OA, Nikitina VN (2007) Macrophytobenthos of the Caucasian Black Sea coast near Tuapse and Great Sochi. Bullet Saint Petersburg Univ Ser 3(2. Bilogia):22–33

Lonsdale DJ (1981) Regulatory role of physical factors and predation for two Chesapeake Bay copepod species. Mar Ecol Prog Ser 5:341–351

Makkaveeva EB (1969) The order of clawed donkeys, or anisopods—Anizopoda (Tanaidacea). In: Determinant of the fauna of the Black and Azov Seas, vol 2. Sciences'. Dumka, Kyiv, pp 402–408

Malej A, Tirelli V, Lučić D, Paliaga P, Vodopivec M, Goruppi A, Shiganova TA (2017) *Mnemiopsis leidyi* in the northern Adriatic: here to stay? J Sea Res 124:10–16. https://doi.org/10.1016/j.seares.2017.04.010

Maltsev VI (1957) Some parasitological aspects of the introduction of the Far Eastern pilengas (Mugil soiuy) in the Azov-Black Sea basin. Thesis. dokl. conf. of young scientists, Vladivostok, pp 49–54

Marambio M, Franco I, Purcell JE, Canepa A, Guerrero E, Fuentes V (2013) Aggregations of the invasive ctenophore Mnemiopsis leidyi in a hypersaline environment, the Mar Menor lagoon (NW Mediterranean). Aquat Invasions 8(2):243–224

Marinov T (1977) Polychaeta. Polychaeta worms—Fauna of Bulgaria, vol 6. Belsu Publishing House. Academy of Sciences, Sofia, p 258

Marinov T (1990) Zoobenthos of Bulgarskya sector na Cherno more (Zoobenthos of the Bulgarian sector of the Black Sea). Bulgar. Acad. Sci. Publ, Sofia, p 196. (in Bulgarian)

May SP, Burkholder JM, Sandra E, Wikfors SH, Gary DF (2010) Effects of the toxic dinoflagellate Alexandrium monilatum on survival, grazing and behavioral response of three ecologically important bivalve molluscs. Harmful Algae 9:281–293. https://doi.org/10.1016/j.hal.2009.11.005

Mayer AG (1912) Ctenophores of the Atlantic coast of North America, vol 162. Publ. Carnegie Inst, Washington, pp 1–58

Mianzan HW (1999) Ctenophora. In: Boltovskoy D (ed) South Atlantic Zooplankton. Backhuys Publishers, Leiden, pp 561–573

Mianzan HW, Mari N, Prenski B, Sanchez F (1996) Fish predation on neritic ctenophores from the Argentine continental shelf: a neglected food resource? Fish Res 27:69–79

Mianzan HW, Martos P, Costello JH, Guerro R (2010) Avoidance of hydrodynamically mixed environments by Mnemiopsis leidyi (Ctenophora: lobata) in open sea populations from Patagonia, Argentina. Hydrobiologia 645:113–124

Mikaelyan AS (1997) Long –term variability of Phytoplankton communities in Open Black Sea in relation to environmental changes. In: Ozsoy MA (ed) Sensivity to change: Black Sea, Baltic Sea and North Sea, vol 27. NATO ASI Series. Kluv. Acad. Publ. S.2, pp 105–116
Milchakova NA (2002) New species of macrophytes of the Black Sea. Ecol Sea T.62:19–24. (in Russian)
Milchakova NA (2003) Macrophytobenthos. In: Eremeev VN, Gaevskaya A (eds) The current state of biodiversity of the coastal waters of the Crimea (Black Sea sector) ECOSI-Hydrophysics. Sevastopol, p 152
Mills CE, Sommer F (1995) Invertebrate introductions in marine habitats: two species of hydromedusae (Cnidaria) native to the Black Sea, *Maeotias inexspectata* and *Blackfordia virginica*, invaders in San Francisco Bay. Mar Biol 122:279–288
Mills EL, Leach JH, Carlton JT, Cekor LS (1993) Exotic species in the Great Lakes: a history of biotic crises and anthropogenic introductions. J Great Lakes Res 19(1):19–54
Mills CE, Pugh PR, Harbison GR, Haddok HD (1996) Medusae, Siphonophores and Ctenophores of the Alboral Sea, North Western Mediterranean. Sci Mar 60(1):145–163
Minicheva GG (2015a) New invader in the Black Sea: Kelp *Chorda tomentosa* Lyngb. Int J Algae 17(3):219–224
Minicheva G (2015b) Finding of alien brown macroalgae *Chorda tomentosa* Lyngb. in the Ukrainian Black Sea coast. J Black Sea/Mediterranean Environ 21(2):227–231
Minicheva GG, Eremenko TI (1993) Algological finds in the northwestern part of the Black Sea. (Algological discoveries in the northwestern Black Sea). Algologiya T.3(4):83–86
Mironov GN (1954) Nutrition of planktonic predators Noctiluca nutrition. Tr Sevast Biol 8:12
Mironov GN (1967) Nutrition of planktonic predators. Nutritional requirements and daily rations of Aurelia aurita L. Biology and distribution of plankton of the South Seas. Nauka, M, pp 124–137
Mironov GN (1971) Biomass and distribution of jellyfish Aurelia aurita (L.) according to trawl catches in 1949–1962 in the Black Sea. Marine Biology, vol 24. Naukova dumka, Kiev, pp 49–69
Mirsoyan ZA, Martynuk ML, Vayzun EV (2006) Current pattern of Ctenophores Beroe Ovata and Mnemiopsis leidyi development in the Sea of Azov. Collection of articles o AzNIIRKH. The main problems of fishery and conservation of water bodies of Azov-Black Sea Basin (2004–2005). Rostov-On-Don, pp 136–148. (in Russian)
Moiseev PA (1989) Biological resources of the World Ocean. Agropromizdat, M, p 368
Moncheva S, Petrova-Karadjova V, Palasov A (1995) Harmful algae bloom along the Bulgarian Black Sea coast and possible patterns of fish and zoobenthic mortalities. In: Lassus P, Arzul G, Denn E, Gentien P (eds) Harmful Marine Algal Blooms. Lavoisier Publ. Inc., pp 193–198
Moncheva S, Doncheva V, Kamburska L (2000) On the long-term response of harmful algal blooms to the evolution of eutrophication off the Bulgarian Black Sea Coast: are the recent changes a sign of recovery of the ecosystem—the uncertainties. In: Report on IX Intern. Conf. “Harmful Algal Blooms”, February 5–11, vol 7, Hobart, Tasmania
Mordukhai-Boltovskoi FD (1960) The Caspian fauna in the Azov-Black Sea basin. USSR Academy of Sciences, ML, p 288
Mordukhai-Boltovskoi FD (1969a) Crustacea. In: Vodyanitskii VA (ed) Determinant of the fauna of the Black and Azov seas, vol 1, Invertebrates edn. Naukova Dumka, Kiev, 151 p
Mordukhai-Boltovskoi FD (1969b) Crustacea. In: Vodyanitsky VA (ed) Guide of the fauna of the Black and Azov seas. Vol. 1. Invertebrates. Naukova dumka, Kiev, p 151
Mordukhai-Boltovskoi FD (ed) (1969c) The identification book of the Black Sea and the Sea of Azov Fauna. V. 2 (Artropoda: Cladocera, Calanoida, Cyclopoida, Monstrilloida, Harpacticoida, Ostracoda, Cirripedia, Malacostraca, Decapoda). Naukova Dumka Publ, Kiev, p 536. (in Russian)
Mordukhai-Boltovskoi FD (1972) The guide of the fauna of the Black and Azov seas, vol 3. (supplement). Naukova dumka, Kiev, pp 307–325
Morozova-Vodyanitskaya NV (1954) Phytoplankton of the Black Sea. Part 2. TR Sevastopol Biol Stat 14:11–19

Muller GF (1986) Review of the hitherto recorded species of Crustacea Decapoda from th Bosphorus, the Sea of Marmara and the Dardanelles. Cerrcetari Marine Ircm N 19:109–130

Muratov MV, Neprochnov YP, Trimonis ES (1980) The history of the development of the Black Sea basin. In: Neprochnov YPM (ed) Geological history of the Black Sea based on the results of deep-sea drilling. Nauka, pp 184–190. (In Russian)

Murina VV (2003) Meroplankton of the Black Sea: the history of study, modern problems. Marine Environ J II(3):41–50

Mutlu E (1999) Distribution and abundance of ctenophores and their zooplankton food in the Black Sea. II. *Mnemiopsis leidyi.* Mar Biol 135(4):603–614

Mutlu E, Bingel F (1999) Distribution and abundance of their zooplankton food.1. *Plerobrachia pileus.* Mar Biol 135:589–601

Nadolinsky VP (2004) Structure and assessment of stocks of aquatic biological resources in the northeastern part of the Black Sea. Ph.D. Thesis. 171 pp

Nesterova DA (2006) Open areas. In: The northwestern part of the Black Sea: biology and ecology. Naukova dumka, Kiev, pp 175–184

Nicolay K (1986) Erosaria spurca (Linné, 1758). La Conchiglia 18(208–209):29

Niermann U (2004) *Mnemiopsis leidyi*: distribution and effect on the Black Sea ecosystem during the first years of invasion in comparison with other gelatinous blooms. In: Dumont H, Shiganova TA, Niermann U (eds) The Ctenophore *Mnemiopsis leidyi* in the Black, Caspian and Mediterranean seas and other aquatic invasions, NATO science series: IV earth and environmental sciences. Kluwer Academic Publishers, Dordrecht, pp 3–32

Nordsieck F (1973) A new Neritidae in the Mediterranean. La Conchiglia 5:4

Ohtsuka S, Shimono T, Hanyuda T et al (2018) Possible origins of planktonic copepods, *Pseudodiaptomus marinus* (Crustacea: Copepoda: Calanoida), introduced from East Asia to the San Francisco: estuary based on a molecular analysis. Aquat Invasions 13(2):221–230. pp

Ojaveer H, Leppakoski E, Olenin S, Riccardi A (2002) Ecological impact of Ponto-Caspian invaders in the Baltic Sea, European inland waters and the Great Lakes: an inter-ecosystem comparison. In: Leppokoski E (eds) Invasive aquatic species of Europe. Kluwer Acad. pub, pp 412–425

Oliveira OMP (2007) The presence of the ctenophore *Mnemiopsis leidyi* in the Oslofjorden and considerations on the initial invasion pathways to the North and Baltic Seas. Aquat Invasions 2: 185–189

Oliveira OMP, Migotto AE (2006) Pelagic ctenophores from the Sao Sebastiano Channal, south-eastern Brasil. Zootaxa 1183:1–26

Oliveira OMP, Miranda TP, Araujo EM, Ayón P, Cedeño-Posso CM, Cepeda-Mercado A et al (2016) Monograph census of Cnidaria (medusozoa) and ctenophora from South American marine. Water S A (Pretoria) 4194(1):001–256

Olson MM (1987) Zooplankton. In: Heck KL Jr (ed) Ecological studies in the Middle Reach of Chesapeake Bay. Springer-Verlag, New York, pp 38–81

Ozgur E, Ozturk B (2008) A population of the alien jellyfish *Cassiopea andromeda* (Forsskal, 1775) (Cnidaria, Scyphosoa, Phizostomea) in the Oludeniz lagoon Turkey. Aquat Invasion 4(3): 423–428

Ozturk B, Ozturk AA (1996) On the biology of the Turkish straits system // CIESM Special Series No 2. Bulleten de I'institut oceanographique. Monacco N special 17:205–221

Papaconstantinou C (1987) Distribution of the Lessepsian fish migrants in the Aegean Sea. Biologia Gallo-Hellenica 13:15–20

Pashkov AN (2001) Ichthyofauna of coastal shelf of the Black Sea in Polyhaline waters, Cand. Sci. (Biol.) dissertation. VNIRO, Moscow, p 263

Pasternak AF (1983) Seasonal dynamics of abundance and biomass of zooplankton in the inshore waters of the north Caucasus. In: Sorokin YI, Vedernikov VI (eds) Seasonal variability of the Black Sea plankton. Nauka, Moscow, pp 139–177. (in Russian)

Pautova LA, Silkin VA, Lifanchuk AV (2012) New for the Black Sea species *throndsenii* (Marino, Montresor & Zingone) (Bacillariophyta): field observation and experimental studies. Int J Algae T.22(2):121–134

Pereladov MV (1988) Some observations for biota of Sudak Bay of the Black Sea. In: III all-Russian conference of marine biology. Naukova Dumka, Kiev, pp 237–238. (in Russian)

Petranu A (1997) Black Sea biological diversity. Romania U.N. Publications, New York, p 314

Petrova-Karadjova VJ (1990) Monitoring of the blooms along Bulgarian Black Sea coast. Rapp Comm Int Mer Medit T.32. V.1:209

Pitsyk GK (1979) Systematic composition of phytoplankton. In: Fundamentals of biological productivity of the Black Sea. Naukova dumka, Kiev, pp 63–70

Polikarpov IG, Saburova MA, Manzhos LA, Pavlovskaya TV, Gavrilova NA (2003) The Black Sea microplankton biodiversity in the coastal zone near Sevastopol (2001-2003). In: Eremeev VN, Gaevskaya AV (eds) Modern state of coastal waters biodiversity of Crimea (Black Sea sector). Ecosea-hydrophysica, Sevastopol, pp 16–43. (In Russian with English summary)

Por FD (1978) Lessepsian migration. The influx of Red Sea biota into the Mediterranean by way of the Suez Canal. Springer-Verlag/New York Publ, Berlin/Heidelberg/New York, pp 88–194

Prodanov K, Mikhailov K, Daskalov G, Maxim K, Chashchin A, Arkhipov A, Shlyakhov V, Ozdamar E (1997) Environmental impact on fish resources in the Black Sea. In: Ozsoy E, Mikaelyan A (eds) Sensivity of North Sea, Baltic Sea and Black Sea to antropogenic and climatic changes. Kluwer Academic Publishers, pp 163–181

Proshkina-Lavrenko AI (1971) On the flora of the Black Sea diatoms. In: Vodyanitsky VA (ed) Problems of marine biology. Naukova dumka, Kiev, pp 41–48

Pryakhin YV, Volovik SP (1997) The results of acclimatization of pilengas in the Sea of Azov. In: Makarov EV, Volovik SP, Tyutina Yu E (eds) The main problems of fisheries and protection of fisheries reservoirs of the Azov-Black Sea basin. Collection of scientific papers. AzNIIRKH, Rostov-on-Don, pp 204–210

Purcell JE, Shiganova TA, Decker MB, Houde ED (2001) The ctenophore *Mnemiopsis* in native and exotic habitats: U.S. estuaries versus the Black Sea basin. Hydrobiologia 451:145–176

Ramsak A, Malej A, Venko K (2012) Comparative phylogeography of meroplanktonic species, Aurelia spp. and Rhizostoma pulmo (Cnidaria: Scyphozoa) in European Seas. Hydrobiologia 690(1):69–80. https://doi.org/10.1007/s10750-012-1053-9

Rass TS (1992) Fish resources of the Black Sea and their changes. Oceanology 32(2):293–302

Rass TS (1993a) Ichthyofauna of the Black Sea and some stages of its history. In: Ichthyofauna of the Black Sea bays in conditions of anthropogenic impact. Naukova dumka, Kiev, pp 6–16

Rass TS (1993b) Ichthyofauna of the Black Sea and some stages of its history. In: Ichthyofauna of the Black Sea bays under anthropogenic influence. Naukova dumka, Kiev, pp 6–16

Rass TS (2001) The Black Sea region and its productivity. Vopr Ichthyol 41(6):742–749

Reeve MR, Syms MA, Kremer P (1989) Growth dynamics of a ctenophore (*Mnemiopsis)* in relation to variable food supply. I. Carbon biomass, feeding, egg production, growth and assimilation efficiency. J Plankton Res 11(3):535–552

Reusch T, Bolte S, Sparwell M, Moss A, Javidpour J (2010) Microsatellites revea origin and genetic diversity of Eurasian invasions by one of the world's most notorious marine invader, *Mnemiopsis leidyi* (Ctenophora). Mol Ecol 19:2690–2699

Revkov NK (2003) Taxonomic composition of the benthic fauna of the Crimean Black Sea coast. In: Yeremeev VN, Gaevskaya AV (eds) The current state of biodiversity of the coastal waters of the Crimea (Black Sea sector). NAS of Ukraine, InBUM. Sevastopol: ECOSI-Hydrophysics, pp 209–229

Rield R (1983) Fauna and Flora des Mittelmeers. Verlag Paul Parey, Hamburg/Berlin, p 836

Riisgård HU, Goldstein J (2014) Jellyfish and ctenophores in Limfjorden (Denmark)—mini-review, with recent new observations. J Mar Sci Eng 2(4):593–615

Sabia L, Uttieri M, Pansera M, Souissi S, Schmitt FG, Zagami G, Zambianchi E (2012) First observations on the swimming behaviour of *Pseudodiaptomus marinus* from lake Faro. Biol Mar Mediterr 19(1):240–241. pp

Salekhova LP et al (1987) The composition of the ichthyofauna in the area of the Karadag State Reserve (Black Sea). Vopr Ichthyol 27(6):898–905

Salekhova LP, Kostenko NS, Vronsky A (1989) Ichthyofauna. In: Koradag Nature Kiev Naukova dumki, pp 242–252

Salsky VO (1964) On the acclimatization of the Far Eastern grass shrimp *Pandalus kessleri* (P. latirostris) in the Khadzhibeysky estuary. Odessa Biol 5:107. (in Ukrainian)

Selifonova ZP (2009) *Oithona brevicornis* Giesbrecht (Copepoda, Cyclopoida) in harborages of the northeastern part of the Black Sea shelf. Inland Water Biol 2:30–32. https://doi.org/10.1134/S1995082909010052

Selifonova ZP (2011) *Amphorellopsis acuta* (Ciliophora: Spirotrichea: Tintinnida)—new species in the Black Sea. Mar Ecol J 10(1):85. (In Russian with English summary)

Senicheva MI (1971) Composition and quantitative development of Phytoplankton of Neritic Zone in the area of Sevastopol' during autumn–winter period of 1968–1969, *Biologiya Morya*. Kiev 24:3–11

Senicheva MI (2001) New and rare Diatom and Dinoflagellate algae species of the Black Sea. Ekologiya Morya 62:25–29

Senichkina LG (1973) Phytoplankton of clear and waste polluted waters in the area of Yalta. Biologiya Morya 28:135–150

Senichkina LG (1993) Changes in the structure of phytoplankton in local areas of the sea under the influence of anthropogenic factors. Plankton of the Black Sea. Naukova dumka, Kiev, pp 32–55

Seravin LN (1994) Revision of the species composition of the genus Mnemiopsis (order Lobata). Zool J 73(1):9–33

Seravin LN, Shiganova TA, Luppova NE (2002) The history of studying the ctenophore Beroe ovata (Ctenophora, Atentaculata, Beroida) and some structural features of its Black Sea representative. Zool J 81(10):1193–1201

Sergeeva NG (1992) Characteristics of bottom communities of the Yalta Bay in conditions of anthropogenic impact. In: Long-term changes in the zoobenthos of the Black Sea. Kyiv Sciences' Dumka, pp 138–170

Sergey S, Belich TV, Sadogurskaya SA (2023) Invasion of the alien species *Bonnemaisonia hamifera* Hariot in Coastal Phytocenoses near the Southern Coast of Crimea (the Black Sea). Inland Water Biol 16(1):81–86. https://doi.org/10.1134/S1995082923010145

Sergin M, Kideys AE (2010) Ongoing process of "Mediterranezation" in the Black Sea. In: CIESM workshop monographs. Climate forcing and its impact on the Black Sea biota, vol 39, pp 61–64

Settepassi F (1967) Atlante malacologico. Molluschi viventi nel Mediterraneo, vol 1. Corporazione Atti Grafiche, Roma

Shadrin NV (2000) New settlers in the Azov and Black Seas: reasons and consequences? In: New settlers in Russian European Seas. Apatyty, pp 76–90. (in Russian)

Shalovenkov NN (2020) Trends in the introduction of alien species of zoobenthos into the Black Sea. Russ J Biol Invasions 1:72–80

Shiganova TA (1993) Ctenophore *Mnemiopsis leidyi* and ichthyoplankton in the Sea of Marmara in October of 1992. Oceanology 33(6):900–903

Shiganova TA (2009) Non-native species in the Seas of Eurasia. D. Sci. Thesis. A. N. Severtsov Institute of Ecology and Evolution. 643 pp. Available from: http://www.sevin.ru/dissertations/gidrobiol/49

Shiganova TA (2011) Review of the status of invasive species with special focus on the most invasive species Mnemiopsis leidyi (A.Agassiz, 1865) and their effect on the Caspian ecosystem. CASPECO, p 123

Shiganova TA (2020) Adaptive strategies of *Mnemiopsis leidyi* A. Agassiz 1865 in different environments of the Eurasian seas. Mar Pollut Bull 161:111737. https://doi.org/10.1016/j.marpolbul.2020.111737

Shiganova TA, Abyzova GA (2021) Revision of Beroidae Eschscholtz, 1825 systematics and distribution in the Southern seas of Europe based on genetic and morphological approaches. Zool J Linnean Soc zlab021:1–26. https://doi.org/10.1093/zoolinnean/zlab021

Shiganova TA, Bulgakova YV (2000) Effects of gelatinous plankton on Black Sea and Sea of Azov fish and their food resources. ICES J Mar Sci 57:000–000. https://doi.org/10.1006/jmsc.2000.0736. Available online at http://www.idealibrary.com

Shiganova TA, Kazmin AS (2024) Variability of mesozooplankton, abiotic factors and ecological status of marine environment in the north-eastern part of the Black Sea. Mar Environ Res 202: 106805. https://doi.org/10.1016/j.marenvres.2024.106805

Shiganova TA, Malej A (2009) Native and non-native ctenophores in the gulf of Trieste, northern Adriatic Sea. J Plankton Res 31(1):62–72

Shiganova TA, Ozturk B (2010) Trend on increasing Mediterranean species arrival into the Black Sea. In: Briand F (ed) CIESM workshop monographs No 39. Climate forcing and its impact on the Black Sea Biota. CIESM, Monaco, pp 75–93

Shiganova TA, Bulgakova YV, Volovik SP, Mirzoyan ZA, Dudkin SI (2000a) The new invader Beroe ovata and its impact on the ecosystem of the Azov-Black Sea basin in August-September 1999. In: Rostov-on-Don VSP (ed) Ctenophore Mnemiopsis leidyi (A.Agassiz) in the Azov and Black Seas and the consequences of its introduction. AzNIIRKH, pp 432–449

Shiganova TA, Bulgakova YV, Sorokin PY, Lukashev YF (2000b) The results of research of the new Beroe ovata alien in the Black Sea. Izv AN Ser biol 2:248–256

Shiganova TA, Bulgakova YV, Volovik SP, Mirzoyan ZA, Dudkin SI (2001a) In: Hydrobiologia Purcel JE, Graham WM, Dumont HJ (eds) A new invader, Beroe Ovata Mayer 1912 and its effect on the ecosystems of the Black and Azov Seas, vol 451. Kluwer Ac. Pub, pp 87–197

Shiganova TA, Mirzoyan ZA, Studenikina EA, Volovik SP, Siokou-Frangou I, Zervoudaki S et al (2001b) Population development of the invader ctenophore Mnemiopsis leidyi in the Black Sea and other seas of the Mediterranean basin. Mar Biol 139:431–445

Shiganova TA, Sapognikov VV, Musaeva EI, Domanov MM, Bulgakova YV et al (2003) Factors that determine pattern of distribution and abundance *Mnemiopsis leidyi* and its effect on ecosystem in the Northern Caspian. Oceanology 43:2003

Shiganova TA, Dumont HJ, Mikaelyan A, Glazov DM, Bulgakova YV, Musaeva EI (2004a) Interaction between the invading ctenophores Mnemiopsis leidyi A. Agassiz and Beroe ovata Mayer 1912, and their influence on the pelagic ecosystem of the northeastern Black Sea. In: Dumont H, Shiganova TA, Niermann U (eds) The Ctenophore Mnemiopsis leidyi in the Black, Caspian and Mediterranean seas and other aquatic invasions. NATO science series: IV. Earth and environmental sciences, vol 35. Kluwer Academic Publishers, Dordrecht, pp 33–70

Shiganova TA, Dumont HJ, Sokolsky AF, Kamakin AM, Tinenkova D, Kurasheva EK (2004b) Population dynamics of Mnemiopsis leidyi in the Caspian Sea, and effects on the Caspian ecosystem. In: Dumont H, Shiganova TA, Niermann U (eds) The Ctenophore *Mnemiopsis leidyi* in the Black, Caspian and Mediterranean Seas and other aquatic invasions. NATO ASI series: 2. Environment, vol 2004. Kluwer Acad. Pub, pp 71–111

Shiganova T, Christou ED, Bulgakova JV, Siokou-Frangou I, Zervoudaki S, Siapatis A (2004c) Study on the distribution and biology of the invader M. leidyi in the northern Aegean Sea, comparison with indigenous species Bolinopsis vitrea. In: Dumont H, Shiganova TA, Niermann U (eds) The Ctenophore Mnemiopsis leidyi in the Black, Caspian and Mediterranean Seas and other aquatic invasions. NATO ASI Series: 2. Environment, vol 2004. Kluwer Acad. Pub, pp 113–135

Shiganova TA, Christou ED, Siokou-Frangou I (2007a) First finding of alien species *Beroe ovata* Mayer 1912 in the Aegean Sea. Mediterr Mar Sci 8(1):5–14

Shiganova T, Bulgakova Y, Siokou-Frangou I, Christou ED, Malej A (2007b) Changes in ctenophore Mnemiopsis leidyi phenology and population dynamics linked to environmental variability and climate forcing in recipient areas of the Southern Seas. In: PICES/GLOBEC symposium on climate variability and ecosystem impacts on the North Pacific: a basin-scale synthesis. Hiroshima

Shiganova TA, Musaeva EI, Lukasheva TA, Stupnikova AN, Zas'ko DN, Anokhina LL, Sivkovich AE, Gagarin VI, Bulgakova YV (2012) Increase in findings of Mediterranean non-native species in the Black Sea. Rossiiskii Zhurnal Biologicheskikh Invasii 3.61:99. https://doi.org/10.1134/S2075111712040042

Shiganova TA, Legendre L, Kazmin AS, Nival P (2014a) Interactions between invasive ctenophores in the Black Sea: assessment of control mechanisms based on long-term observations. 507:111–123. https://doi.org/10.3354/meps10806

Shiganova TA, Riisgård HU, Ghabooli S, Tendal OS (2014b) First report on Beroe ovata in an unusual mixture of ctenophores in the Great Belt (Denmark). Aquat Invasions 9(1):111–116

Shiganova TA, Stupnikova A, and. Stefanova K. (2015) Genetic analyses of non-native species *Oithona davisae* Ferrari FD & Orsi, 1984 in the Black Sea. BioInvasions Rec 4(2):91–95. https://doi.org/10.3391/bir.2015.4.2.04. [details]

Shiganova TA, Sommer U, Javidpour J, Molinero JC, Malej A, Kazmin AS, Isinibilird M, Christou E, Siokou-Frangou I, Marambio M, Fuentes V, Mirsoyan ZA, Gülsahin N, Lombard F, Lilleyj MKS, Angel DL, Galil BS, Bonnet D, Delpy F (2019a) Patterns of invasive ctenophore Mnemiopsis leidyi distribution and variability in different recipient environments of the Eurasian seas: a review. Mar Environ Res 152:104791. https://doi.org/10.1016/j.marenvres.2019.104791

Shiganova TA, Alekseenko E, Kazmin AS (2019b) Predicting range expansion of invasive ctenophore *Mnemiopsis leidyi* A. Agassiz 1865 under current environmental conditions and future climate change scenarios. Estuar Coast Shelf Sci 227:106347. https://doi.org/10.1016/j.ecss.2019.106347

Shurova NM (2006) The appearance of the Atlantic oligochaete Tubificoides benedeni (Annelida, Oligochaeta) in the Black Sea and the features of its distribution on the northwestern shelf. Bull Zool 40(5):453–455

Siapatis A (2014) Maximum entropy modelling of potential habitat of the invasive ctenophore Mnemiopsis Leidyi in the Mediterranean Basin. In: Pitois S, Shiganova T (eds) . Report of the Joint CIESM/ICES Workshop on Mnemiopsis Science (JWMS). CIESM—The Mediterranean Science Commission, pp 53–59

Simakova UV, Smirnov IA (2019) Proceedings of the VI International scientific and practical conference

Skurikhina LA, Yuri K, Anton C, Pan'kova M (2001) Study of two species of mussels, *Mytilus trossulus* and *Mytilus galloprovincialis* (Bivalvia, Mytilidae), and their hybrids in Peter the Great Bay of the Sea of Japan with the use of PCR markers. Russ J Genet 37(12):1448–1451. https://doi.org/10.1023/A:1013264400526

Slynko YV, Kiyashko VI, Yakovlev VN (2000) List of fish species and fishes of the Volga basin. Catalog of plants and animals of the Volga basin. Yaroslavl, pp 252–277

Soldatova IN (1961) The influence of conditions of different salinity on the bivalve mollusk Teredo navalis. Proc JOHN 49:162–179

Sorarrain D (1998) Cambios estacionales en la biomasa de organismos gelatinosos en relacio'n con otros zoopla'ncteres en la bahı'a Samborombo'n. MSc Thesis. Facultad de Ciencias Exactas y Naturales, Departamento de Ciencias Marinas, Universidad Nacional de Mar del Plata. Mar del Plata, Argentina, p 35

Sovinsky VK (1904) An introduction to the study of the fauna of the Ponto-Caspian-Aral Sea basin, considered from the point of view of an independent zoo-geographical province. Notes Kiev Soc Natl 18:497

Starobogatov YI (1972) Class shellfish Loricata//The determinant of the fauna of the Black and Azov Seas. Kyiv Sciences' A thought 3:60–65

Steele JH (1970) Marine Food Chains. University of California Press, p 150. ISBN 978-0-520-013 97-1

Storsberg J (1997) Ein weiterer Fund von *Acteocina mucronata* (Philippi, 1849) im ostlichen Mittelmeer (Gastropoda, Cylichnidae). Sch Mal Haus Natur-Cismar 10:7–9

Studenikina EI, Volovik SP, Miryozan IA, Luts GI (1991) The ctenophore Mnemiopsis leidyi in the sea of Azov. Oceanology 3:722–725

Sullivan LJ, Gifford DJ (2004) Diet of the larval ctenophore *Mnemiopsis leidyi* A. Agassiz (Ctenophora, Lobata). J Plankton Res 26:417–431

Svetlichny L, Hubareva E (2014) Salinity tolerance of alien copepods *Acartia tonsa* and *Oithona davisae* in the Black Sea. Sci Rep 461:201–208. https://doi.org/10.1016/j.jembe.2014.08.012

Svetlichny LS, Abolmasova GI, Hubareva ES, Finenko GA, Bat L, Kideys AE (2004) Respiration rates of Beroe ovata in the Black Sea. Mar Biol 145:585–593

Svetovidov AN (1964) Fish of the Black Sea. Nauka, Moscow, p 550

Swanberg N (1974) The feeding behavior of *Beroe ovata*. Mar Biol 24(1):69–76

Tarkan AN, Isnibilir M, Tarkan AS (2005) Seasonal variation of the zooplankton composition and abundance in the Istabbul Strait. Pak J Biol Sci 8(9):1327–1336

Temnykh A, Nishida S (2012) New record of the planktonic copepod *Oithona davisae* Ferrari and Orsi in the Black Sea with notes on the identity of "*Oithona brevicornis*". Aquat Invasions 7: 425–431. https://doi.org/10.3391/ai.2012.7.3.013

Tendal OS, Jensen KR, Riisgård HU (2007) Invasive ctenophore *Mnemiopsis leidyi* widely distributed in Danish waters. Aquat Invasions 2:272–277

Tenekides NS (1989) On a collection of shells from the Greek Seas. Protopapa Press, Athens, p 187. (in Greek)

Terenko GV (2017) New diatoms genus *Chaetoceros* Ehrenberg in the Ukrainian Black Sea water. In: Proceedings of the XIV congress of the Ukrainian Botanical Society (Kyiv, 25–26 April, 2017) Kyiv, pp 112–113. (In Ukranian)

Terenko LM, Terenko GV (2000) Species diversity of the plankton phytocenosis in the Odessa Bay of the Black Sea. Ekologia moryaT 52:56–59. (in Russian)

Terentev AS (2024) Biocoenosis of *mya arenaria* (Linnaeus, 1758) in the kerch strait (Azov and Black Sea basin) https://doi.org/10.29039/2413-1733-2024-38-101-112

Thompson TE, Crampton DM (1984) Biology of *Melibe fimbriata*, a conspicuous opistobranch Mollusc of the Indian Ocean, which has now invaded the Mediterranean Sea. J Moll Stud 50: 113–121

Tringali L, Villa R (1990) Rinvenimenti malacologici dale coste Turche (Gastropoda, Polyplacophora, Bivalvia). Notiziario del CISMA 12:33–41

Tsikhon-Lukanina EA, Reznichenko OG (1991) Features of nutrition of various-sized individuals of mnemiopsis in the Black Sea. Oceanology 31(3):442–446. (in Russian)

Tsokur AG (1988) Fisrt find of eggs of *Centracantus cirrus* Rafinesque (Centracanthidae) in the Black Sea. Vopr Ikhtiol 28:329–330

Unal E, Shmeleva AA, Zagorodnyaya J, Kıdeyş AE (2000) Zooplankton structure and Copepod species of the Sea of Marmara in spring 1998. In: Marmara Denizi 2000 Sempozyumu Bildiriler Kitabı, İstanbul. Turkish Marine Research Foundation, pp 450–460. (in Turkish)

Uysal Z, Kideys AE, Semichkina L, Georgiva L (1998) Phytoplankton patches formed along the southern Black Sea cost in spring and summer 1996. In: Ivanov L, Ozuz T (eds) Ecosystem modeling as a menegment tool for the Black Sea. Kluwer Acad. Pub. NATO series, pp 151–170

Valkanov A (1936) The brackish waters of Bulgaria. 2nd attempt at their hydrographic and biological research. Annual meeting of the University of Sofia. Facul Phys Math 32(3): 209–341. (in Bulgarian)

Van Ginderdeuren K, Hostens K, Hoffman S, Vansteenbrugge L, Soenen K, De Blauwe H et al (2012) Distribution of the invasive ctenophore Mnemiopsis leidyi in the Belgian part of the North Sea. Aquat Invasions 7(2):163–169

Van Walraven L, Langenberg VT, Van der Veer HW (2013) Seasonal occurrence of the invasive ctenophore *Mnemiopsis leidyi* in the western Dutch Wadden Sea. J Sea Res 82:86–92

Vardala-Theodorou GE (1999) The occurrence of Indo-Pacific molluscan species *Fulvia fragilis* (Forsska, 1775) and *Bulla ampulla* L. 1758 in Elefsis Bay. Newsl Hell Zool Sol Fasc 31:10–11

Velikova V, Moncheva S, Petrova D (1999) Phytoplankton dynamics and red tides (1987–1997) in the Bulgarian Black Sea. Water Sci Technol 39(8):27–36

Vershinin AO (2008) Phytoplankton of north–eastern Black Sea: composition, annual changes in community structure, and the problem of invasive species. In: Materials of Sci. Conf. and VII Marine biology school "Modern problems of algology" (Rostov–on–Don, 9–13 June, 2008). Rostov–on–Don, pp 403–415

Vershinin AO, Orlova TY (2008) Toxic and harmful algae in coastal waters of Russia. Okeanologiya 48(4):568–582

Vershinin AO, Moruchkov AA, Leighfield T et al (2005) Potentially toxic algae in northeast Black Sea coastal phytoplankton in 2000–2002. Oceanology 45(2):240–248. (in Russian)

Vinogradov ME, Shushkina EA, Musaeva EI, Sorokin PY (1989) A new settler in the Black Sea—combtail Mnemiopsis leidyi (Agassiz). Oceanology 29(2):293–299

Vinogradov ME, Sapozhnikov VV, Shushkina EA (1992) Ecosystem of the Black Sea. Nauka, Moscow, p 112

Vinogradov ME, Shushkina EA, Anokhina LL et al (2000) Massive development of the Beroe ovata Eschscholtz ctenophore off the northeastern coast of the Black Sea. Oceanology 40(1): 52–55

Vinogradov AK, Bogatova YI, Sinegub IA (2012) Ecosystems of the seaports of the Black Sea-Azov basin. Astroprint, Odessa, p 528

Vodyanitsky VA (1930) On the question of the origin of the fauna of the Black Sea. The works of the Novorossiysk Biol stations 1(4):3–34

Walter TC (1986) New and poorly known Indo-Pacific species of *Pseudodiaptomus* (Copepoda: Calanoida), with a key to the species groups. J Plankton Res 8(1):129–168

Whitehead PJP, Bauchot ML, Hureau JC, Nielsen J, Tortonese E (1986) Fishes of the Northeastern Atlantic and the Mediterranean, vol III. UNESCO, Paris, p 458

Zagorodnaya YA, Kolesnikova EA (2003) On the problem of the penetration of alien copepod species into the Black Sea. Evolution of marine ecosystems under the influence of intruders and artificial mortality of fauna. Abstracts of reports. Rostov-on-Don, pp 80–81

Zagorodnyaya YA (2002) *Oithona brevicornis* in the Sevastopol Bay: is it a single event of a new invader in the Black Sea fauna? Morskoy Ekologicheskiy Zhurnal (Marine Ecol J) 61:43

Zaika VE, Andryushchenko AA (1969) Taxonomic diversity of phyto- and zooplankton of the Black Sea. Hydrobiol J V(3):12–19

Zaika VE, Kiselyova MI, Mihaylova TV (1992) Many years of changes in zoobentos. Nakhova Dumka, Kiev, p 248

Zaitsev YP (1998) Marine hydrobiological research of the National Academy of Sciences of Ukraine in the 90s of the XX century. Shelf and coastal reservoirs of the Black Sea. Hydrobiol J 34(6):3–21

Zaitsev YP (2006) Introduction to the ecology of the Black Sea. Even, Odessa, p 224c

Zaitsev Y, Mamaev V (1997) Biological diversity in the Black Sea: a study of change and decline.—Black Sea environmental series, vol 3. United Nations Publ, New York, p 208

Zaitsev YP, Ozturk B (eds) (2001) Exotic species in the Aegean, Marmara, Black, Azov and Caspian Seas, vol 8. Published by Turkish Marine Research Foundation, Istanbul, pp 1–26

Zaitsev YP, Polishchuk LN (1984) Outbreak of the Aurelia aurita (L) jellyfish population in the Black Sea. Ecol Sea 17:35–46

Zaitsev YP, Prokopenko V (1989) The delta world. The Danube floodplains. Lighthouse, Odessa, p 142

Zaitzev YP, Alexandrov BG (1998) Biological diversity, vol 7. UN Publ, Ukraine/NY, p 351

Zenetos A, Cinar ME, Pancucci-Papadopoulou MF, Hermelin JR, Furnari J, Andaloru F, Bellou N, Streftaris N, Zibrovius H (2005) Annotated list of marine alien species in the Mediterranean with records of the worst invasive species. Mediterr Mar Sci 6/2:1–56

Zenkevich LA (1963) Biology of the Seas of the USSR. M. USSR Academy of Sciences, p 739

Zenetos A (1994) Scapharca demini (Piani, 1981): first finding in the North Aegean Sea. La Conchiglia 271:37–38

Zevina GB (1994) Biology of marine fouling. Ed. Moscow University, 135

Zevina GB, Poltarukha OP (1999) Barnacles (Cirripedia, Thoracica) of the Black Sea. Bull Mosc Soc Nat Biol Ser 104, part 1:30–39
Zeybek N, Guner H, Aysel V (1986) The marine algae of Turkiye proceed. In: 5th OPTIMA meeting, Istanbul, pp 169–197
Zuev GV, Melnikova EB (2003) Ichthyofauna of the Black Sea coast of Crimea. Ecological diversity of fauna. In: Eremeev VN, Gaevskaya AV (eds) The current state of biodiversity of the coastal waters of the Crimea (Black Sea sector). EKOSI-Hydrophysics Sevastopol, pp 380–424

Chapter 5
The Sea of Azov

Abstract A wide range of issues, related to the non-native species in the Sea of Azov, is being considered including main features of the physical geography, hydrology, and hydrochemistry, origins and composition of the flora and fauna, non-native biota composition, *Mnemiopsis leidyi* population and its effect on the ecosystem, and *Beroe ovata* and its contribution to the ecosystem recovery. Special attention is paid to environment deterioration in the Sea of Azov due to both natural (water warming due to the climate change) and anthropogenic interference (regulation of the rivers' runoff and water withdrawal, industrial construction, irrigation systems, reservoirs construction, pollution etc.) and its effect upon the biota and fish stocks. One of the most disastrous consequences of massive freshwater runoff decrease was the salinization of the sea, which, in turn, facilitated migration of marine species of the Mediterranean or Atlantic origin from the Black Sea. It is noted that the changes in salinity in the Sea of Azov occurred in the form of pronounced phases of salinization/desalination.

5.1 Main Features of the Sea of Azov Physical Geography, Hydrology, and Hydrochemistry

Main physical and geographical features The Sea of Azov is an inland shelf sea in Eastern Europe connected to the Black Sea by the narrow (minimal width about 4 km) and shallow (minimal depth 3–5 m) Kerch Strait and sometimes regarded as a northern extension of the Black Sea (Fig. 1.1, Introduction; Fig. 5.1a). In ancient times, the Sea of Azov did not exist, and the Don River flowed into the Black Sea in the area of the modern Kerch Strait. It is assumed that the filling of the Sea of Azov occurred around 5600 BC. According to the hypothesis (Ryan and Pitman 1999), around 5600 BC, there was a large-scale catastrophic rise in the level of the Black Sea, which may have served as the historical basis for the legends about the Noah's Flood. An earthquake is considered to be the reason for the breakthrough of waters from the Mediterranean Sea into the previously closed Black Sea. Before this cataclysm, the level of the Black Sea was obviously below the general level of the world ocean.

T. A. Shiganova et al., *Non-Native Species in the Ponto-Caspian Seas*,
Monographiae Biologicae 96, https://doi.org/10.1007/978-3-032-18640-9_5

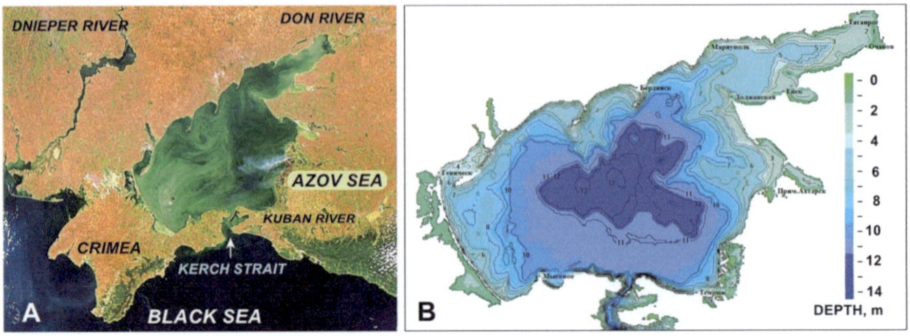

Fig. 5.1 (**a**) the Sea of Azov (optical satellite image); (**b**) bathymetric map of the Sea of Azov

The Sea of Azov is connected to the Atlantic Ocean through the Black, Marmara, Aegean, and Mediterranean seas. Due to its distance from the ocean, the Sea of Azov is the most continental sea on the planet. The modern Sea of Azov extends between 47°17′–45°16′ N and 33°36′–39°21′ E, is about 340 km long and 135 km wide, and has an area of about 37,600 km^2. With the depth varying between 0.9 and 14 m (average depth is 7 m), the Azov is the world's shallowest sea (Fig. 5.1b).

Although more than 20 rivers flow into the Sea of Azov, it is largely affected by the Don and Kuban rivers (Fig. 5.1a), which account for more than 90% of water inflow. The contribution of the Don is about twice that of the Kuban. The Kuban River flows into Temryuk Bay. The Kuban delta is located at the southeast, on the east side of the Kerch Strait. It is over 100 km long and covers a vast flooded area with numerous channels. Because of the spread, the delta has low contrast in satellite images (Fig. 5.1a) and is hardly visible in the map. The Don flows from the north into the large Taganrog Bay. Rivers bring sand and silt, which in turn form numerous bays, limans, and narrow spits. Because of these deposits, the sea bottom is relatively smooth and flat, with the depth gradually increasing toward the center (Fig. 5.1b). Typical values of the annual inflow and outflow of water to the sea, averaged over the period from 1923 to 1985, are as follows: inflow from the Black Sea 36–38 km^3 and outflow 53–55 km^3 (Kosarev et al. 2007). Thus, about 17 km^3 of low-salt water yearly is outflowing from the Azov Sea to the Black Sea.

Since the Sea of Azov is relatively small and nearly surrounded by land, its climate is continental with cold winters and hot and dry summers. In autumn and winter, the weather is affected by the Siberian Anticyclone. Easterly and northeasterly winds with a speed of 4–7 m/s (sometimes up to 15 m/s) provide transport of cold and dry air from Siberia. The mean monthly temperature in January ranges from -1 to -5 °C; during northeasterly storms, it may fall down to -25 to -27 C. In the spring to summer period, warm and fairy weather with weak winds prevails. In July, the mean monthly temperature over the entire sea equals 23–25 °C, while its maximum values reach more than 30 °C. In this season, especially in the spring, Mediterranean cyclones often pass over the sea; they are accompanied by westerly and southwesterly winds with speeds of 4–6 m/s and sometimes by gusts.

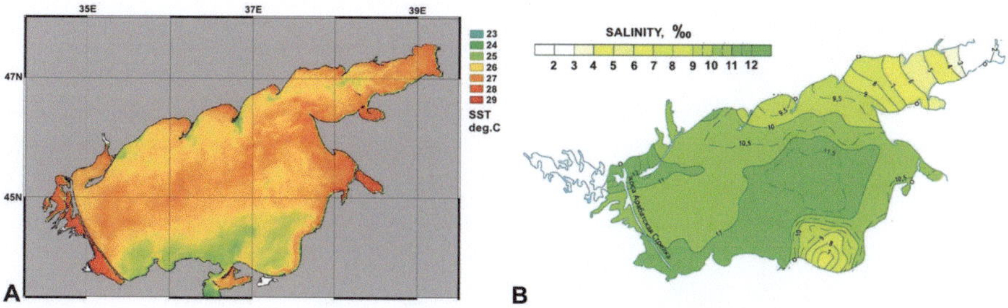

Fig. 5.2 Summer sea surface temperature (IR satellite image) (**a**) and sea surface salinity (**b**) in the Sea of Azov

Hydrology Because of its shallowness, the Sea of Azov reveals pronounced seasonal variability of the water temperature. In winter, over the most part of the basin the sea surface temperature is 0–1 °C; only in the Kerch Strait area it may rise up to 2–3 °C. The shallowness and low salinity of the sea make it vulnerable to freezing during the winter. Fast ice bands ranging from 7 km in the north to 1.5 km in the south can occur temporarily at any time from late December to mid-March. In summer, the temperature is more or less homogeneous over the entire sea being equal to 25–27 °C (Fig. 5.2a). Near the coasts, the temperature may exceed 30 °C.

Due to the shallow waters of the sea, wind and convective mixing reaches the bottom and vertical thermal stratification is practically absent; usually, the temperature difference between surface and bottom is less than 1 °C. Still, during summertime periods under weak winds, the seasonal thermocline may be formed preventing the vertical water exchange.

Because of the massive river inflow and limited exchange with the Black Sea, water in the sea has low salinity with more or less homogeneous spatial distribution (Fig. 5.2b). Prominent horizontal gradients of salinity are observed only in Taganrog Bay (confluence of the Don River) and in Temryuk Bay (confluence of the Kuban River) (Figs. 5.1a and 5.2b). In the open sea, the salinity ranges from 10‰ to 12‰; seasonal salinity variations no exceed 1‰.

In regard to the subject of this book, the long-term variations of the main abiotic environmental factors (i.e., water temperature and salinity) are more relevant. Interannual variability of water temperature and salinity in the Sea of Azov in 1970–2020 is shown at Fig. 5.3.

As a result of global warming, the temperature of the Sea of Azov has been gradually rising since the mid-1980s (Fig. 5.3). Linear trend of temperature rise, calculated from NOAA-CIRES-DOE (US National Oceanic and Atmospheric Administration-Cooperative Institute for Research in Environmental Sciences-Department of Energy) Twentieth Century Reanalysis data, is about +0.06 °C/year (Fig. 5.3), which is consistent with the results of other authors, based on different data sources (e.g., Berdnikov et al. 2019; Ginzburg et al. 2021). The total temperature increase during this period was about 2 °C.

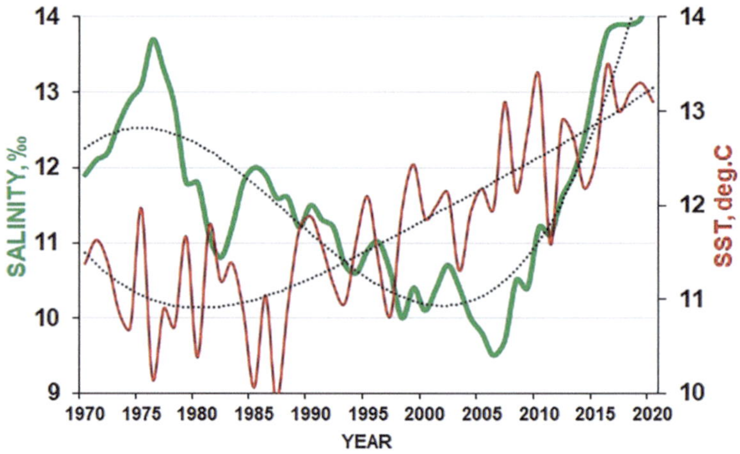

Fig. 5.3 Interannual variability of annual averages of sea surface temperature (°C, red line; data from NOAA-CIRES-DOE twentieth Century Reanalysis v.3, obtained at https://psl.noaa.gov/data/atmoswrit/timeseries/index.html) and salinity (‰, green line; in situ data from Berdnikov et al. 2023) in the Sea of Azov. Dotted lines—polynomial approximations of time series. Calculations and plotting by authors

Regarding the salinity, after its extreme increase up to 13.8‰ in 1976, salinity began to decline and dropped to 9.5‰ by 2006; then, it switched to new phase of salinization with even faster rate than during previous one in 1965–1976 (Fig. 5.3). After 2020, salinity of the Azov continued to rise, reaching record-high 14.97‰ in 2021 (Zhukova et al. 2023). One of the main reasons for the Azov Sea salinization after 2007 assumed to be the long-lasting lack of water in the Don and the Kuban rivers' basins. However, this factor alone is not enough to explain both the salinity decrease in the last quarter of the twentieth century and its rapid increase in 2007–2020. Evaporation rate variability, associated with the large-scale atmospheric forcing variations, also may contribute to the long-term salinity changes (Berdnikov et al. 2023; Zhukova et al. 2023).

The horizontal circulation in the whole thickness of the sea is mainly wind generated. Wind-driven drift currents produce a rise of the sea level off the coast, resulting in compensatory flows. Discharge currents occur in the pre-estuarine areas of the Don and Kuban rivers. In general, averaged resulted circulation pattern is cyclonic (counterclockwise) with typical speed values within 5–25 cm/s (Fig. 5.4). It is well expressed in winds stronger than 5 m/s. In the weak winds, the pattern of currents is not so consistent and is characterized by significant variability of direction and speed. Synoptic dynamic situation differs from averaged pattern and may reveal formation of eddies and filaments, visualized on optical satellite images due to the high number of green algae (Fig. 5.1a).

Water exchange between the Black and Azov seas through the Kerch strait is of importance for species transport between the seas. Under certain external forcing conditions, low-salinity waters from the Sea of Azov flow into the northeastern part

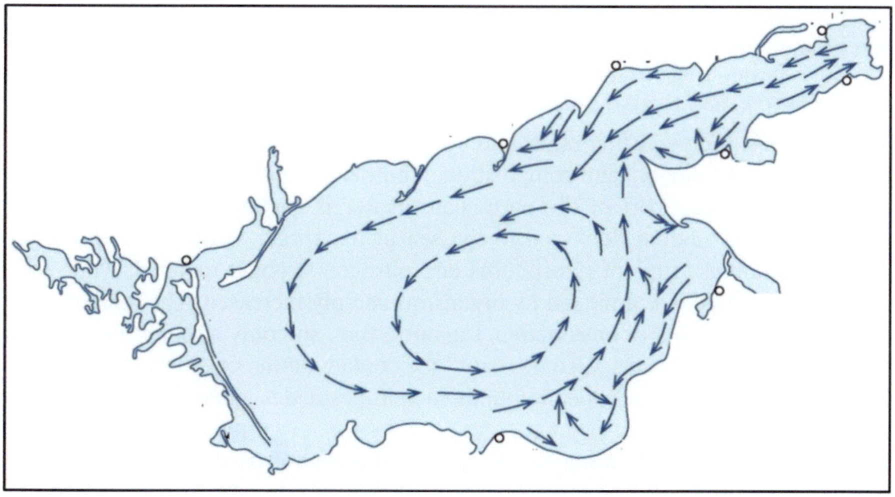

Fig. 5.4 General circulation pattern in the Sea of Azov

of the Black Sea through the narrow Kerch Strait and form a surface-advected buoyant plume. Water flow in the Kerch Strait also regularly occurs in the opposite direction, which results in the spreading of a bottom-advected plume of saline and dense waters from the Black Sea into the Sea of Azov. Recent study (Zavialov et al. 2020) indicated that water exchange in the Kerch Strait is governed by the wind-induced barotropic pressure gradient. As a result, water flow through the shallow and narrow Kerch Strait is a one-way process. Southward transport events are induced by moderate and strong northeasterly wind forcing. In this case, wind stress causes the transport of brackish waters from the southern part of the Sea of Azov, which results in the formation of the buoyant plume in the Black Sea. The northward water transport in the Kerch Strait, oppositely, was registered after a relaxation of strong northeasterly winds that results in a reverse of the barotropic pressure gradient along the strait and triggers inflow from the Black Sea to the Sea of Azov. Thus, strong northeasterly wind plays a restricting role in this process because the intense wind-induced southward surface flow of waters from the Sea of Azov occupies the whole water column in the shallow northern part of the Kerch Strait.

Hydrochemistry Before the early 1950s, under the natural water regime, the Sea of Azov was distinguished by its extremely high biological productivity. The riverine runoff delivered enormous amounts of nutrients (mineral compounds of nitrogen, phosphorus, and silicium), 70–80% of which were supplied during the spring flood period, providing abundant development of phyto- and zooplankton. Along with the good heating, low salinity, sufficient oxygen saturation, long vegetation period, and rapid cycling of nutrients, these factors provided conditions favorable for ichthyofauna.

The regulation of the Don (in 1952) and Kuban (in 1973) rivers and removal of a part of their runoff to fill reservoirs had negative aftereffects like decrease in the riverine runoff and the flood volume and reduction of the spawning areas. Formation of oxygen-deficient zones in the near-bottom layer intensified and in 1987 presence of hydrogen sulfide was first registered.

At present, the amount and composition of nutrients supplied to the sea drastically changed. The major part of the particulate matter is captured in the Tsimlyansk Reservoir; their amount delivered to the sea in the spring sharply decreased, the supply of mineral forms of phosphorus and nitrogen strongly reduced, while their organic forms hardly assimilated by organisms sharply increased. The nutrients that reach the sea are mainly consumed in Taganrog Bay, and only a minor part of them are passed to the open sea. Also increased the contamination of riverine and marine waters with various chemical compounds such as pesticides, phenols, and, at places, with oil products.

The abovementioned ecological changes resulted in a sharp drop in the biological productivity of the sea. The feeding base for fishes decreased by several times and the total catches significantly declined, mostly at the expense of valuable fish species.

The dissolved oxygen regime in the sea, in general, has not changed after the regulation of riverine runoff. High oxygen content is usually observed in autumn and winter due to increased oxygen solubility at low temperatures, active photosynthesis by cold-loving plankton species, and the low oxygen consumption by living organisms.

In this section, information from Dobrovolsky and Zalogin (1982), Kosarev et al. (2007), https://www.britannica.com/place/Sea-of-Azov, and https://en.wikipedia.org/wiki/Sea_of_Azov have been partially used.

5.2 Origins and Composition of the Sea of Azov Flora and Fauna

The Sea of Azov is home to a species with diverse origins, resulting in a composition of distinct faunistic groups—Mediterranean, Ponto-Caspian, and freshwater assemblages (Berg 1949; Zenkevich 1963).

The freshwater and brackish-water communities are primarily composed of Ponto-Caspian relicts. These include 54 species from 32 genera (Mordukhai-Boltovskoi 1969), which inhabit the river mouths and lagoons of the Sea of Azov. The marine community, in contrast, is made up of Mediterranean flora and fauna and contains very few species. For example, in the Sea of Azov, only three Mediterranean Coelenterata species out of 200, 12 mollusk species out of 1457, one crab species out of 51, and eight pelagic Copepoda species out of 300 are found.

The free-living invertebrate fauna of the Sea of Azov comprises 350 species (Mordukhai-Boltovskoi 1969; Zevina 1967). These figures, however, do not account for Turbellaria and Nematodes. When considering only the thoroughly studied groups, the Sea of Azov's fauna is 3.5 times less diverse than that of the Black Sea.

The Sea of Azov not only supports fewer species than the Black Sea but also shows notable changes in the composition of its animal communities. Many Mediterranean species, especially those originating from the Atlantic, are now missing. For example, the Phylum Echinodermata and Kinorhyncha, as well as the Order Pantopoda and Tanaidacea, and several other groups—such as Rotifera, Siphonophora, Ctenophora (*Pleurobrachia pileus*), Sipunculida, Loricata, Scaphopoda, and Appendicularia (*Oikopleura dioica*)—which are found in the Black Sea, even if only as single species, do not occur in the Sea of Azov. Other taxonomic categories are represented by only one or two species, including Nemertini, Porifera, Cirripedia, Bryozoa, and Ascidia. On the other hand, the subphylum Crustacea has seen the smallest decline in diversity, continuing to be the dominant group in the Sea of Azov and making up almost half of its fauna in terms of species number.

The Sea of Azov's most biodiverse groups are amphipods, copepods, and fishes, with Polychaeta and mollusks also being abundant (Mordukhai-Boltovskoi 1969). The vitality of the Sea of Azov's biological communities relied on three principal elements: (i) sufficient influx of nutrient-rich river water and a rapid nutrient turnover rate; (ii) the sea's characteristically low salinity; and (iii) extensive support for fish reproduction—spawning grounds for just migratory and semi-migratory species exceeded 600,000 hectares, while marine fish occupied the entire sea. The synergistic effect of these conditions was intensified by the sea's shallow depth and its specific geographic position.

From the mid-twentieth century onward, the Sea of Azov has experienced escalating anthropogenic pressures that have negatively affected its marine ecosystems. The key factors causing disruption are (1) increasing sea temperatures since the mid-1980s as a result of global warming; (2) fluctuating cycles of salinization and desalinization (after a sharp rise to 13.8‰ in 1976, salinity began to decrease, dropping to 9.5‰ by 2006; a new salinization phase then commenced, advancing more swiftly than the previous 1965–1976 period; post-2020, salinity levels kept rising, reaching a historic high of 14.97‰ in 2021); and (3) incursions of damaging non-native species.

Following this, the structure of indigenous species categorized by taxonomic groups is analyzed.

5.2.1 Phytoplankton

A total of 605 species, varieties, and forms of true and facultative planktonic algae have been documented in the Sea of Azov and Taganrog Bay. Diatoms and green algae are the most numerous in terms of species count. Blue-green and pyrophytic algae also exhibit significant species diversity, whereas euglenic and yellow-green algae constitute approximately 5% of the total species (Studenikina et al. 1999). The dominant algal representatives in the Sea of Azov are primarily planktonic in nature.

Limited water clarity inhibits the growth of submerged aquatic vegetation. Phytoplankton production is fundamental to sustaining the entire heterotrophic community of the Sea of Azov. The hydrological conditions impart unique characteristics to the phytoplankton, particularly within the lagoons. The sea's shallowness and the resulting rapid warming enable algae to thrive throughout almost the entire water column. Phytoplankton development and species composition in the open sea are comparable to those found in the coastal zone. The Sea of Azov is notable for its intense and prolonged algal blooms, high levels of particulate organic matter, and frequent occurrences of oxygen deficiency. At the same time, the salt composition of the Sea of Azov—specifically the ratio of total salt content to chlorinates—is similar to that of oceanic water. This accounts for the dominance of marine species, such as Pyrophytes (Pyrrophyta) and Diatomaceous algae (Bacillariophyta). Water exchange via the Kerch Strait introduces assemblages of algae and animals from the Black Sea into the Sea of Azov. Nevertheless, the sea's highly variable salinity permits only euryhaline species to survive and proliferate. The ability of algae to tolerate salinity fluctuations determines their spatial distribution throughout the Sea of Azov and Taganrog Bay.

The Sea of Azov region, with an average salinity of 11–12‰, supports three distinct ecological algal assemblages: marine, brackish-water-marine, and brackish-water types. In Taganrog Bay, where salinity ranges from 3–4 to 9–10‰, the algal flora is composed of freshwater-brackish-water species, including blue-green algae (Cyanobacteria), green algae (Chlorophyta), and diatomaceous algae (Studenikina et al. 1999).

Rising salinity levels in the sea have led to substantial shifts in the composition of phytoplankton communities. Notable alterations have also been observed in phytoplankton productivity, evident both in the distribution patterns of phytoplankton and in an overall decline in biomass (Studenikina et al. 1999).

5.2.2 *Zooplankton*

The zooplankton community in the Sea of Azov is derived from multiple sources. It includes species from freshwater, relict brackish-water, Ponto-Caspian, and marine groups (Mirzoyan et al. 2000).

In the eastern section of Taganrog Bay, where salinity levels range from 0.5 to 4‰, the dominant organisms are freshwater and brackish-water Cladocera and Copepoda. Rotifers are particularly abundant, especially the armored species *Brachionus plicatilis*, *Keratella curdata*, and *Asplanchna*, which thrive in both fresh and brackish aquatic environments. Moving to the central part of Taganrog Bay, salinity increases to 3–7‰, and the plankton composition becomes more varied: alongside brackish-water and freshwater species, marine organisms are also present. In the western portion of the bay, which opens directly into the sea, the zooplankton is almost entirely composed of marine fauna, including *Acartia clausi*, *Centropages ponticus*, meroplankton, larvae of *Amphibalanus improvisus*, as well as larvae of Gastropoda, Bivalvia, and Polychaeta.

Variations in the salinity levels of Taganrog Bay can cause changes in the distribution of species. For example, when salinity rises, marine organisms tend to move into the eastern parts, whereas during freshening phases, brackish-water and freshwater species spread into the western sections of the bay.

In the main body of the Sea of Azov, zooplankton consists of a limited number of marine species, primarily copepods. The dominant forms include *Acartia clausi*, *Centropages ponticus*, and meroplankton, along with rotifers from the genus *Synchaeta*. In the past, small numbers of *Paracartia latisetosa* (Krichagin 1873) were also recorded. During times of increased salinity, an influx of planktonic species from the Black Sea occurs (Mordukhai-Boltovskoi 1972).

5.2.3 Zoobenthos

The benthic community of the Sea of Azov exhibits limited species richness (Studenikina et al. 1998). Its bottom-dwelling fauna consists of worms, crustaceans, benthic protists, coelenterates, and mollusks. Mollusks alone account for 60–98% of the total biomass of bottom invertebrates. The composition and biomass of this community fluctuate considerably, influenced by a mix of biotic and abiotic conditions. Salinity, the gas regime, and sediment characteristics are among the most significant abiotic factors. Zoobenthos serves as a crucial food source for fish. Competition for habitat also occurs among the various benthic species.

In Taganrog Bay, the zoobenthos community includes freshwater, relict brackish-water, and marine groups. Prior to runoff regulation, the bay was densely populated by Ponto-Caspian species such as *Hypanis colorata*, *Dreissena polymorpha* (Pallas 1771), and *Hyraniola kowalevskyi*, with some extending into the northeastern Sea of Azov. Marine species like *Nereis succinea* and *Cerastoderma lamarcki* were found near the Sea of Azov boundary and partially in the western bay area. During phases of increased salinity, marine species became widespread in the western and central bay, while Ponto-Caspian species were restricted to fresher zones (Studenikina et al. 1998).

5.2.4 Ichthyofauna

The Sea of Azov, including Taganrog Bay, and the northeastern Black Sea host an ichthyofauna consisting of 183 species and subspecies. These are distributed across 112 genera, 55 families, and 22 divisions. Among these, 50 species are considered rare, 19 are classified as vulnerable and face a risk of extinction, and the sturgeon *Huso gueldenstaedtii* is believed to have vanished entirely. In total, records indicate

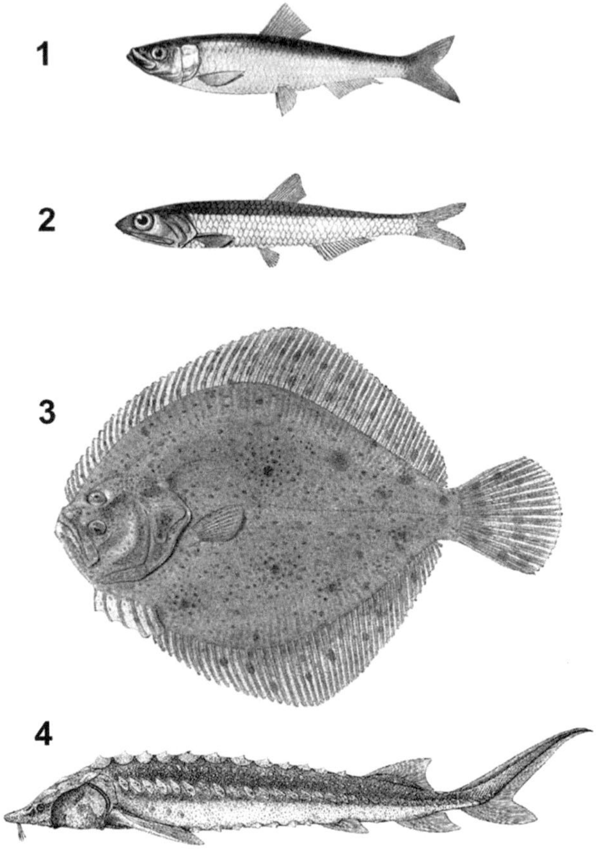

Fig. 5.5 Key commercial fish species found in the Sea of Azov. (1) kilka *Clupeonella cultriventris* (Nordmann 1840); (2) Azov Sea anchovy *Engraulis maeoticus* (Pusanov and Zeeb 1926); (3) turbot *Scophthalmus maeoticus* (Pallas 1814); (4) Sturgeon *Huso gueldenstaedtii* (Brandt and Ratzeburg 1833)

39 marine species, eight freshwater species, 14 species of anadromous and catadromous migrants, and 42 species that inhabit brackish-water environments (Volovik and Chikharev 1998a, b) (Fig. 5.5).

During the period from 1975 to 1977, when salinity levels were exceptionally elevated (especially in the southern region of the sea, where readings frequently reached 15‰), the Sea of Azov saw the arrival of species beyond the typical seasonal migrants such as the anchovy *Engraulis maeoticus* (Pusanov and Zeeb 1926) and *Engraulis encrasicolus* (Linnaeus 1758). Among these additional species were the garfish Belone euxini (Günther 1866), the mullet *Mugil cephalus*, *Chelon auratus* (Risso 1810), *Alticus saliens* (Forster 1788), the atherina *Atherina boyeri* (Risso 1810), the whiting *Merlangius merlangus* (Linnaeus 1758), the picarel *Spicara smaris* (L), and others. These species were either uncommon or had not previously been documented in the Sea of Azov.

A further category includes the bluefish *Pomatomus saltatrix*, the Black Sea plaices Brill *Scophthalmus rhombus* (L) and *Scophthalmus maeoticus* (Pallas 1814), the blue stingray *Dasyatis pastinaca*, the spiny dogfish *Squalus acanthias*, the Brown meagre *Sciaena umbra* (L), and the shi drum *Umbrina cirrosa* (L). Also present are the Black Sea salmon *Salmo trutta* (L), the mackerel *Scomber scombrus* (L), the wrasse *Symphodus ocellatus* (L), the blenny *Parablennius zvonimiri* (Kolombatovic 1892), the blanket bullhead *Aphia minuta*, Mediterranean sand smelt *Atherina hepsetus* (L), the thick pipefish *Syngnathus variegatus*, and various other species.

For the first time in the Sea of Azov, the corkwing *Symphodus cinereus* (Bonnaterre 1788), the rock hoppers *Parablennius Miranda* (Ribeiro 1915), the sand goby *Pomatoschistus* minutus (Pallas 1770) and *Gobius niger*, the puntazzo *Puntazzo puntazzo*, and the Mediterranean shore rockling *Gaidropsarus mediterraneus* were encountered. At the same time, all the previously mentioned species (both rare and newly observed) were detected in limited quantities, primarily in the southern portion of the sea. Only *Gobius niger*, within 2 years of entering the new basin, became widespread not just in the southern but also in the northern areas (Volovik and Dakhno 1983). Therefore, during periods of increased salinity in the Sea of Azov, its fish community can be naturally augmented by immigrants from the Black Sea. These species typically inhabit the northeastern and northwestern regions of the Black Sea and are capable of surviving in water temperatures below 3–5 °C.

5.3 Non-native Biota Composition of the Sea of Azov

The Sea of Azov hosts a diverse array of non-native biota, including various species of fish, invertebrates, and aquatic plants. These organisms have been introduced through human activities such as shipping, aquaculture, and the construction of canals. Notable invasive species include the comb jelly *Mnemiopsis leidyi*, which has significantly impacted local fisheries by consuming zooplankton and fish larvae. Other examples are the rapa whelk *Rapana venosa* and the zebra mussel *Dreissena polymorpha*, both of which compete with native species for resources and alter habitat structures. Climate change and increased maritime traffic continue to facilitate the spread of these non-native species, posing ongoing ecological and economic challenges to the region.

First of all, it is important to note that the species, establishing in the Sea of Azov, have to be euryhaline, eurythermal, euryoxygen, and stenobathno-shallow water species to survive specific environmental conditions of the sea. All non-native species entered the Sea of Azov from the Black Sea via the Kerch Strait. Some of the invaded species negatively affected the ecosystem, while others enriched flora and fauna (Mirzoyan et al. 2000; Bronfman and Khlebnikov 1985; Volovik 1985; Shiganova 2009). Part of the Black Sea species penetrated the Sea of Azov with the currents. Others get during seasonal migrations to the Sea of Azov for spawning and feeding. In the years with the increased water flow from the Black Sea, marine species of Mediterranean origin may reach the Sea of Azov, sometimes spreading over the entire basin (Fig. 5.6).

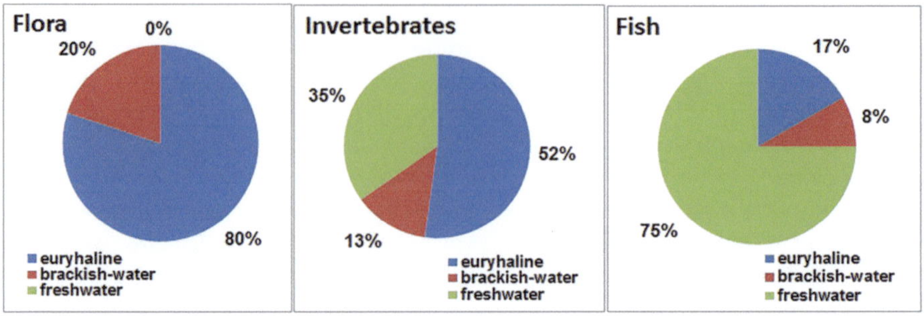

Fig. 5.6 Proportions of the euryhaline, brackish-water, and freshwater among the flora, invertebrates, and fish in the Sea of Azov

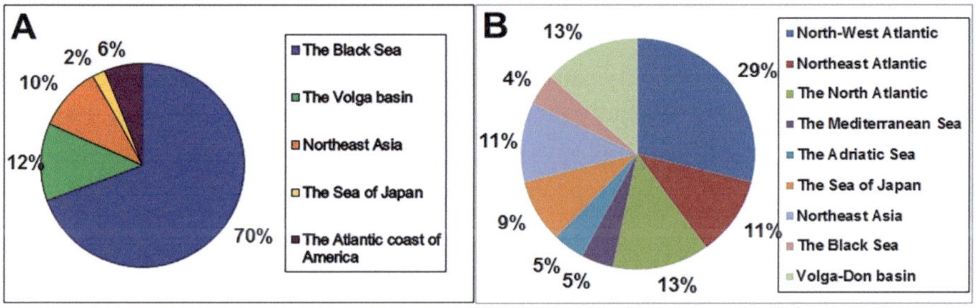

Fig. 5.7 Regions where the species originated within the Sea of Azov (**a**) and regions from which Black Sea species migrated into the Sea of Azov (**b**)

Indeed, every non-native species found in the Sea of Azov has arrived from the Black Sea, carried by currents, in ships' ballast water, or through hull fouling. For the Black Sea itself, all these species are considered established non-natives, with the exception of the mussel *Mytilus galloprovincialis*, the black goby *Gobius niger*, the shipworm *Teredo navalis*, and possibly one species of macrophyte algae, *Ectocarpus caspicus*.

The Sea of Azov has been colonized by species originating from the Black Sea and the Volga basin, as well as by organisms introduced through aquaculture from Northeastern Asia and the Sea of Japan (Fig. 5.7a). The Black Sea species that successfully migrated and established populations in the Sea of Azov themselves trace their origins to a variety of regions (Fig. 5.7b), including the North Atlantic, the Mediterranean Sea, the Adriatic Sea, the Sea of Japan, Northeast Asia, and the Volga-Don basin. Additionally, a small number of species are indigenous to the Black Sea (Fig. 5.7b).

This section consolidates existing data regarding the non-native species composition of multiple systematic groups that have been introduced and have become established in the Sea of Azov. The compilation covers microplankton, phytoplankton, ferns, macrophytes, zooplankton, gelatinous plankton, benthos, and fish (Tables 5.1, 5.2, 5.3, 5.4, 5.5, and 5.6).

Table 5.1 Microplankton

Species	Donor region	Vector	First record	Status of establishment	References
Bacteria *Vibrio cholerae* (El Tor strain)	Southeast Asia	Bacillus carrying	1970 Kerch	Established	Bulgurkov (1968)
Phylum Ciliophora **Class** Spirotrichea Butschli **Fam.** Tintinnidae (Claparede and Lachmann 1858) ***Eutintinnus lusus-undae*** (Entz 1885) Kofoid and Campbell 1939	Mediterranean Sea	Shipping	2002	Established	Kreneva (2003)

A cholera epidemic, triggered by the El Tor strain of the disease-causing bacterium *Vibrio cholerae* (Table 5.2), was documented in Kerch in 1970. Concurrently, an outbreak took place in the Odessa region (Bulgurkov 1968). From that period onward, El Tor outbreaks have periodically emerged in the Black Sea to the Azov area (Shadrin 2000).

The species *Eutintinnus lusus-undae*, which originates from the Mediterranean Sea, was introduced into the Black Sea in 2001, where it expanded its range extensively (Polikarpov et al. 2003) and subsequently (in 2003) entered the Sea of Azov (Kreneva 2003).

The Black Sea serves as a phytoplankton source for the Sea of Azov. Among the introduced species are *Melosira moniliformis*, *Cerataulina bergonii*, *Pseudo-nitzschia seriata*, *Coscinodiscus radiatus*, *Chaetoceros rigidus*, *C. affinis*, and *Pseudosolenia fragilissima*, whose spread is restricted by salinity levels. As a result, marine stenohaline organisms are mainly found in the Kerch Strait and appear only occasionally in the phytoplankton, whereas euryhaline species are distributed throughout the whole basin (Studenikina et al. 1999). The majority of incoming species have a neutral effect on the native ecosystem. However, the arrival of certain species—such as *Alexandrium tamarense*, *A. ostenfeldii*, *Dinophysis acuminata*, *Pseudo-nitzschia pungens*, and *Heterosigma akashiwo*—is harmful because they can trigger toxic algal blooms (Kovaleva 2002). The earliest recorded non-native species was the diatom *Pseudosolenia calcar-avis* (=*Rhizosolenia calcar-avis*). It entered the Black Sea around 1910 and reached the Sea of Azov in 1924 (Table 5.2) (Усачев 1928). Today, *P. calcar-avis* frequently becomes the dominant phytoplankton species in the Sea of Azov during warmer months (Studenikina et al. 1999). It emerges in April–March and continues to grow through the summer. Due to its large size, it is not easily consumed by Azov fish or invertebrates and tends to outcompete native algae (Proshkina-Lavrenko 1963).

The Sea of Azov supports just two macrophyte species: the non-native brown algae *Ectocarpus caspicus* and *Desmarestia viridis* (Table 5.3) (Stepanyan 2003). Initially, *E. caspicus* was thought to be a recent arrival from the Caspian Sea, but it was later identified as a species from the northwestern Black Sea. *Desmarestia*

Table 5.2 Phytoplankton

Species	Donor region	Vector	First record	Status of establishment	References
Class Centrophyceae Order Bacillariophyta Fam. Solanaceae ***Pseudosolenia calcar-avis*** (M. Schultze) Sentstrom 1986 (=*Rhizosolenia calcar-avis* *(M. Schultze)*	North Pacific	Shipping	1924	Established	Usachev (1928)
Bacteriastrum hyalinum Lauder *B. delicatulum* Cleve Fam. Hemiaulaceae *Cerataulina pelagica* (Cleve) Hendey (=*C. bergoii* (H. Peragallo) Schutt	The Mediterranean Sea? North Atlantic?	Shipping?	1980s	Established	Studenikina et al. (1999)
Fam. Chaetocerotaceae ***Chaetoceros socialis*** *Laud*	North Atlantic	Shipping?	1980s	Established	Studenikina et al. (1999)
Fam. Leptocylindraceae ***Leptocylindrus danicus*** **Cl.**	North Atlantic?	Shipping?	1980s	Established	Studenikina et al. (1999)
Thalassiosira nana *Lohmann 1908*	North Atlantic	Shipping?	1930s Coastal waters of Sevastopol	Established	

Table 5.3 Macrophytes

Species	Donor region	Vector	First record	Status of establishment	References
Desmarestiales Desmarestiaceae *Desmarestia viridis* (Muller) (Lamouroux,1813)	North Atlantic, Adriatic Sea	Penetration with currents, shipping?	1980s	Established	Stepanyan (2003)
Fucophyceae Ectocarpales Сем. Ectocarpaceae *Ectocarpus caspicus* (Henckel)	From the north-western part of the Black Sea?	Penetration with currents, shipping?	1980s	Established	Stepanyan (2003)

viridis, first observed in Odessa Bay in 1990 (Minicheva and Eremenko 1993), subsequently spread along the northwestern Black Sea coast and entered the Sea of Azov.

The hydromedusa *Blackfordia virginica*, which thrives in low-salinity environments, has been established in the Sea of Azov since the 1930s (Mordukhai-Boltovskoi 1968). In certain areas, *Blackfordia virginica* populations can grow to significant densities. These medusae consume zooplankton and may compete with fish larvae for food resources.

In 1961, *Bougainvillea (Paragonimus) megas* was identified in the Kerch Strait (Zevina et al. 1963). Although it is common, this species cannot endure low temperatures (Zevina et al. 1963).

Mnemiopsis leidyi was first observed near the Kerch Strait in August 1988. Since then, it has entered the Sea of Azov each spring or summer via currents, leading to population surges in summer or early autumn. It dies off when temperatures fall below 4 °C, typically by late October or November (Studenikina et al. 1991). A second ctenophore, *Beroe ovata*, which preys on *M. leidyi*, arrived in the Sea of Azov between September and October 1999 (Mirzoyan et al. 2002; Shiganova et al. 2000a).

During the period of increased salinity, the Black Sea Polychaeta community in the Sea of Azov consisted of *Aricia cuvieri*, *Nerinides cantabra*, *Eteone picta*, *Harmothoe imbricata*, *Pectinaria koreni*, *Phyllodoce tuberculata*, *Nerinides tridentata*, *Pygospio elegans*, *Polydora ciliata*, and *Spio filicornis* (Stark 1955; Kiseleva 1987). When salinity levels later decreased to 10.6‰, only the final two species, *Polydora ciliata* and *Spio filicornis*, remained (Frolenko 2000). Another introduced polychaete, *Mercierella (Ficopomatus) enigmatica*, was documented in the Kerch port prior to 1960 (Lebedev 1961). At present, its population is limited and it has no significant influence on the ecosystem.

The barnacle *Amphibalanus improvisus* (Darwin 1854) was first documented in the Black Sea during the nineteenth century. Today, it is a common fouling organism in the Sea of Azov. In Taganrog Bay, its peak biomass can reach 7 kg.m^{-2} (Partalii

Table 5.4 Planktonic invertebrates

Species	Donor region	Vector	First record	Status of establishment	References
Phylum Coelenterata **Order** Hydrozoa Fam. Campanulinidae *Blackfordia virginica* (Mayer 1910)	Atlantic coast of North America Chesapeake Bay	Shipping fouling or penetration with currents through the Kerch Strait	1930s	Established	Mordukhai-Boltovsky (1968)
Order Leptolida Fam. Bougainvillidae *Bougainvillea (Paragonimus) megas* (Kinne,1956)	Chesapeake Bay	Shipping fouling or penetration	1950s	Established	Naumov (1968)
Ctenophora **Order** Lobata Esch **Fam.** Mnemiidae **Genus** *Mnemiopsis* L. Agassiz *Mnemiopsis leidyi* A. (Agassiz,1865)	Atlantic coastal waters and estuaries of North America	Annual reintroduction with currents through the Kerch Strait	1988 Southern part of the Sea of Azov	Established	Studenikina et al. (1991) and Shiganova et al. (2000a)
Order Beroida Esch. Fam. Beroidae **Genus** *Beroe* (Browne 1756) *Beroe ovata* (Bruguière 1789)	Atlantic coastal waters and estuaries of North America	Annual reintroduction with currents through the Kerch Strait	1999 Southern part of the Sea of Azov	Established	Shiganova et al. (2000b) and Mirzoyan et al. (2002)
Class Crustacea **Order** Copepoda Fam. Acartiidae *Acartia tonsa* (Dana,1849)	Atlantic coastal waters and estuaries of North America	Penetration with currents through the Kerch Strait	After 2000	Established	Gubanova (2003)

1980). While it degrades ecological conditions, its larvae are consumed by planktivorous fish, and its empty shells offer refuge for benthic invertebrates. Certain species, including the amphipods *Gammarus locusta*, *Stenothoe monoculoides*, and *Jassa ocia*, even reproduce within these shells.

In 1956, the predatory rapa whelk *Rapana venosa* was observed in the southern Sea of Azov near the Kerch Strait. Its further spread was limited by the area's low salinity (Kostyuchenko and Nazarenko 1960). Regarding bivalve mollusks, two shipworm species, *Teredo navalis* and *Lyrodus pedicellatus*, colonized the sea

Table 5.5 Benthic species

Species	Donor region	Vector	First record	Status of establishment	References
Phylum Entoprocta Class Kamptozoa Fam. Urnatellidae *Urnatella gracilis* (Leidyi,1851)	Fresh waters of North America and then found in Belgian fresh waters	Shipping (fouling) or penetration with currents through the Kerch Strait	1954–1962 in lower reaches of the Don, 1962 in Taganrog Bay	Established	Mordukhai-Boltovsky (1968)
Class Polychaeta Order Serpulimorpha Fam. Serpulidae *Mercierella (Ficopomatus) enigmatica* (Fauvel 1923)	English Channel, North Sea, French coastal waters	Shipping (fouling)? Penetration from the Black Sea?	Before 1960 Kerch port	Established	Lebedev (1961)
Phylum Mollusca Class Gastropoda Fam. Muricidae *Rapana venosa* (Valenciennes 1846)	Sea of Japan	From the Black Sea, spreading through the Kerch Strait	1956	Established	Kostyuchenko and Nazarenko (1960)
Order Discopoda Fam. Littorinidae *Potamopyrgus jenkinsi* (Smith,1889)	Thames Estuary, Sweden waters (?)	Shipping, spreading through the Kerch Strait	1960s	Established	Mordukhai-Boltovsky (1968)
Order Nudibranchia Fam. Crambidae *Doridella obscura* (Verrill 1870)	Atlantic coastal waters of North America	Shipping (fouling) Penetration	1986–1991 Kerch Strait	Established	Sinegub (1994)

(continued)

Table 5.5 (continued)

Species	Donor region	Vector	First record	Status of establishment	References
Phylum Mollusca **Class Bivalvia** **Order** Venerida Fam. Myidae *Mya arenaria* (Linne.,1758)	The North Sea? Atlantic coastal waters of Europe or North America?	Shipping (fouling) penetration	1966–1967	Established	Mordukhai-Boltovsky (1968)
Order Mytilida Fam. Anadaridae *Anadara inaequivalvis* (Bruguiere 1789)	The northern part of the Adriatic Sea, there from the Pacific Ocean	Shipping (fouling) penetration	1989–1990	Established	Chikharev et al. (1994) and Zolotarev (1996)
Fam. Mytilidae *Mytilus galloprovincialis* Lmk.		Penetration	1950s	Established	Stark (1960)
Fam. Teredinidae *Teredo navalis* (Linne 1758)	Northern Atlantic or Pacific	Shipping (fouling) penetration	1953–1955	Established	Rybnikov and Nikolaeva (1963)
Class Crustacea **Order** Cirripedia Сем. Balanidae *Amphibalanus improvisus* (Darwin 1854)	Atlantic coastal waters of North America	Larvae—penetration with currents adults—fouling of vessels	Nineteenth century	Established	Mordukhai-Boltovsky (1968)
Fam. Mysidae Genus *Paramysis* *Paramysis agigensis* (Bacescu 1940)	Lower reaches of the Don Lower reaches of the Don Lower reaches of the Don	Penetration, ballast water?	2000 2000 2000	Established Established Established	Daneliya (2003)

Paramysis intermedia *Paramysis lacustris* (Czerniavsky 1882) *Limnomysis benedeni* (Czerniavsky 1882) *Katamysis warpachowskyi*	Lower reaches of the Don Lower reaches of the Volga		2000 2000	Established Established	
Class Crustacea **Order** Decapoda Fam. Xanthidae Dutch crab *Rhithropanopeus harrisii* (Maitland 1874)	Zuider See Bay (North Sea), previously from the Atlantic waters of North America	Shipping, spreading through the Kerch Strait	1960s	Established	Mordukhai-Boltovsky (1968)
Fam. Portunidae Blue crab *Callinectes sapidus* (Rathbun 1930)	Atlantic coastal waters of North America	Shipping, spreading through the Kerch Strait	1967 Kerch Strait	Established	Bulgurkov (1968)
Fam. Varunidae (Grapsidae) *Eriocheir sinensis* (Milne-Edwards 1853)	Introduced in Europe (the North Sea) in 1912 from the estuaries of the East Asia and the coastal waters of the South China and Yellow Seas	Via the rivers?	1998 Northeastern part of the Sea of Azov	Established	Murina and Antokolsky (2001)

Table 5.6 Fish species

Species	Donor region	Vector	First record	Status of establishment	References
Pisces **Class** Osteichthyes **Order** Mugiliformes Fam. Mugilidae *Planiliza haematocheilus* (Temminck and Schlegel 1845)	Estuaries of northern part of the Sea of Japan	Intentional introduction in the Black and Azov seas	1979–1984 juveniles released in Dairy Estuary in 1979, breeding herd was formed, release into the sea began in 1984	Established	Pryakhin and Volovik (1997)
Fam. Cyprinidae *Pseudorasbora parva* (Schleg)	Far Eastern rivers	Delivered along the way with *Hypophthalmichthys molitrix and Mylopharyngodon piceus*	1980s Lower reaches of the Don	Established	Nadtoka and Abramenko (1998)
Order Cypriniformes Fam. Cyprinidae *Hypophthalmichthys nobilis* (Richardson 1 845)	Rivers of China	Intentional introduction	1980s The Don Delta, reservoirs, brackish coastal waters of the Sea of Azov	Established	Volovik and Chikharev (1998a, b)
Order Cypriniformes Fam. Cyprinidae *Hypophthalmichthys molitrix (Valenci-ennes 1844)*	Rivers of China	Intentional introduction	1980s The Don Delta, reservoirs, brackish coastal waters of the Sea of Azov	Established, but the population is supported by artifi-cial reproduction	Volovik and Chikharev (1998a, b)
Order Cypriniformes Fam. Cyprinidae *Mylopharyngodon piceus* (Richardson)	Rivers of China	Intentional introduction	1980s The Don Delta, reservoirs	Established, but the population is supported by artifi-cial reproduction	Volovik and Chikharev (1998a, b)

Order Cypriniformes Fam. Cyprinidae *Ctenopharyngodon idella (Valenciennes 1844)*	Rivers of China	Intentional introduction	1980s The Don Delta, reservoirs	Established, but the population is supported by artificial reproduction	Volovik and Chikharev (1998a, b)
Ictiobus cyprinellus (Valenciennes)	North America	Intentional introduction	1980s Reservoirs	Established	Volovik and Chikharev (1998a, b)
Ictiobus bubalis (Rafinesque)	North America	Intentional introduction	1980s Reservoirs	Established	Volovik and Chikharev (1998a, b)
Ictiobus niger (Rafinesque 1819)	North America	Intentional introduction	1980s Reservoirs	Established	Volovik and Chikharev (1998a, b)
Carassius auratus (Linnaeus 1758)	Amur River	Incidental introduction	1930s	Established	Abramenko et al. (1997)
Order Cyprinodontiformes Fam. Poeciliidae *Gambusia holbrooki* (Girard 1859)	The coast of Italy, the Adriatic Sea, there from America, desalinated areas of the Black Sea?	Shipping, for the purpose of mosquito control	1970s Kuban Basin	Established	Volovik and Chikharev (1998a, b)
Fam. Oryziatidae *Oryzias latipes* (Temminck and Schlegel 1846)	Fresh waters of Japan	Brought by aquarists, probably penetrated from the Black Sea	1970s Kuban Basin	Established	Volovik and Chikharev (1998a, b)
Gobiidae *Gobius niger* Liner	Black Sea	Penetrated from the Black Sea	1970s Penetrated during salinization		Volovik and Dakhno (1983)

from the Black Sea between 1953 and 1955 (Rybnikov and Nikolaeva 1963; Iljin 2008). Under the present low-salinity environment, only *Teredo navalis* persists, and its population surges can threaten wooden underwater structures.

Another bivalve, the brackish-water clam *Mya arenaria*, was noted in the 1960s (Morduhai-Boltovskoi 1968). It has since become a significant part of the benthic community, particularly within low-oxygen sediment layers (Studenikina and Frolenko 2003; Frolenko 2006).

Following April 1989, the species *Anadara inaequivalvis* (Chikharev et al. 1994) expanded its range across the Sea of Azov. It is frequently found within *Cerastoderma* biocenosis but can also establish its own distinct community across extensive seabed regions. The proliferation of *Anadara inaequivalvis*, along with that of *M. arenaria*, is considered beneficial, as both mollusks inhabit biotopes with low oxygen levels that are unsuitable for other species (Frolenko and Dvoryainova 1998). These species hold commercial promise, containing up to 40% palatable meat (Studenikina and Frolenko 2003). The Black Sea mussel *Mytilus galloprovincialis* colonized the Sea of Azov in the late 1950s, coinciding with a period of rising salinity. Today, this mussel is the dominant benthic organism.

The crab *Rhithropanopeus harrisii* stands as one of the enduring non-native species in the Sea of Azov; it was formerly present even within the freshwater sections of the lower Don River, functioning as a food source for gobies, flatfishes, turbots, and sturgeons. Consequently, its introduction proved relatively advantageous (Morduhai-Boltovskoi 1968). A different crab species, *Callinectes sapidus*, emerged in 1967 within the Kerch Strait and is currently found there in limited numbers. A small number of individuals belonging to the Chinese mitten crab *Eriocheir sinensis* were initially discovered in 1968 in the northeastern region of the Sea of Azov (Murina and Antokolsky 2001). Five freshwater Ponto-Caspian mysids have been recently documented in the Sea of Azov, the delta of the Don River, and in the Tsimlyansk reservoir (Daneliya 2003).

The Taman Bay and the northern Kerch Strait serve as the northern distribution limit for numerous fish species. During phases of strong Black Sea water inflow, these species can extend their range further, thereby enriching the Sea of Azov's fauna (Volovik and Chikharev 1998a, b). In the years 1975–1976, a period of elevated salinity in the Sea of Azov, several species beyond the typical seasonal migrants temporarily entered its waters. These included the garfish *Belone belone euxini*, the mullets *Mugil cephalus* (Linnaeus 1758), *Chelon auratus* (Risso 1810), and *Chelon saliens* (Risso 1810), the atherine *Atherina boyeri* (Risso 1810), the Black Sea whiting *Merlangius merlangus* (Linnaeus 1758), the picarel *Spicara smaris* (Linnaeus 1758), and others, such as the bluefish *Pomatomus saltatrix* (Linnaeus 1766), the Black Sea flounder *Scophthalmus rhombus* (Linnaeus 1758), and turbot *Scophthalmus maeoticus* (Pallas 1814), along with the stingray *Dasyatis pastinaca* (Linnaeus 1758), the Spiny dogfish *Squalus acanthias*, the brown meagre *Sciaena umbra*, and the shi drum *Umbrina cirrosa*. Also recorded are the Black Sea salmon *Salmo trutta* (Linnaeus 1758), the mackerel *Scomber scombrus* (Linnaeus 1758), the wrasse *Symphodus ocellatus* (Linnaeus 1758), the blenny *Parablennius zvonimiri* (Kolombatovic 1892), the transparent goby *Aphia minuta* (Risso 1810), the sea smelt *Atherina hepsetus*, and the thickly snouted pipefish *Syngnathus variegatus* (Pallas 1814), among others. Species newly identified in the Sea of Azov include the corkwing *Symphodus cinereus* (Bonnaterre 1788), the mystery blenny *Parablennius incognitus* (Bath 1968) and *Parablennius sanguinolentus* (Pallas 1814), the sand goby *Pomatoschistus minutus* (Pallas 1770), the Black goby *Gobius niger* (Linnaeus 1758), the puntazzo *Diplodus puntazzo* (Walbaum 1792), and the Mediterranean catfish *Gaidropsarus mediterraneus* (Linnaeus 1758).

While these species (both rare and first recorded) were encountered in limited quantities, *Gobius niger* quickly became abundant (Volovik and Dakhno 1983). This suggests that under conditions of rising salinity, the Sea of Azov could be naturally populated by Black Sea immigrants capable of surviving temperatures below 3–5 °C. In such a scenario, the total number of fish species might rise to 140–150 (Volovik and Chikharev 1998a, b).

5.3.1 Deliberate Introduction of Commercially Valuable Fish Species

The majority of intentionally introduced species were benthic, while the rest were pelagic (Volovik and Chikharev 1998a, b). Among them, 15 species came from the fresh and brackish waters of the Far East and North America. The most successful introduction was the mullet *Planiliza haematocheilus* (Temminck and Schlegel 1845), previously known as *Liza haematocheilus* (*Mugil soiuy*). This species exhibits high tolerance to varying salinity and oxygen levels, feeds on detritus, and efficiently utilizes organic matter (Volovik and Pryahin 1997). It has now colonized nearly the entire Sea of Azov. Annually, a portion of its spawning population migrates from the Sea of Azov to the Black Sea. In recent years, mullet catches have been rising, establishing it as a key commercial fish in both the Black and Azov Seas.

To advance aquaculture in ponds and lagoons, species such as bighead carp (*Hypophthalmichthys nobilis*, Richardson 1845), silver carp (*Hypophthalmichthys molitrix*, Valenciennes 1844), black amur (*Mylopharyngodon piceus*, Richardson 1846), and grass carp (*Ctenopharyngodon idella*, Valenciennes 1844) were introduced. While silver carp and grass carp have expanded into rivers and lagoons, they reproduce exclusively in pond environments (Volovik and Chikharev 1998a, b).

Alongside the silver carp and grass carp, the stone moroko, *Pseudorasbora parva* (Temminck and Schlegel 1846), was also introduced. It is currently found in estuaries, rivers, and the Taganrog Bay (Nadtoka and Abramenko 1998). The fry and juveniles consume zooplankton and plant matter, whereas adults feed on benthic organisms and fish eggs. This species competes with both juvenile and adult planktivorous and benthivorous Cyprinidae, as they inhabit the same environments. The stone moroko is recognized as a highly invasive fish, spreading quickly throughout Europe.

The Far East goldfish, *Carassius auratus* (Linnaeus 1758), was unintentionally introduced with stocked fish (Abramenko et al. 1997) and has now spread near the Don River's mouth, appearing in the brackish waters of the Sea of Azov. It has outcompeted the endemic carp, *Carassius carassius* (Linnaeus), in Taganrog Bay, as well as other species. The Japanese medaka, *Oryzias latipes*, is an aquarium species that likely spread from the Black Sea, where it was released by aquarium hobbyists. It is now found in the Kuban River delta and in the brackish environments of the Sea of Azov (Volovik and Chikharev 1998a, b).

The mosquitofish, *Gambusia holbrooki*, was introduced into the Kuban River during the 1970s to help control mosquito populations.

The status of nine additional introduced fish species remains unclear (Volovik and Chikharev 1998a, b).

5.3.2 Environmental Degradation in the Sea of Azov and Its Impact on Marine Life

Historically, the Sea of Azov held the second most abundant fish populations in the Ponto-Caspian region, surpassed only by the Caspian Sea. In the recent past, each hectare of its waters yielded approximately 80 kilograms of fish, with valuable species such as sturgeons, pike perches, breams, and sea roaches accounting for half of that catch. The total annual harvest from this relatively small sea amounted to 300,000 tons. This remarkable productivity was supported by three key conditions: (1) substantial river inflow, which delivered plentiful nutrients; (2) relatively low salinity levels; and (3) the naturally high productivity of the fish stocks themselves.

However, starting in the mid-twentieth century, the environmental health of the Sea of Azov began to decline due to a combination of natural causes and human activities. The rise in water temperature associated with global warming (discussed in Sect. 5.4.1) was a key natural factor. Simultaneously, the intensive growth of the agro-industrial sector around the sea led to a major degradation of the ecosystem.

Primary human-induced factors that negatively impacted the environment included the regulation of river flows and water extraction, industrial and urban development, the building of irrigation networks and reservoirs within watersheds, and the expansion of transport infrastructure. A particularly damaging outcome of the substantial reduction in freshwater inflow was the sea's increasing salinity. This salinization, in turn, encouraged the migration of marine species from the Black Sea that originated in the Mediterranean or Atlantic. It is important to highlight that the shifts in the Sea of Azov's salinity manifested as distinct phases of increasing and decreasing salt content (refer to Sect. 5.4.1).

A clear illustration of salinity's impact on marine life is the altered development patterns of the harmful scyphozoan jellyfish species *Rhizostoma pulmo* and *Aurelia aurita*. Periods of increased salinity in the Sea of Azov have coincided with the mass proliferation of these jellyfish. Their initial documented appearance occurred in the

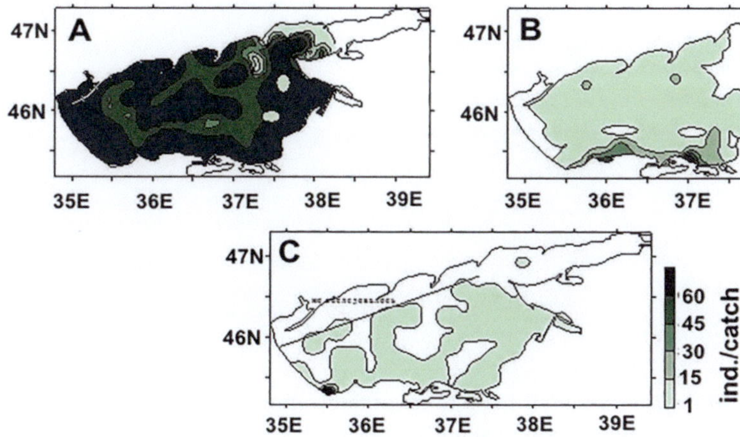

Fig. 5.8 Jellyfish *Rhizostoma pulmo* abundance across varying salinity stages (ind./catch). (**a**) 1976, 14.04‰; (**b**) 1983, 11.53‰; (**c**) 2018, 14.25‰. (Adapted from Mirzoyan et al. (2019) (Modified by authors))

early 1970s, corresponding with a salinity level of 13.6‰. A significant surge in their abundance and biomass was subsequently observed during 1975–1976, when salinity was approximately 13‰ (Fig. 5.8a).

Starting in the 1980s, a trend of desalination led to a decline in scyphozoan jellyfish populations (Fig. 5.8b). From 1981 to 2017, these species were only found sporadically. A notable resurgence was recorded in 2018, when salinity levels surpassed 13.3‰, marking the first substantial increase in jellyfish numbers since the 1980s. Despite the similar salinity conditions, the population outbreak in 2018 was considerably less intense than the one witnessed in the 1970s (Fig. 5.8c).

Consequently, it is highly pertinent to conduct a comparative analysis of the distribution, abundance, biomass, and age structure of *Rhizostoma pulmo* and *Aurelia aurita* populations from past and present salinization phases to identify the factors responsible for the differing developmental rates of these jellyfish.

The scaphoid jellyfish species *Rhizostoma pulmo* and *Aurelia aurita* form a characteristic macroplankton community in the Black Sea, distinguished by their yellow bodies. This complex thrives in the Sea of Azov when a specific set of environmental conditions align: elevated salinity and temperature, abundant food resources, and a scarcity of competing species. During the 1970s, these factors converged perfectly, leading to an intensive proliferation of jellyfish. They populated the Sea of Azov in enormous quantities, spreading across almost the entire basin, including the Taganrog Bay. The population peak, estimated at approximately 7.0 billion individuals, was documented in 1978, whereas the highest biomass was recorded 2 years earlier, in 1976. An illustration of a significant *Rhizostoma pulmo* bloom along the Azov Sea coastline is presented in Fig. 5.9.

However, in the contemporary phase of salinization, the growth of scaphoid jellyfish takes place amid intense food competition from the recent invasive species,

Fig. 5.9 Aggregation of *R. pulmo* at the coast of the Sea of Azov (the Taman Bay area). (Photo by T. Shiganova)

the ctenophore *Mnemiopsis leidyi*. Despite favorable salinity and temperature conditions, the extremely limited food availability places the jellyfish in a state of poor trophic security, leading to significantly less vigorous development compared to earlier salinization periods. In 2018, jellyfish numbers reached only 0.01 billion individuals, with biomass recorded at approximately 0.014 million tons—figures that are two orders of magnitude lower than during the jellyfish expansion in the Sea of Azov in the 1970s.

5.3.3 A Brief Overview of the Impact of Non-native Species on Ecosystem

This section offers a concise overview of both the detrimental and beneficial outcomes resulting from non-native species invasions in the Sea of Azov ecosystem. The analysis excludes the specific impacts of the invasive ctenophore *M. leidyi* and its predator *B. ovata*, as these will be examined thoroughly in subsequent sections. It is important to note that certain invasive species can simultaneously produce both negative and positive effects. Since the nineteenth century, the count of non-native species in the Sea of Azov has increased exponentially, reaching a total of 46 species by the period 2000–2005 (Fig. 5.10).

The following key examples of invaders' impacts are notable:

1. Within the microplankton community, the El Tor strain of *Vibrio cholerae* poses the greatest threat to human health. This pathogen can survive in coastal marine environments and transmit infections to populations in nearby terrestrial areas. Initial epidemic outbreaks of cholera attributed to the El Tor strain were documented in 1970 along the Black Sea coastline, particularly near the Kerch and Odessa regions.
2. *Tintinnid ciliates* serve as a crucial macroplanktonic connection within aquatic food webs. They prey on planktonic algae and cyanobacteria, capable of ingesting as much as 27% of the yearly primary production in coastal regions, which diminishes the base of the food supply. Concurrently, tintinnids

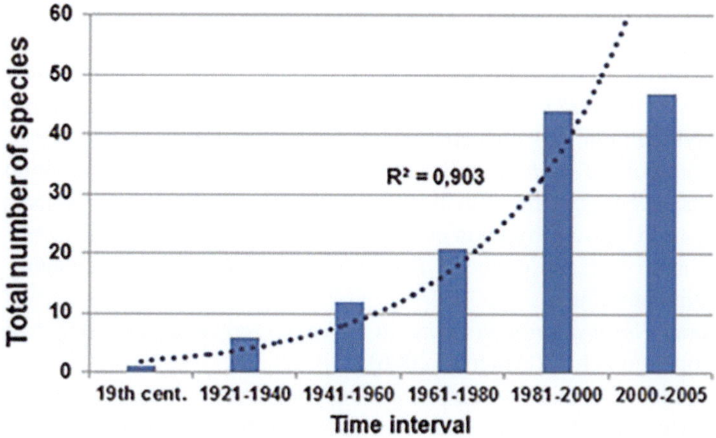

Fig. 5.10 Rise in the count of nonindigenous species within the Sea of Azov. Dotted line—exponential approximation (accuracy of the approximation $R^2 = 0.9336$)

themselves represent a significant seasonal dietary element for meso- and macrozooplankton, including copepods (Capriulo and Carpenter 1983), large zooflagellates, and fish larvae.

3. The majority of phytoplankton species play a vital role in the ecosystem by serving as a primary food source for zooplankton. However, the diatom alga *Pseudosolenia calcar-avis*, when it experiences massive proliferation, can displace more beneficial native species of trophic phytoplankton. Additionally, certain phytoplankton species (e.g., *Alexandrium tamarense*, *A. ostenfeldii*, *Dinophysis acuminata*, *Pseudonitzschia pungens*, *Heterosigma akashiwo*) are detrimental to the ecosystem, as they have the potential to generate toxic blooms (Kovaleva 2002).
4. The gelatinous hydromedusa *Blackfordia virginica* is a damaging invasive organism, extensively found in brackish environments across the World Ocean. Its medusae consume zooplankton and may act as competitors to fish larvae. The ctenophore *Mnemiopsis leidyi* is another detrimental invader, which caused severe disruption to the Sea of Azov ecosystem following its introduction. However, the arrival of its predator, *Beroe ovata*, led to a progressive restoration of the ecosystem. This topic will be examined in greater depth in the subsequent sections. The copepod *Acartia tonsa* is a beneficial species, serving as a food source for planktivorous fish.
5. The detrimental effects of the fouling organism *Amphibalanus improvisus* are evident on the hulls of vessels, within pipelines, on piers and dams, and across seabed areas colonized by their communities. Conversely, their free-swimming larval stages represent a major component of the coastal meroplankton diet.
6. The benthic organisms *Mya truncata* (L) and *Anadara inaequivalvis* (L) commonly populate low-oxygen zones that are unsuitable for other

bottom-dwelling species. They serve as an important food supply for benthophagous fish, and their larval stages are eaten by small pelagic fish. Furthermore, *Mya truncata* (L) is a type of edible shellfish consumed by humans. *Anadara inaequivalvis* (L), a marine bivalve mollusk belonging to the Myidae family, is also appropriate for aquaculture.

7. The shipworm *Teredo navalis* is too large to be eaten by zooplankton. During blooms, it poses a threat to the Sea of Azov ecosystem by displacing native species and causing hypoxia near the seabed (Studenikina et al. 1999). Furthermore, and of greater concern, *Teredo navalis* is capable of damaging wooden hydraulic structures.
8. The crab *Rhithropanopeus harrisii* has emerged as a new food resource for benthophagous fish (Frolenko 2006).
9. The mullet species *Planiliza haematocheilus* has become a commercially significant fish in the Sea of Azov, a region where native mullet populations are sparse and do not reproduce naturally (Volovik and Chikharev 1998a, b).
10. In 1947, the gastropod *Rapana venosa* (Valenciennes) was first recorded in the Novorossiysk region. This predatory species feeds on oysters, mussels, and various other bivalves (Drabkin 1953). More recently, however, *Rapana* has gained recognition as a prized delicacy.
11. *Pseudorasbora parva* (Amur chebachok), along with carp and white cupid—species originally introduced for aquaculture—spread into the Black Sea through canal and river networks, eventually reaching the Sea of Azov. Today, *Pseudorasbora parva* is considered a widespread weed invasive fish.

5.4 *The Mnemiopsis leidyi* population in the Sea of Azov Has Significantly Influenced the Region's Ecosystem

5.4.1 Development of Mnemiopsis leidyi *Population*

Mnemiopsis leidyi was initially introduced into the Sea of Azov from the Black Sea in August 1988, as documented by Studenikina et al. (1991). A representative specimen of *M. leidyi* from the Sea of Azov is illustrated in Fig. 5.11.

Since then, it has entered the Sea of Azov annually, establishing a breeding population that dies off entirely when temperatures fall below 4 °C (Mirzoyan et al. 2000). The presence of *M. leidyi* in the Sea of Azov, along with its distribution and population dynamics, is influenced by a combination of biotic and abiotic factors. These include (i) the synoptic processes occurring over the Azov-Black Sea basin, such as prevailing wind patterns and temperature conditions; (ii) the density of *M. leidyi* individuals in the Black Sea and the strength of their influx; and (iii) the availability of food resources in the Sea of Azov.

The timing of *M. leidyi's* emergence in the Sea of Azov is influenced by the prevailing wind patterns during spring. Long-term studies have identified two

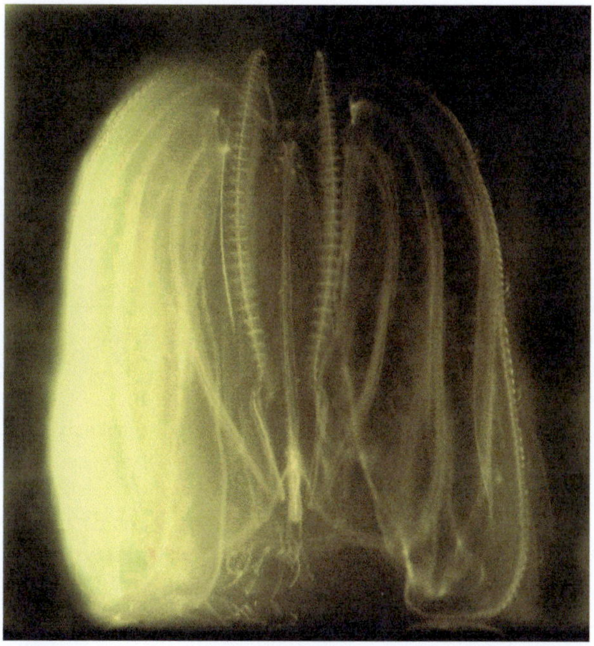

Fig. 5.11 *Mnemiopsis leidyi* from the Sea of Azov. (Photo by T. Shiganova)

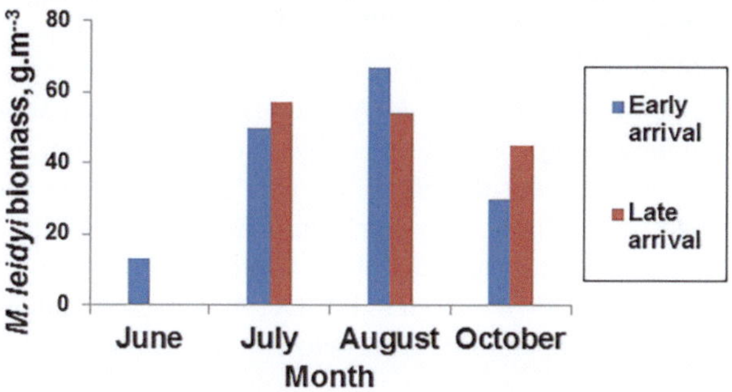

Fig. 5.12 Seasonal development patterns of *M. leidyi* differ significantly depending on whether its arrival occurs early or late in the year

distinct patterns in the seasonal population dynamics of this invasive species: an early or a late arrival of *M. leidyi* from the Black Sea (Fig. 5.12). Dominant southerly winds promote an earlier influx of a significant number of individuals from the Black Sea, typically occurring in April–May. Conversely, when easterly and northerly winds are more frequent, the springtime inflow of ctenophores is reduced. In certain situations, offset phenomena can even lead to a contraction of areas already colonized by the species. Under these conditions, a delayed development of the *M. leidyi* population is observed.

The highest levels of *M. leidyi* abundance and biomass occur during the late summer and autumn months (Fig. 5.12). Individuals of *M. leidyi*, which grew up in the Sea of Azov, are smaller than the Black Sea ones. Maximal length of an adult

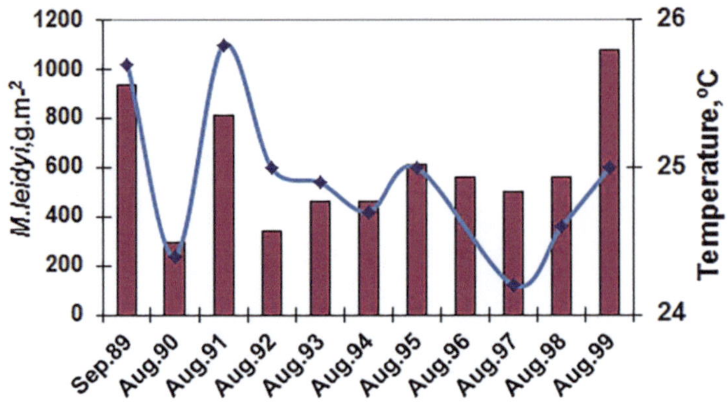

Fig. 5.13 Interannual variability of *M. leidyi* seasonal maximum of biomass (g.m^{-3}, bars) and sea surface temperature (°C, blue line) in the Sea of Azov

M. leidyi, born in the Sea of Azov, may reach 65 mm, while an average size is about 20 mm. The peak dimensions of *M. leidyi* are typically observed during August and September, influenced by the temperature of the surface water (Fig. 5.13).

5.4.2 Mnemiopsis leidyi *Reproduction*

In the Sea of Azov, *M. leidyi* begins reproduction soon after its appearance, if the zooplankton concentration is high enough. Later on, *M. leidyi* gradually reproduces and extends over the entire sea area except for the eastern part of Taganrog Bay, where the salinity is lower than 3‰. Reproduction reaches its peak in August (in case of early arrival) or in September–October (in case of late arrival). The reproduction slows down in late October; in November, the entire population extinct when the temperature drops below 4 °C (Mirzoyan et al. 2000).

5.4.3 Mnemiopsis leidyi *Fecundity*

During experimental observations, the highest reproductive rate in the open sea was documented for a 30 mm specimen (1250 eggs ind.$^{-1}$ day^{-1}) and for a 72 mm specimen (1210 eggs ind^{-1} day^{-1}). The lowest recorded fecundity was 50 eggs ind^{-1} day^{-1}, observed in a 50 mm individual. The largest specimens collected from Yasinskii Bay did not produce any eggs. The computed average fecundity in the open sea was 897 ± 393 eggs ind^{-1} day^{-1}, while in Yasinskii Bay, it was 600 ± 455 eggs ind^{-1} day^{-1}, both measured at a consistent water temperature of 25 °C (Shiganova and Mirzoyan, unpublished).

5.4.4 *The Impact of* Mnemiopsis leidyi *on Zooplankton*

In the Sea of Azov, *M. leidyi* mirrored its effects in the Black Sea, but with greater intensity and speed. This heightened effect is attributed to the Sea of Azov's smaller area, shallow depth, and high zooplankton productivity. The explanation lies in the ctenophore's capacity for extremely high grazing rates on virtually all plankton groups and species, ranging from the smallest organisms like rotifers and copepod nauplii to the largest plankton representatives. A critical factor is that the sea's shallowness makes the entire water column of zooplankton accessible to *M. leidyi*. Furthermore, the predatory activity of *M. leidyi* directly influences the seasonal development patterns of the zooplankton community (Fig. 5.14).

During the spring season, when *M. leidyi* is not yet present, zooplankton populations experience active growth. If *M. leidyi* arrives early, this growth phase is limited to just one spring month (May). However, when *M. leidyi* appears later, the spring zooplankton peak extends over a longer period, continuing into the early summer (Mirzoyan et al. 2000). Due to a decline in zooplankton consumption by planktivorous fish—whose numbers dropped significantly following the introduction of *M. leidyi* (Luts et al. 1997; Rogov 1996). The plankton maximum became twice as intense compared to periods before *M. leidyi* was established.

The second phase of zooplankton development is marked by the arrival of *M. leidyi*. From June through late autumn, zooplankton faces intense predation pressure from *M. leidyi*. As a result, both the abundance and biomass of zooplankton drop to extremely low levels (Fig. 5.14); by July, these values are several orders of magnitude lower, and in some areas, zooplankton disappears entirely by August. Throughout the summer, only 10% of the annual plankton production remains,

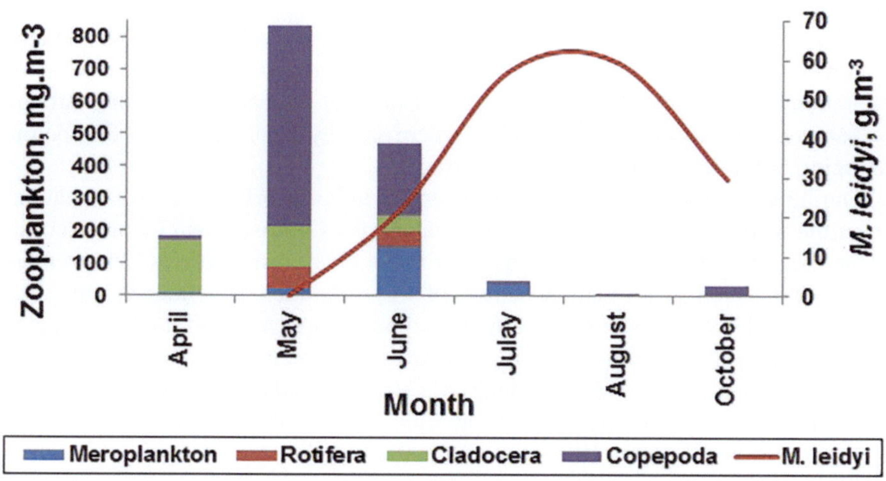

Fig. 5.14 Seasonal variability of zooplankton (bars) and *M. leidyi* (red line) biomass (mg.m^{-3}) in the Sea of Azov

whereas in the early 1980s, before *M. leidyi* appeared, this figure stood at 56% and was evenly distributed across the summer months.

Consequently, after the arrival of *M. leidyi*, the seasonal pattern of zooplankton in the Sea of Azov no longer includes a summer peak. This shift impacted the feeding efficiency of planktivorous fish and altered their life cycle (Mirzoyan et al. 2000).

5.4.5 Impact of **Mnemiopsis leidyi** *on Zoobenthos*

Following the introduction of *M. leidyi*, its initial prey consisted of meroplankton species, particularly pelagic larvae. Meroplankton can account for as much as 90% of the total zooplankton biomass and is heavily preyed upon by *M. leidyi*, potentially altering larval settlement rates. The dominant portion of meroplankton is made up of Bivalvia larvae. As a result, their populations were the most severely impacted by *M. leidyi*.

Bivalve mollusk reproduction in the Sea of Azov begins in May and June. Because their larvae can remain in the water column for more than a month, the second generation is also likely affected by *M. leidyi*. In such cases, the number of settled juvenile mollusks measuring under 2 mm in July and August, and less than 8 mm in October, becomes extremely low (Studenikina and Frolenko 2000). In areas with the highest concentrations of *M. leidyi*, bottom biocenoses were nearly devoid of mollusk juveniles, indicating a significant decline in their reproductive success.

5.4.6 Impact of **Mnemiopsis leidyi** *on Fish*

The Azov anchovy, *Engraulis maeoticus* (Pusanov and Zeeb 1926), represents the dominant small pelagic fish species in the region. It undertakes an annual seasonal migration into the Sea of Azov from the Black Sea, typically from April through November. This species inhabits zones where salinity levels exceed 7–8‰ before returning to the Black Sea for overwintering.

During February and March, or occasionally in late April, the anchovy that has spent the winter in the near-bottom waters along the Caucasian and Crimean coasts of the Black Sea becomes active. It ascends to the surface and initiates its movement toward the Kerch Strait. The schedule of this migration is influenced by solar radiation and water temperature. In the southern spawning grounds, surfacing generally takes place by mid-March, whereas in northern areas, it occurs in early April. The migration proceeds along a narrow coastal strip, typically within 1.5–2.0 km of the shoreline. The rate at which the shoals travel is determined by the pace of water warming (Rogov et al. 2000).

After entering the Sea of Azov through the Kerch Strait, the anchovy spreads throughout the water body. The highest fish densities are typically located in the northern, northeastern, and western areas, whereas the central and southeastern zones show the lowest concentrations. The thermal properties of the water and the existing zooplankton biomass are the primary factors controlling the fish's distribution and numbers (Rogov et al. 2000).

In terms of its feeding habits, the anchovy is a filter feeder, and its diet in the Sea of Azov is dominated by seston organisms (Kornilova 1960). The variety of organisms it consumes is generally determined by the seasonal makeup of the plankton community. Prior to the introduction of *M. leidyi*, the anchovy's trophic conditions were adequately broad. For the first half of its feeding season, it fully met its nutritional needs by consuming between 30% and 90% (averaging 50%) of the spring and summer zooplankton production (Mirzoyan et al. 1998).

Throughout the main summer feeding period, anchovy and kilka continue to feed exclusively on plankton, though without depleting it entirely. These points to a stable "zooplankton-planktivorous fishes" food web, enabling efficient feeding within shared marine foraging grounds and supporting relatively high population levels (Mirzoyan et al. 1998).

Following the *M. leidyi* invasion, shifts in dietary composition became evident by August. By this period, the diet had shifted to low-energy items such as phytoplankton, meiobenthos (including worms, larvae, and juvenile crabs), and mysids, as roughly 90% of the traditional food source (Copepoda species) had virtually vanished. The intestinal fullness index recorded was 25‰ (Mirzoyan et al. 1998). Consequently, the latter part of the anchovy's feeding season took place under even more challenging trophic circumstances. The sea lacked adequate feeding areas. Zooplankton biomass fell below the critical threshold, measuring only 2–77 $mg{\cdot}m^{-3}$, whereas the minimum required for effective feeding by planktivorous is at least 100 $mg{\cdot}m^{-3}$ (Rogov et al. 2000).

After the arrival of *M. leidyi*, the body length of the fish by late August (the premigration phase) became 10–15 cm shorter compared to the period before the invasion. Additionally, the average body weight and fat content of fingerlings declined. These changes signal the poor physical state of the young fish and their lack of readiness for autumn migration and overwintering, contributing to a higher rate of natural mortality during their first year. The overall degradation of zooplankton conditions linked to the *M. leidyi* invasion ultimately led to a severe reduction in the stocks and catches of Azov anchovy (Fig. 5.15).

The Azov kilka population exhibits notable variations in its abundance compared to generational yields. These fluctuations are distinctly evident; in certain instances, the most robust generation's abundance surpasses that of the weakest by as much as 50 times. Consequently, during productive years, over 80% of the population is composed of under-yearlings, whereas in less productive years, the share of older age groups—particularly 2-year-old individuals—rises, accounting for up to 75% of the total population. The abundance of 3-year-old fish is generally low. Additionally, kilka biomass undergoes substantial interannual changes.

A decline in the kilka stock and a reduction in its quality have been clearly observed since 1988, following the introduction of the ctenophore *M. leidyi*. The fat content in 2- and 3-year-old individuals reached its lowest recorded levels during this period, ranging from 8.1% to 13.3%, compared to 24.1% to 29.3% in 1987. In

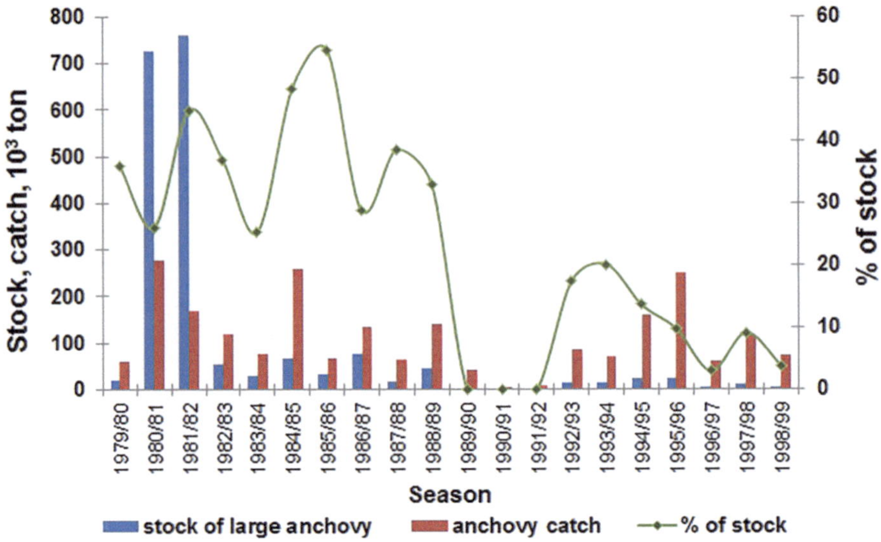

Fig. 5.15 Azov anchovy stock levels (red bars), catch quantities (blue bars), and the proportion of catch relative to stock (green line). (Compiled from Rogov et al. 2000; plotted by T. Shiganova)

the years that followed, fat content stabilized; however, current peak values no longer reach the average levels seen prior to the appearance of *M. leidyi* (Rogov et al. 2000).

The reproductive efficiency of kilka also declined due to the deterioration of feeding conditions. A shortage in the forage base during the summer led to the division of the kilka population into two distinct groups with different qualities. Those individuals that had already spawned moved to the sea to feed but encountered unfavorable nutritional circumstances. This was because of following the introduction of *M. leidyi*, the biomass of zooplankton during the peak feeding months for kilka (July–August) became extremely limited. As a result, by autumn, under ongoing food scarcity, kilka accumulate minimal fat reserves, with 60–65% of the population having a fat content no greater than 13% of their body mass. This poorly nourished and weakened segment of the population then skips the subsequent spawning season.

The invasion of *M. leidyi* has led to severe declines in fishing productivity. Kilka catches have plummeted by factors ranging from several tens to several hundreds. Fixed-net kilka fishing has become nearly insignificant. Due to stock depletion and delayed maturation in parts of the population that skip spawning, the biomass of spawners at reproductive sites did not exceed 50 thousand tons. Catch per unit effort dropped by more than four times, and total fixed-net catches fell by more than eight times (Fig. 5.16).

Consequently, the arrival of the ctenophore *M. leidyi* compromised the food supply for planktivorous fish during their feeding seasons, leading to a decline in the quality of anchovy and kilka. This nutritional shortfall and poor physiological

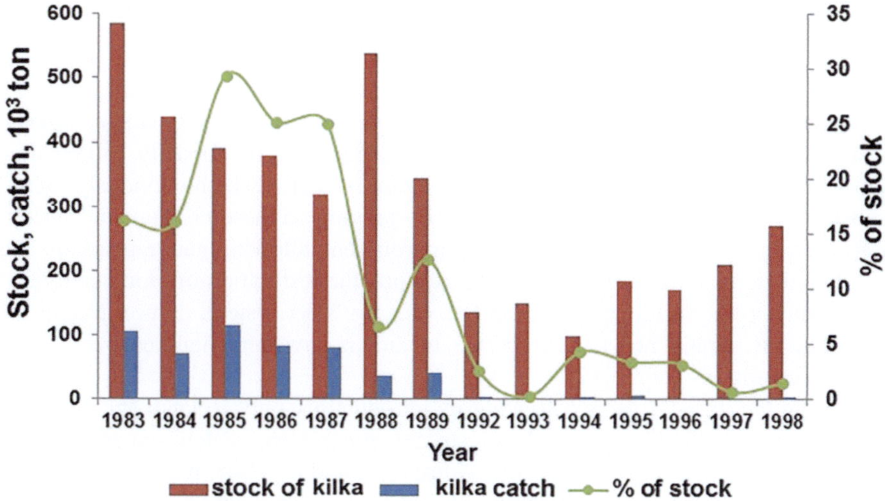

Fig. 5.16 Stock (red bars), catch (blue bars), and percentage of catch from stock (green line) of Azov kilka. (Compiled after Rogov et al. 2000; plotted by T. Shiganova)

state contribute to the low reproductive success of these key pelagic species, ultimately causing a twofold reduction in fish stocks and a tenfold drop in commercial catches (Rogov et al. 2000).

No specific research has been conducted on the effects of *M. leidyi* on phytoplankton. Nevertheless, a rise in primary production exceeding 1.5 times was observed following the arrival of *M. leidyi* between 1988 and 1998 (Aleksandrova et al. 2000).

In summary, it is important to emphasize that the Sea of Azov offered more favorable conditions for the development of the *M. leidyi* population than the Black Sea. This is primarily due to higher summer temperatures and a greater biomass of its primary food source, zooplankton. However, because of colder winters, *M. leidyi* can only inhabit the Sea of Azov during the warmer months. In this sea, *M. leidyi* achieves a higher abundance compared to the Black Sea, with a recorded maximum of 21,020 ind. m^{-3} (Mirzoyan et al. 2000) nearly three times greater than the peak abundance in the Black Sea (7600 ind. m^{-2}) (Volovik and Chikharev 1998a, b). Despite this, the biomass of *M. leidyi* in the Sea of Azov is considerably lower (1089 g m^{-2}) than in the Black Sea (4600 g m^{-2}), which is attributed to the notably smaller individual size and correspondingly lower individual weight of the ctenophores in the Sea of Azov.

Thus, in the Sea of Azov, a cascading effect of *M. leidyi* was observed both up the trophic web and downward to the lower trophic levels. Moreover, due to the small size, shallow waters, and high productivity of the sea, the impact level was even more significant than in the Black Sea.

5.5 *Beroe ovata* in the Sea of Azov and Its Contribution to the Ecosystem Recovery

A species of comb jelly has played a notable role in the region's ecological restoration. This gelatinous organism is recognized for its capacity to consume other invasive ctenophores, particularly *Mnemiopsis leidyi*, which previously caused substantial disruptions to local food webs. By preying on these problematic species, *Beroe ovata* helps rebalance the marine environment, allowing native plankton and fish populations to recover. Its presence supports the restoration of biodiversity and contributes to the stabilization of the Sea of Azov's ecosystem, demonstrating the potential for certain non-native species to aid in environmental recovery under specific conditions (Fig. 5.17).

During the initial 4 years following its introduction, only isolated specimens of *B. ovata* were sporadically found in restricted zones across a few sea regions. By 2003, however, *B. ovata* had become widespread, and by 2005, it had established a large, persistent population. This population comprised individuals from all life stages, including larvae that had not been observed in earlier periods. Larval densities ranged from 0.5 to 1.0 individual per cubic meter, with average concentrations reaching 5.0 ind.m^{-3} in areas of high aggregation. These findings confirm that *B. ovata* successfully acclimated to its new surroundings, reproducing and

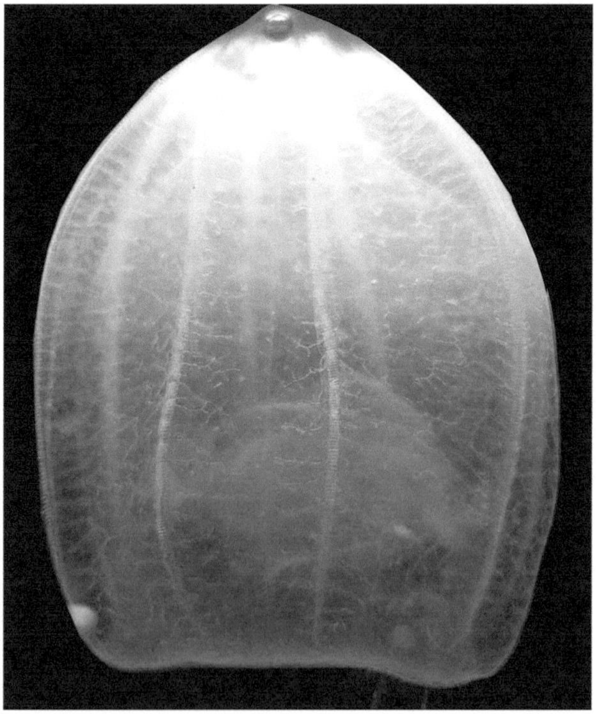

Fig. 5.17 An example of the Sea of Azov *Beroe ovata* with the swallowed *Mnemiopsis leidyi* inside

forming a vigorously feeding community. Estimates indicate that it consumes between 73% and 99% of the *M. leidyi* population.

During the late autumn season, *B. ovata* disappears and remains absent from the Sea of Azov during the winter and spring. Therefore, it can be considered similar to *M. leidyi* a seasonal migrant that reestablishes its population annually. The specific pattern of *M. leidyi* development in the Sea of Azov after 2005 is linked to a notable reduction in its inflow from the Black Sea. This, in turn, resulted from the decline of the *M. leidyi* population in the Black Sea due to control by *B. ovata*. This explains the diminished delivery of *M. leidyi* in the southern Sea of Azov and the lengthening of the initial concentration period. Consequently, delayed development of the *M. leidyi* population became more common.

Recent developments in *M. leidyi* populations show that their initial abundance has decreased by three orders of magnitude compared to periods when *B. ovata* was absent. This decline has resulted in a slower initial population growth rate and an extended developmental period. When *M. leidyi* inflow is reduced—similar to scenarios involving its late arrival—zooplankton productivity remains relatively high during the first 2–3 months. This allows pelagic fish, especially anchovies, to produce substantial numbers of eggs and larvae due to increased trophic resources. In such years, *M. leidyi* typically spreads rapidly across the entire basin during the more active second phase of its development cycle, multiplying its abundance and biomass to levels that fall within the range of interannual variations.

During the autumn season, the growth rate of the *M. leidyi* population progressively slowed. By October, isolated zones began to emerge in the southern and southeastern regions of the sea, where *M. leidyi* numbers were minimal or entirely absent. In 2005, similar areas were also documented in the central and western parts of the sea. The formation of these regions with lower *M. leidyi* concentrations generally coincided with the arrival of *B. ovata* in the same areas.

Over the past 7 years in Taganrog Bay, *M. leidyi* has not been observed during the month of June. In July, it was missing in both 2001 and 2004. During the remaining years, *M. leidyi* was found exclusively in the western section. By August, the species had extended its presence only to the central boundary of the basin. Only in the years noted when population development was most intense did its range expand to include the traditional zones of the western and central areas. The eastern part of the bay was an exception, as salinity levels below 3‰ prevented the spread of *M. leidyi*. In recent times, the ongoing desalination of Taganrog Bay has led to a further reduction in the summer habitat area of *M. leidyi*. In previous years, the abundance and biomass of *M. leidyi* in Taganrog Bay declined due to the invasion of *B. ovata*.

Therefore, it is evident that the Sea of Azov sustains two distinct seasonal populations of ctenophores, whose growth is influenced by particular biotic and abiotic conditions. The relationship between these two ctenophore species adheres to a predator (*B. ovata*)-prey (*M. leidyi)* dynamic. *B. ovata* exerts direct biological regulation over the development of *M. leidyi* (Fig. 5.18). The efficiency of this regulatory mechanism improves over time, contributing to the progressive recovery of food supplies and fish populations.

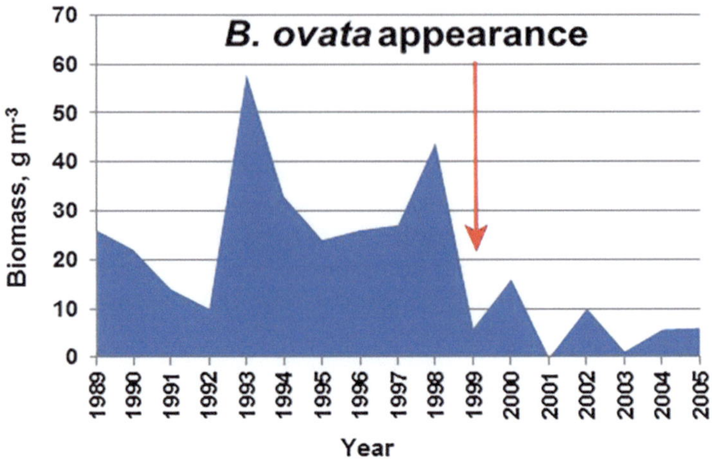

Fig. 5.18 Yearly fluctuations in the biomass (g.m^{-3}) of *M. leidyi* within the southern region of the Sea of Azov proper from 1999 to 2005. (Adopted from Mirzoyan et al. 2006; modified by T. Shiganova). The red arrow indicates the period when *B. ovata* first appeared

References

Abramenko MI, Kravchenko OV, Velikoivanenko AE (1997) The genetic structure of populations in the diploid-triploid complex of the silver crucian carp Carassius auratus gibelio in the basin of the Lower Don. Quest Ichthyol 37(1):62–67

Aleksandrova ZV, Baskakova TG, Romova MG (2000) Features of the hydrochemical regime and production of primary organic matter in the ecosystem in the modern period. In: Volovik SP (ed) Ctenophore Mnemiopsis leidyi (A.Agassiz) in the Azov and Black Seas: biology and consequences of invasion. AzNIIRKH, Rostov-on-Don, pp 173–188. (in Russian)

Berdnikov SV, Dashkevich LV, Kulygin VV (2019) Climatic conditions and hydrological regime of the sea of azov in the XX – early XXI centuries. Aquatic Biores Environ 2(2):7–19. http://journal.azniirkh.ru, www.azniirkh.ru, ISSN 2618-8147 print, ISSN 2619-1024 online

Berdnikov SV, Kulygin VV, Dashkevich LV (2023) Reasons for Rapid Increase of Water Salinity in the Sea of Azov in the twenty-first Century. Phys Oceanogr 30(6):714–730

Berg LS (1949) Fishes of fresh waters of the USSR and neighboring countries. Parts 1,2,3, 4th edn. USSR Academy of Sciences, Moscow. (in Russian)

Bronfman AM, Khlebnikov EP (1985) The Sea of Azov. In: Fundamentals of reconstruction. L.: hydrometeorological unit. 272 pp

Bulgurkov KI (1968) Callinectes sapidus rathbun in the Black Sea // izv. Res Dev 9:97–99

Chikharev AS, Frolenko LN, Rekov YI (1994) The bivalve mollusk of Anadara sp. (Cunearca cornea) a new non-native species in the Sea of Azov. Fisheries. No. 3, pp 35–37

Daneliya ME (2003) Mysids (Crustacea: Mysidacea) of the Sea of Azov Basin. The abstract of Ph. D. thesis. St.-Petersburg. 23 p

Dobrovolsky AD, Zalogin BS (1982) Seas of the USSR. Publishing House of Moscow State University. 192 p

Frolenko LN (2000) Polychaetes of the Sea of Azov and features of their development. In: Makarov EV, Volovik SP, Tyutina Yu EC (eds) The main problems of fisheries and protection of fisheries reservoirs of the Azov-Black Sea basin. AzNIIRKH, Rostov-on-Don, pp 62–64

Frolenko LN (2006) Assessment of bottom biocinosis in the Sea of Azov. In: Voynov NV (ed) Basic problems of fisheries and protection of fishery reservoirs in the Azov-Black Sea basin (2004–2005). AzNIIRKH, Rostov-on-Don, pp 77–87

Ginzburg AI, Kostianoy AG, Serykh IV, Lebedev SA (2021) Climate Change in the Hydrometeorological Parameters of the Black and Azov Seas (1980–2020). Oceanology 61:745–756. https://doi.org/10.1134/S0001437021060060

Gubanova AD (2003) Acartia tonsa is a new species of Copepoda in the Black Sea. In: Eremeev VN, Gaevskaya AV (eds) Current state of biodiversity in the coastal waters of Crimea (Black Sea sector). ECOSI-Hydrophysics, Sevastopol, pp 99–103NAS of Ukraine, InBUM

Kiseleva MI (1987) Changes in the composition and distribution of polychaete worms in the Sea of Azov. Hydrobiol J 23(2):40–46

Kornilova VP (1960) Biology and fish farming of the Azov anchovy (Engraulis encrasicholus maeoticus pus). Proc AzCherNIRO 18:50–73

Kosarev AN, Kostianoy AG, Shiganova TA (2007) The Sea of Azov. In: Kostianoy AG, Kosarev AN (eds) The Black Sea environment. The handbook of environmental chemistry, vol 5Q. Springer, Berlin, Heidelberg. https://doi.org/10.1007/698_5_091

Kostyuchenko RA, Nazarenko VV (1960) Rapana in the Sea of Azov. Nature 12:107–109. Inst Oceanol. USSR Academy of Sciences. 49:118–135 pp

Kreneva KV (2003) A new species of shell ciliates for the fauna of the Sea of Azov. In: Evolution of marine ecosystems under the influence of intruders and artificial mortality of fauna. Abstracts of reports. Rostov-on-Don, 20–21 pp

Lebedev EM (1961) Fouling vessels in the fouling of the Sea of Azov and the Kerch Strait. Tr Inst Oceanol USSR Acad Sci 49:118–135

Mirzoyan ZI, Volovik SP, Kukhta MA, Naboka TV, Lutz GI, Rogov SF (1998) Characteristics of nutrition and provision of food for pelagic fish of the Sea of Azov. In: Collection of scientific papers of the AzNIIRHa "The main problems of fisheries and protection of fisheries reservoirs of the Azov-Black Sea basin" (1996–1997). Rostov–on-Don, pp 58–67

Mirzoyan ZA, Volovik SP, Kornienko GG, Dudkin SI, Logichevskaya TV (2000) Biology of ctenophore *Mnemiopsis leidyi* in the Sea of Azov. In: Volovik SP (ed) Ctenophore *Mnemiopsis leidyi* (A.Agassiz) in the Sea of Azov and the Black Sea and its effects on ecosystems, Rostov-on-Don, pp 101–144. (in Russian)

Mirzoyan ZI, Volovik SP, Martynyuk ML (2002) Features of Beroe ovata population development in the Azov-Black Sea basin. In: Caspian Floating University No. 3. KaspNIRKh, Astakhan, pp 105–112

Mirzoyan ZI, Martynyuk ML, Viazun EV (2006) Features of the development of ctenophores *Beroe ovata* and *Mnemiopsis leidyi* in the Sea of Azov in the modern period. In: Proceeding of scientific AZNIIRKH "The main problems of fisheries and the protection of fisheries reservoirs in the Azov-Black Sea basin" (2004–2005), Rostov-on-Don, pp 136–148

Mirzoyan ZA, Martynyuk ML, Khrenkin DV, Afanasyev DF (2019) Development of the scyphozoan jellyfish rhizostoma pulmo and aurelia aurita populations in the Sea of Azov. Aqua Bioresour and Habit 2(2):27–35. (in Russian)

Mordukhai-Boltovskoi FD (1969) Crustacea. In: Vodyanitskiy VA (ed) Determinant of the fauna of the Black and Azov Seas, vol 1. Invertebrates. Naukova Dumka: The Caspian fauna in the Azov-Black Sea basin, M. N., L. N.: USSR Academy of Sciences, Kiev. 288 p. (in Russian)

Mordukhai-Boltovskoi FD (1972) Guide of the Black and Azov Seas Fauna. Naukova Dumka, Kiev. (in Russian)

Murina VV, Antokolsky AG (2001) Chinese crab *Eriocheir sinensis* – an invader in the Sea of Azov. Ecol Sea 55:37–39

Naumov DV (1968) Hydrozoan class Hydrozoa (Class Hydrozoa). Definition of the fauna of the Black and Azov Seas, vol 1. Naukova Dumka, Kiev, pp 56–70

Polikarpov IG, Saburova MA, Manzhos LA, Pavlovskaya TV, Gavrilova NA (2003) The Black Sea microplankton biodiversity in the coastal zone near Sevastopol (2001–2003). In: Eremeev VN,

Gaevskaya AV (eds) Modern state of coastal waters biodiversity of Crimea (Black Sea sector). Ecosea-hydrophysica, Sevastopol, pp 16–43. (In Russian with English summary)
Rogov SF (1996) Dynamics of the population of the Azov Hamsa in 1986–1992. In: Collection of scientific tr. Azniirkha, Rostov-on-Don, pp 174–178
Rogov SF, Lutz GI, Volovik SP (2000) Biology and adaptation of hamsa and sprat in connection with the introduction of ctenophores. In: Volovik SP (ed) Ctenophore Mnemiopsis leidyi in the Azov and Black Seas: biology and consequences of invasion. AzNIIRKH, Rostov-on-Don, pp 234–294
Ryan WB, Pitman WC (1999) Noah's flood: The new scientific discoveries about the event that changed history. Simon and Schuster, New York. 319 p. ISBN 0-684-85920-3 ISBN 978-0-684-85920-0
Rybnikov PP, Nikolaeva GG (1963) Settlement of *Teredo navalis* L. (Teredinidae, Mollusca) larvae on a tree in the Black Sea. Tr Inst Oceanol 70:179–185
Shadrin NV (2000) New settlers in the Azov and Black Seas: reasons and consequences? In: New settlers in Russian European Seas. Apatyty, pp 76–90. (in Russian)
Shiganova TA (2009) Non-native species in the Southern seas of Eurasia. D.Sc. Thesis, A.N. Severtzhev Institute of ecology and evolution RAS, Moscow. 642 pp. (in Russian)
Shiganova TA, Bulgakova YV, Sorokin PY, Lukashev YF (2000a) The results of research of the new Beroe ovata alien in the Black Sea. Izv AN Ser Biol 2:248–256
Shiganova TA, Bulgakova YV, Volovik SP, Mirzoyan ZA, Dudkin SI (2000b) The new invader Beroe ovata and its impact on the ecosystem of the Azov-Black Sea basin in August–September 1999. In: Volovik SP (ed) Ctenophore Mnemiopsis leidyi (A.Agassiz) in the Azov and Black Seas and the consequences of its introduction. AzNIIRKH, Rostov-on-Don, pp 432–449
Sinegub IA (1994) On the expansion of the range of the nudibranch *Doridella obscura* Verrill in the Black Sea. Hydrobiol J 30(3):107–109
Stark IN (1960) Annual and seasonal dynamics of benthos in the Sea of Azov. Proc AzNIIRH 1(1): 167–231
Stepanyan OV (2003) Biological pollution of the Azov and Black Seas. Macroalgae invasion. The evolution of marine ecosystems under the influence of intruders and artificial mortality of fauna. Theses. Rostov-on-Don, pp 32–34
Studenikina EINS, Volovik NPN, Mirzoyan ZNANG, Luts IN (1991) Ctenophore Mnemiopsis leidyi in the Sea of Azov. Oceanology 31(6):722–725
Studenikina EI, Volovik SP, Tolokonnikova I, Frolenko LN, Selivanova EV (1998) Current characteristics of benthic communities in the Sea of Azov. In: Makarov EV, Volovik SP, Tyutina Yu E (eds) The main problems of the fishery and preservation of fish stocks of the Azov-Black Sea basin, Rostov-on-Don. 67 pp. (in Russian)
Studenikina EI, Aldakimova AY, Gubina GS (1999) Phytoplankton of the Sea of Azov under Anthropogenic Impact. AzNIIRKH, Rostov-on-Don. 175 pp. (in Russian)
Usachev PI (1928) About phytoplankton of the Sea of Azov. In: The diary of Vses. Botan. Congress. Leningrad, pp 64–70
Volovik SP (1985) Productivity and problems of ecosystem management in the Sea of Azov. Diss. for the scientific step.Doctor of Sciences. Rostov-on-Don, 563 pp
Volovik SP, Chikharev AS (1998a) Antropogenic changes of ichthyofauna of the Azov basin. In: Makarov EV, Volovik SP, Tyutina Yu E (eds) The main problems of the fishery and preservation of fish stocks of the Azov-Black Sea basin, Rostov-on-Don. (in Russian)
Volovik SP, Chikharev AS (1998b) Anthropogenic transformation of the ichthyofauna of the Azov basin. In: Makarov EV, Volovik SP, Tyutina Yu E (eds) Basic problems of fisheries and protection of fishery reservoirs of the Azov-Black Sea basin. AzNIIRKH, Rostov-on-Don, pp 7–22
Volovik SP, Dakhno VD (1983) On species composition of the Azov ichthyofauna in conditions of salinity increasing. Abstract on results of research of AzNIIRKH for 25 years. Rostov–on-Don (in Russian)

Zavialov I, Osadchiev A, Sedakov R, Barnier B, Molines J-M, Belokopytov V (2020) Water exchange between the Sea of Azov and the Black Sea through the Kerch Strait. Ocean Sci 16: 15–30. https://doi.org/10.5194/os-16-15-2020

Zenkevich LA (1963) Biology of the seas of the USSR. USSR Academy of Sciences, Moscow. 739 pp. (in Russian)

Zevina GB (1967) Distribution of bryozoans and entoprocta in the Sea of Azov. Hydrobiol J 3(1) (in Russian)

Zhukova SV, Mirzoyan AV, Shishkin VM, Podmareva TI, Lutynskaya LA, Taradina EA, Burlachko DS, Karmanov VG (2023) Possible scenarios for the formation of the continental runoff and the salinity of the Azov Sea, taking into account the current and future trends in climate change. Aqua Bioresour Environ 6(4):7–30. http://journal.azniirkh.ru, www.azniirkh.ru. ISSN 2618-8147 print, ISSN 2619-1024 online

Zolotarev V (1996) The Black Sea Ecosystem Changes related to the introduction of new Mollusc species. Mar Ecol 17(1–3):227–236

Chapter 6
The Caspian Sea

Abstract A detailed overview of the challenges linked to the introduction of non-native species into the Caspian Sea is provided. The discussion covers these key areas: principal characteristics of the Caspian Sea's physical geography, hydrology, and hydrochemistry; the origins and makeup of its native flora and fauna; the composition of non-native biota and its ecological consequences; the population dynamics of *Mnemiopsis leidyi* and its impact on the ecosystem; the role of *Beroe ovata* as a biological control agent for *M. leidyi*; and anthropogenic pressures affecting the Caspian Sea's plant and animal life. Throughout the twentieth century, environmental conditions in the Caspian Sea declined significantly, primarily because of sea level fluctuations related to river runoff management and increasing pollution. The reconnection of the Black and Azov seas with the Caspian Sea following the completion of the Volga-Don Canal in 1952 allowed zoobenthic fouling organisms and macrophytes to spread into the Caspian. By the early 1980s, with the adoption of ship ballast tanks, phytoplankton and zooplankton also began arriving from the Black Sea. The establishment of most non-native species led to profound transformations within native biological communities. The Caspian Sea example illustrates how the rapid proliferation of gelatinous carnivorous invaders can swiftly disrupt the entire ecosystem's functioning.

6.1 Key Physical, Hydrological, and Hydrochemical Attributes of the Caspian Sea

Main Physical and Geographical Features The Caspian Sea is the planet's largest enclosed body of water, entirely landlocked and situated thousands of kilometers from any ocean. It exhibits limnological characteristics and stands as the Earth's sole limnoocean (Dumont 1998). Located within the Eurasian continent between 47°07′–36°33′ N and 46°43′–54°50′ E, it extends approximately 1200 km from north to south, with a mean width of 320 km and a total area of roughly 386,400 km^2. Its surface rests about 27 m below global mean sea level. The sea is bordered by the Caucasus Mountains to the west, the expansive steppes of Central Asia to the east, the fertile plains of Southern Russia to the north, and the mountainous Iranian

T. A. Shiganova et al., *Non-Native Species in the Ponto-Caspian Seas*,
Monographiae Biologicae 96, https://doi.org/10.1007/978-3-032-18640-9_6

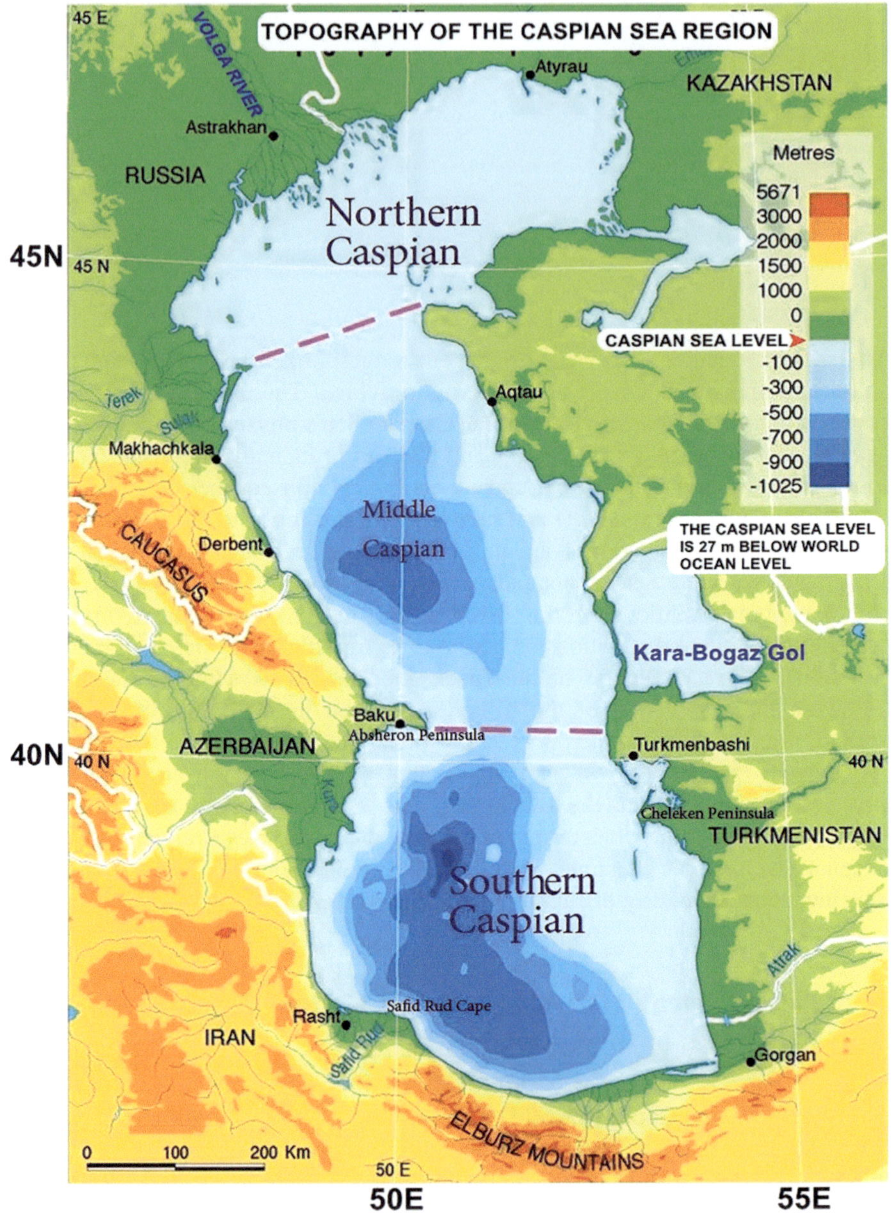

Fig. 6.1 Bathymetric chart of the Caspian Sea's depths and the landforms of its adjacent regions

Plateau and Elburz ridge to the south (Fig. 6.1). Its primary source of freshwater is Europe's longest river, the Volga, which enters at the shallow northern extremity and contributes more than 70% of the total river inflow.

The Caspian Sea is typically divided into three separate geographical zones: the Northern, Middle, and Southern Caspian (Fig. 6.1). The boundary between the Northern and Middle sections is the Mangyshlak Threshold, while the Apsheron Threshold separates the Middle and Southern regions. Water depths vary from 5 to 7 m in the Northern Caspian to 700–800 m in the Middle Caspian, reaching 700–900 m in the Southern Caspian, which contains the sea's deepest point at 1025 m. Isolated basins divided by the Apsheron Threshold are found in both the Middle and Southern Caspian. The continental shelf, defined as areas shallower than 100 m, makes up about 62% of the sea's total area, and the average depth across the entire sea is 208 m.

The Caspian Sea's water level undergoes significant long-term variations driven by climatic factors that influence the equilibrium between water inputs—such as river inflow and rainfall—and outputs, primarily evaporation. Throughout the twentieth century, a consistent declining trend was observed in these level changes, with the sea dropping by 3.2 m over a 75-year period, reaching a low of −29 m in 1977. This represented the lowest recorded level in five centuries. Consequently, the sea's surface area contracted by over 40,000 km^2, an area larger than that of the Sea of Azov. Starting in 1978, the sea level began to rise rapidly, climbing to approximately −27 m relative to global ocean levels by 1996. This was followed once more by a gradual decline, as illustrated in Fig. 6.2.

Owing to its considerable north-south elongation, the Caspian Sea spans multiple climatic zones. Its northern section experiences a temperate continental climate, while the western shore enjoys a temperate warm climate. The southwestern and

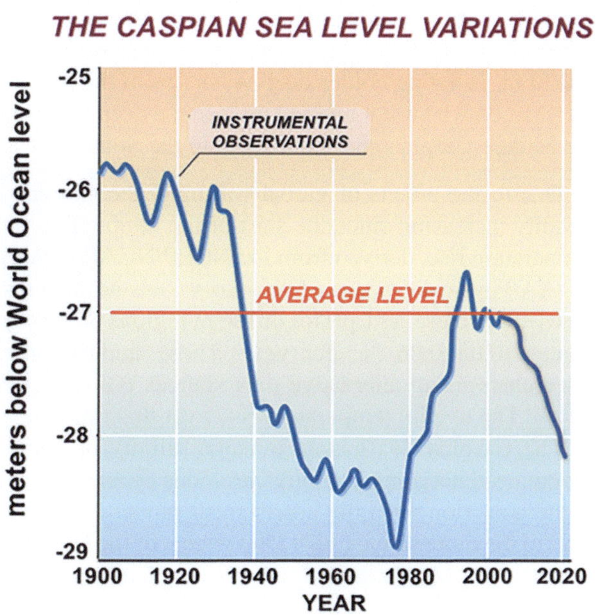

Fig. 6.2 Fluctuations in the Caspian Sea's water levels from 1900 to 2020

southern coasts fall within subtropical zones, and the eastern shore is characterized by a desert climate.

During winter, weather patterns over the Northern and Middle Caspian are shaped by Arctic continental air masses, whereas the Southern Caspian is frequently affected by southern cyclones. Conditions in the west tend to be unstable and wet, while the east remains dry. In summer, the western and northwestern areas come under the influence of extensions from the Azores atmospheric high-pressure system, and the southeastern regions are dominated by the Iran-Afghanistan low-pressure area. This combination results in consistently dry, warm weather.

Prevailing winds over the sea are from the north and northwest (up to 40%) and from the southeast (around 35%). Average wind speeds range from approximately 6 to 9 m/s. The fierce northern storms known as "Baku Nords" can reach velocities of 20–25 m/s.

The lowest mean monthly air temperatures, around −10 °C, occur in January and February in the northeastern parts, dropping to as low as −30 °C in the harshest winters. In the southern regions, temperatures during these months average 8–12 °C. By July and August, the average monthly temperature across the entire sea area is 25–26 °C, with peaks of up to 44 °C recorded on the eastern coast. The northern portion of the Caspian Sea typically freezes over in winter.

Hydrology During January and February, the surface water temperature in the Northern Caspian Sea approaches freezing, increasing progressively southward to reach about 11 °C near the Iranian coast. In the summer months, surface temperatures generally warm to 23–28 °C across most areas. An exception is the eastern shelf of the Middle Caspian, where coastal upwelling occurs in July and August, causing surface temperatures to fall to 12–17 °C (Fig. 6.3). In winter, strong convective mixing prevents any vertical thermal stratification. During summer, a seasonal thermocline develops below the warmed surface layer at depths of 20–30 m, isolating colder deep waters from the warmer surface layer. Throughout the year, bottom water temperatures in the deep basins remain stable at 4.5–5.5 °C in the Middle Caspian and 5.8–6.5 °C in the Southern Caspian.

Due to the effects of global warming, the Caspian Sea's temperature has been steadily increasing since the start of the 1980s (Fig. 6.4). Positive linear trends for temperature rise, derived from the NCEP/NCAR reanalysis dataset (accessed via the NOAA/Physical Sciences Laboratory website at https://psl.noaa.gov/cgi-bin/data/timeseries/timeseries1.pl) for different regions of the sea during summer, are approximately 0.05–0.06 °C per year. These findings align with the work of other researchers using alternative data sources (Ginzburg et al. 2005; Kostianoy et al. 2019). The overall temperature rise over this timeframe has been around 1.8–2 °C.

The Caspian Sea has an average salinity of approximately 12.8‰. The most pronounced horizontal salinity variations occur in the Northern Caspian, particularly at the transition from the nearly fresh (under 1‰) river estuaries of the Volga and Ural to the more saline (10–11‰) waters of the Middle Caspian. Salinity differences between the Middle and Southern Caspian are minimal, with a gradual increase

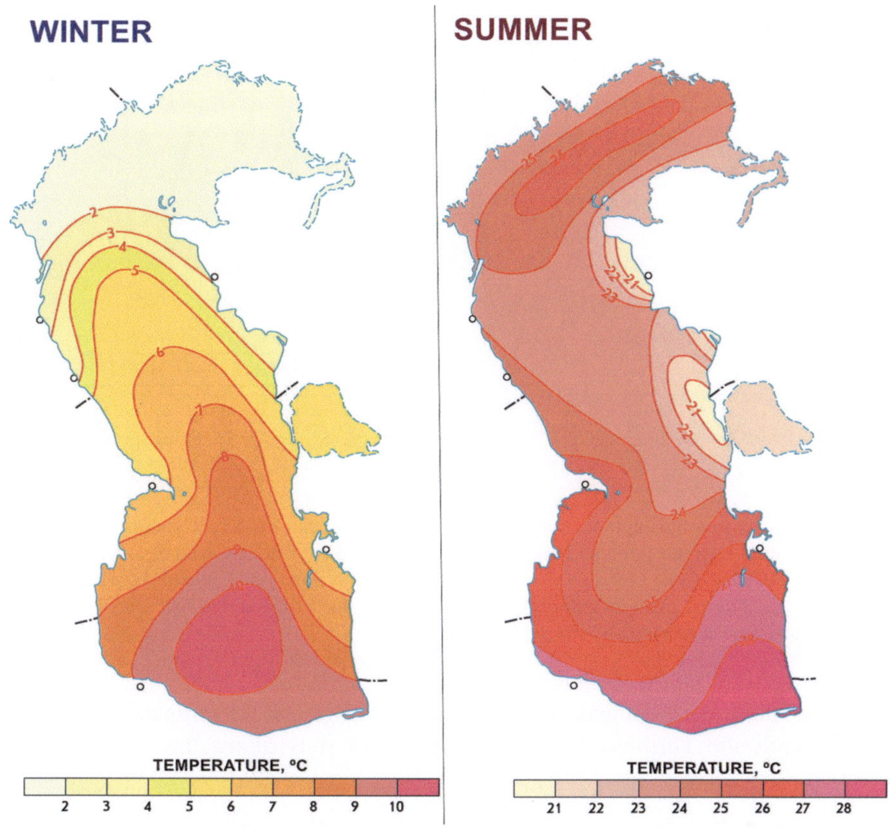

Fig. 6.3 Winter and summer sea surface temperatures in the Caspian Sea

observed from the northwest toward the southeast (Fig. 6.5). In the open waters, salinity remains relatively consistent with depth.

The overall circulation within the Caspian Sea is driven by wind stress, spatial variations in water density, the Coriolis effect, coastline shape, and seabed topography. In the northern region, river discharge and water level changes also play a significant role. The interplay of these elements produces a complicated, predominantly cyclonic (counterclockwise) circulation system. This system features two main cyclonic gyres—one in the Middle Caspian and another in the South Caspian–along with numerous smaller eddies (Fig. 6.6).

Typical wind-driven current velocities range from 15 to 20 cm/s, with maximum speeds reaching up to 70 cm/s. In the Northern Caspian, the dominant winds generate a flow that travels along the northwestern shoreline toward the southwest. Within the Middle Caspian, this current converges with the western segment of the regional cyclonic gyre and proceeds to follow the western coastline. Near the Absheron Peninsula, the current splits. One segment moves into the open sea and enters the Middle Caspian's cyclonic circulation, while the nearshore portion

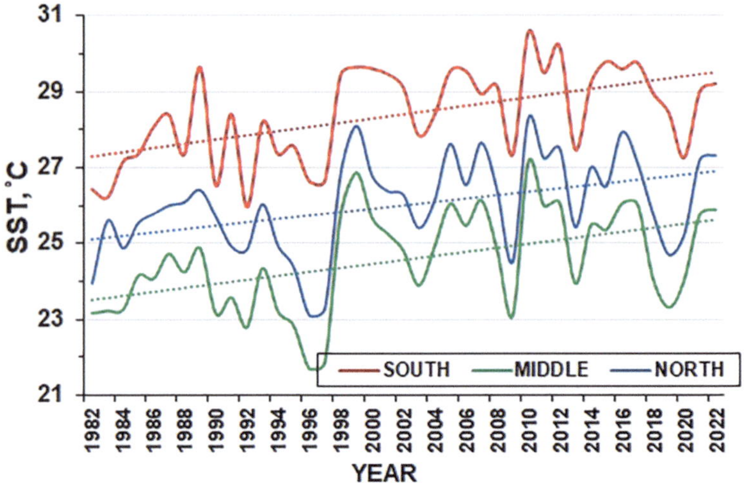

Fig. 6.4 Year-to-year fluctuations in summer sea surface temperature (SST, °C) across the northern (blue line), middle (green line), and southern (red line) regions of the Caspian Sea. Data sourced from the NCEP/NCAR reanalysis dataset (https://psl.noaa.gov/cgi-bin/data/timeseries/timeseries1.pl). Dotted lines indicate the linear trends for each time series. All calculations and visualizations were performed by the authors

continues along the Southern Caspian's coasts before turning northward. There, it merges with the northward coastal current that runs along the entire eastern seaboard.

Hydrochemistry The current salt makeup of the Caspian Sea's water stems from the transformation of its original oceanic water due to continental runoff, following its separation from the World Ocean around 10,000 years ago. Figure 6.7 illustrates the salt composition across different Caspian Sea regions in comparison to that of the World Ocean. Compared to the World Ocean's salt composition, the Caspian Sea's waters show reduced levels of cations (Na, K, Ca, Mg), which originate from denudation processes.

Due to the increased impact of continental runoff, the concentration of chloride anions in the Caspian Sea is 3.5 times lower (and up to seven times lower in its northern region) compared to oceanic water (Fig. 6.7). In the Middle and South Caspian, chloride anion levels are nearly identical, measuring 5.44 and 5.46 g kg^{-1}, respectively. In the North Caspian, near the mouth of the Volga River, the concentration drops to 0.09 g kg^{-1}. For the same reason, the Caspian Sea exhibits elevated concentrations of sulfate (SO_4) and bicarbonate (HCO_3) anions (Fig. 6.7).

Given that hydrocarbonates make up over 40% of the ionic composition in river discharge entering the Caspian Sea, its waters exhibit carbonate oversaturation, primarily in the form of calcium hydrocarbonate. As river water transitions into seawater, calcium hydrocarbonate converts to carbonates, which then settle out of the water column into the underlying sediments. This precipitation is particularly

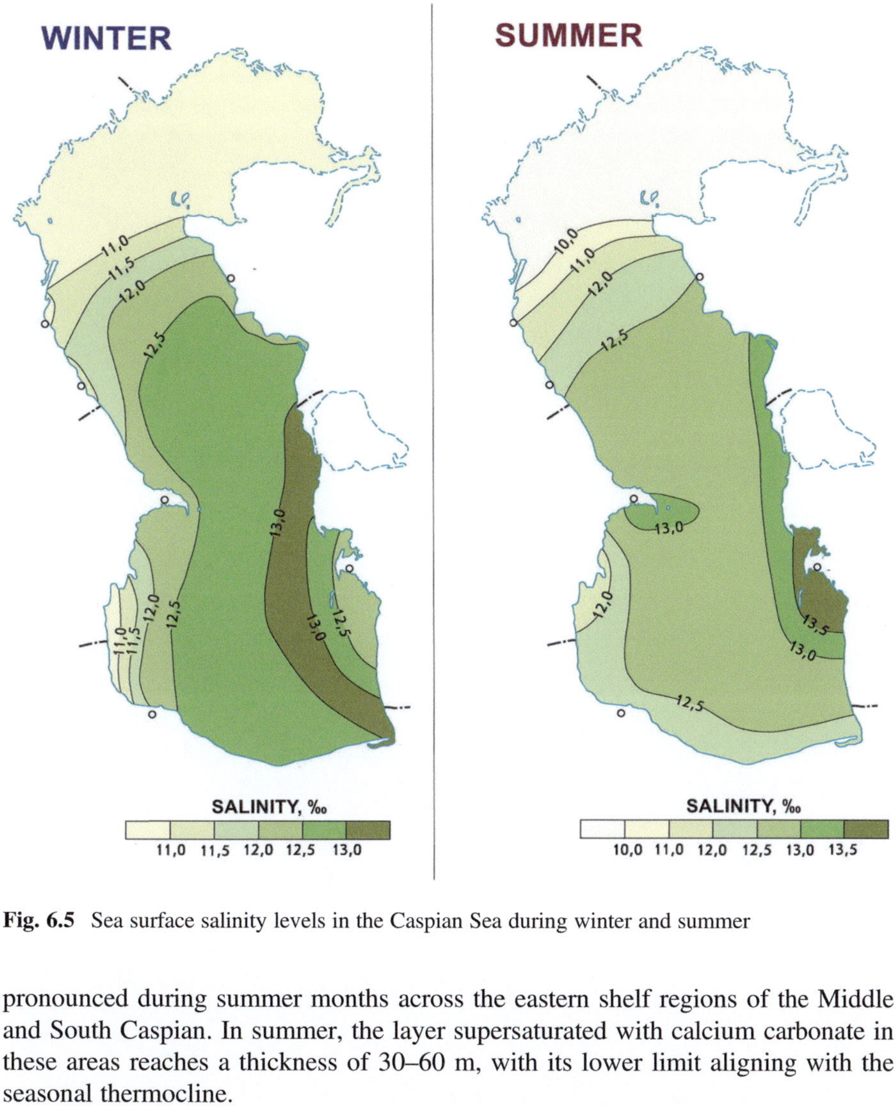

Fig. 6.5 Sea surface salinity levels in the Caspian Sea during winter and summer

pronounced during summer months across the eastern shelf regions of the Middle and South Caspian. In summer, the layer supersaturated with calcium carbonate in these areas reaches a thickness of 30–60 m, with its lower limit aligning with the seasonal thermocline.

Within the Middle and Southern Caspian Sea, which constitutes the majority of the Caspian's water volume, the proportion of key cations (excluding Na and K) relative to chloride anion levels is markedly greater than in the global ocean. Magnesium concentrations are twice as high, calcium is three times greater, SO_4 is fourfold higher, and HCO_3 levels are elevated by a factor of 5.5.

The oxygen conditions in the shallow Northern Caspian Sea contrast markedly with those in its deeper regions. In the shallowest areas, where depths are under 10 m, the vertical differences in dissolved oxygen, temperature, and salinity are minimal all year round. For the entire water column in the Northern Caspian, the highest dissolved oxygen levels (8–11 ml l^{-1}) occur during the colder months from

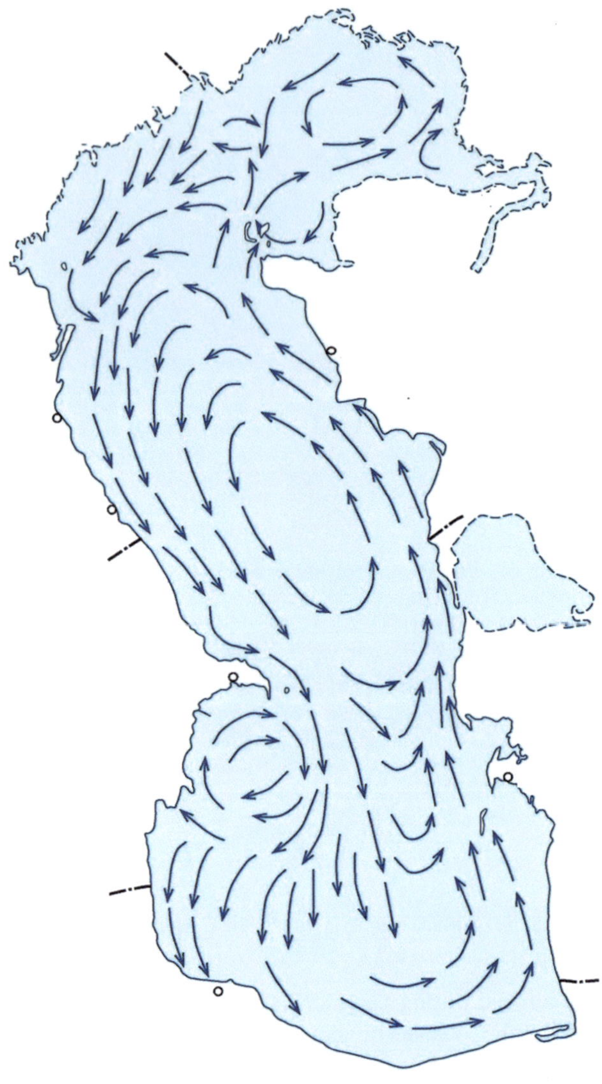

Fig. 6.6 General circulation pattern in the Caspian Sea

October to April. In winter, this is due to lower temperatures, while in spring, it results from phytoplankton growth and associated photosynthesis. During spring, as intense warming and freshwater influx from river floods begin, a pycnocline develops in the outer, deeper shelf zone (10–20 m). This density layer limits vertical mixing and leads to a reduction in dissolved oxygen near the seabed. By August, near-bottom oxygen concentrations are 15–20% lower than surface levels. Nevertheless, oxygen saturation in bottom waters generally remains above 80%, which is above the hypoxic threshold.

During the winter months, the dissolved oxygen levels in the upper mixed layer across the entire Caspian Sea are nearly in equilibrium with the atmosphere, reaching

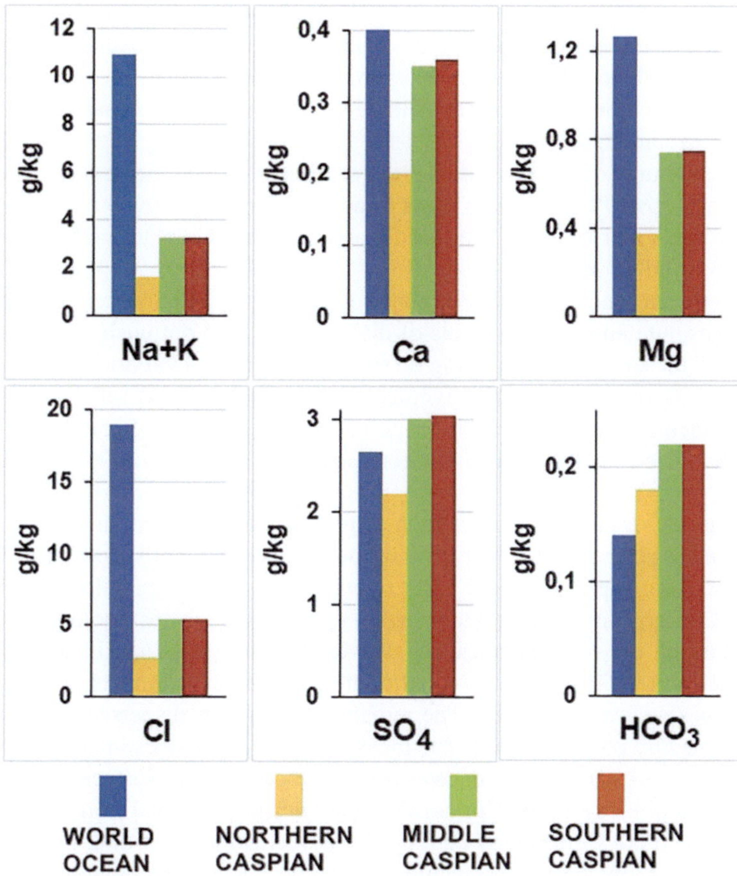

Fig. 6.7 Concentrations of major ions (g/kg) in the global ocean and across the primary regions of the Caspian Sea. Information sourced from Tuzhilkin et al. (2005), with visual representation prepared by the authors

approximately 100% saturation. In the Middle Caspian, the absolute oxygen content is relatively uniform in the upper 100–200 m due to winter vertical convective mixing, ranging from about 7 to 9 ml l^{-1}. In the Southern Caspian, the upper quasi-homogeneous layer is shallower, around 50 m thick in winter, and oxygen concentrations are slightly lower (7–7.5 ml l^{-1}) compared to the Middle Caspian, primarily because of higher water temperatures.

With the arrival of spring, intensive photosynthesis in the top 20-m layer of the Caspian Sea's deep-water regions leads to a peak in absolute oxygen concentration, exceeding winter values by 0.2–0.3 ml l^{-1}. This can cause oxygen supersaturation to reach 110% and occasionally climb as high as 130–135%.

In summer, the highest absolute (6.0–6.5 ml l^{-1}) and relative (105–110%) surface concentrations of dissolved oxygen are found in areas with the most vigorous phytoplankton photosynthetic activity, notably the Northern Caspian and the upwelling zone along the eastern coast of the Middle Caspian. Below the seasonal thermocline, from 30 to 100 m deep, oxygen saturation drops to its lowest summer levels, falling to around 80% in the Middle Caspian and approximately 65% in the Southern Caspian.

As temperatures decline in autumn, the seasonal thermocline deepens and eventually breaks down. This process leads to an increase in dissolved oxygen concentrations in the upper layers of the Middle and Southern Caspian, with levels rising between 6.5 and 7.5 ml l^{-1}.

Within the intermediate and deep zones of the Middle and Southern Caspian Sea, both the absolute and relative levels of dissolved oxygen diminish from the northwest toward the southeast, exhibiting no significant seasonal variation. In the Middle Caspian's 200–600-m layer, the average yearly absolute (and relative) dissolved oxygen content declines with depth from 5.5 ml l^{-1} (70%) to 3.5 ml l^{-1} (50%). For the Southern Caspian, the equivalent decrease is from 4.0 ml l^{-1} (60%) to 2.5 ml l^{-1} (40%).

The primary origin of nutrients such as phosphorus, nitrogen, and silicon in the Caspian Sea is river inflow. Over 95% of the nitrogen and phosphorus entering the sea comes from the Volga and Ural rivers, which discharge into the Northern Caspian. Consequently, the northern region exhibits the greatest biological productivity. Most of the mineral nutrients delivered to the sea—specifically, 96% of phosphorus and 98% of nitrogen—settle into the bottom sediments of the shallow coastal zones near river mouths. The remaining portion is dispersed across the sea and consumed by phytoplankton. The most pronounced changes in nutrient concentrations as one moves away from river estuaries occur in the Northern Caspian.

Peak nutrient levels in the Northern Caspian Sea are recorded from May to June, coinciding with the spring flood period. The lowest amounts of their mineral forms occur during the low-water season (July–September), when phytoplankton growth is at its most intense. Between June and August, concentrations of the most unstable mineral nutrients—specifically phosphorus and nitrite nitrogen—in the Northern Caspian can drop by several multiples.

Within the Middle and Southern Caspian Sea, the spatial and seasonal patterns of mineral nutrient availability are primarily governed by biochemical cycles—specifically production and decomposition—alongside the physical transport of water through currents and turbulent mixing. Photosynthetic activity dominates the surface layer down to 30–50 m, while organic matter breakdown prevails in deeper waters, leading to a progressive rise in nutrient levels with increasing depth. During winter, when photosynthetic production is minimal and advective-diffusive mixing is strongest, nutrient concentrations reach their peak in these regions. Conversely, in spring and autumn, a shift in the balance between biochemical and hydrodynamic processes causes nutrient levels in the sunlit euphotic zone to drop to near-zero values.

Overall, the coastal zones of the Caspian Sea are distinct from its open waters due to higher concentrations of hydrochemical parameters—such as dissolved oxygen, pH, and nutrients—and a strong reliance on the quantity and chemical makeup of river inflows. Conversely, the hydrochemical conditions in the deep-sea regions of the Caspian Sea are primarily shaped by biochemical processes (production and decomposition) and dynamic mechanisms (advection and diffusion). Nevertheless, even in these offshore areas, fluctuations in the delivery of materials via river discharge remain the dominant driver of long-term changes.

Throughout the twentieth century, the Caspian Sea exhibited a cyclical shift between two distinct hydrochemical regimes. The first regime was prevalent prior to the mid-1930s and reemerged after the 1970s, coinciding with periods of increased river inflow. This phase was marked by diminished dissolved oxygen levels near the seabed (reaching hypoxic conditions) and lower nutrient concentrations in surface waters. The second type of hydrochemical regime, which peaked during the 1960s and 1970s, occurred under conditions of minimal river discharge. This period displayed the opposite pattern, characterized by anomalies in the previously mentioned features that were reversed in nature.

This section incorporates data partially sourced from Dobrovolsky and Zalogin 1982; Tuzhilkin et al. 2005; https://www.britannica.com/place/Caspian-Sea; and https://en.wikipedia.org/wiki/Caspian_Sea.

6.2 Origins and Composition of the Caspian Sea Flora and Fauna

This and the following sections examine different facets of how non-native species invasions affect the Caspian Sea's biological communities, drawing in part on the review conducted by Shiganova et al. (2019a).

This section focuses exclusively on the most recent phase of the Caspian Sea's geological development. The contemporary flora and fauna of the Caspian evolved during a prolonged period of isolation and fluctuating salinity levels, which remain lower than those of the ocean. Consequently, the Caspian Sea has emerged as a focal point for the development of a brackish-water biological community. Furthermore, species that entered the Caspian during the postglacial era, following its reconnection with other seas, subsequently diversified into endemic species and subspecies. Although species diversity is relatively low, the Caspian Sea exhibits high biological productivity, particularly in its northern regions. Strong solar radiation and substantial river inflows have led to elevated nutrient concentrations in the water, resulting in correspondingly high productivity. The substantial fish populations in the Caspian Sea are the product of both this high productivity and the distinctive composition of its fauna (Kasymov 1987).

Today's Caspian Sea is divided into three distinct sections: the northern, middle, and southern basins. Each possesses its own unique hydrophysical characteristics and biological diversity (refer to Sect. 6.5.1). The middle and southern basins host the greatest concentration of endemic Caspian species, which trace their origins to ancient marine life. These are the deepest regions of the sea, areas that never fully desiccated despite climatic shifts over past geological eras. A critical volume of water has perpetually persisted here, sustaining biological continuity. Nonetheless, during periods of regression linked to rising salinity, the population of marine organisms typically declined (Aladin et al. 2002).

The Northern Caspian accounts for one-third of the total surface area of the Caspian Sea yet holds only about 1% of the basin's total water volume—a contrast explained by its extreme shallowness. Salinity in this region varies and is typically two to three times lower than in the Middle Caspian, largely due to substantial freshwater inflow from rivers. The Volga River, which enters the Caspian in the north, contributes roughly 80% of the sea's total river runoff. Because of the lower salinity, species common in the main body of the Caspian are seldom found here. Instead, the Northern Caspian hosts its own distinct brackish-water community, noted for considerable biodiversity. Certain species migrate into the area to spawn, and many freshwater organisms live exclusively here, unable to survive in the higher salinity of the Middle and Southern Caspian. This section of the sea is exceptionally shallow, with depths generally not surpassing 10 m. Its shallowness has led to repeated drying during past geological epochs, causing local species either to perish or to endure unfavorable periods within the deltas and estuarine reservoirs of Caspian rivers.

Therefore, it is inaccurate to treat the Caspian Sea's biodiversity as uniform across its entire basin. Variations in hydrophysical conditions across its three distinct sections have led to significant differences in biological diversity. The northern region possesses the most abundant fauna and flora. Biodiversity is somewhat reduced in the Middle and Southern Caspian, with the lowest levels found in the Kara-Bogaz-Gol Bay. Furthermore, a complete understanding of species distribution throughout the Caspian requires including the organisms living within its river deltas (Aladin et al. 2002).

Based on their origins, the species inhabiting the Caspian Sea can be categorized into four distinct groups (Derzhavin 1956; Zenkevich 1963; Kasymov 1987).

6.2.1 The Autochthonous Complex

This assemblage is predominant, comprising 513 species (Kasymov 1987), which represents 43.8% of the total fauna and flora in the Caspian Sea. Indigenous species account for approximately 50% of fish, over 60% of crustaceans, more than 80% of mollusks, and 14 phytoplankton species (Kasymov 1987). According to Mordukhai-Boltovskoy (1960), these species are endemic to the Caspian Sea or to the broader Ponto-Caspian region. Indigenous species in the Black and Azov Seas also encompass those inhabiting the low-salinity waters near river estuaries. In contrast, within the Caspian Sea, the autochthonous or Ponto-Caspian complex species are also capable of living in the saline waters of the main sea basin.

The species within this complex can only persist within a narrow range of salinity variation. Experimental research by Karpevitch (1975) demonstrated that this fauna cannot endure high salinity conditions, particularly when compared to the

Ponto-Caspian species currently found in the Ponto-Azov area (Mordukhai-Boltovskoy 1960; Zenkevich 1963).

6.2.2 The Arctic Complex

Species originating from the Arctic are frequently identified as glacial relicts. Their forebears entered the Caspian Sea from the north during the postglacial era, specifically when the New Euxinic basin formed approximately 0.019–0.009 million years ago, a time marked by melting ice sheets and rising sea levels. This allowed Arctic species to colonize the Caspian Sea (Zenkevich 1963). Presently, the Arctic complex consists of 14 species, representing just 1.2% of the Caspian Sea's total fauna (Kasymov 1987). It currently encompasses Copepoda (*Limnocalanus macrurus grimaldii* (Guerne, 1886)), Isopoda (*Saduria entomon* (Linnaeus, 1758)), Amphipoda (*Pseudalibrotus caspius* (Grimm) G.O. Sars, 1896), *Onisimus platyceras* (G.O. Sars, 1896), *Gammaracanthus caspius* (Sars, 1896), *Monoporeia affinis* (Lindström, 1855), Mysidacea *Mysis caspia* (G.O. Sars 1895), *M. macrophthalma* (G.O. Sars 1895), *M. macrolepis* (G.O. Sars), and *M. amblyops* (G.O. Sars, 1907) (Kasymov 1987)) and fishes (Salmonidae (whitefish, i.e., belorybitza) *Stenodus leucichthys* (Güldenstädt, 1772) and Caspian salmon or trout Salmo caspius (Kessler, 1877). The polychaeta *Manayunkia caspica* (Annenkova, 1928) and the seal *Pusa caspica* (Gmelin, 1788) are also part of this complex (Zenkevich 1963).

6.2.3 The Mediterranean Complex

The species, which trace their origins to the ancient Black and Mediterranean Seas and are classified as Mediterranean or Atlantic settlers, entered the Caspian Sea via the Kuma-Manych Depression and later spread eastward through the Uzboy River roughly 13,000 years ago. The following species colonized the region through this route: the seagrass *Zostera noltii* (Hornemann, 1832); the bivalve mollusk *Cerastoderma glaucum* (Bruguière, 1789); the polychaete *Fabricia sabella* subspecies caspica (Zenkevich 1963); the bryozoan *Amathia imbricata*, *Amathia imbricata* (Adams, 1800); and the fishes *Atherina boyeri* (Black Sea), the needle fish *Syngnathus abaster* (1827), and the goby *Knipowitschia caucasica* (Zenkevich 1963).

6.2.4 The Freshwater Complex

Species originating from freshwater environments constitute the predominant population in the Northern Caspian, having first entered the region during the late Tertiary era. The introduction of freshwater organisms into the Caspian Sea occurred repeatedly throughout periods of reduced salinity. These species continue to migrate into the sea today, a process that will be examined further in the section detailing contemporary flora and fauna. Freshwater species account for roughly 19.5% of the total nonparasitic animal life and represent the most significant portion of the Caspian Sea's phytoplankton. This category also encompasses migratory and semi-anadromous fish species. They arrived via river inflows along the eastern and western shores of the Northern Caspian. This group of fish includes sturgeons, salmons, pike, catfish, carp, and perch. Additionally, it comprises various invertebrates (Oligochaeta, Turbellaria, Chironomidae, Rotifera, Cladocera, Copepoda) that inhabit the desalinated zones, primarily in the Northern Caspian and the pre-estuarine regions of the Middle and Southern Caspian (Kasymov 1987).

These combined ecosystems collectively harbor over 300 endemic species (Kasymov 1987; Dumont 2000; Khlebovich 1963). The endemic animal life is exceptionally rich and varied. It comprises four species of Spongia, 29 species of Turbellaria, three species of Nematodes, two species of Rotifera, two species of Oligochaeta, one species of Polychaeta, 19 species of Cladocera, three species of Ostracoda, 23 species of Copepoda, 14 species of Mysidae, one species of Isopoda, 41 species of Amphipoda, seven species of Cumacea, one species of Decapoda, two species of Hydracarina, 99 species of Mollusca, 54 species of fish, and one species of mammal. Regarding the plant life, 14 Caspian phytoplankton species from the Bacillariophyta group are also endemic (Proshkina-Lavrenko and Makarova 1968).

6.2.5 Current State of Flora and Fauna in the Caspian Sea

Currently, the Caspian Sea exhibits relatively low species diversity: its species richness is 2.5 times lower than that of the Black Sea (Zenkevich 1963). This is primarily due to the fact that the modern flora and fauna of the Caspian evolved under conditions of prolonged isolation and periodically fluctuating salinity levels, which are lower than oceanic. Consequently, the Caspian Sea has developed a distinctive brackish-water fauna (Birshtein 1968; Mordukhai-Boltovskoy 1962).

Later studies have indicated an increase in the number of known metazoan species in the Caspian Sea: Zenkevich (1963) documented around 450 species, Kasymov (1987) reported about 950 species.

A key characteristic of the Caspian Sea, resulting from its extended isolation, is the exceptionally high proportion of endemic species. This feature defines the unique nature of the Caspian's biota.

6.2.6 Phytoplankton

Currently, the phytoplankton community of the Caspian Sea consists of 808 species (Table 6.1). The phytoplankton of the Northern Caspian exhibits typical estuarine characteristics and is distinct from that found in the Middle and Southern Caspian (Proshkina-Lavrenko and Makarova 1968).

The indigenous phytoplankton of the Northern Caspian includes 448 freshwater species, accounting for 55% of the total phytoplankton species in the entire basin (Table 6.1). Of these, 239 species are found exclusively in the lower Volga River, 163 species occur only in the Northern Caspian, and 61 species are present in both the lower Volga and the estuarine areas. Chrysophyta is restricted to the Northern Caspian, as are the majority of Xanthophyta and Chlorophyta species. Additionally, 98 widespread euryhaline organisms (some of which are cosmopolitan) are present across all three regions.

Fourteen members of the Oscillatoriales order (families Schizothrichaceae Elenkin, 1949, Phormidiaceae Anagnostidis and Komárek, 1988, and Pseudanabenaceae (Lemmermann) Komárek, 1974), two species from the order Nostocales, and a single representative from the Oocystaceae family are found exclusively in the Southern Caspian region. Merely fourteen species of Bacillariophyta in the Caspian are considered endemic (Proshkina-Lavrenko and Makarova 1968). The proportions of freshwater, brackish, and marine species are detailed in Table 6.1.

6.2.7 Microphytobenthos

Microphytobenthos comprises 129 diatom species and 55 blue-green algae species, which primarily inhabit solid surfaces and are associated with the fouling biotope, along with specific green algae like *Enteromorpha* and *Cladophora*. These green algae develop filamentous colonies and more complex formations that can anchor to the substrate.

6.2.8 Zooplankton

Based on research by Kasymov (1987), a total of 315 zooplankton species and subspecies have been documented in the Caspian Sea. In the Northern Caspian region, approximately 200 species are found. Among these, Ciliata represent the most numerous group, comprising over 70 species. They are followed by Rotifera (more than 50 species), Diplostraca (formerly known as Cladocera) (over 30 species), and Copepoda (more than 20 species), which show comparatively lower diversity (Yablonskaya 1985; Kasymov 1987).

Table 6.1 Taxonomic composition and ecological condition of the indigenous phytoplankton in the Caspian Sea

Phylum	Classes	Orders	Families	Genera	Species	Environment						Notes
						Freshwater	%[a]	Brackish	%[a]	Marine	%[a]	
Cyanophyta	–	3	15	44	147	59	40	82	56	2	1.4	
Chrysophyta	1	2	5	8	20	19	100	0	0	0	0	
Bacillariophyta	2	16	35	48	260	87	34	158	62	10	4	14 endemic brackish-water species
Xanthophyta	1	1	3	4	13	12	92	1	8	0		
Cryptophyta	1	–	1	3	13	8	61	5	99	0	0	
Dinophyta	1	4	8	18	44	13	30	16	37	14	33	
Euglenophyta	1	1	3	7	40	25	71	15	24	0	0	
Chlorophyta	4	13	30	84	271	225	85	40	15	0	0	
Total	11	40	90	216	808	448	55	317	39	26	6	

[a] % from the total number of species in type

Within the Middle and Southern Caspian Sea, the most plentiful native Copepoda species are found in deep-water zones where salinity measures 12–13‰. These include the endemic species *Eurytemora minor* (Behning, 1938), *E. affinis* (Poppe, 1880), *E. grimmi* (Sars G.O., 1897), and *Heterocope caspia* (Sars, 1897), which are present year-round. The thermophilic *E. minor* is the dominant species in the Southern Caspian across all seasons and in the Middle Caspian during the summer. Alongside these native species, Arctic Copepoda such as *Limnocalanus macrurus grimaldii* (Guerne, 1886) and four species of planktonic Mysidacea—*Mysis caspia* (G.O. Sars, 1895), *M. macrophthalma* (G.O. Sars, 1895), *M. macrolepis* (G.O. Sars), and *M. amblyops* (G.O. Sars, 1907)—also inhabit the deep-sea regions (Kasymov 1987).

Species originating from the Mediterranean are found across the Caspian basin, with their highest concentrations in coastal zones. The most frequently encountered is *Calanipeda aquae-dulcis* (Krichagin, 1873) (Aladin et al. 2002).

In the western regions of the Middle and Southern Caspian, Cladocera are the dominant zooplankton group, making up 25–55% of all zooplankton species. This assemblage includes the endemic forms *Polyphemus exiguus* (Sars, 1897), *Podonevadne trigona* (G.O. Sars, 1897), and *Cornigerius maeoticus hircus* (Sars, 1902), along with nine representatives of the family Cercopagidae (Sars 1897, 1902; Mordukhai-Boltovskoi 1964, 1968). Also present in the Middle and Southern Caspian are widespread euryhaline Cladocera species: *Podonevadne angusta* (Sars, 1902), *P. camptonyx* (Sars, 1897), *Evadne anonyx* (Sars, 1897), *E. prolongata* (Behning, 1938), and *Cornigerius bicornis* (Zernov, 1901).

Within the Diplostraca group, the invasive and detrimental *Cercopagis pengoi* (Ostroumov, 1891), originally from the Black and Caspian Seas, deserves attention. It has spread extensively across reservoirs and was introduced to the Baltic Sea and North American lakes via ballast water. This species is listed among the world's 100 most damaging invasive organisms in the Global Invasive Species Database.

Common inhabitants of the shallow coastal zones, where both salinity and temperature fluctuate widely, are *Calanipeda aquae-dulcis* and *Heterocope caspia*. According to Sars G.O., 1897, *C. aquae-dulcis* stands out as the most euryhaline crustacean species in the Caspian Sea. It is found with equal frequency in the freshwater zones near river estuaries and in the more saline bays; it reproduces continuously throughout the year, except during the ice-covered period in the North Caspian, and feeds on phytoplankton and bacteria. *Heterocope caspia* thrives within a more limited salinity range. In the Caspian Sea, it exists primarily as a near-bottom form, inhabiting mainly the freshened regions and, to a lesser extent, the Middle and South Caspian. Its reproduction occurs during the warmer months. In the North Caspian, it is absent from phytoplankton samples from November through April.

The zooplankton community also features members of the Harpacticoida group, which comprises two endemic, six widespread, and two brackish-water species, as well as the Cyclopoida group, with four endemic and 22 freshwater species.

Additionally, meroplankton—primarily made up of bivalve larvae and benthic crustacean larvae—adds to the overall biodiversity of these zooplankton assemblages.

6.2.9 Gelatinous Species

The Caspian Sea hosts a limited number of indigenous Cnidaria species. Within the Hydrozoa group, the marine organism *Moerisia pallasi* (Derzhavin, 1912) is found in the Northern and Middle regions of the Caspian Sea (Derzhavin, 1956).

6.2.10 Zoobenthos

A defining characteristic of the benthic fauna is the prevalence of autochthonous Caspian species, frequently comprising Caspian endemics that have been integrated into endemic genera or subgenera. This assemblage includes Caspian endemics along with relicts from the tertiary sea's fauna, which existed approximately five million years ago. Notable groups encompass the Hydrobiidae (Stimpson, 1865), with species such as *Clathrocaspia isseli* (Logvinenko and Starobogatov, 1969), two species of Oligochaeta, four species of Polychaeta, leeches (Hirudinea), 14 species of Mysidae, one species of Isopoda, 41 species of Amphipoda, seven species of Cumacea, one species of Decapoda, and 99 species of Mollusca—among them six species of Dreissenidae (Gray, 1840). Additionally, the fauna includes 29 species of Turbellaria, three species of Nematodes, one species of Polychaetes, four species of Spongia (Demospongiae), and pearl weed (Bryozoa) (Kasymov 1987; Dumont 2000; Khlebovich 1963).

The second category encompasses freshwater organisms that colonized the Caspian during its period of reduced salinity, subsequently adjusting to brackish conditions. This assemblage is comparatively limited, featuring only a handful of oligochaete (Oligochaeta) species, the freshwater mollusks *Dreissena polymorpha* (Pallas, 1771) and *D. bugensis* (Andrusov, 1897), along with certain microfauna such as turbellarians (Turbellaria), ostracods (Ostracoda), and benthic rotifers (Rotifera) (Kasymov 1987).

The third group is composed of Arctic species that entered the Caspian Sea from northern regions during the postglacial era, approximately 10–12 thousand years in the past. This group contains Polychaeta, Mysidacea, and Amphipoda (Aladin et al., 2002).

A majority of benthic organisms inhabit the seafloor, either on its surface or within the sediment (epi- and endobenthos). This group includes periphyton (fouling) representatives such as sponges (Demospongiae), bryozoans (Bryozoa), polychaete worms (Polychaeta), bivalves (Bivalvia), *Dreissena* spp., ciliates (Ciliata), as well as nektobenthic shrimps (Mysidacea) and planktobenthic copepods (Copepoda), cladocerans (Diplostraca), and rotifers (Rotifera) (Aladin et al., 2002).

The indigenous fouling community of the Caspian Sea is comparatively limited, comprising species such as *Dreissena polymorpha* (Pallas, 1771), *D. elata elata* (Andrusov, 1897), *Cordylophora caspia* (Pallas, 1771), *Chelicorophium curvispinum* (G.O. Sars, 1895), and *Chelicorophium robustum* (G.O. Sars, 1895).

These organisms colonize all appropriate benthic surfaces, including stones, shell debris, pilings, and submerged timber (Aladin et al., 2002).

In order to finalize the assessment of Caspian invertebrates, it is also essential to acknowledge 325 parasitic species, which belong to a variety of taxa—ranging from protists to crustaceans. The majority of these parasites are associated with specific hosts, and their evolutionary development is tied to these relationships. While most are widely distributed species linked to freshwater hosts, certain groups have given rise to endemic forms within the Caspian region (Karpinsky 2002).

The Distribution of Benthic Organisms across the Seabed Is Influenced by Several Key Factors

- Salinity (especially in the Northern Caspian).
- Distribution and population size of the primary benthic consumers, specifically benthivorous fish species.

When it comes to salinity levels, the zoobenthic community in the Caspian Sea can be categorized into four distinct ecological groups:

1. Freshwater species: Gastropoda and Bivalvia, Oligochaeta, and Chironomidae larvae, found in estuarine environments where salinity levels are between 0 and 2‰.
2. Brackish and freshwater species: Oligochaeta, Bryozoa, Chironomidae larvae, and members of the native Caspian fauna such as Crustacea (*Pterocuma sowinskyi* (Sars, 1894)), Polychaeta (Ampharetidae), Mollusca (*Hypanis vitrea bergi* Starobogatov 1972), and *Dreissena polymorpha*. These organisms primarily inhabit areas with salinity conditions ranging from 0–2 to 7‰. Certain crustaceans are euryhaline and can survive across a broader spectrum of salinity and depth (Romanova 1960, 1979).
3. Brackish-water species flourish in salinity ranges from 3–5 to 10–11‰. Numerous relict Caspian mollusks are part of this category, confined to the Northern Caspian Sea, such as *Didacna trigonoides* (Pallas, 1771) and *Dreissena rostriformis grimmi* (Andrusov, 1890). Additional forms consist of freshwater and brackish Amphipoda, which include 41 endemic species, along with the Cumacean *Stenocuma diastyloides* (G.O. Sars, 1897) (Starobogatov 1972, 1994).
4. Marine species originating from the Mediterranean: Bivalvia *Cerastoderma glaucum* (Bruguière, 1789) and salt-adapted variants of the relict Caspian complex, primarily inhabiting the Middle and Southern Caspian, including the Mollusca *Didacna profundicola* (Logvinenko and Starobogatov, 1966) and *Didacna longipes* (Grimm, 1877). These forms thrive at salinities of 8–13‰.

6.2.11 Macrophytes and Sea Grasses

The Caspian Sea hosts 104 macrophyte species, distributed across 27 families from three phyla. The species distribution is detailed in Table 6.4. The most diverse families are Cladophoraceae (Wille, 1884 (20 species)), Characeae (S.F. Gray,

Table 6.2 Taxonomic composition of the native Caspian Sea macrophytes

Phylum	Families	Species	Freshwater	Brackish	Marine, brackish	Environment is unknown
Chlorophyta	15	64	8	15	15 (+2 parasites)	26
Pheophyta	4	13		4	6	3
Rhodophyta	8	27		9	8	10

1821 (18 species)), Ulvaceae (J.V. Lamouroux ex Dumortier, 1822 (14 species)), and Ectocarpaceae (C. Agardh, 1828 (8 species)). The foundational element of the Caspian algal flora is formed by the Cladophoraceae family (Table 6.2).

Macrophytobenthos is a key component of the coastal ecosystem in the Caspian Sea. Macroalgae and seagrasses deliver a steady supply of organic material to support heterotrophic life while also creating essential habitats for invertebrates and photophilic fish. Shifts in aquatic vegetation significantly influence the structure, distribution, and abundance of benthic communities. In the Northern and Middle Caspian, the phytobenthos is particularly sensitive to fluctuations in sea level and broader environmental changes.

Across the Caspian Sea, 11 macrophyte species are widespread. In the coastal waters of Kazakhstan within the Northern Caspian, 38 species have been documented, with additional species found in the wetlands of transitional zones and estuaries.

In the eastern region of the Caspian Sea, only eight species of higher plants have been documented. These are all angiosperms: *Zostera noltei* (Hornemann, 1832), *Potamogeton pectinatus*, *Stuckenia pectinata* (L.) (Börner, 1912), *Ruppia spiralis* (Linnaeus ex Dumortier 1827), *R. maritima* (Linnaeus, 1753), and *Najas marina* L. The dominant species are *Zostera noltei*, *Myriophyllum spicatum* L., and *Ceratophyllum demersum* L. (Aladin et al. 2002).

6.2.12 Ichthyofauna

As noted by Derzhavin (1956) and Zenkevich (1963), the Caspian Sea's ichthyofauna comprises 78 species (Table 6.3). The highest concentration of endemic species occurs among herrings and gobies, though they are also present in other fish groups. Berg classified the Caspian as a distinct ichthyo-geographic subregion. The speciation of fish in the Caspian Sea is linked to its geological past. Cycles of marine regression and transgression, along with alternating phases of salinization and desalinization, have driven the emergence of new races, subspecies, and species.

Fish progress through three primary phases in their life cycle: spawning, feeding, and overwintering. For freshwater species, these stages often overlap and involve short-range migrations. Semi-anadromous fish feed in marine waters before moving into rivers to spawn and overwinter. Anadromous species, which reside in the sea,

Table 6.3 Ichthyofauna family-level species composition in the Caspian Sea (based on Derzhavin 1956; Zenkevich 1963)

Family	Number of species
Petromyzonidae (Bonaparte, 1831)	1
Acipenserilae (Bonaparte, 1831)	5
Clupeidae (Cuvier, 1816)	9
Salmonidae (Jarocki or Schinz, 1822)	2
Esocidae (G. Cuvier, 1817)	1
Cyprinidae (Rafinesque, 1815)	15
Cobitidae (Swainson, 1838)	2
Siluridae (Rafinesque, 1815)	1
Gadidae (Rafinesque, 1810)	1
Gasterosteidae (Bonaparte, 1831)	1
Syngnathidae (Bonaparte, 1831)	1
Atherinidae (Risso, 1827)	1
Persidae (Rafinesque, 1815)	4
Gobiidae (Cuvier, 1816)	12

travel into rivers solely for spawning. Pelagic fish remain in the sea throughout their entire lives. Some undertake extensive migrations, while others are confined to local regions. Most fish in the Caspian Sea are found in coastal zones at depths between 50 and 75 m (Aladin et al. 2002).

Among anadromous fish, sturgeons—which spawn in the rivers of the Caspian basin—are the most prized for human consumption. They journey hundreds of kilometers upstream to reproduce, favoring pebbly and sandy substrates for spawning. The construction of dams on rivers (except the Ural River) has blocked sturgeon access to many spawning sites, hindering their reproduction.

Currently, sturgeons spawn in the Ural River at locations up to 800 kilometers from its mouth. Due to siltation, the area of available spawning grounds has declined from 1700 hectares in the late 1970s to 1100 hectares in the early 1990s, and this reduction continues.

Sturgeons (Acipenseridae) The Caspian sturgeon stands out as the most prominent and well-known species. Fossil evidence indicates that these ancient fish have been present for more than ten million years, with a presence in the Caspian Sea spanning at least 5–6 million years, showing little evolutionary change (Bogachev 1928). Today, they are considered both endemic species and living relics.

Six species of sturgeon inhabit the Caspian Sea. These include five sturgeon types: the migratory beluga *Huso huso* (Linnaeus, 1758), the Russian sturgeon *Huso gueldenstaedtii* (Brandt and Ratzeburg, 1833) with its subspecies (the Persian sturgeon *Acipenser persicus* Borodin, 1897), the stellate sturgeon *Huso stellatus* (Pallas, 1771) with its South Caspian form (the stellate sturgeon *Platichthys stellatus* (Pallas, 1787)), the ship sturgeon *Huso nudiventris* (Lovetsky, 1828), and the sterlet *Huso ruthenus* (Linnaeus, 1758), which resides in the Volga River basin (Ivanov 2000).

Belugas, Russian sturgeons, and stellate sturgeons primarily feed and grow in the North Caspian during the warmer months, from April to October. As water temperatures drop in the Northern Caspian, these sturgeon species move southward into warmer waters.

Russian sturgeons consume benthic crustaceans such as Cumacea, Amphipodae, and Mysidacea. The diet of juveniles additionally includes clam worms and crabs (comprising 23 to 40%). As they mature, the proportion of mollusks—particularly abra and crabs—rises, and the intake of small fish like common kilka and gobies also increases significantly. In later growth stages, mollusks become the dominant food source. The Persian sturgeon has a dietary spectrum similar to that of the Russian sturgeon (Yablonskaya 1985).

In the diet of young stellate sturgeon, benthic crustaceans are also predominant. As they grow, worms begin to appear in their food intake. Mature stellate sturgeon additionally feed on mollusks and fish such as gobies or common kilka. Adults primarily consume crabs, kilka, gobies, and abra. Worms make up only approximately 2% of their diet (Yablonskaya 1985).

Starting in the 1980s, a sharp decline in sturgeon catches commenced. This was driven by (i) prior overfishing and (ii) a reduction in the number of sexually mature fish due to diminished natural reproduction linked to the loss of available spawning grounds. By the early 1990s, catches had fallen by half and continued to drop. In recent years, sturgeon harvests have not surpassed a few thousand tons annually.

Currently, two main types of measures are being implemented to protect sturgeon populations in the Caspian Sea: (1) strict catch quotas and (2) artificial breeding and release of juveniles into the wild. Regrettably, most of the declared sturgeon fishing regulations are not followed. It is evident that the existing system for monitoring sturgeon fisheries and enforcing rules and prohibitions needs restructuring and stronger enforcement (Ivanov 2000).

6.2.13 *Marine Fish*

Overall, 39 distinct marine or marine-brackish-water species and subspecies are recognized, though the classification of three remains contested. With the exception of *Sander marinus* (Cuvier, 1828), all are endemic to the Caspian Sea. The status of *S. marinus* itself is uncertain, however, as the evolutionary connections between sea sander populations in the Black and Caspian Seas have not yet been examined. Only a small number of native brackish-water fishes share a direct evolutionary lineage with species adapted to higher salinity in the Caspian. Exceptions include the Caspian Atherina, *Atherina boyeri* (Risso, 1810), and the plump-cheeked needlefish, *Syngnathus abaster* (Risso, 1827). These latter species are more closely related to typical Mediterranean fish fauna.

Commercial marine fish species are found throughout the Caspian Sea, though their distribution patterns vary. Certain marine species can reproduce, grow, and feed in the shallow northern region of the sea, including the common kilka *Clupeonella caspia* (Svetovidov, 1941), the Caspian shad *Alosa braschnikowi* (Borodin, 1904), and other *Alosa* spp.

Kilka fish serve as the principal commercial species for every nation bordering the Caspian Sea. Prior to the year 2000, the anchovy kilka, *Clupeonella engrauliformis* (Borodin, 1904), accounted for 70–90% of the total kilka harvest in the Caspian (Ivanov 2000). This warm-water species resides in the Middle and Southern regions of the Caspian Sea. During the warmer months from May to December, when water temperatures range from 15 to 25 °C, it reproduces in the surface layer (Paritsky 2014).

The populations of big-eyed kilka *Clupeonella grimmi* (Kessler, 1877) are found primarily in the open waters of the Middle and Southern Caspian Sea at depths exceeding 40–50 m. During winter and spring, the fish gather in schools in the northwestern and eastern sections of the Southern Caspian, migrating toward the northwest regions of the Middle and Southern Caspian by autumn. This species tends to avoid surface waters, inhabiting depths between 20 and 200 m and sometimes descending as deep as 500 m. In winter, schools are typically observed at 80–150 m; in spring and summer, they rise to 40–60 m. By autumn, they return to deeper waters around 100–150 m. Spawning is extended, occurring mainly from January to April and again from October to November. Larvae and juveniles inhabit the upper water layers and feed on zooplankton.

The common kilka, *Clupeonella caspia*, is a euryhaline species capable of thriving across a broad spectrum of salinity levels. Its abundant schools populate the Northern Caspian, where they reproduce in the shallow, fresher waters. This species entered the Caspian Sea via the Kumo-Manych depression and is now distributed throughout the sea, predominantly inhabiting coastal areas at depths of 10–60 m. The kilka undertakes migrations into the lower sections of the Volga, Ural, and Terek rivers. Since 1964, the common kilka has expanded its range within the Volga River system, establishing itself in the river's reservoirs and forming resilient populations. Consequently, it has effectively become an invasive species in these areas.

Most marine species in the Caspian Sea lack a distinct wintering phase, as the favorable climate supports the continuous growth of prey organisms, enabling fish to feed actively year-round. The Middle and Southern Caspian serve as crucial feeding grounds for various fish species.

Nonetheless, the typical wintering habitats for multiple kilka species—common, anchovy, and big eyed—are situated along the Turkmen shoreline of the Southeastern Caspian. Substantial commercial populations of prized herring varieties and mullet species are found in Turkmenistan's offshore waters. Twenty-six fish species inhabit the Turkmenbashi Gulf, among them gobies, sand smelts, herrings, mullets, and pipefish; the Caspian roach *Rutilus caspicus* (Yakovlev, 1870) is the most numerous, while *Salmo caspius* (Kessler, 1877), the Russian sturgeon *Huso gueldenstaedtii* (Brandt and Ratzeburg, 1833), and the stellate sturgeon *Acipenser stellatus* (Pallas, 1771) are uncommon (Aladin et al., 2002).

6.2.14 *Caspian Seal* **Pusa Caspica** *(Gmelin, 1788)*

The Caspian seal, *Pusa caspica* (Gmelin, 1788), stands as the sole mammal inhabiting the Caspian Sea and is an endemic species, living exclusively in its brackish waters. It represents the smallest member of the *Pusa* genus, having entered the Caspian from the Arctic during the Quaternary period and becoming landlocked following the retreat of the continental ice sheet.

Its range encompasses not only the coastal areas but also the many rocky islands and drifting ice formations. These seals undertake seasonal migrations into the deltas of the Volga and Ural Rivers. Considerable seal populations are found on islands in the eastern sector of the Northern Caspian and on the sandy spits of the Southern Caspian. Nonetheless, the greatest concentrations are observed along the edge of the ice cover in the Northern Caspian during their winter breeding period. In 1993, the Caspian seal was classified as a vulnerable species by the IUCN (International Union for Conservation of Nature) and was subsequently added to the IUCN Red List. The population has declined dramatically from an estimated 1.5 million individuals at the start of the twentieth century to roughly 360,000–400,000 by the close of the 1980s.

6.2.15 *Anthropogenic Impacts on the Caspian Sea Flora and Fauna*

During the 1900s, the ecological state of the Caspian Sea declined significantly. This was primarily caused by changes in sea level linked to the management of river flows, along with pollution from sources like hydrocarbons and phenols (Kosarev 2006).

Furthermore, the connection between the Caspian Sea and the Black and Azov seas was reestablished with the completion of the Volga-Don Canal in 1952. This allowed zoobenthic fouling organisms and aquatic plants to enter the Caspian by attaching to ship hulls. By the early 1980s, following the widespread use of ballast tanks on vessels, phytoplankton and zooplankton species from the Black Sea also began arriving in the Caspian (Shiganova 2010).

Consequently, the Caspian Sea experienced impacts from both accidental invasions and intentional introductions of non-native species. While some species rapidly established themselves, others appeared only briefly before disappearing. The inventory of non-native species continues to evolve quickly, a common trend in regions influenced by global transformations.

Ongoing climate change is also reshaping the Ponto-Caspian ecosystem. Given that abiotic factors—such as air and water temperature, wind patterns, and atmospheric humidity—profoundly affect the ecosystem, analyzing their fluctuations remains essential.

6.3 Non-native Biota Composition in the Caspian Sea

The introduction of non-native species and shifts in native biodiversity within the Caspian Sea have occurred, with limited exceptions, across three distinct phases (Fig. 6.8).

The initial phase involved a systematic, extensive stocking effort, launched during the 1930s as part of a stock enhancement initiative. This program was designed to boost the populations of commercially valuable fish species or their prey within the Caspian Sea (Fig. 6.8).

Following these introductions, only two fish species—the mullets *Chelon saliens* (Risso, 1810) and *Chelon aurata* (Popov, 1930) from the Black Sea—along with two benthic organisms (the polychaete *Hediste (Nereis) diversicolor* (O.F. Müller 1776) and the bivalve *Abra segmentum* (Récluz, 1843)) from the Sea of Azov, have managed to establish themselves successfully in the Caspian Sea. Two shrimp species, *Palaemon adspersus* (Rathke, 1836) and *P. elegans* (Rathke, 1836), as well as the crab *Rhithropanopeus harrisii* (Gould, 1841), were brought from the Black Sea at the same time as the deliberate mullet introductions. An unintended consequence was the accidental introduction of fish parasites (Karpevitch 1975).

The most notable exception was the introduction of the bivalve *Mytilaster lineatus* (Gmelin, 1791). It reached the Caspian Sea as fouling on small vessels, arriving at Baku harbor in 1919 (Bogachev 1928).

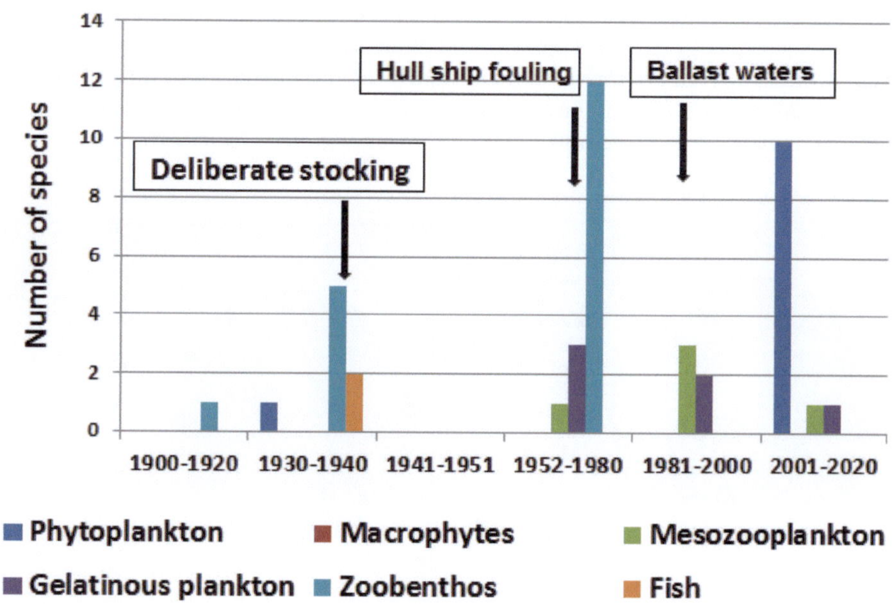

Fig. 6.8 Primary pathways and temporal patterns of species introduction into the Caspian Sea

The second phase commenced with the construction of the Volga-Don Canal in 1952. During this period, there was a substantial rise in the number of non-native species introduced via hull fouling. This development established a significant new pathway for biological invasions. Additionally, the Volga River is now linked to the Baltic basin through the Volga-Baltic Waterway. Collectively, these elements created "invasion corridors" into the Ponto-Caspian basin (Panov et al. 1999) (refer to Fig. 2.2 in Chap. 2). At that time, zoobenthic species and macrophytes were transported from the Black Sea on ships' hulls through fouling.

The third phase commenced in the early 1980s, following the installation of ballast tanks on vessels, which led to the introduction of pelagic phytoplankton and zooplankton species via ballast water. Species of phytoplankton and mesozooplankton originating from the Black Sea, such as Copepoda and Diplostraca, are now documented in the Caspian Sea. These introduced species have subsequently displaced native Caspian populations (Fig. 6.9).

Every non-native species that arrived and became established during the indicated timeframes originated from the Black Sea, with some being intentionally introduced from the Sea of Azov (Fig. 6.9).

The initial group consists of 37 species, all of which are widespread euryhaline species from the Black Sea. These species have Atlantic-Mediterranean origins but have resided in the Black Sea for an extended period, adapting to its comparatively low salinity. This group includes phytoplankton (nine species), macrophytes (nine species), zooplankton (four species) comprising Copepoda (one species) and Diplostraca (three species), gelatinous plankton (one species), benthos (12 species) including Amphinoda (one species), Bivalvia (four species), Cirripedia (one species), Decapoda (three species), Polychaeta (one species), and fish (two species).

The subsequent category comprises 11 brackish-water species from the Black Sea, which evolved to thrive in low-salinity environments prior to colonizing compatible regions of the Caspian Sea. This group encompasses the Hydrozoa *Odessia maeotica* (Ostroumoff, 1896), the amphipods *Gammarus aequicauda*

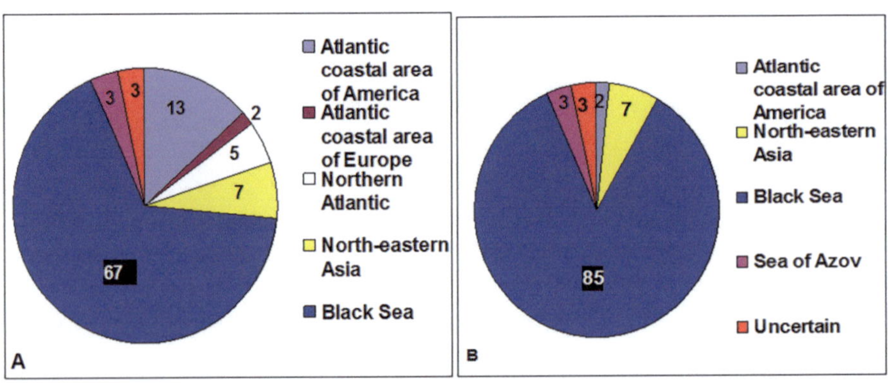

Fig. 6.9 Proportion (%) of introduced species in the Caspian Sea. (**a**) the original routes of species introduction; (**b**) species originating from the Black Sea, including its non-native arrivals

(Martyinov, 1931) and *Iphigenella shablensis* (Carausu, 1943), the mollusks *Monodacna colorata* (Eichwald, 1829), *Hypanis colorata* (Ménétries, 1832), *Dreissena rostriformis bugensis* (Andrusov 1897), *Lithoglyphus naticoides* (Pfeiffer), and *Tenellia adspersa* (Nordmann, 1845), the kamptozoan *Barentsia benedeni* (Foettinger, 1887), and the pearlwort *Conopeum seurati* (Canu, 1928), and *Electra*, *Einhornia crustulenta* (Pallas, 1766) (Karpinsky et al. 2006).

Alongside the indigenous species of the Black Sea, several invasive organisms from that region have also been established in the Caspian Sea. These introductions total nine species. A number of these invaders first reached the Black Sea from the Atlantic shores of North America, such as the comb jelly *Mnemiopsis leidyi* (Aggasiz, 1885) and its natural predator, the comb jelly *Beroe ovata* (Bruguière, 1789). The copepod *Acartia tonsa* (Dana, 1849) and the hydrozoan *Calyptospadix cerulea* (Clarke, 1882) also followed this route. Additional species originated from the European Atlantic coast, including the polychaete worm *Ficopomatus enigmaticus* (Fauvel, 1923), the crab *Rhithropanopeus harrisii* (Gould, 1841), the hydrozoan *Blackfordia virginica* (Mayer, 1910), and two diatom species: *Rhizosolenia calcar-avis* (M. Schultze, 1858) and *Cerataulina pelagica* (Morozova – Vodyanitskaya 1948).

In 1999, divers identified specimens of *Mnemiopsis leidyi* and *Aurelia aurita* near the border between the Middle and Southern Caspian Sea (Ivanov et al. 2000; Shiganova et al. 2001). Genetic studies confirmed that *Mnemiopsis leidyi* originated from the Black Sea, which itself had received the species from the Gulf of Mexico (specifically Tampa Bay, Florida). The populations of *M. leidyi* in the Black and Caspian Seas display comparable allelic diversity and a low FST value, indicating a close genetic relationship between these groups. Additionally, specimens gathered from the Northern and Southern Caspian show minimal variation across populations (Ghabooli et al. 2011).

Several species have migrated into the Northern Caspian from the brackish and freshwater habitats of the Volga River and its reservoirs. Among these are the quagga mussel *Dreissena bugensis* (Andrusov, 1897) (Orlova et al. 1999) and, within the brackish waters of the Volga delta, the gastropod *Lithoglyphus naticoides* (Pfeiffer).

Five species of freshwater fish were introduced concurrently through aquaculture activities originating from East Asia.

The origins of other non-native species in the Caspian, such as the pearlwort *Lophopodella carteri* (Hyatt, 1866) and the three-spine stickleback *Gasterosteus aculeatus* (Linnaeus, 1758), remain ambiguous. These are regarded as cryptogenic species (of uncertain origin) (Shiganova 2010).

To date, 64 nonindigenous species have become established in the Caspian Sea, excluding four freshwater fish species introduced for commercial aquaculture. These organisms originally inhabited estuaries, reservoirs, and the brackish waters of the North Caspian (Karpevitch 1975). Certain species appear only occasionally. The origins of the remaining four species remain uncertain. The status of *Amphibalanus eburneus* (Gould, 1841) is unclear. Therefore, aside from a limited number of freshwater invertebrates intentionally brought from the Atlantic coastal region of

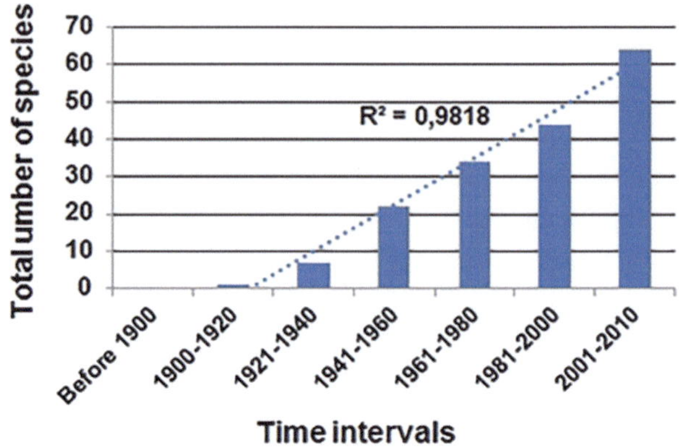

Fig. 6.10 A rise in the count of invasive species within the Caspian Sea has been observed since the start of the twentieth century. The dotted line indicates the linear trend (approximation accuracy $R^2 = 0.9818$).

America and northeastern Asia, along with two species deliberately introduced from the Sea of Azov, all established non-native species—primarily euryhaline—were introduced from the Black Sea (Fig. 6.9). These introduced euryhaline species have outcompeted native biota in both the brackish Northern Caspian and the more saline Middle and Southern Caspian. Marine euryhaline non-native species have colonized the Middle and Southern Caspian, frequently displacing indigenous species. Brackish and freshwater species have established themselves in the Northern Caspian, though those most tolerant to salinity have managed to spread into the Middle and Southern Caspian. The count of invasive species in the Caspian Sea has increased linearly since the start of the twentieth century (Fig. 6.10).

6.3.1 Non-native Species Impact on the Caspian Sea Ecosystem

6.3.1.1 Non-native Phytoplankton Species Impact

Within the Northern Caspian region, phytoplankton species, whether indigenous or introduced, primarily originate from estuarine brackish-water environments. Conversely, in the Middle and Southern Caspian areas, euryhaline marine neritic species are dominant (Proshkina-Lavrenko and Makarova 1968). A compilation of non-native phytoplankton species is provided in Table 6.4.

The diatom species *Rhizosolenia* (syn. *Pseudosolenia*) calcar-avis was unintentionally introduced into the Caspian Sea in 1934, coinciding with the intentional release of mullet fish. Following its introduction, this species spread quickly throughout the Caspian, displacing the native diatom *Rhizosolenia fragilissima f. fragilissima* (Bergon, 1903) and the dinoflagellate *Prorocentrum cordatum* (Ostenfeld) (J.D. Dodge, 1975) in regions where these species were previously dominant. Consequently, the total phytoplankton biomass in the area increased twofold. From the 1930s onward, *R. calcar-avis* came to constitute 80–92% of the

Table 6.4 Non-native species of phytoplankton

Species	Donor region	Vector	First record	Status of establishment	References
Bacillariophyta *Rhizosolenia calcar-avis* (Schultze, 1858)	Black Sea	During deliberate stocking of gray mullets	1934	Established	Usachev (1948)
Bacillariophyta *Pseudo-nitzschia seriata* (Cleve) (H. Peragallo, 1899)	Black Sea	Ballast waters	2004	Established	Shiganova et al. (2005)
Bacillariophyta *Pseudo-nitzschia pseudodelicatissima* (Hasle) (Hasle, 1993)	Black Sea	Ballast waters	2004	Established	Shiganova et al. (2005)
Bacillariophyta *Chaetoceros peruvianus* (Brightwell, 1856)	Black Sea	Ballast waters	2004	Established	Shiganova et al. (2005)
Bacillariophyta *Tropidoneis lepidoptera* (Gregory) (Cleve, 1894)	Black Sea	Ballast waters	2004	Established	Shiganova et al. (2005)
Bacillariophyta *Cerataulina pelagica* (Cleve) (Hendey, 1937)	Black Sea	Ballast waters	2008	Established	Pautova et al. (2015)
Coccolithophore *Braarudosphaera bigelowii* (Gran and Braarud) (Deflandre, 1947)	Black Sea	Ballast waters	2004	Established	Shiganova et al. (2005)
Coccolithophore *Emiliania huxleyi* (Lohmann) (Hay and Mohler, 1967)	Black Sea	Ballast waters	2004	Established	Shiganova et al. (2005)
Dinophyta *Akashiwo sanguinea* (K. Hirasaka) (Gert Hansen and Moestrup, 2000)	Black Sea	Ballast waters	2004	Established	Shiganova et al. (2005)
Dinophyta *Protoperidinium crassipes* (Kofoid) (Ballech 1974)	Black Sea	Ballast waters	2004	Established	Shiganova et al. (2005)
Dinophyta *Gonyaulax polygramma* (F. Stein, 1883)	Black Sea	Ballast waters	2012	Established	Pautova et al. (2015)

phytoplankton biomass in the Middle and Southern Caspian. As a large diatom, *R. calcar-avis* is not a suitable food source for zooplankton in the Caspian. Instead, the algae sink to the seafloor, where they are broken down by bacteria and incorporated into the benthic food chain. Its establishment triggered a significant shift in energy flow within the ecosystem (Levshakova and Sanina 1973).

In the third phase, beginning in the early 1980s, multiple phytoplankton species were introduced via ballast water from the Black Sea (Table 6.4). These included two dinoflagellate species, native to the Black Sea (*Akashiwo sanguinea* and

Protoperidinium crassipes), several diatom species (*Pseudo-nitzschia seriata*, *Pseudo-nitzschia pseudodelicatissima*, *Chaetoceros peruvianus*), and, for a limited period, the planktonic *Tropidoneis lepidoptera*. All of these species achieved broad distribution across the region (Shiganova et al. 2005; Pautova et al. 2015). In the Middle Caspian, the coccolithophore *Braarudosphaera bigelowii* and the globally widespread *Emiliania huxleyi* were documented in 2004 (Shiganova et al. 2005). *Emiliania huxleyi* is now extensively distributed and frequently forms blooms in the Black Sea (Mikaelyan et al. 2013).

Among introduced species, the diatom *Pseudo-nitzschia seriata* and the dinoflagellate *Akashiwo sanguinea* are known to be toxic. The majority of these species now impact the Caspian ecosystem and frequently become dominant, forming blooms whose biomass can account for over 90% of the total phytoplankton biomass. A nine-year phytoplankton study (2004–2012) in the Middle Caspian revealed shifts in species composition and community structure following the introduction of non-native species. Blooms of the following algae influenced the phytoplankton community in the deep area (Derbent Depression) of the Middle Caspian: diatoms (in 2004, 2005, and 2010), dinoflagellates (in 2006), and the non-native *Emiliania huxleyi* (2.4×10^3 cells·l^{-1}) (Pautova et al. 2015). Consequently, the makeup of the dominant phytoplankton species altered. Rather than the large-cell non-native diatoms *Pseudosolenia calcar-avis* and *Rhizosolenia fragilissima f. fragilissima* (Bergon, 1903), which were dominant from the 1960s to the 1980s, the non-native *Pseudo-nitzschia seriata*, *Cerataulina pelagica*, and *Chaetoceros peruvianus* now prevail during spring-summer-autumn bloom events in the Middle Caspian (Pautova et al. 2015). Additionally, a winter bloom of *Cerataulina bergonii* (Ostenfeld, 1903) was first observed in February 2008 in the Middle Caspian at a temperature of 11.4 °C, with concentrations reaching 10^6 cells·l^{-1}, constituting more than 90% of phytoplankton biomass in the 0–50 m layer. In deeper layers (50–120 m), a bloom of *Pseudo-nitzschia seriata* (Cleve) (H. Peragallo, 1899) (classified as a harmful species) was recorded, with an abundance of approximately 10^6 cells·l^{-1}. In effect, these blooms consisted of a single species, making up over 90% of the total phytoplankton biomass.

Subsequently, during May–June 2012, two distinct phytocenoses were recorded in the Middle and Southern Caspian, separated by the lower limit of the seasonal thermocline. The warm-water community was dominated by the large centric diatom *Coscinodiscus perforatus* (Ehrenberg, 1844), along with the dinoflagellates *Prorocentrum cordatum* (Ostenfeld) (J.D. Dodge, 1976) and the non-native species *Gonyaulax polygramma* (F. Stein, 1883). At a single station near the Apsheron peninsula, at a depth of 23 m corresponding to the upper thermocline boundary, the potentially toxic non-native *Gonyaulax polygramma* was observed with a concentration of 48×10^3 cells·l^{-1} and a biomass of 2289.6 mg·m^{-3}. Its contribution to the total phytoplankton biomass reached 81.43%, and to the biomass of Dinoflagellata, it was 90.6%. Above and below this specific layer, the concentration of this species was significantly lower. This distribution was likely linked to particular hydrological conditions (Pautova et al. 2015).

Peak phytoplankton biomass levels in the deep-water zones of the Middle and Southern Caspian Sea matched those found in the layer above the seasonal thermocline (up to 400 $mg.m^{-3}$) and below it (up to 570 $mg.m^{-3}$). The cold-water community was dominated by the pennate non-native diatom *Pseudo-nitzschia seriata* (Cleve) (H. Peragallo, 1899), which is classified as harmful. Beneath the thermocline, this species accounted for 96% of the total phytoplankton biomass, attaining its highest concentrations (400–570 $mg.m^{-3}$) close to the upper limit of the pycnocline. Other algal species were exceptionally scarce (Pautova et al. 2015).

Nine years (2004–2012) of phytoplankton study in the Middle Caspian demonstrated that the cold-water phytoplankton community, living below the thermocline, was a remnant of the winter-spring blooms. The species, constituting those blooms (*Pseudo-nitzschia seriata*, *Cerataulina pelagica*, *Chaetoceros peruvianus*), were non-natives, which suppressed native ones due to their higher plasticity (Pautova et al. 2019).

A nine-year investigation (2004–2012) of phytoplankton in the Middle Caspian Sea revealed that the cold-water community existing beneath the thermocline was a leftover from the winter-spring blooms. The species forming these blooms—*Pseudo-nitzschia seriata*, *Cerataulina pelagica*, and *Chaetoceros peruvianus*—were non-native and outcompeted indigenous species because of their greater adaptability (Pautova et al. 2019).

Consequently, beginning in the late 1990s, shifts in diatom composition toward smaller, non-native species such as *Cerataulina pelagica*, *Pseudo-nitzschia seriata*, and *Chaetoceros peruvianus* during phytoplankton blooms have contributed to a reduction in overall phytoplankton biomass. As a result, from May to November, the biomass of diatom blooms has ranged from 0.15 to 0.2 g C m^{-3}, which is approximately half the levels recorded in the 1970s and 1980s, a period when *Rhizosolenia calcar-avis* was the dominant species. This alteration in the dominant diatom species, coupled with the growing importance of picoplankton, has led to a decline in primary production within the Middle Caspian Sea (Pautova et al. 2015).

Between 2000 and 2010, the proliferation of algal blooms in the Caspian Sea contributed to the development of hypoxic areas. Nevertheless, comprehensive information regarding the scale of this hypoxia—including its frequency, duration, and geographic spread—as well as its related impacts, remains unavailable. Owing to limitations in nitrogen and phosphorus, combined with diminished grazing pressure, the composition of phytoplankton species underwent a transformation: the proportion of small-celled microalgae rose, whereas large-celled algae, which had been predominant for many years (such as *Rhizosolenia calcar-avis*), declined (Ardabieva et al. 2005). The marked decrease in *Rhizosolenia calcar-avis* concentration led to a substantial rise in silicium levels within the euphotic zone (Pautova et al. 2015).

In the Northern Caspian Sea, the sizable *R. calcar-avis* and *Spyrogira* gave way to smaller diatoms from the genera *Navicula*, *Nitzschia*, *Cyclotella*, and *Chaetoceros*, along with the green algae *Scenedesmus*, *Ankistrodesmus*, and *Binuclearia* and the blue-green algae *Merismopedia* (Pautova et al. 2015).

Relative to the 1990s, the 2000s saw a greater variety of algal species in the Middle and Southern Caspian Sea (Ardabieva et al. 2005; Polyaninova 2007).

Diatoms were the predominant group, with their numbers nearly doubling since 1998. Small-celled algae, including *Cyclotella caspia* (absent in WoRMS), *Cyclotella laevissima* (Van Goor, 1920), *Pleurosigma elongatum* (W. Smith, 1852), and *Actinocyclos ehrenbergii* (Ralfs, 1861), along with species from the genera *Nitzschia* and *Navicula*, became dominant within the diatom community. The abundance of *Rhizosolenia calcar-avis* decreased by a factor of 13.5. Non-native species from the Azov-Black Sea region, such as *Pseudo-nitzschia seriata* and *Cerataulina pelagica*, as well as *Tropidoneis lepidoptera* and *Chaetoceros peruvianus* (first recorded in 2004, Shiganova et al. 2005), became widespread in the Middle and South Caspian (Pautova et al. 2019).

The non-native dinoflagellate *Gonyaulax polygramma* forms summer blooms in the eastern Middle Caspian under wind-driven upwelling conditions. Under these circumstances, its biomass can attain 16 g m^{-3} (Pautova et al. 2019).

The influence of blue-green and green algae, particularly from the genus *Oscillatoria* (Cyanophyceae), has grown more significant relative to the 1990s. Among these, *Oscillatoria* sp. and *Jaaginema subtilissimum* (Kützing ex Forti) (Anagnostidis and Komárek, 1988) (first recorded in 2001) developed more extensively. In 2001 and 2002, the quantity of blue-green and green algae substantially surpassed the typically dominant diatoms and dinoflagellates in this region (Southern Caspian). During 2001 and 2005, anomalous summer blooms exhibiting unusual patchiness occurred across large sections of the Caspian Sea for approximately 4–5 months. Since 2005, blue-green algae have continued to proliferate in substantial amounts, while green algae were diverse but present in relatively low abundance (Pautova et al. 2015).

An extraordinary proliferation of toxic cyanobacteria (*Nodularia spumigena* Mertens ex Bornet and Flahault, 1888 and *Nodularia harveyana* Thuret ex Bornet and Flahault, 1886) was documented in Iranian coastal waters from mid-August through late September 2005 (refer to the image of the bloom in the Southern Caspian region near the Iranian shoreline in Fig. 6.11), despite these species

Fig. 6.11 A bloom of *Nodularia spumigena* was observed along the Iranian coastline in September 2005. (Photograph from Roohi et al. 2010)

previously being found only in the Northern Caspian. The affected area covered 20 km^2. Based on a sampling effort on September 20, 2005, conducted by an Iranian research team, *Oscillatoria* sp. was also identified in significant amounts alongside *Nodularia* sp. The cell density of *Nodularia* sp. reached 189×10^6 cells m^{-3} in the upper 7 m and $10{,}069 \times 10^6$ cells m^{-3} in the upper 20 m. The mean cyanobacterial abundance and biomass in the top 7 and 20 m were 5829×10^6 cells m^{-3} (with *Nodularia* sp. contributing 512×10^6 cells m^{-3}) and a biomass of 1655 mg m^{-3}, respectively (Fig. 6.11) (Roohi et al. 2010). *Nodularia spumigena* generates compounds that are toxic to humans and fish, including hepatotoxins and neurotoxins (Kruger et al. 2009).

In subsequent years, these species' blooms occurred in other parts of the Southern Caspian (Akhuntov et al. 2021). It is likely that these unusual summer blooms have been linked to the effects of *Mnemiopsis leidyi* proliferation.

6.3.2 Non-native Zooplankton Species Impact

Black Sea euryhaline mesozooplankton species were transported to the Caspian Sea via ballast water (Table 6.5). These organisms are originally Atlantic-Mediterranean species that had long inhabited the Black Sea, where they acclimated to conditions of comparatively low salinity. The majority are broadly dispersed throughout the Black Sea.

Within the Copepoda, the most notable invasive species in the Caspian Sea is *Acartia tonsa* (Dana 1849). It was initially recorded as *A. clausi* (Giesbrecht, 1889) (Kurasheva and Abdulaeva 1984) and later reidentified as *Acartia tonsa* by Shiganova et al. (2005).

A. tonsa was introduced into the Black Sea from North Atlantic coastal waters and subsequently transported to the Caspian Sea. Following its establishment in the Caspian Sea in 1981, *Acartia tonsa* displaced native copepods (Shiganova et al. 2004b) and emerged as the dominant species, especially after the invasion of *Mnemiopsis leidyi* (Fig. 6.12), serving as a food source for kilka (Zabralieva et al. 2006). Even prior to the *M. leidyi* invasion in 1986, *A. tonsa* attained densities of up to 3100 ind.m^{-3} in the Southern Caspian and supplanted the native *Eutemora grimmi* in the Middle and Southern Caspian. Furthermore, *A. tonsa* also replaced another abundant species, *Calanipeda aquae-dulcis*, in coastal regions. Before the *A. tonsa* invasion (1971–1981), *C. aquae dulcis* concentrations ranged from 201 to 610 ind.m^{-3} in the western Caspian and from 238 to 1013 ind.m^{-3} in the eastern Caspian. After *A. tonsa* became established in 1981, *C. aquae dulcis* numbers declined due to this displacement. Consequently, from 1982 to 1993, *C. aquae dulcis* abundance varied between 15 and 844 ind.m^{-3} in the western Middle Caspian and between 58 and 344 ind.m^{-3} in the eastern Middle Caspian. In the Southern Caspian, following the proliferation of *A. tonsa*, *C. aquae dulcis* abundance fell from 563 to 0.3 ind.m^{-3} in the west and from 968 to 68 ind.m^{-3} in the east (Fig. 6.12).

Table 6.5 Non-native zooplankton species

Species	Donor region	Vector	First record	Status of establishment	References
Copepoda					
Acartia tonsa (Dana, 1849)	Black Sea First introduced in the Black Sea from north American waters	Shipping ballast waters	1981	Established	Kurasheva and Abdulaeva (1984), reidentified by Shiganova et al. (2001)
Copepoda					
Oithona similis (Claus, 1866)	Black Sea	Shipping ballast waters	2004	?	Shiganova et al. (2005)
Cirripedia					
Amphibalanus improvisus (Darwin, 1854)	Black Sea or Azov Sea	Shipping hull fouling	1955	Established	Derzhavin (1956)
Amphibalanus eburneus (Gould, 1841)		Shipping hull fouling	1955	? Fate Unknown	Zevina (1965)
Diplostraca					
Pleopis polyphemoides (Leukart, 1859)	Black Sea	Shipping Ballast waters	1958	Established	Mordukhai-Boltovskoy (1962)
Diplostraca					
Penilia avirostris (Dana, 1849)	Black Sea	Shipping Ballast waters	1985	Established	Aladin et al. (2002); Shiganova, unpublished
Diplostraca					
Podon intermedius (Lilljeborg, 1853)	Black Sea	Shipping	1985	Established	1985 Kurasheva and Abdulaeva (1984)
Copepoda					
Calanus euxinus (Hulsemann, 1991)	Black Sea	Shipping	2004	? Fate is now unknown	2004 Shiganova et al. (2005)
Chaetognatha					
Parasagitta setosa (Muller, 1847)	Holomeso-plankton	Shipping Ballast waters	2004	? Fate is now unknown	2004 Shiganova et al. (2005)

While both *E. grimmi* and *C. aquae dulcis* are plant-eating species, *A. tonsa* proves to be the more effective competitor for nutritional resources (Tatarintseva et al. 2000). Research conducted in the Black Sea indicates that *A. tonsa* exhibits high reproductive capacity and generates resting eggs capable of enduring adverse environmental circumstances (Gubanova 2003). Furthermore, this species deposits

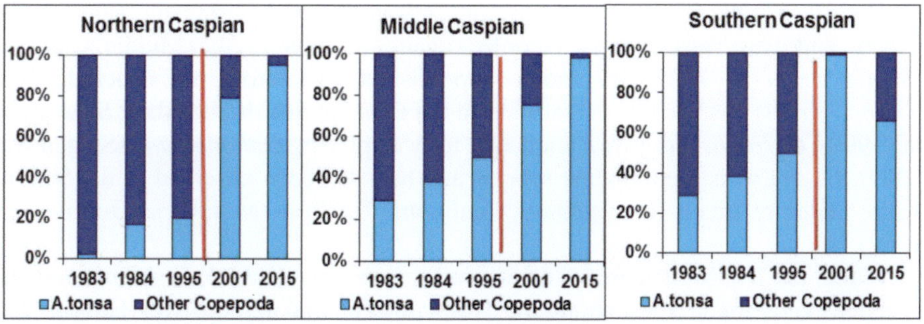

Fig. 6.12 Proportion of *Acartia tonsa* within total Copepoda abundance during summer months across key subregions of the Caspian Sea (data for 1983–1995 sourced from Tatarintseva et al. 2000; data for 2001–2004 sourced from Shiganova et al. 2004b; drawing by T. Shiganova). Vertical red lines mark the period of *M. leidyi* introduction

its eggs freely into the water column rather than retaining them in an attached sac. The seasonal population patterns of *A. tonsa* in the Caspian Sea contrast with those observed in the Black and Azov seas: it maintains a year-round presence in the Middle and Southern Caspian regions (Tatarintseva et al. 2000), whereas in the Black and Azov seas, its development is restricted to warmer periods (Gubanova 2003).

From 2001 to 2004, *A. tonsa* emerged as the sole copepod species capable of surviving following the *Mnemiopsis leidyi* invasion. It accounted for 75–99% of the zooplankton in late summer (Fig. 6.12) and 90% of the kilka diet (Shiganova et al. 2004b). Recent research shows a rise in the proportion of *A. tonsa* within Copepoda abundance (Fig. 6.12) across all regions of the Caspian Sea. During summer zooplankton surveys conducted by the Caspian Scientific Research Institute of Fisheries (CaspNIRKh, Astrakhan, Russian Federation) in 2014 and 2015 in the deep zone of the Northwestern Caspian, nauplii and young copepodites represented 100% of the copepod population. *Calanipeda aquae-dulcis* was previously subdominant (around 50%), but only in early summer before *M. leidyi* had spread to the Northern Caspian. The abundance of *A. tonsa* reached 2700 ind.m^{-3} with a biomass of 23.3 mg.m^{-3}, whereas other Copepoda constituted just 5% of abundance and 14% of biomass (Fig. 6.12).

In May 2004, multiple specimens of Black Sea zoo- and phytoplankton species, not previously documented in the Caspian, were gathered in the Middle Caspian. A recently ingested and undigested specimen of *Calanus euxinus* (copepodite stage V; 2.5 mm long) was discovered in the gut of a kilka. The plankton sample also contained the chaetognath *Parasagitta setosa* (Table 6.5). It remains uncertain whether these species have become established in the Caspian Sea (Shiganova et al. 2005). *Calanus euxinus* and *Parasagitta setosa* are cold-water Black Sea species that undertake daily vertical migrations, moving to the lower boundary of the oxygenated zone (Vinogradov et al. 1992). Consequently, the likelihood of *C. euxinus* and *P. setosa* establishing themselves in the Caspian Sea is relatively low.

The non-native copepod *Oithona similis* was detected during the same 2004 survey and was subsequently recorded again in 2008 in the Middle Caspian (Shiganova et al. 2005; Shiganova, unpublished). Therefore, we conclude that *Oithona similis* has become established in the Caspian Sea. In the Black Sea, since the 1990s, *O. similis* has grown abundant, fully displacing *Oithona nana* (Shiganova 2009). *O. similis* is a euryhaline species and has also been observed in the Sea of Azov under relatively low salinity conditions (9.07‰) (Mordukhai-Boltovskoy 1969).

Within the Diplostraca group, the nonindigenous species *Pleopis polyphemoides* is extensively found throughout the Caspian Sea, making up over half of the cladoceran community (Polyaninova et al. 2003; Shiganova et al. 2004b; Kamakin, unpublished data from 2014–2015). Research conducted in the Northwestern Caspian during June 2014 and 2015 revealed that the non-native *P. polyphemoides* was dominant, reaching an abundance of 92 individuals per cubic meter and a biomass of 0.4 mg per cubic meter, followed by the native *Podonavadne trigor* at 48 individuals per cubic meter and 0.7 mg per cubic meter of biomass.

In June 2015, an investigation in the Middle Caspian Sea, along a transect from Derbent to Cape Peschanyi, showed relatively limited zooplankton diversity. Among copepods, *A. tonsa* was the most numerous species (568.2 individuals per cubic meter), accounting for 98% of all Copepoda. The native *C. aqua edulcis* followed, with a significantly lower density of 7.23 individuals per cubic meter. Within Diplostraca, the non-native *Pleopis polyphemoides* was most abundant, with 188.4 individuals per cubic meter and a biomass of 7.23 mg per cubic meter, representing 90% of all Cladocera. Another subdominant group consisted of larvae from the non-native barnacle *Amphibalanus improvisus* (Fig. 6.13). Species richness was relatively low: in the central area, it included 13 species; in the western part, nine species; and in the eastern section, ten species during the survey period.

During concurrent surveys in the Southern Caspian in June 2015 (along a transect from Kurinsky Stone Island to Ogurchinsky Island), mesozooplankton species richness was limited to just ten species. The community was primarily composed of Copepoda, which accounted for 65–66% of both total abundance and biomass, with *A. tonsa* being the dominant species. Larvae of the barnacle *Amphibalanus*

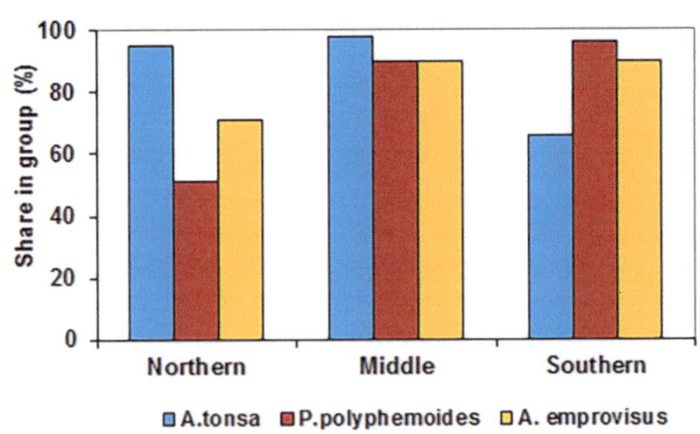

Fig. 6.13 Proportion of introduced species within meso- and meroplankton taxonomic groups across various Caspian Sea regions in June 2015. (Data source: A. Kamakin; drawing by T. Shiganova)

improvisus made up 18% of the total zooplankton abundance and biomass. Within the Cladocera, *Pleopis polyphemoides* was predominant, representing 90% of total Diplostraca biomass and 12% of the overall zooplankton biomass (Fig. 6.13).

Among meroplanktonic species, the larvae of the non-native barnacle *Amphibalanus improvisus* were the most numerous. In June 2015, they constituted 71% of the total meroplankton in the Northern Caspian and 90% in the Middle and Southern Caspian (Fig. 6.13).

The Black Sea Diplostraca species *Podon intermedius* (Table 6.5) was first documented in the Middle Caspian in 1985 at concentrations of 2–6 ind.m^{-3}. *P. intermedius* gradually expanded into the coastal zones of the Middle and Southern Caspian, though it did not reach the Northern Caspian. In the Southern Caspian, it was detected in the 0–100 m depth layer. By 1989, its abundance had risen to 17 ind. m^{-3} in certain locations, but it later became less prevalent. Nevertheless, it formed a significant component of the diet for anchovy and common kilka (Polyaninova 2007).

A different non-native Diplostraca species, *Penilia avirostris*, was documented on multiple occasions in the Caspian Sea prior to 2003, but not in subsequent years. However, specimens of *Penilia avirostris* were observed again in 2008 in the nearshore waters of the Middle Caspian near Makhachkala (Shiganova, unpublished).

Following the initial proliferation of *M. leidyi* in 2001, the zooplankton community structure in the Caspian Sea started to become less complex. After the species spread to the Northwestern Caspian, the zooplankton diversity index, calculated using the Shannon-Weaver method, fell from 8.8 in June to 1.2 by October. In the Middle Caspian, it declined from 9 to 1.24 and in the Southern Caspian from 5 to 0.85. A comparison of species diversity data from August 2001 with that from August 1998 revealed the index had decreased sixfold in the Middle Caspian and ninefold in the Southern Caspian (Shiganova et al. 2004a, b). In May 2004, when *M. leidyi* was not present, the diversity index in the presence of *A. tonsa* was 0.118 at depths of 50 and 100 m. Conversely, in coastal waters where *A. tonsa* was absent, the diversity index reached 1.425.

Based on zooplankton metrics, specifically the integrated EQR index, and the classification system for marine waters that uses zooplankton characteristics (WFD) and the Shannon-Weaver index, the entire Caspian Sea is assessed as having a "Bad ecological status."

Therefore, prior to the invasion of *M. leidyi*, the primary prey for kilka consisted of Caspian copepods, most notably *Eurytemora grimmi* and *Heterocope caspia* (G.O. Sars, 1897). Following the *M. leidyi* invasion, both the population numbers and biomass of these native copepods declined substantially. Consequently, the introduced species *A. tonsa* emerged as the predominant food source for all kilka varieties. As early as the 1990s, *A. tonsa* accounted for up to 90% of the stomach contents in anchovy kilka and up to 77% in common kilka (Elizarenko 1992).

During the summer of 2003 and the autumn of 2004, extensive transect surveys were conducted across the western regions of the Middle and Southern Caspian Sea. Research involving 350 stomach content analyses of both anchovy kilka

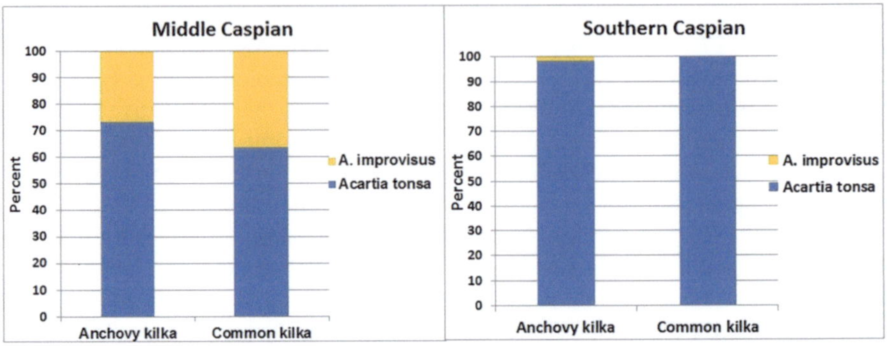

Fig. 6.14 Proportion of *Amphibalanus improvisus* and *Acartia tonsa* (%) in the stomach contents of anchovy and common kilka within the middle (summer) and southern (autumn) regions of the Caspian Sea (data from Zabralieva et al. 2006; drawing by T. Shiganova)

(*Clupeonella engrauliformes*) and common kilka (*Clupeonella delicatula*) revealed that their digestive tracts contained only two non-native species: the copepod *Acartia tonsa* and the larvae of the barnacle *Amphibalanus improvisus*. *Acartia tonsa* constituted the greatest proportion of the ingested material (Zabralieva et al. 2006). Hence, after the *M. leidyi* invasion, merely two non-native plankton species were left as accessible food resources for kilka in the Middle and Southern Caspian Sea (Fig. 6.14).

6.3.3 Non-native Gelatinous Species Impact

Based on existing knowledge, the Caspian Sea's ecosystem does not include any indigenous gelatinous plankton species. Non-native gelatinous species started to emerge shortly after the completion of the Volga-Don canal, transported via ship ballast water originating from the Black Sea. These introduced species encompassed the hydrozoan *Blackfordia virginica* (Mayer, 1910)—originally native to North American estuaries and subsequently established in the Black Sea—and the Black Sea Ponto-Caspian species *Odessia maeotica* (Ostroumoff, 1896), which is native to the brackish waters of the Black Sea's northwestern region. Their initial documented occurrences were in 1956 and 1960, respectively (Naumov 1968; Logvinenko 1959). Subsequently, the fouling hydrozoan *Calyptospadix cerulea* (Clarke, 1882), which arrived in the Black Sea from North America or Europe, was similarly introduced into the Caspian Sea through the same vector (Zevina 1968) (Table 6.6).

In 1999, the jellyfish species *Aurelia aurita* (alternatively identified as *Aurelia* sp. by Ramsak et al. 2012) and the ctenophore *Mnemiopsis leidyi* were documented in the Middle Caspian Sea (Ivanov et al. 2000). *M. leidyi* was introduced from the North American coast to the Black Sea and subsequently entered the Caspian Sea via ballast water, a pathway confirmed by molecular studies (Ghabooli et al. 2011).

Table 6.6 Non-native gelatinous species

Species	Donor region	Vector	First record	Status of establishment	References
Hydrozoa					
Blackfordia virginica (Mayer, 1910)	Black Sea Native for estuaries of the North America	Ship's hull fouling	1936 Black Sea 1956 Caspian	Established	Naumov (1968)
Hydrozoa					
Odessia maeotica (Ostroumoff, 1896) (former *Moerisia maeotica*) (Ostroumoff, 1896)	Black Sea	Ship's hull fouling	1960	Established	Logvinenko (1959)
Hydrozoa					
Calyptospadix cerulea (Clarke, 1882) *(former Bougainvillea megas* (Kinne 1956)	Black Sea Native North America or Europe	Ship's hull Fouling	1962	Established	Zevina (1965)
Scyphozoa					
Aurelia aurita (L, 1758) (or *Aurelia* sp.)	Black Sea	?	1999	?	Ivanov et al. (2000)
Ctenophora					
Mnemiopsis leidyi (A. Agassiz, 1865)	Black Sea Native for the North America	Ballast waters	1999	Ballast waters	Ivanov et al. (2000)
Ctenophora					
Beroe ovata (Bruguière, 1789)	Black Sea Native for the North America	Ballast waters	2019	Ballast waters	Roohi et al. (2022)

Aurelia aurita (or *Aurelia* sp.) is widely distributed in the Black Sea. This species originates from the temperate boreal region of the Atlantic Ocean (Mordukhai-Boltovskoy 1969). Recent genetic research indicates that the Black Sea *Aurelia aurita* falls within the clade Borealis (Ramsak et al. 2012), while populations in the Mediterranean should be provisionally classified as *Aurelia* sp. pending comprehensive global molecular analysis.

All members of the *Aurelia* sp. group exhibit a complex reproductive strategy involving both sexual and asexual phases. Each distinct life stage—medusa, planula, scyphistoma, ephyra, and adult medusa—possesses its own unique morphology and ecological niche.

During the year 2000, *Aurelia* sp. ephyrae were identified within plankton samples collected from the Middle Caspian Sea. In subsequent years, no medusae or ephyrae were observed. Nevertheless, in 2005, an adult *Aurelia* sp. individual was discovered near the Turkmenistan coastline (V. Nazarenko, pers. comm.). Furthermore, a 2008 survey in the eastern Middle Caspian, at depths of 30–73 m, documented abundant scyphistomae of *Aurelia* sp., with densities ranging from 100 to 10,000 individuals per square meter (Korsun et al. 2012). Adult medusae, however, were not observed, likely due to the low salinity of the water. For instance, we have documented the presence of *Aurelia* sp. in the Sea of Azov under similarly low salinity conditions (approximately 10.6–11‰): the species was introduced into the Sea of Azov via Black Sea waters and could only survive during periods of increased salinity. Consequently, *Aurelia* sp. experienced intensive development in the Sea of Azov during the 1970s when salinity levels were between 12.6 and 13.3‰. In 2018, marking the first significant occurrence since the 1980s, a notable increase in the presence and development of *Aurelia* sp. occurred in the Sea of Azov at a salinity of 13.4‰ . Therefore, the maximum salinity levels in the Caspian Sea—12.6‰ in the Middle Caspian and 13‰ in the Southern Caspian—are insufficient to support the permanent establishment of *Aurelia* sp.

6.3.4 Non-native Zoobenthos Species Impact

Within the Caspian Sea, zoobenthos is predominantly composed of autochthonous species, with only a limited number of exceptions (Zenkevich 1963). The initial non-native organisms to successfully colonize the Caspian Sea were benthic species. A list of these non-native zoobenthic species is provided in Table 6.7.

The mussel *Mytilaster lineatus* (Gmelin, 1791) became the initial unintentional invader, having been transported from the Black Sea as hull fouling in 1919. It progressively spread throughout the entire sea, with the exception of the low-salinity regions in the Northern Caspian. Between 1934 and 1938, its biomass surged dramatically from 11% to 42% of the total benthic biomass, reaching up to 7 kg/m^2 in certain locations (Brotskaya and Netsengevch 1941). *M. lineatus* establishes extensive colonies at depths reaching 100 m on rocky substrates and coarse sediments. The highest biomass is found at depths of 1–25 m along the eastern shoreline of the Middle Caspian (Karpinsky 2002).

This same ecological niche was previously occupied, until around 1938, by the endemic Caspian species *Dreissena elata* and *D. caspica*. However, the first postwar investigations in 1955–1957 revealed that they had been entirely displaced by *M. lineatus*. The replacement of the two *Dreissena* species can be attributed not only to the more specialized filtering apparatus of *M. lineatus* but also to its superior tolerance for low oxygen levels—a condition exacerbated by its own dense populations—to which the oxygen-sensitive *Dreissena* species could not adapt (Logvinenko 1965).

The establishment of *M. lineatus* also inflicted considerable harm on native populations of Cardiidae and *Cerastoderma lamarcki*. These native bivalves were

Table 6.7 Non-native species of zoobenthos

Species	Donor region	Vector	First record	Status of establishment	References
Mollusca (Bivalvia)					
Mytilaster lineatus (Gmelin, 1791	Black Sea	Fouling of the boats	1919	Established	Brotskaya and Netsengevch (1941)
Mollusca (Bivalvia)					
Hypanis colorata (Ménétries, 1832)	Azov-Black Sea	Shipping hull fouling	1994	Established	Orlova et al. (1999)
Mollusca (Bivalvia)					
Dreissena rostriformis bugensis (Andrusov, 1897) (subspecies) *Abra segmentum* (Récluz, 1843)	Sea of Azov	Stocked deliberately	1947–1948	Established	Karpevitch (1975)
Mollusca (Bivalvia)					
Monodacna colorata (Eichwald, 1829)	Azov-Black Sea	Shipping hull fouling	1959	Established	Saenkova (1960)
Mollusca (Gastropoda)					
Lithoglyphus naticoides (C. Pfeiffer, 1828) *Tenellia adspersa* (Nordmann, 1845)	Black, Azov seas	Shipping Hull fouling Hull fouling	1971 1989	Established ?	Pirogov (1972) and Antsulevitch and Starobogatov (1990)
Crustacea (Cirripedia)					
Amphibalanus improvisus (Darwin, 1854) *Amphibalanus eburneus* (Gould, 1841)	Black, Azov seas Black, Azov seas	Shipping Hull fouling Shipping Hull fouling	1955 1956	Established ?	Derzhavin (1956) and Zevina and Starostin (1961)
Crustacea (Amphipoda)					
Corophium volutator (Pallas, 1766) *Gammarus*	Black, Azov seas Black,	Ships hull fouling Ships hull fouling	1950–1960 <1994 <1969	? Established Established	Birshtein (1968); Piatakova and Tarasov (1996);

(continued)

Table 6.7 (continued)

Species	Donor region	Vector	First record	Status of establishment	References
aequicauda (Martynov, 1931) *Iphigenella shablensis* (Carausu, 1943)	Azov seas	Ships Hull fouling			and Mordukhai-Boltovskoy (1969)
Crustacea (Decapoda)					
Rhithropanopeus harrisii (Gould, 1841) *Eriocheir sinensis* H. (Milne Edwards, 1853)	Black Sea Via rivers and canals	Deliberately Simultaneously with stocked Via rivers and canals	1958 1990s	Established Established	Nebolsina (1959) Slynko et al. (2000)
Crustacea (Decapoda)					
Palaemon adspersus (Rathke, 1836) *P. elegans* (Rathke, 1837)	Black Sea	With stocked deliberately of gray mullets	1930–1934	Established	Lavrov-Navozov (1939)
Bryozoa					
Conopeum seurati (Canu) *(Membranipora crustulenta)* *Electra crustulenta* (Pallas, 1766) *Lophopodella carteri* (Hyatt, 1866)	Black, Azov seas Black, Azov seas Black, Azov seas	Ships hull fouling Ships hull fouling Ships hull fouling	1958 1950s	Established Established Established	Zevina and Kuznetsova (1965); Abricosov (1959); and Abricosov and Kosova (1963)
Entoprocta					
Barentsia benedeni (Foettinger, 1887)	Black, Azov seas	Ships Hull fouling	1962	Established	Zevina (1968)
Polychaeta					
Hediste diversicolor (O.F. Müller, 1776) *Ficopomatus enigmaticus* (Fauvel, 1923)	Azov Sea Black, Azov seas	Stocked deliberately Ships Hull fouling	1958 1958	Established Established	Hartman (1960) and Bogoroditsky (1963)

effectively "smothered" as *M. lineatus* settled on their shells near the siphons (Romanova 1960).

During the 1970s, following the full colonization of the Caspian Sea by *M. lineatus*, its population stabilized, accounting for 40% of the total benthic biomass. However, after the invasion of *M. leidyi*, the proportion of *M. lineatus* in the overall benthic biomass increased slightly at depths of 20–50 m, especially in summer, reaching 45.7% (Koncheva 2004). In the summer of 2005, the biomass of *M. lineatus* was recorded at 47.63 $g{\cdot}m^{-2}$ at 10 m and 248.5 $g{\cdot}m^{-2}$ at 25 m, representing 4.6% and 86.2% of total biomass, respectively (Zabralieva et al. 2006). Adult *M. lineatus* serves as a food source for roach, bream, and gobies in the Northern Caspian and for zander, sturgeon, and stellate sturgeon in the Middle Caspian. Nevertheless, their contribution to sturgeon diets is considerably lower compared to other benthic species; moreover, they are inaccessible to sturgeon in rocky habitats. The planktonic larvae of *M. lineatus* are consumed by zooplanktivorous fish (Karpinsky 2002). Following the *M. leidyi* invasion, the abundance of native edible zoobenthos declined, while the role of *M. lineatus* in sturgeon feeding did rise.

Consequently, *M. lineatus* expanded its range throughout the entire Caspian Sea, substantially grew in number, and emerged as the predominant species within the Caspian zoobenthic community. It also provided an extra, though nonessential and intermittent, food supply for sturgeon. Additionally, its larvae could be consumed by zooplanktivorous fish. Nonetheless, it displaced native Bivalvia species that historically constituted more crucial dietary components for sturgeons. This invasion has consequently degraded the nutritional quality of the benthos in the Middle and Southern Caspian regions (Karpevitch 1975; Karpinsky et al. 2006).

Two shrimp species, *Palaemon adspersus* (Rathke, 1837) and *P. elegans* (Rathke, 1837), were inadvertently introduced from the Black Sea alongside intentional gray mullet stockings. They proliferated across the Caspian, with biomass occasionally attaining 150 g/m^2. *P. elegans* became especially numerous in the southeastern sector of the sea (Lavrov-Navozov 1939). These shrimps are detritivorous and phytophagous, with abundant food sources available. They serve as a valuable supplementary food for benthophagous animals, including sturgeons and seals, and also hold considerable commercial importance (Karpevitch 1975).

The polychaete worm *Hediste (Nereis) diversicolor* (O.F. Müller, 1776) (Fig. 6.15) and the bivalve *Abra segmentum* (Récluz, 1843) are now considered classic cases of successful intentional introductions, implemented to enhance the food supply for commercially important fish populations, particularly sturgeons. These species were released into an available ecological niche occupied by detritus feeders, which inhabit the sediment and do not compete with the indigenous fauna of the Caspian Sea (Zenkevtch and Birshtein Ya 1934).

Hediste diversicolor was introduced to the North Caspian between 1939 and 1941 from the Sea of Azov and was later discovered in significant quantities within sturgeon digestive tracts in 1944 (Spassky 1945). This species quickly established itself across the Northern Caspian and soon spread into the western regions of the Middle Caspian, eventually reaching the Southern Caspian. It subsequently became

Fig. 6.15 *Hediste (Nereis) diversicolor*

Fig. 6.16 *Abra segmentum*

a dietary component for all benthic-feeding fish, particularly sturgeons (Hartman 1960; Khlebovich 1963).

The bivalve *Abra segmentum* (Fig. 6.16) was transported from the Sea of Azov during 1947–1948 (Karpevitch 1975).

Abra segmentum quickly started to rival *Cerastoderma lamarcki* not just for the soft muddy substrates favored by both species but also for the planktonic detritus that serves as their food source. In this regard, *Abra segmentum* is a filter feeder, whereas *Cerastoderma lamarcki* gathers detritus from the sediment surface (Romanova 1979).

Subsequently, *Hediste diversicolor* expanded widely across the entire sea, excluding the eastern coast and zones with comparatively low benthic biomass, and became a dominant species. Its habitat resembles that of *Abra segmentum* (silt, silty sand, sand, silty shells). The seasonal variation of *A. segmentum* is consistent and predictable. Fluctuations in its biomass are solely caused by sturgeon grazing. Consequently, benthic biomass in the ecosystem ranges from 0.04 to 51.7 $g{\cdot}m^{-2}$, averaging 10.7 $g{\cdot}m^{-2}$, while the biomass of *H. diversicolor* varies from

0.02 to 25.2 g·m^{-2}, with an average of 5.6 g·m^{-2}. *A. segmentum* was the second most abundant species in terms of biomass and numbers in the Caspian Sea (Karpinsky 2002).

Both *Hediste diversicolor* and *Abra segmentum* developed substantial biomass in muddy seabed regions, emerging as favored prey for Russian and stellate sturgeons. This significantly boosted the benthic food supply in the Northern Caspian and along the western shoreline, a key feeding ground for sturgeon during their spring and autumn migrations (Fig. 6.17).

Hediste diversicolor emerged as a preferred prey for starred sturgeon, whereas other sturgeon species favor *Abra segmentum*. In recent times, as sturgeon populations have declined, the abundance of *Hediste diversicolor* has risen. The invasion of *Mnemiopsis leidyi*, especially in the Southern Caspian Sea, has subsequently reversed this trend once more.

Following the invasion of *M. leidyi*, the two species deliberately introduced as edible benthic resources—*Abra segmentum* and *Hediste diversicolor*—retained significant value as food sources in the Middle Caspian. In the eastern coastal waters of the Southern Caspian, even during winter, *H. diversicolor* and *A. segmentum* were commonly found, accounting for 86% and 62% of the total zoobenthos, respectively. The biomass of *A. segmentum* rose with depth, peaking at 41.2 g m^{-2}. The mollusks *Cerastoderma lamarki* and *A. segmentum*, along with the polychaete *H. diversicolor*, contributed the largest proportions. Crustaceans were nearly absent, largely due to predation by benthic fish and sturgeons (Koncheva 2004).

During summer in the Middle Caspian, benthic species diversity reached 48 species, supported by the influx of benthic invertebrates. In the western section, the non-native *H. diversicolor* was predominant, making up 81% of Polychaeta, while *A. segmentum* constituted 52% of mollusks. In the eastern region, the dominant edible benthic species was *H. lineatus* (50%). The greatest density of *A. segmentum* was observed south of Makhachkala, where it represented 93% of edible benthic fauna. In the eastern Southern Caspian during summer, benthic richness included 25 species with a total biomass of 23.2 g·m^{-2}. Among Polychaeta, *H. diversicolor* was dominant, comprising up to 92% (Koncheva 2004).

The sole true benthic predator, the crab *Rhithropanopeus harrisii* (Goud, 1841), entered the Caspian Sea at the same time as the intentional introduction of gray mullets from the Black Sea, where it had itself been brought from the Atlantic coast of North America (Kobyakova and Dolgopolskaya 1969). First recorded in 1958 (Nebolsina 1959), it established itself in the Northern Caspian before expanding across the shallow zones of the Middle and Southern Caspian (Reznchenko 1967; Karpinsky 2002; Zolota and Feneva 2018). Mature crabs consume *H. diversicolor*, bivalve mollusks (*M. lineatus*, Cardiidae, *Dreissena* spp., *A. segmentum*), oligochaetes, and dead fish, while juveniles feed on corofiids. Its broad, polyphagous diet, which overlaps with that of benthivorous fish, suggests it could compete with sturgeon species. Conversely, the crab serves as prey for Russian and stellate sturgeons (Reznchenko 1967). Prior to the crab's arrival, sturgeons were the only predators feeding on Caspian benthos (Karpinsky 2002).

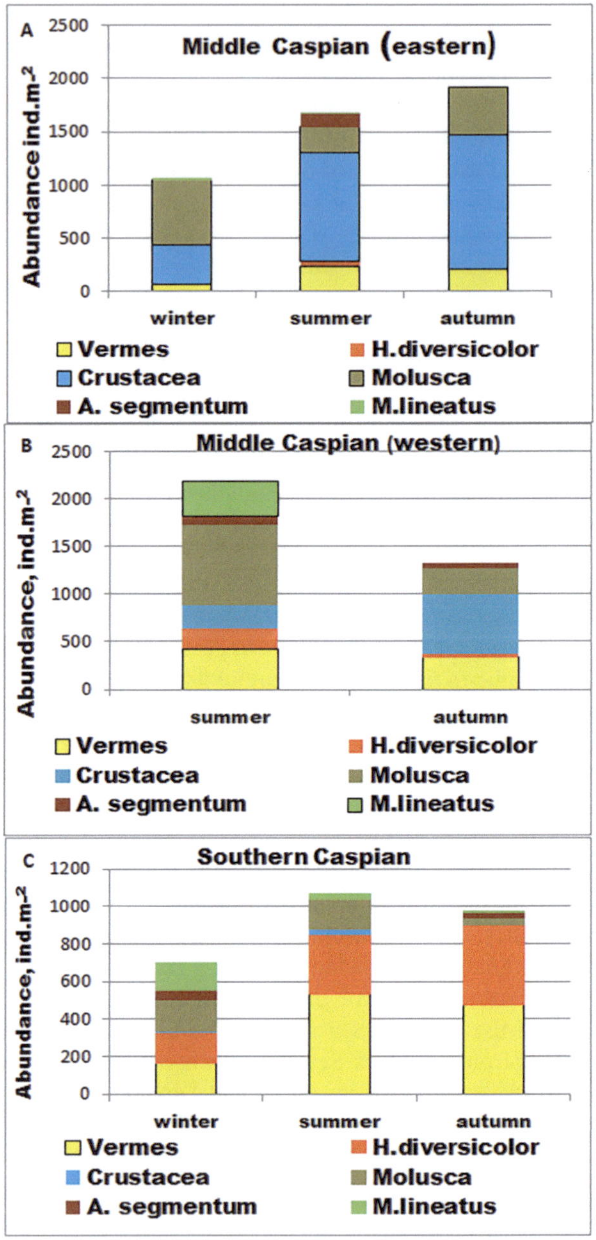

Fig. 6.17 Abundance of the non-native zoobenthos species in 2003. (Data from Koncheva 2004; drawing by T. Shiganova)

Two barnacle species, *Amphibalanus improvisus* and *A. eburneus*, were first identified in the Caspian Sea between 1955 and 1956 (Derzhavin 1956; Zevina and Starostin 1961). It is likely they were the initial invaders, arriving through the Volga-Don Canal attached to the hulls of ships. Their protective shells enable them to endure harsh environmental conditions. *A. improvisus* is a stationary filter feeder that lives on solid surfaces in areas with strong water flow. It emerged as the predominant species and formed a significant portion of the fouling biomass. *A. eburneus* was initially documented in the Krasnovodsk Bay and around Ogurchinsky Island but has not been reported in subsequent Caspian Sea surveys. *A. improvisus* is a problematic fouling organism, damaging vessels and port infrastructure. However, its larvae serve as a food source for zooplanktivorous fish.

Following the barnacles' establishment, several other species appeared, including the bryozoans *Conopeum seurati*, *Electra crustulenta* (Abricosov 1959), *Lophopodella carteri* (Abricosov and Kosova 1963), the entoproct *Barentsia benedeni* (Foetinger, 1887) (Zevina 1968), and the polychaete worm *Ficopomatus enigmaticus* (*Mercierella enigmatica*) (Fauvel, 1923) (Karpinsky et al. 2006).

The polychaete worm *Ficopomatus enigmaticus* was first documented in 1958, quickly establishing itself and beginning to reproduce in Krasnovodsk Bay between 1958 and 1961. Its biomass peaked at 30 kg.m-2 (Bogoroditsky 1963), though it later became exceptionally scarce (Karpinsky 2002).

Subsequently, several other species were introduced into the Caspian Sea. These included the bivalve mollusk *Monodacna colorata* (Saenkova 1960), the amphipods *Corophium volutator* (Pallas, 1766) (Birshtein 1968), *Gammarus aequicauda* (Martyinov 1931), and *Iphigenella shablensis* (Carausu, 1943) (Grigorovich et al. 2003), along with the bryozoan *Conopeum seurati* (Canu) (also known as *Membranipora crustulenta*) (Karpinsky et al. 2006). The gastropod *Lithoglyphus naticoides*, which serves as an intermediate host for multiple trematode parasite species, colonized the Volga Delta in 1971 (Pirogov 1972).

Later records include the nudibranch sea slug *Tenellia adspersa* in the Southern Caspian (Antsulevitch and Starobogatov 1990) and the quagga mussel *Dreissena rostriformis bugensis* (Andrusov, 1897) in the North Caspian (Orlova et al. 1999). Two amphipod species, *Gammarus aequicauda* and *Iphigenella shablensis*, were also identified in the Northern Caspian (Grigorovich et al. 2003).

While many of these species play a minor role in the Caspian ecosystem, certain non-native organisms such as *Hediste diversicolor*, *Abra segmentum*, *Mytilaster lineatus*, and the crab *Rhithropanopeus harrisii* are crucial for sturgeon nutrition (Molodtsova and Polianinova 2009).

Overall, the introduction of benthic species has substantially altered the trophic organization of the Caspian zoobenthos. Several non-native species have taken over the ecological (trophic) niches of native ones, displacing them due to greater adaptability and competitive success in foraging. Only *Abra segmentum* and *Hediste diversicolor*, which are mobile collectors of sediment, inhabit the biotope of silty soils—an area found in shallow zones and continually supplied with fresh detritus. This niche was previously partly occupied by mobile, burrowing sestonphages like *Cerastoderma lamarcki*. As this ecological niche was largely vacant, these newcomers actively colonized it, exploiting the abundant detritus reserves on the sediment surface (Karpinsky 2002).

6.3.5 Non-native Macrophyte Species Impact

The Caspian macrophytobenthos consists of brown, red, and green algae, along with Charophyta and Angiospermae (seagrasses and higher plants) (Milchakova 2003). More than 120 species of macrophytes have been documented in the Caspian Sea.

Notable non-native macrophytes (Table 6.8), likely introduced via ship fouling and first observed in the late 1950s, include the green algae *Ulvella parasitica*, *Ulvella leptochaete*, *Ulva flexuosa*, and *Monostroma latissimum*; the brown algae *Ectocarpus siliculosus*, *Litosiphon laminariae*, and *Entonema oligosporum*; and the red algae *Colaconema daviesii*, *Ceramium diaphanum*, and *Polysiphonia variegata*. All red algae species originate from the Black Sea (Milchakova 2011). In Crimean coastal waters, *Ceramium diaphanum* and *Polysiphonia variegata* are the dominant species, whereas *Acrochaetium daviesii* occurs as an accompanying species. *C. diaphanum* rapidly became dominant in the Northern Caspian following its introduction (Zevina 1994). The ecological impact of these species within the Caspian ecosystem remains unstudied.

6.3.6 Non-native Fish Species Impact

Following repeated and intentional efforts to stock commercial fish species starting in the 1930s, just two types of gray mullet have successfully established populations: the sharp-nosed mullet *Chelon saliens* (Risso, 1810) and the golden gray mullet *Chelon auratus* (Risso, 1810), both originating from the Black Sea (Babayan 1957) (Table 6.9).

Chelon saliens and *Chelon auratus* are currently found along the eastern shoreline of the Middle and Southern Caspian Sea. These mullet species have also been occasionally recorded in the Northern Caspian. Their spawning grounds are situated in the central regions of the Middle and Southern Caspian. Spawning is prolonged, occurring over a 6-month period from June through November. The sharp-nose mullet exhibits a faster growth rate and now makes up the bulk of mullet catches in the Caspian Sea. A geographical distinction exists in the distribution of these species: the singil tends to occupy more northern waters, primarily inhabiting the Middle Caspian, whereas the sharp-nose is found further south, mainly in the Southern Caspian. Furthermore, when residing in coastal areas, the singil typically lives at greater depths than the sharp-nose (Karpevitch 1975). An analysis of growth rates, fertility, and maturation reveals that both species grow considerably faster in the Caspian Sea, particularly the singil, and reach maturity sooner than their Black Sea counterparts. Singil attains sexual maturity at 4 years (sometimes three), while sharp-nose matures at 3 years. In the Black Sea, singil reaches maturity between 4 and 6 years of age. Caspian mullets also demonstrate higher fecundity compared to Black Sea populations (Karpevitch 1975). Presently, neither species contributes

Table 6.8 Non-native macrophytes species

Species	Donor region	Vector	First record	Status of establishment	References
Chlorophyceae					
(green algae) *Ulvella parasitica* (Oltmanns) (R. Nielsen, C.J. O'Kelly and B. Wysor, 2013)	Black Sea	Shipping fouling or ballast waters	1950s	Established	Zenkevich (1963)
Chlorophyceae					
(green algae) *Ulvella leptochaete* (Huber) (R. Nielsen, C.J. O'Kelly and B. Wysor, 2013)	Black or Azov Sea	Shipping fouling or ballast waters	1950s	Established	Zevina (1958)
Chlorophyceae					
(green algae) *Ulva flexuosa* (Wulfen, 1803)	Black or Azov Sea	Shipping fouling or ballast waters	1950s	Establishes	Zevina (1958)
Chlorophyceae					
(green algae) *Monostroma latissimum* (Wittrock, 1866)	Black or Azov Sea	Shipping fouling or ballast waters	1950s	Establishes	Zevina (1958)
Phaeophyceae					
(brawn algae) *Ectocarpus siliculosus* (Dillwyn) (Lyngbye, 1819)	Black or Azov Sea	Shipping fouling or ballast waters	1950s	Shipping fouling or ballast waters	Zevina (1958)
Phaeophyceae					
(brawn algae) *Litosiphon laminariae (Lyngbye)* (Harvey, 1849)	Black or Azov Sea	Shipping fouling or ballast waters	1950s	Shipping fouling or ballast waters	Zevina (1958)
Rhodophyta					
(red algae) *Colaconema daviesii* (Dillwyn) (Stegenga, 1985)	Black Sea	Shipping fouling or ballast waters	1950s	Shipping fouling or ballast waters	Zevina (1958)
Rhodophyta					
(red algae) *Ceramium diaphanum* (Lightf) (Roth 1806)	Black Sea	Shipping fouling or ballast waters	1950s	Shipping fouling or ballast waters	Kireeva and Schapova (1957)
Rhodophyta					
(red algae) *Polysiphonia variegata* (C.Ag.) Zanard	Black Sea	Shipping fouling or ballast waters	1950s	Shipping fouling or ballast waters	Kireeva and Schapova (1957)

Table 6.9 Non-native fish species, recorded in the Caspian Sea

Species	Donor region	Vector	First record	Status of establishment	References
Chelon saliens (Risso, 1810) *Chelon auratus* (Risso, 1810)	Black Sea Black Sea	Stocked deliberately Stocked deliberately	1930–1934 1930–1934	Established Established	Babayan (1957)
Ctenopharyngodon idella (Valenciennes, 1844	East Asia	Aquaculture	Since 1955	Established	Karpevitch (1975)
Hypophthalmichthys nobilis (Richardson, 1845) *Hypophthalmichthys molitrix* (Valenciennes, 1844)	East Asia	Aquaculture	Since 1955	Established Established	Karpevitch (1975)
Ictiobus cyprinellus (Valenciennes, 1844)	East Asia freshwater fish	Aquaculture	Since 1955	Established	Karpevitch (1975)
Hemiculter leucisculus (Basilewsky, 1855)	East Asia weed fish	Simultaneously with aquaculture	Since 1955	Established	Karpevitch (1975)
Pseudorasbora parva (Temminck and Schlegel, 1846)	East Asia weed fish	Simultaneously with aquaculture	Since 1955	Established	Karpevitch (1975)
Oryzias latipes (Temminck and Schlegel, 1846)	East Asia weed fish	Simultaneously with aquaculture	Since 1955	Established	Gvozdev and Mitrofanov
Gasterosteus aculeatus (Linnaeus, 1758)	Origin is not known weed fish	Origin is not known	2001	Established	Shakirova and Atamuratova (2001)
Channa argus spp.	East Asia Weed fish	Simultaneously with aquaculture	2001	Established	Shakirova and Atamuratova (2001)
Gambusia holbrooki (Girard, 1859)	Kazakhstan consumer of mosquito larvae	Penetrated from rivers	1963	Established	Zenkevich (1963)
Anguilla anguilla (Linnaeus, 1758)	Freshwater eel	Penetrated	1960s	Established	Slynko et al. (2000)

significantly to overall fishery catches, except in the South Caspian, where they remain relatively abundant, especially *Chelon auratus* (Fazli et al. 2022).

The practice of raising freshwater herbivorous fish from East Asia in ponds began in 1955. Consequently, only the grass carp (*Ctenopharyngodon idella*) established self-sustaining populations beyond the ponds, specifically in the freshened zones of the Northern Caspian Sea. Furthermore, freshwater species introduced to the Caspian basin are now present in the Volga Delta and the low-salinity regions of the Northern Caspian. These include the bighead carp (*Hypophthalmichthys nobilis*), the silver carp (*Hypophthalmichthys molitrix*), the bigmouth buffalo (*Ictiobus cyprinellus*), and others, which are maintained through artificial propagation (Karpevitch 1975). The initial releases of these fish occurred in the Volga Delta. Later, they were reared in aquaculture facilities within the Ural, Volga, and Kura river basins. Both intentional and accidental discharges of these fish into the wild have occurred from these farms (Karpevitch 1975; Ivanov 2000).

During the 1960s, the European eel (*Anguilla anguilla*) was deliberately stocked into Lake Seliger, and since the 1970s, it has been captured throughout the Volga River and in the low-salinity areas of the Northern Caspian (Slynko et al. 2000). Additionally, it was recorded in Azerbaijan in 1964. In certain years, the eel has also been observed in the coastal waters of Iran.

In addition to the white amur and silver carp, various weed fish species were sometimes introduced, including the common sawbelly *Hemiculter leucisculus*, the Amur chebachok *Pseudorasbora parva*, the snakehead *Channa argus* spp. (Shakirova and Atamuratova 2001), and the medaka *Oryzias latipes*. Another weed fish, the three-needle stickleback *Gasterosteus aculeatus*, was identified in Turkmenian waters in 1981. Its source remains unclear. This species has since proliferated extensively in the region, being observed in the coastal areas of Hasan-Kuli and the Kara-Bogaz-Gol zone. Spawning activity was noted in the lower sections of the Atrek River (Shakirova and Atamuratova 2001). Subsequently, it was also detected along the Iranian coastline (Panin et al. 2005). Although lacking commercial value, this euryphagous fish competes with the young of numerous other species. It also preys on the eggs and larvae of commercially important fish, adversely affecting their populations. One documented instance recorded a single three-needle stickleback consuming 74 fry measuring 6 mm in length over 5 h, followed by another 62 fry after 2 days. Furthermore, it is notable that when this fish enters hydrotechnical systems, it can impair their operational efficiency (Shakirova and Atamuratova 2001).

Gambusia holbrooki entered the Caspian Sea via Kazakhstan's rivers in 1934. It currently inhabits the Caspian Sea near river estuaries. Originally introduced deliberately into the Black Sea to manage mosquito populations, Gambusia consumes mosquito larvae and was used in the swampy water bodies of the Black Sea basin (Zenkevich 1963).

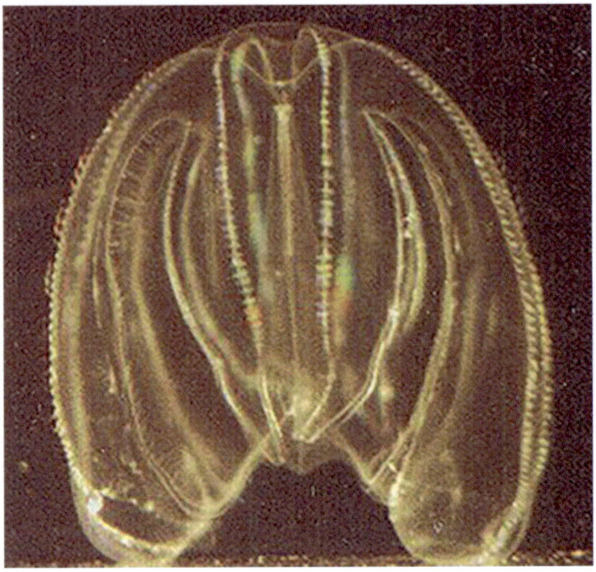

Fig. 6.18 Adult *Mnemiopsis leidyi* from the Caspian Sea. Length is 32 mm. (Photo by T. Shiganova and A. Kamakin)

6.4 *Mnemiopsis leidyi* Population in the Caspian Sea and Its Effect on the Ecosystem

The ctenophore *Mnemiopsis leidyi* was first documented in the Middle Caspian Sea in 1999 (Ivanov et al. 2000). As previously noted, *M. leidyi* was introduced from the Atlantic coast of North America into the Black Sea (refer to Chap. 3). It subsequently reached the Caspian Sea via ballast water. Molecular studies have confirmed its Atlantic North American origin (Ghabooli et al. 2011). A specimen of Caspian Sea *M. leidyi* is displayed in Fig. 6.18.

By the year 2000, *M. leidyi* had colonized the entire Caspian Sea in regions where salinity exceeded 4.3‰ (Shiganova et al. 2003). Once established, it exhibited a seasonal pattern of propagation, involving both source and sink regions. Its primary habitat (the source area) is the Southern Caspian, where it persists year-round because temperatures never fall below 10 °C—conditions suitable for both survival and reproduction. From the Southern Caspian, it spreads seasonally to other sections of the sea (the sink areas). The development of *M. leidyi* begins with water warming in the Southern Caspian during late winter to early spring, extending northward into the Middle and Northern Caspian as temperatures there become optimal. It lives, grows, and reproduces in these areas during the warm season and perishes with the onset of winter cooling.

Water temperature and the availability of food, specifically zooplankton concentrations, are the primary drivers influencing the population size of *M. leidyi*. The initial winter aggregations of *M. leidyi* form in the Southern Caspian Sea, where the winter temperature is a critical determinant for the population size in the subsequent year. From the Southern Caspian, *M. leidyi* moves into the western region of the Middle Caspian by May, expands throughout that area during June and July, and arrives in the Northern Caspian by late July to early or mid-August. The species' abundance and the extent of its spread are governed by water temperature, as well as the direction and speed of wind-driven currents. The prevalence of westerly and southwesterly winds facilitates an earlier northward advance of *M. leidyi*. These wind-driven currents also influence the scope of its distribution within the Northern Caspian, particularly in zones where salinity exceeds 4.3‰ (Shiganova et al. 2003, 2004b).

M. leidyi typically appears in the western and central regions of the Northern Caspian Sea by August. Yet, in the eastern sector of the Northern Caspian, it was not observed until 2008. On those rare occasions, only a few specimens in poor condition were discovered. This could be attributed to the elevated levels of suspended particles and the low density of zooplankton (Shiganova et al. 2003). In 2008, *M. leidyi* expanded into the northeastern zones of the Northern Caspian for the first time. Its population density reached 3–40 ind.m^{-3} at depths of 3–8 m, where salinity measured 7–9‰; by 2009, its abundance had grown, ranging from 3–64 ind. m^{-3} at salinities of 4.4–8.3‰. In 2010, the population of *M. leidyi* doubled relative to 2009, and in coastal waters, it attained densities of 150–232 ind.m^{-3} at salinities between 9 and 11.4‰. In these coastal zones, the individuals were smaller and more numerous in shallower waters (3–5 m) compared to deeper areas (>6 m depth) (Shiganova 2011).

Beginning in 2008, a decline in the abundance of *M. leidyi* in the Southern Caspian Sea commenced, attributed to rising temperatures reaching up to 30 °C, conditions which are unfavorable for the species (Shiganova et al. 2004b). At the same time, *M. leidyi* populations grew in the Middle and Northern Caspian, likely due to overall water warming and the consequent earlier arrival of suitable thermal conditions. Consequently, *M. leidyi* arrives in the Middle and Northern Caspian earlier in the season, typically by July. Furthermore, *M. leidyi* expanded into the northeastern Caspian, an area where it was not found prior to 2008. Its population there peaked during the exceptionally warm year of 2010.

M. leidyi is a zooplanktivorous ctenophore whose primary diet consists of edible zooplankton, fish eggs, and larvae. Therefore, its introduction into the Caspian Sea had catastrophic effects on the zooplankton community. By the 2000s, when it had colonized the entire sea, zooplankton biomass had dropped by a factor of 5 to 20 across all taxonomic groups, a trend that intensified in subsequent years. The estimated daily predation rate by *M. leidyi* during its peak summer development significantly exceeds the available zooplankton biomass, leading to periods of starvation for the ctenophore (Shiganova et al. 2004b).

Consequently, the Environmental Quality Report (EQR) index for *M. leidyi* biomass and abundance (Cochrane et al. 2010) has been rated as "Bad" across all regions in recent years, despite a decline in its biomass within the Southern Caspian Sea.

6.4.1 M. leidyi *Impact on Trophic Levels of the Caspian Ecosystem*

M. leidyi exerts a cascading influence across every trophic tier of the ecosystem, extending from holozooplankton, meroplankton, and ichthyoplankton up to zooplanktivorous fish species—most notably kilka—and onward to predatory fish, sturgeons, and seals. This cascade also extends downward from the zooplankton community to phytoplankton and microplankton. Ultimately, it leads to a reconfiguration of matter and energy flows throughout all trophic levels and alters the balance of production and decomposition processes within the marine environment (Fig. 6.19) (Shiganova et al. 2004b).

Planktivorous fish serve as the primary dietary rivals to *M. leidyi* and experienced significant impacts following its invasion. During 2000–2001, as *M. leidyi* proliferated, a catastrophic decline occurred in the catches of anchovy kilka, *Clupeonella engrauliformis* (Fig. 6.20). This outcome is partly due to the anchovy kilka being a warm-water species that inhabits the surface layers of the Middle and Southern Caspian Sea from May to December for spawning—a period that coincides with the

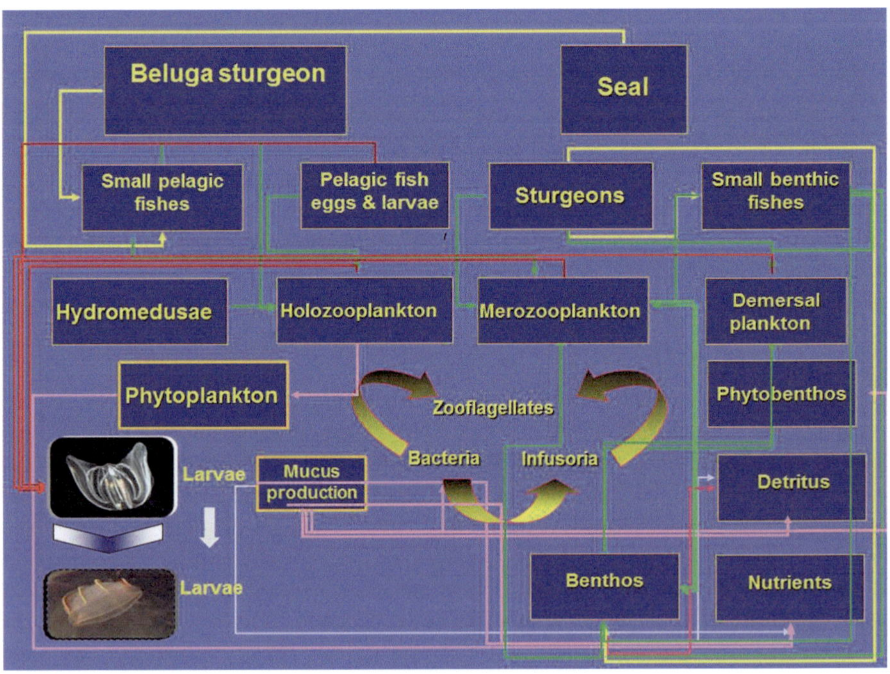

Fig. 6.19 Diagram of *M. leidyi* cascade effect across all ecosystem trophic levels prior to the arrival of *B. ovata*: red lines indicate direct consumption by *M. leidyi*; green lines represent up effect; pink lines signify down effect. Bold arrows denote the microbial loop, generated by mucus from both *M. leidyi* and *B. ovata*. The influence of *B. ovata* upon its establishment: direct predation by adults on *M. leidyi* adults and by larvae on *M. leidyi* larvae

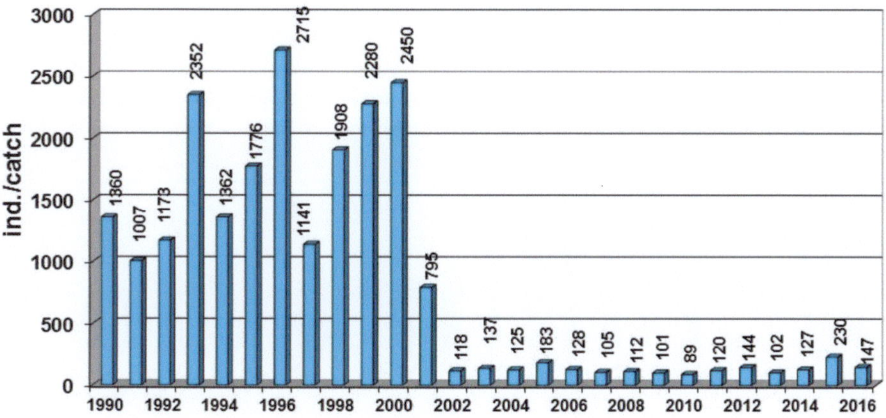

Fig. 6.20 Annual average anchovy kilka (*Clupeonella engrauliformis*) research catches recorded by the Russian Federation from 1990 to 2016 (Kamakin et al. 2018)

peak development of *M. leidyi*. Consequently, the spawning and growth phases of anchovy kilka align closely with those of *M. leidyi*, leading to intense competition for zooplankton resources between *M. leidyi* and both adult and larval anchovy kilka. Moreover, because the embryonic and postembryonic development of anchovy kilka takes place in surface waters, its eggs and larvae are directly preyed upon by *M. leidyi* (Paritsky 2014).

The populations of big-eyed kilka *Clupeonella grimmi* were negatively impacted by *M. leidyi* across all life stages, from eggs and larvae to juveniles and adults. The larvae and juveniles inhabit the upper water layers, feeding on zooplankton, and experienced significant declines due to competition with *M. leidyi* as well as predation by *M. leidyi* on their eggs and larvae (Paritsky 2014).

Common kilka *C. caspia* was less affected, as it can migrate into the brackish and even freshwater zones of the Northern Caspian Sea, where it feeds on brackish and freshwater plankton (Zabralieva et al. 2006; Roohi et al. 2010). Consequently, common kilka started to dominate the kilka catches by Russia. Accordingly, the proportion of common kilka in Russian catches reached 67.0%, anchovy kilka accounted for 32.7%, and big-eye kilka made up only 0.8% (Fig. 6.21) (Paritsky 2014).

The common kilka emerged as the most prevalent species within the kilka group across all surveyed regions, a fact clearly illustrated by the recorded Iranian catch data (Fig. 6.22).

Prior to the arrival of *M. leidyi*, mollusks served as the primary food source for Russian sturgeon. Both juvenile and adult sturgeon also consumed Caspian kilka species, including anchovy, common, and big-eyed kilka (Polyaninova et al. 2003). Kilka played a crucial role in the diet of sturgeon populations within the Middle and Southern Caspian Sea. In these regions, kilka constituted up to 40% of the sturgeon's food intake. For both the Russian sturgeon (*Huso gueldenstaedtii*) and the stellate sturgeon (*Huso stellatus*), a significant portion of their diet, reaching 20–40%,

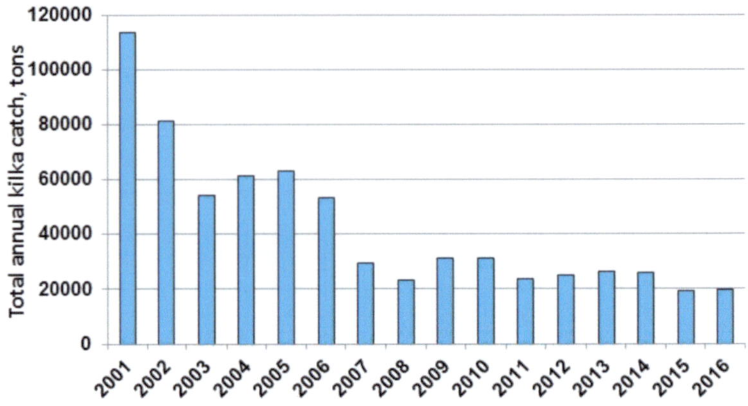

Fig. 6.21 Annual Russian catches of kilka from 2001 to 2016 (data sourced from A. Kamakin)

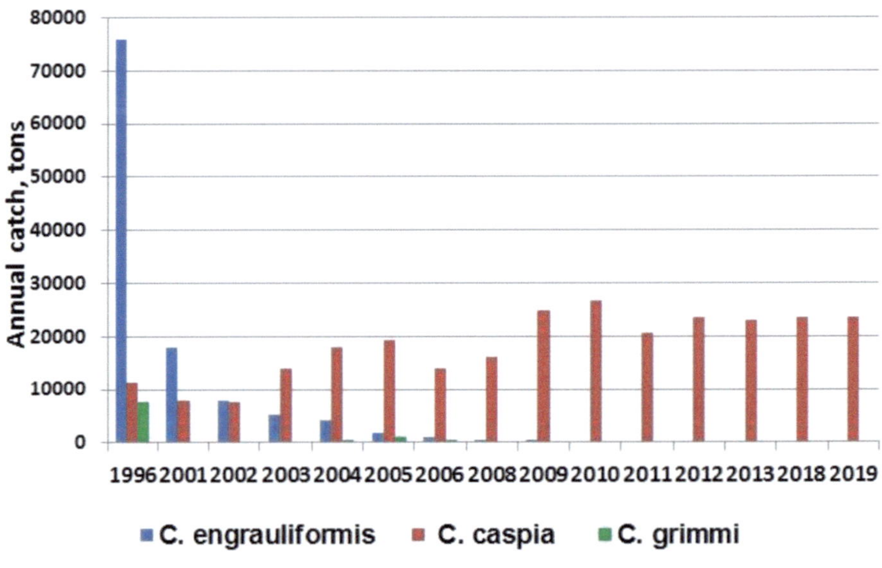

Fig. 6.22 Iranian catches of kilka species: anchovy kilka *Clupeonella engrauliformis*, big-eye *C. grimmi*, common kilka *Clupeonella caspia* (data provided by A. Roohi)

consisted of kilka during the winter and spring seasons (Molodtsova and Polyaninova 2009). After the year 2000, overall kilka stocks declined, leading to a corresponding decrease in their presence in sturgeon diets. Only the common kilka was subsequently documented in the diets of the Russian sturgeon, stellate sturgeon, and beluga *Huso huso* (Linnaeus, 1758) (Molodtsova and Polyaninova 2009; Kamakin et al. 2018).

The influence of *M. leidyi* on sturgeon populations is multifaceted, complicating precise assessment of its total effect. Nevertheless, a comparison of recent data with records from before the *M. leidyi* invasion shows a decline across all sturgeon stocks, alongside a worsening of their morphological characteristics (Kamakin et al. 2018).

The Caspian seal, *Pusa caspica*, is the sole mammal endemic to the Caspian Sea. The effects of the *M. leidyi* outbreak were especially severe for Caspian seals during the initial invasion years (2000–2002), due to a sharp drop in anchovy and big-eyed kilka stocks, which had constituted roughly 60% of the seals' diet. Consequently, the average seal weight fell by 10% in 2001 relative to 2000, and the number of breeding females plummeted. In 2001, total barrenness reached 79.8%, with only 10.1% of females being pregnant.

After 2002, seals started migrating to the Northern Caspian to feed, shifting their diet from anchovy and big-eyed kilka to common kilka. Common kilka accounted for 40.59% of the seal diet in shallow waters and 86.5% in deeper areas, as seals intensified their feeding migrations into Northern Caspian regions where *M. leidyi* cannot survive. This dietary shift led to an improvement in the overall condition of the seal population and their reproductive success (Khuraskin et al. 2006).

Therefore, a bottom-up influence propagated through the ecosystem's trophic levels, manifesting as (1) elevated nutrient levels that boost phytoplankton growth; (2) grazing by zooplankton and meroplankton, which serve as prey for planktivorous fish like kilka and herring; (3) a decline in kilka and other planktophagous fish stocks, alongside a reduction in zoobenthic species with planktonic life stages, leading to a consequent decrease in benthic fish populations; and, ultimately, (4) diminished food availability for sturgeons and seals, which depend on benthic organisms and planktivorous fish. Conversely, top-down regulation of these trophic levels caused a significant drop in zooplankton abundance. This reduced grazing pressure on phytoplankton, promoting its proliferation and thereby raising chlorophyll concentrations. The mucus excreted by *M. leidyi* further alters the marine environment and lower trophic levels, impacting the microbial loop—comprising bacteria, pico-phytoplankton, nano-autotrophic/heterotrophic flagellates, microphytoplankton, chlorophyll a, primary production (PP), and micro-zooplankton (Fig. 6.19) (Shiganova et al. 2004b, 2019a, b).

Consequently, a rise in blue-green algae was noted during the periods when *M. leidyi* proliferated, resulting in detrimental algal blooms (e.g., see Fig. 6.11).

Within the Caspian Sea, solely the most resilient nonindigenous species, alongside certain native ones such as common kilka and a few others, demonstrated sufficient adaptability to sustain their populations following the incursion of *M. leidyi* (Shiganova 2009).

The significant influence of *M. leidyi* as it colonized the Caspian Sea was amplified by the lack of natural predatory checks. A known predator of *M. leidyi*, the non-native ctenophore *Beroe ovata*, was first documented in the Southern Caspian as recently as November 2019 (Roohi et al. 2022).

6.5 *Beroe ovata* as a Biological Controller of *Mnemiopsis leidyi* Population

The primary driver behind the rapid proliferation of *M. leidyi* was the lack of any natural predator to control its population. Fortunately, a potential predator, another ctenophore species named *Beroe ovata* (Fig. 6.23), emerged in the Southern Caspian Sea in November 2019. Specimens of this species were first identified during a routine plankton sampling initiative conducted by Iranian researchers (Roohi et al. 2022). The ctenophore *Beroe ovata* was originally introduced into the Black Sea in 1997, likely transported in ballast water (Konsulov and Kamburska 1998). Molecular studies have classified it as a warm-water species originating from the Atlantic coast of North America, similar to the earlier invasive species, *Mnemiopsis leidyi* (Johansson et al. 2018; Shiganova and Abusova 2021). Like many other ctenophore species, *Beroe ovata* is a hermaphrodite and possesses the ability to self-fertilize.

Existing field surveys of *B. ovata* in the Caspian Sea are summarized in Fig. 6.24. Following the arrival of *B. ovata* in the Southern Caspian, the initial field investigation—which also included a concurrent study of *M. leidyi*—was carried out in 2018–2019 along the Iranian coast. *B. ovata* was observed in November 2019 at depths shallower than 5 and 10 m and during October–November at depths above 30 m, occurring after the seasonal peak of *M. leidyi* in July. The highest abundance and biomass of *B. ovata* were documented above 30 m depth, with a mean density of

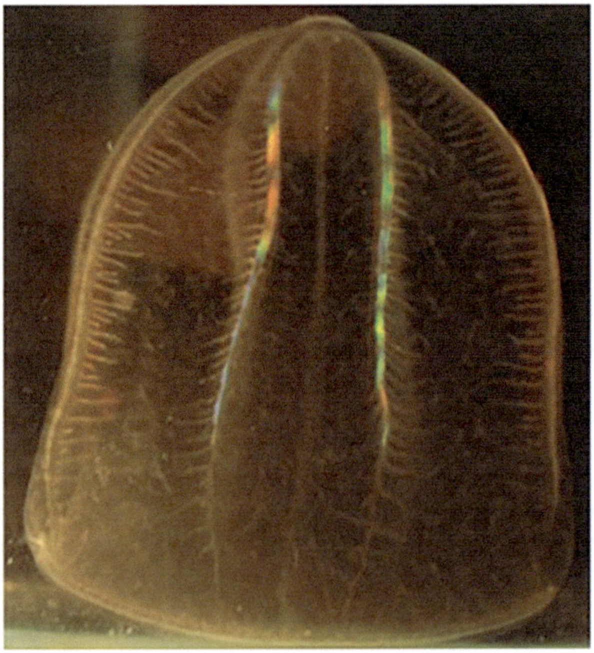

Fig. 6.23 Adult *Beroe ovata* from the Caspian Sea. (Photo by T. Shiganova)

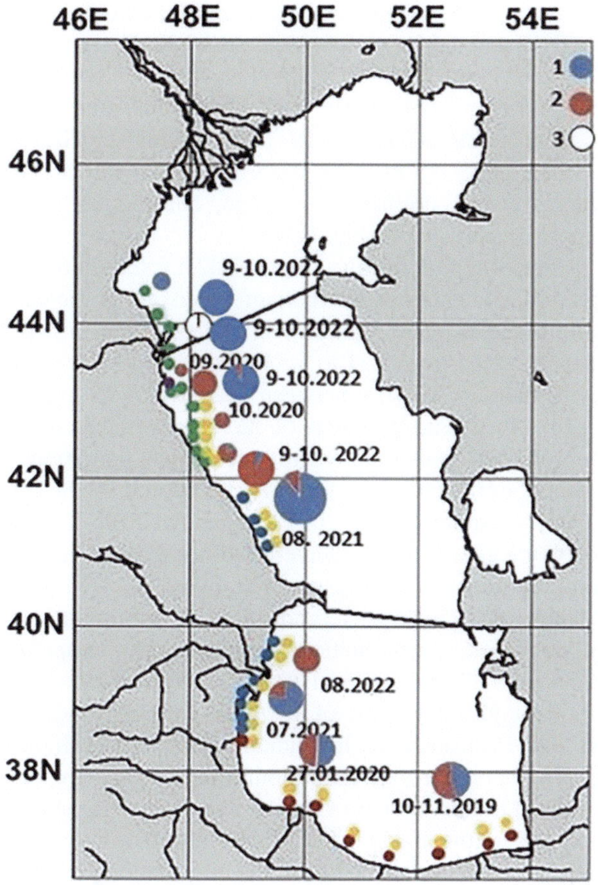

Fig. 6.24 Documented observations of *B. ovata* and *M. leidyi* in the Caspian Sea: read dots, in the Iranian waters in 2019–2020 (after A. Roohi et al. 2022, Bagheri et al. 2021); blue dots, in Azerbaijani waters in 2021–2022; green dots, in the Middle Caspian in 2020 (after Sayapina et al. 2021); violet dot, station off Makhachkala in 2020 (Vostokov et al. 2020). Neighboring yellow dots indicate simultaneous presence of *M. leidyi* during the observations. Circles indicate the ratio of mean abundance of (1) *M. leidyi*, (2) *B. ovata*, (3) blank circle—an absence of both species; labels indicate the date of observation

12.22 ind.m^{-3}. After the emergence of *B. ovata*, the abundance and biomass of *M. leidyi* declined sharply in the winters of 2018 and 2019.

The recorded sizes of the ctenophores were as follows: *M. leidyi* ranged from 5 to 29 mm (with an average of 7.8 ± 5.01 mm), and *B. ovata* measured 3–15 mm (averaging 7.5 ± 4.3 mm). The occurrence of juveniles from both species confirms that reproduction is taking place in the Southern Caspian. It is notable that adult *B. ovata*, like *M. leidyi*, attain a smaller size under the low-salinity conditions of the Caspian Sea compared to those in the Black Sea (Shiganova 2020).

In January 2020, *B. ovata* was documented within aggregations of *M. leidyi* near fish cages in the southwestern Caspian (Bagheri et al. 2021).

During early October 2020, *B. ovata* was observed in the shelf waters off Dagestan. The average concentration reached 22 ind./m^2, with a biomass of 6.15 g/m^2. Larvae and small individuals measuring 5–20 mm dominated the population. *M. leidyi* was not detected in this survey (Vostokov et al. 2020) (Fig. 6.24).

During a November 2020 research expedition conducted by the Southern Center of RAS and the Caspian branch of SIO RAS, significant quantities of *B. ovata* were gathered in the Middle Caspian Sea, specifically near Makhachkala, Derbent, and at the northern edge of the Agrakhan Peninsula (Fig. 6.24). The survey did not document any presence of *M. leidyi* (Ushivtsev 2021; Sayapina et al. 2021).

Consequently, the observed limit of *B. ovata* near the Agrakhan Peninsula likely represents its northernmost range, considering the prevailing salinity levels of approximately 10‰ (Fig. 6.24).

Subsequently, in July–August 2021, a substantial number of *B. ovata* specimens were documented in Azerbaijani waters along the eastern coast of the Middle and Southern Caspian (Akhuntov et al. 2021; Shiganova 2011). Their recorded abundance ranged from 0.31 to 5.5 individuals per cubic meter (averaging 1.93 ind./m^3), with biomass varying between 0.28 and 4.95 grams per cubic meter (averaging 1.74 g/m^3). The presence of larval, juvenile, and adult stages indicated active reproduction under the existing ecological conditions in these regions. In terms of ctenophore abundance, *M. leidyi* accounted for 86.4%, while *B. ovata* constituted 13.6% (Akhuntov et al. 2021; Shiganova 2011).

Our earlier laboratory investigations, designed to determine the distribution range and reproductive parameters of *B. ovata*, showed that it can survive at a minimum salinity of 7‰ and reproduce at a minimum salinity of 10‰ (Shiganova, unpublished). Reproduction commences at a temperature of 19 °C, peaks between 25 and 26 °C, and ceases at 28.2 °C (Shiganova, unpublished). A further constraint is food supply: the larvae of *M. leidyi* serve as food for *B. ovata* larvae, while adult *B. ovata* consume any individuals of *M. leidyi* (Shiganova et al. 2003).

Based on our long-term monitoring of *M. leidyi* distribution, its source region is situated in the Southern Caspian Sea. Following spring warming, *M. leidyi* starts to expand northward into the Middle and Northern Caspian, forming sink areas for reproduction. Synthesizing all the information presented, we assert that *B. ovata* has indeed become established in the Southern and Middle Caspian. We further conclude that the primary zone of occurrence for both *B. ovata* and *M. leidyi* is the Southern Caspian, which functions as their source area.

Therefore, it is likely that *B. ovata* will spread into regions where *M. leidyi* is already established, provided thermal and salinity conditions are suitable (Fig. 6.24). This includes the Southern, Middle, and southern sections of the Northern Caspian, where salinity rises above 10‰ during warmer periods. Both species can persist as long as these environmental parameters remain. Nevertheless, there is a chance that *M. leidyi* populations in the Southern or Middle Caspian could be fully preyed upon by *B. ovata*, preventing their expansion into the Northern Caspian. Surveys conducted in the Middle Caspian near Makhachkala and the Agrakhan Peninsula found only *B. ovata* present, indirectly suggesting that *M. leidyi* individuals there may have been entirely consumed. Should this occur, *B. ovata* itself could eventually vanish from those areas due to a lack of suitable prey.

Summer water temperatures in the Iranian sector of the Southern Caspian can reach 30 °C, a level unfavorable for both *M. leidyi* and *B. ovata*. Consequently, their populations peak and reproduction primarily occurs during the autumn-winter period, mirroring patterns observed in the Mediterranean Sea where high summer temperatures are also limiting. In the Middle and Northern Caspian, the seasonal population dynamics of both ctenophore species resemble those documented in the Black Sea (Shiganova, unpublished).

Over time, this sequence of events is expected to initiate a recovery process for the Caspian ecosystem.

6.5.1 How Non-native Species Did Affect the Caspian Sea Ecosystem: Concluding Remarks

Within the Caspian Sea, nonindigenous species encompass a diverse array of ecological, taxonomic, and functional categories. Native species maintain dominance solely among the fish populations. However, their dietary patterns have been significantly altered following the invasion of *M. leidyi*. At present, the primary food sources for fish—including both planktivorous and benthivorous species—are largely supplied by non-native organisms.

These introduced species, possessing greater competitive abilities, have displaced native counterparts with similar environmental needs. Subsequently, some of these invaders have impacted other trophic levels, ultimately leading to profound shifts in ecosystem dynamics and a substantial reorganization of the original energy pathways. Certain non-native arrivals have occupied the ecological (trophic) niches previously held by native species, supplanting them due to superior adaptability and success in securing food resources. Consequently, the introduction of non-native species has typically yielded adverse outcomes, as less competitive indigenous species have been displaced and, in some instances, nearly eradicated.

The alga *Rhizosolenia calcar-avis* is unsuitable for consumption by pelagic herbivores due to its coarse texture. It contributes to greater phytoplankton biomass, yet diminishes its dietary quality. Following the invasion of *M. leidyi* into the Caspian Sea in 2001, the quantities of blue-green and green algae have substantially surpassed those of the typical diatoms and dinoflagellates. Unusual summer blooms with atypical patchiness can be seen across extensive regions for roughly 4—5 months. Starting in 2005, blue-green algae also expanded significantly in the Southern Caspian. Furthermore, a new taxonomic group, the coccolithophore *Emiliania huxleyi*—known for its extensive blooms in the Black Sea—was recently identified in the Caspian Sea (Shiganova et al. 2005).

Nevertheless, in certain situations, non-native species can serve as a food source for native organisms. For instance, the introduced species *Hediste diversicolor* and *Abra segmentum* have become favored prey for sturgeons. They do not compete with native Caspian species because they utilize reserve resources, specifically the abundant detritus on sediment surfaces that native species do not exploit (Zenkevtch and Birshtein Ya 1934).

The crab *Rhithropanopeus harrisii*, which rivals sturgeons for food resources, also serves as a suitable prey item for them. Even *Mytilaster lineatus*, out of reach for sturgeons on the hard-rock seabed of the Northern and Middle Caspian, contributes to their diet in the Southern Caspian, where it inhabits soft sediments.

The invasive copepod *Acartia tonsa* has displaced key native Caspian copepods, particularly *Eurytemora grimmi* and *Heterocope caspia*, which were fundamental prey for zooplanktivorous fish, mainly kilka. Nevertheless, *Acartia tonsa* itself has now become their primary food source.

In the Southern Caspian, the introduced mullet has grown into a commercially valuable fishery target.

Still, despite these beneficial cases, the arrival of most non-native species has led to significant reorganization within native communities. The most striking instance is the invasion of the ctenophore *Mnemiopsis leidyi*, which triggered profound transformations across the entire ecosystem.

Following the invasion of *M. leidyi*, the Caspian ecosystem's functioning was altered, mirroring earlier events in the Black and Azov seas (Shiganova et al. 2004a, b), with its influence growing over time. In the Caspian Sea, *M. leidyi* achieved greater population densities than in the Black Sea. This rise in abundance persists, and *M. leidyi* started expanding northward sooner than in prior years, colonizing new regions in the Northern Caspian that were previously unsuitable. The transformation of the Caspian Sea ecosystem followed the same pattern as in the Black Sea but to a more extensive degree (Shiganova et al. 2004b; Paritsky 2014; Kamakin et al. 2018).

Nevertheless, *M. leidyi* itself serves as prey for another ctenophore, *Beroe ovata*, which was transported into the Caspian Sea from the Black Sea via ballast water. *B. ovata* is a monotrophic species and can consume only zooplanktivorous ctenophores, such as *M. leidyi*. The precise mechanism of its transfer from the Black Sea to the Caspian remains uncertain, though ballast water is the presumed vector. *B. ovata* was first documented in the Southern and Middle Caspian between 2019 and 2021. Its complete consumption of *M. leidyi* in those regions, which prevented *M. leidyi* from reaching the Northern Caspian for three consecutive years, has created an opportunity for the Caspian ecosystem to rebound.

Consequently, the situation in the Caspian Sea illustrates how the swift spread of gelatinous carnivorous invaders can rapidly disrupt the entire ecosystem's functioning.

References

Abricosov GG (1959) A new invader in the Caspian Sea. Zool Zhurnal 38:17–45. (in Russian)

Abricosov GG, Kosova AA (1963) Finding of new tropical freshwater pearlwort. *Lophopodella carteri* (Bryozoa, Phylactolaemata) in avandelta of the Volga. Zool Zhurnal 42:1724. (in Russian)

Akhuntov MM, Mamedov EV, Dzapharova EE (2021) The state of the plankton community as a result of the appearance of *Beroe ovata* and its impact on the ecosystem of the Caspian Sea. VIII scientific and practical conference. Astrakhan, 22–34 (in Russian)

Aladin NV, Plotnikov IS, Fillipov AA (2002) Invaders in the Caspian Sea. In: Leppakovski E, Gollash S, Olenin S (eds) Invasive aquatic species of Europe, distribution, impacts and management. Kluwer Ac.pub, pp 351–359

Antsulevitch AE, Starobogatov YI (1990) First finding of mollusks order Nudibranchia (= Tritoniiformes) in the Caspian Sea. Zool Zhurnal 69(11):138. (in Russian)

Ardabieva AG, Tatarintseva TA, Terletskaya OV, Morozyuk VV (2005) Phytoplankton of the Caspian Sea in 2004. Fishery studies in the Caspian Sea: results of research in 2004. KaspNIRKH, Astrakhan, pp 100–121. (in Russian)

Babayan KE (1957) Caspian mullets. Zool J 36(10):1505–1513. (in Russian)

Bagheri S, Sayad Bourani M, Babaei H, Roohi A, Ghandi AD (2021) An investigation on the effects of fish farming in marine cages on abundance and structure of *Mnemiopsis leidyi* and *Beroe ovata* (Ctenophora: Lobata) in the southwestern Caspian Sea during 2018–2020. Iran J Fish Sci 20(5). https://doi.org/10.22092/ijfs.2021.351053.0

Birshtein AY (1968) Crustacea. In: Birshtein AY (ed) Atlas of invertebrates of the Caspian Sea. Food Industry, Moscow, pp 118–120. (in Russian)

Bogachev VV (1928) *Mytilaster* in the Caspian Sea. Russian Gydrobiol J 7(8–9):187–188. (in Russian)

Bogoroditsky PV (1963) Mass development of polychaete *Merceriella enigmatica* Fauvel in the Krasnovodsk Bay. Trudy SIO RAS 70:26–28. (in Russian)

Brotskaya VA, Netsengevch MR (1941) *Mytillaster linneatus* distribution in the Caspian Sea. Zool Zh 20(1):79–99. (in Russian)

Cochrane SKJ, Connor DW, Nilsson P, Mitchell I, Reker J, Franco J, Valavanis V, Moncheva S, Ekebom J, Nygaard K, Serrão Santos R, Narberhaus I, Packeiser T, van de Bund W, Cardoso AC (2010) Marine strategy framework directive task group 1 report biological diversity EUR 24337 EN – 2010. http://publications.jrc.ec.europa.eu/repository/bitstream/111111111/13642/1/tg1final.pdf

Derzhavin AN (1956) New invader in the Caspian Sea—barnacles *B. improvisus* Darw. Rep Azerbaijan Acad Sci 12(1):43–47. (in Russian)

Dobrovolsky AD, Zalogin BS (1982) Seas of the USSR. M. Publishing House of Moscow State University, p 192

Dumont HJ (1998) The Caspian Lake: history, biota, structure, and function. Limnol Oceanogr 43: 44–52

Dumont HJ (2000) Endemism in the Ponto-Caspian fauna, with special emphasis on the Onychopoda (Crustacea). Adv Ecol Res 31:181–196

Elizarenko MM (1992) Kilka feeding in summer in the middle Caspian. Biological resources of the Caspian Sea. 1st Intern Conf Astrahan Caspryba: 111–113 (in Russian)

Fazli H, Kaymaram F, Daryanabard R (2022) Effects of fishing and environmental parameters on the commercial bony fish assemblage in the South Caspian Sea. Oceanol Hydrobiol Stud 51(1): 90–99. https://doi.org/10.26881/oahs-2022.1.08

Ghabooli S, Shiganova TA, Zhan A, Cristescu ME, Eghtesadi-Araghi P, MacIsaac HJ (2011) Multiple introductions and invasion pathways for the invasive ctenophore *Mnemiopsis leidyi* in Eurasia. Biol Invasions 13:679–690

Ginzburg AI, Kostianoy AG, Sheremet NA (2005) Sea surface temperature variability. The Caspian Sea environment. In: Kostianoy AG, Kosarev AN (eds) The handbook of environmental chemistry, vol. 5. Springer-Verlag GmbH, pp 59–81

Grigorovich IA, Therriault TW, MacIsaac HJ (2003) History of aquatic invertebrate invasions in the Caspian Sea. Biol Invasions 5:103–115

Gubanova AD (2003) *Acartia tonsa* – a new species of Copepoda in the Black Sea. In: Eremeev VN, Gaevskay AV, NAN Ukraine, IBSS (eds) The present state of biodiversity of the coastal zone of Crimea (Black Sea sector). ECOSI- Hydrophysics, Sevastopol, pp 99–103. (in Russian)

Hartman O (1960) On *Neanthes diversicolor* in the Caspian Sea and increasing of its area of distribution. Zool Zh 39:35–39. (in Russian)

Hillebrand H, Durselen C, Kirschtel D, Pollingher U, Zohary T (1999) Biovolume calculation for pelagic and benthic microalgae. J Phycol 35:403–424

Ivanov VP (2000) Biological resources of the Caspian Sea. KaspNIRH, Astrakhan, p 100

Ivanov VI, Kamakin AM, Ushivtzev VB, Shiganova TA, Zhukova O, Aladin NV, Wilson SI, Harbison GR, Dumont HJ (2000) Invasion of Caspian Sea by the comb jellyfish *Mnemiopsis leidyi* (Ctenophora). J Biol Invasions 2:255–258

Johansson ML, Shiganova TA, Ringvold H, Stupnikova AN, Heath DD, MacIsaac HJ (2018) Molecular insights into the ctenophore genus Beroe in Europe: new species, spreading invaders. J Hered 109(5):520–529. https://doi.org/10.1093/jhered/esy026

Kamakin AM, Chigenkova OA, Zaitsev VF (2009) Distribution and level of *M.leidyi* development in the Caspian Sea in cold seasons. Biodiversity of the Caspian Sea and its coastal ecosystems. Association of Universities of pre-Caspian states. Makhachkala-Guilan 3:92–98. (in Russian)

Kamakin AM, Khodorevskaya RP, Paristkyi YA (2018) Effect of the new invader ctenophore Mnemiopsis leidyi (a. Agassiz, 1865) on the main levels of the Caspian Sea ecosystem. Bulletin ASTU aquatic biological resources and their rational use. Fish Ind 1:35–48. https://doi.org/10.24143/2073-5529-2018-1-35-48. (in Russian)

Karpevitch AF (1975) Theory and practice of aquatic organism acclimatization. Food Industry V. 2, Moscow. (in Russian)

Karpinsky MG (2002) Ecology of the benthos of the middle and South Caspian. VNIRO, Moscow. (in Russian)

Karpinsky MG, Shiganova TA, Katunin DN (2006) Introduced species. In: Kostianoy AG, Kosarev AN (eds) The Caspian Sea environment, The handbook of environmental chemistry. Springer–Verlag, Berlin/Heidelberg, pp 175–190

Kasymov AG (1987) Biota of the Caspian Sea. ELM, Baku, p 156. (In Russian)

Kazmin AS (2021) Multidecadal variability of the hydrometeorological parameters in the Caspian Sea. Estuar Coast Shelf Sci 250:107150. https://doi.org/10.1016/j.ecss.2020.107150

Khlebovich VV (1963) On systematic of the Caspian Nereis. Zool Zh 42:129. (in Russian)

Khuraskin LS, Zakharova NA, Kuznetsov VV, Yanchenkov VP, Khoroshko VI, Artemova AV, Volodina VV (2006) Assessment of the current state of population of Caspian seal in Volgo-Caspian basin and forecast of its commercial catch in 2007. Fishery research in the Caspian. In: Karpuk MI et al (eds) Result of research in 2005. CASPNIRKH, Astrakhan, pp 350–361. (In Russian)

Kireeva MS, Schapova TF (1957) Materials on the systematic composition and biomass of algae and higher aquatic vegetation of the Caspian Sea. Proccedings of IOAN. V.23

Kisilev IA (1969) Plankton of the seas and inland water bodies. V.1. Nauka, Moscow, p 657. (in Russian)

Kobyakova ZI, Dolgopolskaya MA (1969) Order Decapoda. In: Guide of the fauna of the black and Azov seas. V. 2. Naukova Dumka, Kiev, pp 270–306. (in Russian)

Koncheva LA (2004) Characteristic of benthos in the middle and South Caspian. Fishery Research Astrakhan CaspNIIRKH, pp 132–138. (in Russian)

Konsulov AS, Kamburska LT (1998) Ecological determination of the new Ctenophora – *Beroe ovata* invasion in the Black Sea. Trudy Institute of Oceanology, Varna, pp 195–197. (in Bulgarian)

Korsun S, Fahrni JF, Pawlowski J (2012) Invading *Aurelia aurita* has established scyphistoma populations in the Caspian Sea. Mar Biol 159:1061–1069. https://doi.org/10.1007/s00227-012-1886-9

Kosarev AN (2006) Physico-geographical conditions of the Caspian Sea. In: Kostianoy AG, Kosarev AN (eds) The Caspian Sea environment, The handbook of environmental chemistry Part P. Springer–Verlag, Berlin/Heidelberg, pp 5–32

Kostianoy AG, Ginzburg AI, Lavrova OY, Lebedev SA, Mityagina MI, Sheremet NA, Soloviev DM (2019) Comprehensive satellite monitoring of Caspian Sea conditions. In: Barale V, Gade M (eds) Remote sensing of the Asian Sea. Springer, Cham, pp 505–521

Kurasheva EK, Abdulaeva NM (1984) *Acartia clausi* (Calanoida, Acartiidae) in the Caspian Sea. Zool Zh 63(6):931–933. (in Russian)
Lavrov-Navozov NP (1939) The Black Sea shrimp in the Caspian Sea. Zool Zhurnal 18:443. (in Russian)
Levshakova VD, Sanina LV (1973) Summer phytoplankton of the middle Caspian before and after *Rhizosolenia* introduction. VNIRO Proc VNIRO Moscow 80:18. (in Russian)
Logvinenko BM (1959) On finding jelly fish *Blackfordia virginica* in the Caspian Sea. Zool Zhurnal 38:1257. (in Russian)
Logvinenko BM (1965) On changes in the Caspian mollusk fauna genus *Dreissena* after introduction of *Mytilaster lineatus* (Gmel.). Scientific reports of higher school. Biol Sci 4:14. (in Russian)
Lovetskaya AA (1951) Caspian kilkas and their fishery. Pishizdat, Moscow, p 45. (in Russian)
Mikaelyan AS, Zatsepin AG, Chasovnikov VK (2013) Long-term changes in nutrient supply of phytoplankton growth in the Black Sea. J Mar Syst 117:53–64
Milchakova NA (2003) Macrophytobenthos. In: Eremeev VN, Gaevskaya AV (eds) Current state of biodiversity of the coastal waters of Crimea (Black Sea area). ECOSI-Hydrophysics, Sevastopol, p 152. (in Russian)
Milchakova NA (2011) Marine plants of the Black Sea. An illustrated field guide. DigitPrint Press, Sevastopol, p 144. ISBN 978-966-02-5801-3. (in Russian)
Molodtsova AA, Polianinova AA (2009) Feeding of sturgeon, stellate sturgeon and beluga in the Caspian Sea. Fishery Problems 10(40):718–740. (in Russian)
Molodtsova AI, Polyaninova AA, Kashentseva AI, Kamelov AK (2004) The state of sturgeon feeding in the Caspian Sea in 2003. Fishery Research in the Caspian CaspNIRKH, pp 215–225. (in Russian)
Moncheva S., Parr B. (2010) Manual for phytoplankton sampling and analysis in the Black Sea. www.blacksea-commission.org
Mordukhai-Boltovskoy FD (1960) Caspian fauna in the Azov and Black Sea basins. USSR Academy of Sciences, Moscow-Leningrad. (in Russian)
Mordukhai-Boltovskoy FD (1962) Appearance representatives of Mediterranean Polyphemidae in the Caspian Sea. Zool Zh 41(2):289–290. (in Russian)
Mordukhai-Boltovskoy FD (1969) In: Vodyanitsky VA (ed) Crustacea. Guide on identification of the black and Azov seas fauna. V.1 Invertebrata. Naukova Dumka, Kiev, p 151. (in Russian)
Morozova –Vodyanitskaya NV (1948) Phytoplankton of the Black Sea Part 1. Phytoplankton of Sevastopol and total review phytoplankton of the Black SeaIn: Proceedings of Sevastopol biological station. YI: 39–172. (in Russian)
Naumov DV (1968) Phylum Coelenterata. In: Atlas of invertebrates of the Caspian Sea. Food industry, Moscow, pp 43–50. (in Russian)
Nebolsina TK (1959) Crab in the Caspian Sea. Priroda N 6:116. (in Russian)
Orlova MI, Arakelova ES, Komendantov A (1999) On overlap of the habitat *Dreissena bugensis* (Andr.) and *Dreissena polymorpha* (Pall.) in the Volga delta and in shallow waters of the North Caspian. In: Abstracts of conference, Astrakhan, pp 67–69. (in Russian)
Osipov VV (2006) Change of growth rate and life cycles of *Clupeonella curtriventris* Nordman 1840 after its introduction in fresh water ecosystems. PhD thesis. A.N. Severtzhev Institute of ecology and evolution RAS, Moscow, p 125
Panin GN, Mamedov RM, Mitrophanov IV (2005) In: Hubkfryan MG (ed) The current state of the Caspian Sea. Nauka, Moscow, p 358
Panov VE, Krylov PI, Telesh IV (1999) The St. Petersburg harbour profile. In: Gollasch S, Leppäkoski E (eds) Initial risk assessment of alien species in Nordic coastal waters. Nord 1999:8. Nordic Council of Ministers, Copenhagen, pp 225–244
Paritsky YA (2014) Pattern of reproduction of Caspian kilkas. In: Paritsky Yu A, Aseinova AA, Razinkov VP (eds) Current state of bioresources of the inland waters, Reports of All-Russian conference II. M: POLIGRAPF-PLUS. V 2, Borok, pp 430–436. (in Russian)

Pautova LA, Kravchishina MD, Vostokov SV, Zernova VV, Silkin VA (2015) Pattern of vertical structure of summer phytoplankton of deep regions of the Caspian Sea. Rep Russ Acad Sci 462(4):479–483. (in Russian)

Pautova LA, Silkin V, Kravchishina M (2019) The dominanation of invasive species in the present phytoplankton of the Caspian Sea. Rapp Comm Int Mer Médit 42:191

Pautova LA, Kravchishina MD, Silkin VA, Klyuvitkin AA, Artemiev VA, Vazyulya SV, Burenkov VI (2022) Invasive species of diatoms in the autumn phytoplankton of the Caspian Sea: the role in the formation of total biomass and distribution in the salinity gradient. Russ J Biol Invasions. (in press) 15:68

Piatakova GM, Tarasov AG (1996) Caspian Sea amphipods: biodiversity, systematic positions and ecological peculiarities of some species. Int J Salt Lake Res 5:63–79. (in Russian)

Pirogov AI (1972) On finding *Lithogliphus naticoides* in the Volga delta. Zool Zhurnal 51:912. (in Russian)

Polyaninova A.A. (2007). Species invaders in the Caspian Sea and their role in ecosystem. Astrahan, CaspNIRH. 104. (in Russian)

Polyaninova AA, Tatarintseva TA, Teletskaya OV, Tinenkova DK, Petrenko EL, Kochneva LA (2003) Hydrobiological research in the middle and South Caspian after invasion of ctenophore *Mnemiopsis leidyi*. Fishery research in the Caspian. In: Karpuk MI et al (eds) Result for 2002. CaspNIRH, Astrakhan, pp 121–144. (in Russian)

Proshkina-Lavrenko AI, Makarova IV (1968) In: Gollerbakh L (ed) Algae of the Caspian Sea plankton. Nauka, Moscow, p 288. (in Russian)

Ramírez-Romero E, Molinero JC, Sommer U, Salhi N, Yahia OKD, Yahia MND (2020) Phytoplankton size changes and diversity loss in the southwestern Mediterranean Sea in relation to long-term hydrographic variability. Estuar Coast Shelf Sci 235:106574. https://doi.org/10.1016/j.ecss.2019.106574

Ramsak A, Malej A, Venko K (2012) Comparative phylogeography of meroplanktonic species, *Aurelia spp*. and Rhizostoma pulmo (Cnidaria: Scyphozoa) in European seas. Hydrobiologia 690(1):69–80. https://doi.org/10.1007/s10750-012-1053-9

Reznchenko OG (1967) Transatlantic introduction of crab *Rhithropanopeus harrisii*, (Crustacea, Brachyura). Proc SIO RAS. 85:136–177. (in Russian)

Romanova NN (1960) Benthos distribution in the middle and South Caspian. Zool Zhurnal 39:811–825. (in Russian)

Romanova NN (1979) On relationship between *Cerastoderma lamarcki* and *Abra ovata* off the western coast of the middle Caspian. All-USSR meeting on mollusks research. Abstract. Proceedings 6. Leningrad. Nauka: 109–110 (in Russian)

Roohi A, Kideys A, Sajjadi A, Hashemian A, Pourgholam R, Fazli H, Khanari AG, Eker-Develi A (2010) Changes in biodiversity of phytoplankton, zooplankton, fishes and macrobenthos in the South Caspian Sea after the invasion of ctenophore *Mnemiopsis leidyi*. Biol Invasions 2:2343–2361. https://doi.org/10.1007/s10530-009-9648-4

Roohi A, Kideys AE, Jolodar MN, Bandpei MA, Rostami AM (2021) Long-term changes in gelatinous zooplankton, mesozooplankton and kilka fish in the South Caspian Sea: environmental controls and trophic webs interactions. J Oceanogr Mar Sci

Roohi H, Kideys A, Farabi SMV, Rowshan Tabari M, Finenko G, Arashkevitch E4, Bagheri S, Negarestan H, Rostamian M, Rostami H, Pourgholam R, Tamara Shiganova (2022) The non-native ctenophore *Beroe ovata* could result in ecosystem recovery for the Caspian Sea. Iran J Fish Sci. NO: 1401/1155/247/1

Saenkova AK (1960) Colorful Monodacna in the Caspian Sea. Priroda 11:111. (in Russian)

Sayapina VV, Ushivtsev VB, Oleynikova EP, Dosaev FG (2021) The Ctenophore, Beroe ovata Bruguiиre, 1789—the new invader in the ecosystem of the Caspian Sea. Okeanologiia 61(5): 753–758. (In Russian)

Shakirova FM, Atamuratova AH (2001) Arrival of non-native species in the Caspian Sea. Caspian Floating Univ Sci Bull 2:81–84. (in Russian)

Shiganova TA (2009) Non-native species in the south seas of Eurasia. Thesis of D. Sci. A.N. Severtzhev Institute of ecology and evolution RAS, Moscow, p 642. (in Russian)
Shiganova T (2010) Biotic homogenization of inland seas of the Ponto-Caspian. Annu Rev Ecol Evol Syst 41:103–126
Shiganova TA (2011) Determination of the status of invasive species with special attention to the most aggressive invader *Mnemiopsis leidyi* in the Caspian Sea. CaspECO report, p 125
Shiganova TA (2020) Adaptive strategies of *Mnemiopsis leidyi* A. Agassiz 1865 In different environments of the Eurasian seas. Mar Pollut Bull 161:111737. https://doi.org/10.1016/j.marpolbul.2020.111737
Shiganova TA, Abusova GA (2021) Revision of Beroidae Eschscholtz, 1825 systematics and distribution in the southern seas of Europe based on genetic and morphological approaches. Zool. J. Linn Soc. 194(1). https://doi.org/10.1093/zoolinnean/zlab021
Shiganova TA, Kamakin AM, Zhukova OP, Ushivtzev VB, Dulimov AB, Musaeva EI (2001) Invader in the Caspian Sea – ctenophore *Mnemiopsis* and initial results of its effect on the pelagic ecosystem. Oceanology 41(4):542–549. (in Russian)
Shiganova TA, Sapognikov VV, Musaeva EI, Domanov MM, Bulgakova YV et al (2003) Factors that determine pattern of distribution and abundance *Mnemiopsis leidyi* and its effect on ecosystem in the North Caspian. Oceanology 43(5):716–733. (in Russian)
Shiganova TA, Dumont HJD, Mikaelyan AS, Glazov DM, Bulgakova YV et al (2004a) Interaction between the invading ctenophores *Mnemiopsis leidyi* (A. Agassiz) and *Beroe ovata* Mayer 1912, and their Influence on the Pelagic Ecosystem of the Northeastern Black Sea. In: Dumont HJ, Shiganova TA, Niermann U (eds) The Ctenophore Mnemiopsis leidyi in the Black, Caspian and Mediterranean Seas and other aquatic invasions, NATO ASI Series, IV earth and environment sciences, vol 35. Kluwer Academic Publishers, Dordrecht/Boston/London, pp 33–70
Shiganova TA, Dumont HJ, Sokolsky AF, Kamakin AM, Tinenkova D, Kurasheva EK (2004b) Population dynamics of *Mnemiopsis leidyi* in the Caspian Sea, and effects on the Caspian ecosystem. In: Dumont HJD, Shiganova TA, Niermann U (eds) The Ctenophore *Mnemiopsis leidyi* in the Black, Caspian and Mediterranean Seas and other aquatic invasions, NATO ASI Series, IV earth and environment sciences. Kluwer Academic Publishers, Dordrecht/Boston/London, pp 71–111
Shiganova TA, Musaeva EI, Pautova LA, Bulgakova YV (2005) The problem of invaders in the Caspian Sea in the context of the findings of new zoo- and phytoplankton species from the Black Sea. Biolog Bul N 1:78–87
Shiganova TA, Sommer U, Javidpour J, Molinero JC, Malej A, Kazmin AS, Isinibilir M, Christou E, Siokou- Frangou I, Marambio M, Fuentes V, Mirsoyan ZA, Gülsahin N, Lombard F, Lilley MKS, Angel DL, Galil BS, Bonnet D, Delpy F (2019a) Patterns of invasive ctenophore Mnemiopsis leidyi distribution and variability in different recipient environments of the Eurasian seas: a review. Mar Environ Res:104791. https://doi.org/10.1016/j.marenvres.2019.104791
Shiganova TA, Mikaelyan AS, Moncheva S, Stefanova K, Chasovnikov VK, Mosharov SA, Mosharova IN, Slabakova N, Mavrodieva R, Stefanova E, Zasko DN, Dzhurova B (2019b) Effect of invasive ctenophores *Mnemiopsis leidyi* and *Beroe ovata* on low trophic webs of the Black Sea ecosystem. Mar Pollut Bull 141:434–447. https://doi.org/10.1016/j.marpolbul.2019.02.049
Shorygin AA, Karpevich AF (1948) New invaders of the Caspian Sea and their role in the biology of this basin. Krymizdat, Simferopol. (in Russian)
Slynko YV, Kiyashko VI, Yakovlev VN (2000) Catalog of plants and animals of the Volga watershed, Yaroslavl, pp 252–277
Spassky NN (1945) Finding of representative of Annelides in the North Caspian. Zool Zhurnal 24: 23. (in Russian)
Starobogatov YI (1972) The Loricata shellfish class. Determinant of the fauna of the black and Azov seas. Kyiv Sci A Thought 3:60–65

Starobogatov YI (1994) Taxonomy and paleontology//Dreissena polymorpha (Pall.) (Bivalvia, Dreissenidae). Species of fauna of Russia and neighboring countries. Nauka, Moscow, pp 18–46
Tatarintseva TA, Ardabieva AG, Terletskaya OV, Tinenkova DK, Malinovskaya LV, Tarasova LI, Petrenko EL (2000) The Mediterranaen introducers in the plankton and bottom fauna of the Caspian Sea. In: Species introducers in the European seas in Russia, Apatity, p 16. (in Russian)
Thibaut T, Meinesz A (2004) *Caulerpa taxifolia* 18 years of infestation in the Mediterranean Sea. In: Dumont H, Shiganova T, Niermann U (eds) Aquatic invasions in the Black, Caspian, and Mediterranean seas, NATO ASI series, 2. Environment. Kluwer Academic Publishers, pp 287–993
Thompson PA, O'Brien TD, Paerl HW, Peierls BL, Harrison PJ, Robb M (2015) Precipitation as a driver of phytoplankton ecology in coastal waters: a climatic perspective. Estuar Coast Shelf Sci 162:119–129. https://doi.org/10.1016/j.ecss.2015.04.004
Tinenkova DH, Petrenko EL (2004) Zooplankton composition of the Middle and South Caspian in October 2003, Results of research in 2003, Astrakhan, pp 130–131. (in Russian)
Tuzhilkin VS, Goncharov AV (2008) About ventilation of deep waters of the Caspian Sea. Research of oceans and seas. GOIN Proc 211:43–64
Tuzhilkin VS, Katunin DN, Nalbandov YR (2005) Natural chemistry of Caspian Sea waters. In: Kostianoy AG, Kosarev AN (eds) The Caspian Sea environment. The handbook of environmental chemistry, vol 5P. Springer, Berlin/Heidelberg. https://doi.org/10.1007/698_5_005
Usachev PI (1948) Quantitative fluctuations of phytoplankton in the North Caspian. Proc SIO RAS 2:60–88. (in Russian)
Ushivtsev VB (2021) Nauka and culture. Nature of the Caspian Sea. Short communication (in Russian)
Vinogradov ME, Sapognikov VV, Shuskina EA (1992) Ecosystem of the Black Sea. M. Nauka, p 112. (in Russian)
Vostokov SV, Gadgiev AA, Vostokova AS, Rabazanov NI (2020) The ctenophore *Beroe cf. ovata* in the Caspian Sea. The beginning of a new stage in the evolution of the Caspian ecosystem? South Russia Ecolo Develop 15(4):21–35. (In Russian). https://doi.org/10.18470/1992-1098-2020-4-21-35
Yablonskaya EA (1985) Caspian sea. Fauna and biological productivity. Nauka, Moscow, p 276. (in Russian)
Zabralieva TS, Gadjiev RV, Akhundov MM, Gasanov NG, Muradova IT (2006) Characteristic of feeding of kilka in the western coastal areas of the middle and South Caspian under new environmental conditions. In: Karpyuk MI, Mazhnik aY, Katunin DN et al (eds) Fisheries reserch in the Caspian, vol 435. CaspNIRKH, Astrakhan, pp 140–156. (in Russian)
Zenkevich LA (1963) Biology of the seas of USSR. Publisher USSR Academy Sciences, Moscow, p 739. (in Russian)
Zenkevtch LA, Birshtein Ya A (1934) On possible measures for increasing the Caspian and Aral seas productivities. Fishing Ind 3:38. (in ussian)
Zevina GB (1958) New organisms in the Caspian Sea. Priroda 7:79. (in Russian)
Zevina GB (1965) Biocenosis of fouling in the Caspian Sea and their changes associated with the introduction of new organisms. Moscow Nauka: 200–212 (in Russian)
Zevina GB (1968) Phylum Entoprocta. In: Atlas of invertebrate of the Caspian Sea. Food Industry, Moscow, pp 65–67. (in Russian)
Zevina GB (1994) Biology of marine fouling. Moscow University, p 135. (in Russian)
Zevina GB, Kuznetsova IA (1965) The role of shipping in the alteration of the Caspian Sea fauna. Okeanologiya 5:518–527. (in Russian)
Zevina GB, Starostin IV (1961) The qualitative and quantitative changes in the fouling of the Caspian Sea in connection with the navigation in the newly opened Volga-Don Canal. Proc SIO RAS 49:97–107. (in Russian)
Zolota AK, Feneva I Yu (2018) The most dangerous invasive species of Russia (Top-100). Moscow, KMK Scientific Press. Ed. Dgebuadze Yu. I., Petrosyan B.G. Hlyap L.A. 420–426. (in Russian)

Chapter 7
The Aral Sea

Abstract The Aral Sea stands as a stark illustration of how an inland sea-lake can vanish as a result of unsustainable economic exploitation of natural water resources. The sea's deterioration caused permanent alterations to its ecosystem, the loss of formerly abundant valuable fish species, and other devastating outcomes. This analysis examines the history and current condition of the Aral Sea, the species composition prior to its desiccation phase, and the transformations of its biota under human influence. Given that water salinity is a primary environmental factor affecting aquatic life, the sharp increase in the Aral Sea's salinity represents the most significant environmental impact. The reduction in the sea's area and volume has been paralleled by a steady rise in salinity levels. From 1960, salinity climbed from typical levels of 10–12‰ to nearly 140‰ by 2018. Most aquatic organisms that lived in the Aral Sea before the salinization process have vanished from the basin. Fish populations, benthic invertebrates (macrozoobenthos), zooplankton, and macroalgae communities were severely affected. Shifts in species composition_both naturally occurring and human induced—are reviewed across three phases: before salinization, during the salinization period, and after the division of the Aral Sea basin.

7.1 History and Modern State of the Aral Sea

The Aral Sea is widely recognized as a landlocked body of water that is vanishing because of the unsustainable exploitation of its natural water reserves for economic activities. This environmental decline resulted in permanent changes to its ecological system, the loss of once-abundant valuable fish populations, and numerous other damaging effects.

The Aral Sea's geological history encompasses two distinct marine phases, separated by an extensive continental period that spanned the entire Pleistocene epoch. During this lengthy interval, the sea vanished entirely, allowing subaerial processes to reshape its exposed floor. The initial marine stage, representing the Aral Sea's prehistory, extended from the late Pliocene into the Pleistocene. The Aral depression was already a geographical feature then and was inundated during the late

T. A. Shiganova et al., *Non-Native Species in the Ponto-Caspian Seas*,
Monographiae Biologicae 96, https://doi.org/10.1007/978-3-032-18640-9_7

Pliocene by the Akchagylian and Apsheronian transgressions of the Caspian Sea. These incursions created large marine bays, containing freshened water. The second marine stage was comparatively brief and confined to the Holocene. In its early phase, the Aral basin contained a network of water bodies with fluctuating salinity; these periodically desiccated, transforming into solonchaks. Around the mid-Holocene, a significant shift occurred when the Amu Darya River diverted its flow from the Caspian Sea to the Aral, dramatically increasing the water volume entering the basin. This influx led to the formation of a brackish-water marine environment, which was subsequently colonized by *C. glaucum*. The sea's level experienced fluctuations of varying duration and magnitude (reaching up to 20 m), while salinity levels shifted from a few to several tens of pro mille. Following this period, and aside from some relatively short dry intervals between the third- and first-century BC, the combined discharge from the two primary rivers generally sustained a high-water level in the sea until the 1960s. The natural development of the Aral Sea, which commenced approximately 4800 years ago, has been drastically altered since 1960 by the shrinking and degradation of the marine basin. This year signifies the start of a new, anthropogenic phase in its history.

The Aral Sea was previously ranked as the fourth largest inland water body globally, following the Caspian Sea, the Great Lakes, and Lake Victoria. A dramatic reduction in both its surface area and water volume commenced during the latter half of the twentieth century. This was primarily due to the diversion of river water for irrigation from its two principal tributaries: the Syr Darya in the north and the Amu Darya in the south (Fig. 7.1). Between 1961 and 2008, the Aral Sea's area and volume contracted from 66,100 km^2/1066 km^3 to 10,400 km^2/110 km^3. Since 1960, the sea's surface level has fallen by more than 24 meters.

Before 1960, the Aral Sea was conventionally separated into the Large Aral (the principal water area, further divided into western and eastern basins) and the Small Aral (or the northern basin) (Fig. 7.1). In 1989, the Aral Sea separated into two distinct water bodies: the North Aral (Small Aral) and the South Aral (Large Aral). By 2003, the eastern and western basins of the South Aral had also divided (Fig. 7.1).

By 2009, the eastern section had vanished entirely, while the western part had diminished to a narrow band along the former South Aral's western fringe (Fig. 7.1). In later periods, sporadic water influx resulting from substantial snowmelt has occasionally allowed the eastern basin to be partially refilled to a limited extent (Fig. 7.1).

Salinity levels in water represent a key environmental factor affecting aquatic life. For the purposes of this publication, the dramatic increase in the Aral Sea's salinity stands out as the most significant ecological consequence. The reduction in the sea's surface area and total volume has occurred alongside a steady rise in salt concentration. From 1960 onward, salinity climbed from its typical range of 10–12‰ to nearly 140‰ by 2018 (Fig. 7.2). Between 1995 and 1997, salinity surpassed the threshold that is detrimental to most marine organisms. Continued desiccation of the Aral Sea resulted in hypersaline environments. This escalating water salinity transformed all ecosystem communities. Most aquatic species that lived in the

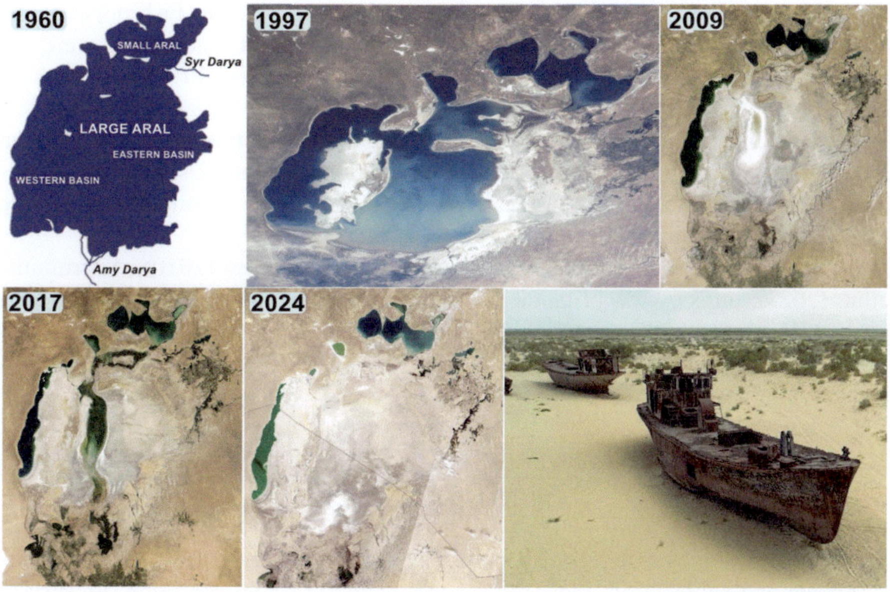

Fig. 7.1 The Aral Sea's dramatic decline. 1960, its original extent; 1997, the division into western, eastern, and northern sections; 2009, the remaining western and northern (Small Aral) water bodies; 2017, a portion of the eastern basin refilled following significant snowmelt in 2015; 2024, continued contraction of the western lakes. Bottom right, vessels stranded in the desert: the destiny of the Aral Sea's fleet

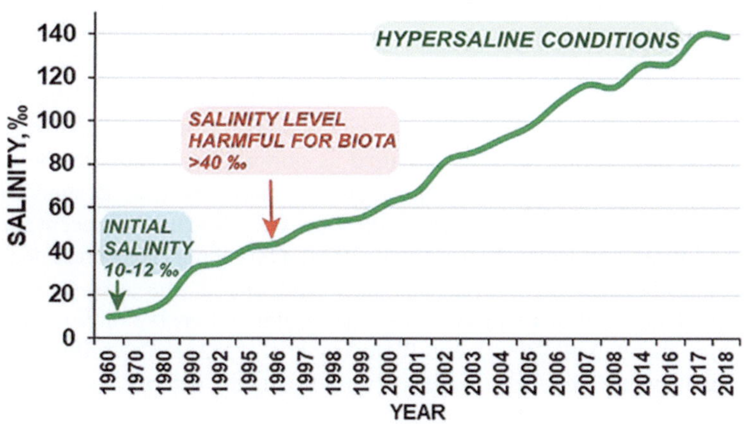

Fig. 7.2 Year-to-year changes in salinity within the Aral Sea's western basin during its desiccation phase. (Information sourced from Kostianoy and Kosarev (2010) and Andrulionis et al. (2025). Figure creation and explanatory notes provided by the authors)

Aral Sea prior to its salinization have vanished from the region. Fish populations, benthic invertebrates (macrozoobenthos), zooplankton, and macroalgae were the most severely impacted. The changes were less pronounced for communities of bottom microalgae (microphytobenthos) and phytoplankton.

Consequently, due to the desiccation resulting from human activity, the Aral Sea has experienced significant transformations, shifting from a brackish, uniform, well-mixed, and biologically rich sea into a hypersaline, highly stratified, and largely lifeless body of water.

In this section, information from Kostianoy and Kosarev (2010), Andrulionis et al. (2025), https://www.britannica.com/place/Aral-Sea, and https://en.wikipedia.org/wiki/Aral_Sea have been partially used.

7.2 The Biological Makeup of the Aral Sea Prior to Its Desiccation Period

From the mid-nineteenth century until 1960, a period marking the start of its shrinkage and salinization, the Aral Sea maintained a stable condition; its surface area, water volume, and salinity levels remained largely unchanged. The biodiversity of the Aral Sea's biological community prior to human impact was relatively modest. Records indicate the existence of approximately 375 phytoplankton species, predominantly from the Bacillariophyta and Chlorophyta groups, alongside 195 invertebrate species (Yablonskaya 1974), and 20 fish species (Nikolsky 1940). However, these figures might not be entirely precise, as certain smaller aquatic organism groups, such as protozoa, were not thoroughly investigated. Furthermore, species permanently inhabiting the freshwater zones at the mouths of the Amu Darya and Syr Darya rivers, or those introduced via river discharge, were often excluded from these counts. Even though these species spread across significant portions of these waters, they were typically classified as non-native to the Aral Sea's fauna (Andreev 1989). The level of endemism was remarkably low, despite the sea's geographic isolation, a fact attributed to the water body's relatively recent formation (Aladin and Plotnikov 1995; Boomer et al. 2000). Only three Harpacticoidea species (Borutzkiy 1974) and no fish species were endemic to the Aral Sea. Detailed summaries of the Aral Sea's biotic composition, with extensive bibliographies, are available in Aladin et al. (2004), and Plotnikov et al. (2023). The following discussion is primarily based on these sources, incorporating the authors' own observations and supplementary information.

The indigenous free-living invertebrate population of the Aral Sea originated from diverse sources. As noted by Yablonskaya (1974), 78% of the free-living invertebrate species were residents of continental freshwater and brackish-water habitats, 17% were species from the Caspian Sea, and a mere 5% were of Mediterranean-Atlantic origin. Consequently, species with freshwater ancestry formed the dominant component of the Aral Sea's fauna.

The variety of native zooplankton species in the Aral Sea was most abundant within its freshwater regions. In areas with typical salinity levels, the range of species was notably more limited. The indigenous zooplankton of the Aral Sea can be categorized into three groups based on their salt tolerance: species that live in the open sea under normal salinity conditions (10–12‰), mainly freshwater species that can withstand slight increases in salinity (including 29 species of Rotatoria, 8 species of Cladocera, and 1 species of Copepoda), and primarily river-dwelling species that migrate into the Aral Sea (Andreev 1989).

The initial species assemblage established the foundation for the Aral Sea's indigenous zooplankton community. Twenty-one Rotatoria species were commonly found across freshwater, brackish, and marine environments (Kutikova 1970); among the seven Cladocera and 22 Copepoda species (Andreev 1980) that constituted this group, only a limited number were present in large quantities. Four Cladocera species—*Cercopagis pengoi aralensis* (M. Boltovskoi), *Evadne anonyx* (G. Sars), *Podonevadne camptonyx* (G. Sars), and *P. angusta* (G. Sars)—are characteristic of the endemic Ponto-Caspian group, which served as the most stable element of the original Cladocera fauna. The remaining three Cladocera species—*Ceriodaphnia reticulata* (Jurine), *Alona rectangula* (G. Sars), and *Moina mongolica* (Daday)—were generally not noted for their abundance, with a few isolated cases (Andreev 1989). *Arctodiaptomus salinus* (Daday) was historically the predominant copepod, its broad presence typifying fauna from inland saline water bodies (Lukonina 1960). The endemic subspecies *Halicyclops rotundipes aralensis* (Borutzky) was found across the entire Aral Sea. Additional copepods included freshwater euryhaline species of Cyclopoida, with *Mesocyclops leuckarti* (Claus) being the most frequently encountered, and 15 brackish and marine species of Harpacticoida, among which *Schizopera aralensis* (Borutzky), *S. reducta* (Borutzky), and *Enhydrosoma birsteini* (Borutzky) were identified as Aral Sea endemics (Borutzkiy 1974). During summer, the most numerous organisms in the Aral Sea's zooplankton were the larvae of bivalves from the genera *Dreissena* and *Hypanis* (Lukonina 1960; Kortunova 1975).

The Aral Sea's native benthic community was primarily composed of freshwater and Caspian species. The macrozoobenthos included roughly 12 species and one mollusk subspecies, six Oligochaeta species, five Crustacea species, and the larvae of more than 25 Chironomidae species (Dengina 1959; Belyanina and Konstantinov 1974). Nevertheless, only a select group were found in significant numbers: the Aral-Caspian bivalve mollusks *Dreissena polymorpha aralensis* (Andrusov), *D. p. obtusecarinata* (Andrusov), *D. caspia pallasi* (Andrusov), *Hypanis minima minima* (Ostroumoff), and *H. m. sidorovi* (Starobogatov); the oligochaetes *Nais elingius* (Müller) and *Paranais simplex* (Hrabe); the ostracod *Cyprideis torosa* (Jones); the amphipod *Dikerogammarus aralensis* (Uljanin); chironomid larvae from the genus *Chironomus* (with *C. behningi* (Goetghebuer) as the most common); and trichopteran larvae of the genus *Oecetis intima* (MacLachlan).

Marine mollusks, specifically *Cerastoderma* spp. and salt-tolerant *Caspiohydrobia* spp., were abundant exclusively in the saline bays along the eastern shoreline (Husainova 1958; Dengina 1959; Yablonskaya 1960). The gastropod

Theodoxus pallasi (Lindholm) was found, although in significantly smaller quantities compared to the bivalves from the *Dreissena* and *Hypanis* genera (Yablonskaya 1960; Yablonskaya et al. 1973). Mollusks accounted for 63% of the zoobenthos biomass, while chironomid larvae made up 33% (Karpevich 1964). The species makeup of benthic communities was heavily influenced by sediment type and the composition of benthic vegetation, with the bivalve *Dreissena* emerging as the dominant organism (Zenkevich 1963).

The Aral Sea's indigenous fish population originally included just 20 species spanning 7 families. Cyprinidae was the most numerous family, comprising 12 species, or 60% of the total fish fauna. The Percidae family was represented by three species, while the Acipenseridae, Salmonidae, Siluridae, Esocidae, and Gasterosteidae families each contributed only a single species. Eurybiontic species formed the dominant group of native fish in the Aral Sea, making up approximately 95% of the fish fauna (Nikolsky 1940).

Just three out of the six native fish species assemblages in the Aral Sea were found within the basin (Nikolsky 1940):

1. Upper quaternary fauna pike (*Esox*) and perch (*Perca*).
2. Aral-Caspian fauna: *Acipenser*, *Rutilus*, *Abramis*, *Aspius*, *Barbus*, and *Pungitius*. All of these species inhabited the marine environment and the downstream sections of the rivers. This collection was characteristic of the Aral Sea's fish fauna and comprised nine species, accounting for 45% of the total fish fauna.
3. Northern immigrants: two groups of northern (mainly Siberian) fish fauna, including
   (a) Stenothermal psychrophilic fishes: Aral salmon *Salmo trutta aralensis*
   (b) Eurythermal limnophilic fishes in the lower reaches of rivers and partially across the Aral Sea: silver crucian *Carassius auratus gibelio*, *Leuciscus idus oxianus*, and ruffe *Gymnocephalus cernuus* (=*Acerina cernua*).

The deep-water region of the Aral Sea was devoid of any known fish species, indicating that its fish fauna originated from the limnophilic fauna of the Amu Darya basin (Nikolsky 1940). Furthermore, the Aral Sea lacked any native pelagic fish that primarily consumed zooplankton and phytoplankton. Nevertheless, certain species, particularly sabrefish, shemaya, roach, and white-eye bream (Nikolsky 1940), supplemented their diet with plankton from the pelagic zone while also feeding on benthos. The principal food items for fish in the pelagial were insect pupae, predominantly from the Chironomidae family, which float on the water surface during mass emergence events, along with their adult forms and the adults of caddis flies. Zooplankton did not constitute a major food resource for the fish community, with the stickleback (*Pungitius platygaster aralensis*) being the only species that relied mainly on zooplankton for sustenance (Nikolsky 1940).

The Aral Sea's original fish population consisted mainly of species that originated in freshwater environments. According to Bervald (1964) and Nikolsky (1940), the availability of suitable conditions for natural reproduction was the primary factor determining the abundance of this fauna. These indigenous fish species typically

Fig. 7.3 Fauna of the Aral Sea. Principal native and introduced species in the Aral Sea's ecological community. *Mesocyclops leuckarti* (**a**) and *Arctodiaptomus salinus* (**b**) were major constituents of the zooplankton, though the latter was eventually displaced by the non-native *Calanipeda aquaedulcis* (**c**). Within the benthic environment, the native *Dikerogammarus aralensis* (**e**) was largely replaced by the invasive shrimp *Palaemon elegans* (**d**) and *Paramysis lacustris* (**f**). Native *bivalves Adacna vitrea* (**i**), *Cerastoderma sp* (**j**), *Dreissena caspia pallasi* (**l**) and *Ecrobia grimmi* (**m**) together with the introduced species *Abra segmentum* (**k**), acted as essential food sources for fish. The non-native polychaete *Hediste diversicolor* (**h**) fulfilled a comparable dietary function. Larvae and pupae of chironomids (**g**) were a critical element in fish diets. Native fish: Sander (**p**), Carp (**s**). Introduced fish: Baltic Herring (**n**), Plaice (**r**). Both the Aral Sea Trout (**o**) and the Aral Sea Sturgeon (**q**) are now considered extinct. Adapted from Plotnikov et al., 2023 (with modifications by the authors)

migrated to freshwater zones to reproduce. Their most successful spawning occurred in the freshwater bays adjacent to river deltas. Key commercial species, including bream, carp, and roach, historically spawned in locations with fluctuating salinity levels. However, embryonic development is known to be impaired at elevated salinity concentrations. Definitive information regarding the maximum salinity levels that permit normal larval development is currently unavailable.

For reference, the most significant native, introduced, and extinct species of the Aral Sea fauna are illustrated in Fig. 7.3.

7.3 Transformations in the Aral Sea's Biological Communities Resulting from Human Influence

The variety of free-living invertebrates and fish species in the Aral Sea has experienced major shifts from the mid-twentieth century into the early twenty-first century due to human influence. These changes stem from deliberate species introductions as well as the basin's desiccation and rising salinity caused by unsustainable water resource management. According to Plotnikov et al. (1991), the recent history of the Aral Sea is marked by three key periods, with the present era characterized by the divergent evolution of two separate lakes.

The initial crisis, stemming from both deliberate and accidental introductions of invertebrate and fish species into the Aral Sea, occurred between 1957 and 1960 and led to relatively minor impacts. However, starting in the 1960s, the main factor shaping the composition of invertebrates and other fauna became the rapidly rising salinity of the Aral Sea, caused by unregulated diversion of river inflows. During this period, the Aral Sea changed from being the world's fourth largest inland water body into the diminished Small (North) Aral and the highly saline remnants of the former Large (South) Aral (Fig. 7.4).

The second period of crisis unfolded between 1971 and 1976, a time when the salinity of the Aral Sea rose beyond 12–14‰. This led to a sharp decline in the overall count of invertebrate and fish species. Freshwater and brackish-water organisms vanished, leaving only euryhaline and marine species to persist.

A third crisis commenced in 1986, as salinity levels climbed past 23–25‰, and continued until the Aral Sea separated into two distinct water bodies—the North Aral (Small Aral) and the South Aral (Large Aral)—in 1989. Throughout this phase, brackish-water species originating from the Caspian region went extinct (Plotnikov et al. 1991). The present era started with the division of the Aral Sea, with both sections inheriting a shared fauna.

The decline of the Aral Sea is clearly illustrated by the trends in commercial fish harvests (Fig. 7.5). For instance, between 1971 and 1976, as the Aral's salinity rose beyond 12–14‰, commercial catches fell dramatically. By 1986, when salinity levels surpassed 23–25‰, the harvests had dropped to virtually zero. At the same time, the variety of fish species diminished from an initial 20 to only the introduced flounder species by 1990 (Fig. 7.5, insert).

In the North Aral Sea, the 2005 construction of the Kokaral Dam (Fig. 7.4) enabled the collection of runoffs, which significantly expanded the freshwater area and enlarged the natural habitat for native commercial fish species. Populations of several highly valued fish rebounded to levels that once again permitted commercial

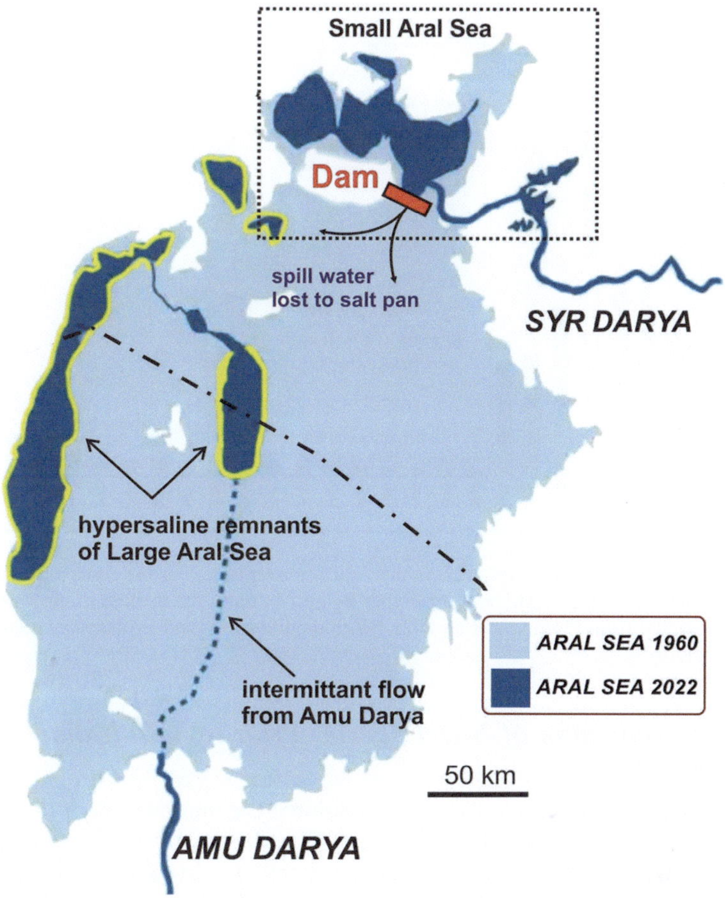

Fig. 7.4 The Aral Sea's surface area has dramatically decreased from its 1960 size to the present. The pale blue shading shows the sea's original extent, while the dark blue color denotes the contemporary water coverage. Yellow outlines delineate the shifting coastlines of the hypersaline remains of the Large Aral Sea. The red box indicates the position of the Kokaral Dam. (Adapted from Plotnikov et al. (2023) (Modified by authors))

fishing. Since 2005, commercial fisheries in the North (Small) Aral have been recovering (Fig. 7.5).

Comprehensive analyses of human impacts on Aral Sea biota, with detailed references, are available in Plotnikov et al. (1991), Aladin et al. (2004), Plotnikov et al. (2012), and Plotnikov et al. (2023). The following summary is generally based on these sources, with the authors' own comments and additions.

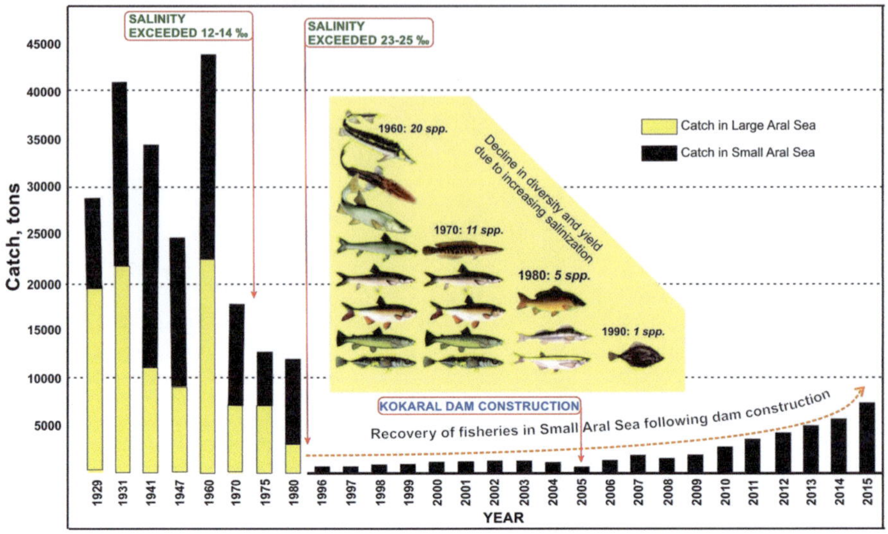

Fig. 7.5 The commercial fishing yields in the Aral Sea over time. For the earliest timeframe, data are limited and intermittent. The accompanying figure illustrates the decline in the variety of fish species. (Adapted from Plotnikov et al. 2023 (With modifications and explanatory notes provided by the authors))

7.3.1 Introduction of Novel Species Prior to Salinization

To enhance fish production, researchers explored the introduction of diverse invertebrate species as potential fish food sources. The plan involved adding—besides the bivalve mollusks *Syndosmya segmentum* (Recluz), also known as *Abra ovata* (Philippi), and *Mytilaster lineatus* (Gmelin) (Zenkevich and Birshtein 1934)—amphipods, shrimp, and crab species from the Caspian Sea, along with mysids and other brackish-water organisms from the Azov-Black Sea basin, the Baltic, and Far Eastern seas (Averintsev 1936; Behning 1936; Zenkevich and Birshtein 1934, 1937; Zenkevich 1947; Ilyin 1954). These proposals were later assessed by Karpevich in multiple publications (Karpevich 1947, 1948, 1953, 1960, 1975, 1986).

The limited zooplankton output in the Aral Sea was insufficient to maintain substantial populations of planktivorous fishes (Yablonskaya and Lukonina 1962). For instance, the calanoid copepod *Arctodiaptomus salinus*, which has a low reproductive rate of one generation annually, made up 70% of the spring zooplankton bloom (Kortunova 1975; Lukonina 1960). In summer months, overall zooplankton concentrations dropped, and rotifers—common in delta regions—were found only in limited quantities in the open waters (Karpevich 1975). These conditions, along with the restricted variety of native species in the Aral Sea, prevented efficient use of all available food sources. Additionally, certain resources were lost from the ecosystem; for example, wind carried adult chironomids into the desert, significantly reducing the organic matter present in the sea (Karpevich 1960, 1975).

Given the anticipated significant hydrological shifts resulting from the regulation of Syr Darya and Amu Darya flows, it was predicted that salinity levels would increase, forcing the native flora and fauna to either adapt or perish. To mitigate these adverse outcomes, strategies were developed to modify the Aral Sea's biodiversity in order to preserve its fishery value (Karpevich 1953). Consequently, the introduction of euryhaline species was recommended, with precautions to avoid adding undesirable organisms. Experts suggested starting with species from lower trophic levels—first phytoplankton, then zooplankton and zoobenthos, and finally fish. Among fish, benthophages and predators could be introduced relatively easily, while planktophage species required a prior expansion of their food supply through the introduction of highly productive planktonic invertebrates (Karpevich 1960, 1975).

Based on an assessment of the Aral Sea's ecosystem and initial field studies, a proposal was made to introduce numerous species of planktonic algae, along with planktonic and benthic invertebrates and fish. The following benthic invertebrates were suggested for introduction: the polychaete worms *Hediste* (= *Nereis*) *diversicolor* (O.F. Müller) and *Nephtys hombergii* (Savigny) in Lamarck, the bivalve mollusks *Monodacna colorata* (Eichwald) and *Syndosmya segmentum*, the cumaceans *Schizorhamphus bilamellatus* (Sars) and *Pterocuma pectinata* (Sowinski), and the amphipods *Corophium nobile* (Sars), *C. curvispinum* (Sars), along with several other species. To reinforce the second pelagic trophic level—planktonic invertebrates—the introduction of the following species was recommended: the highly productive brackish-water and marine crustaceans *Calanipeda aquaedulcis* (Kritchagin), *Heterocope caspia* (Sars), *Acartia clausi* (Giesbrecht), *A. latisetosa* (Kritchagin), *Acanthocyclops vernalis* (Fischer), *Halicyclops sarsi* (Akatova), *Centropages kroyeri* (Giesbrecht), and rotifers from the genus *Synchaeta* (Karpevich 1960, 1975).

However, during the mid-1950s, when aquatic invertebrates and fish species began to be introduced into the Aral Sea, scientifically supported recommendations were disregarded. The possible adverse effects were not taken into account. The adaptation process was launched by adding fish species without first reinforcing the lower trophic levels (Karpevich 1947, 1948, 1953, 1960, 1975).

By accident, several invertebrate and noncommercial fish species were introduced into the Aral Sea alongside valuable commercial fish. These introductions were not only unplanned but also undesirable. Of the six goby species originating from the Ponto-Caspian region, three became successfully established: the bubyr *Knipowitschia caucasicus* (Berg), the monkey goby *Neogobius fluviatilis* (Pallas), and the round goby *N. melanostomus* (Pallas). By the mid-1960s, a sharp rise in the goby population caused a substantial decline in benthic invertebrate numbers (Doroshev 1968; Markova 1972; Karpevich 1975; Yablonskaya et al. 1973). During an unsuccessful effort between 1954 and 1956 to introduce Caspian mullets *Liza auratus* (Risso) and *L. saliens* (Risso), shrimp species were inadvertently brought into the Aral Sea (Husainova 1958). Initially, it was thought that two shrimp species (*Palaemon elegans* (Rathke) and *P. adspersus* (Rathke)) had been introduced, but later studies confirmed that only *P. elegans* (Malinovskaya 1961) had spread across the entire sea (Gavrilov 1970; Karpevich 1975). Additionally, the plankton-feeding atherine *Atherina boyeri caspia* (Eichwald) became naturalized along with the mullets (Karpevich 1975).

Between 1954 and 1956, the Baltic herring *Clupea harengus membras* (Linnaeus) was introduced into the ecosystem. It quickly became naturalized (Konovalov et al. 1958) and was widespread by 1960. Following the introduction of Baltic herring, atherine, and gobies, a significant shift occurred toward the consumption of low-productivity zooplankton as a primary food source. Because their food supplies were not replenished (Karpevich 1960), the Baltic herring and atherine species soon depleted their available food reserves (Karpevich 1975; Yablonskaya and Lukonina 1962; Kortunova 1975). A marked decline was observed in the populations of large planktonic crustaceans, including *Cercopagis pengoi aralensis*, *Moina mongolica*, *Ceriodaphnia reticulata*, and cyclopoids. Notably, the abundance of the copepod *Arctodiaptomus salinus*, a key planktonic species in the Aral Sea, was drastically reduced. Consequently, from 1961 onward, the larvae of bivalve mollusks constituted the majority of the summer zooplankton biomass (Lukonina 1960; Yablonskaya and Lukonina 1962; Kortunova and Lukonina 1970; Kortunova 1975), while the number of small rotifers rose. This scarcity of accessible zooplankton led to a mass die-off of the Baltic herring and atherine populations (Bykov 1964; Kortunova and Lukonina 1970; Kortunova 1975). The introduction of plankton-eating fish also affected the abundance of zoobenthos species—such as planktonic larvae of bivalves, as well as larval chironomids and trichopterans—which were important elements of the Aral Sea's zooplankton community (Lukonina 1960; Husainova 1958).

The mysids *Paramysis* (*Serrapalpisis*) *lacustris* (Czerniavsky), *P. (Mesomysis) intermedia* (Czerniavsky), *P. (Metamysis) ullskyi* (Czerniavsky), and *Paramysis (Paramysis) baeri* (Czerniavsky) were the first invertebrates deliberately introduced into the Aral Sea between 1958 and 1960. In 1958, these invertebrates were released into Bolshoy Sary-Cheganak bay, but they failed to survive due to high salinity levels. Two years later, in 1960, *P. lacustris* and *P. intermedia* were successfully established in the freshwater Karateren bay. By 1964, these invertebrates had expanded across the entire near-delta region of the Syr Darya. Additionally, *P. intermedia* was successfully introduced into the more freshwater Abbas Bay in the southern Aral Sea. In 1964, mysids from the Syr Darya mouth were relocated to the mouth of the Amu Darya (Bekmurzaev 1974; Husainova 1971). By 1965, *P. ullskyi* was observed in the southern parts of the sea, likely having migrated from the Amu Darya water reservoirs. By the late 1960s, mysids had spread throughout the Aral Sea at depths of up to 5–7 meters, with the exception of the more saline Akpetkinsky archipelago. Initially, *P. lacustris* was the dominant species, but over time, *P. intermedia* became the most abundant mysid, making up approximately 90% of their population (Karpevich and Bokova 1970; Kortunova 1970; Karpevich 1975).

By the early 1960s, three additional invertebrate species had been intentionally introduced to the Aral Sea, and one arrived accidentally. Between 1927 and 1963, 17 new fish species appeared in the Aral Sea, significantly increasing fish diversity from 20 to 34 species. However, the composition of commercially important fish species remained largely unchanged. Overall, the negative consequences of these introductions outweighed the positive outcomes.

7.3.2 *The Arrival of New Species During Times of Salinization*

Since the 1960s, the Aral Sea has been steadily shrinking and growing shallower, primarily because of irrigation withdrawals from the Amu Darya and Syr Darya rivers. The rapid anthropogenic desiccation and salinization of the Aral Sea led to initiatives aimed at introducing euryhaline species capable of surviving the rising salinity levels.

During the early 1960s, the polychaete *Hediste diversicolor* and the bivalve *Abra ovata* were introduced into the Aral Sea from the Sea of Azov. *Hediste diversicolor* became abundant starting in 1963, while *Abra ovata* proliferated from 1967 onward. Both species developed into important food sources for benthophagous fishes. In the mid-1960s, an unsuccessful effort was made to introduce the bivalve mollusk *Monodacna colorata* (Karpevich 1975). Between the mid-1960s and early 1970s, several attempts were also undertaken to introduce planktonic invertebrates. Candidates for introduction included two euryhaline copepods—*Calanipeda aquaedulcis* and *Heterocope caspia*. *Calanipeda aquaedulcis* successfully naturalized and, by 1970, had become the dominant zooplankter in the Aral Sea (Kazakhbaev 1974; Andreev 1978), replacing the previously dominant *Arctodiaptomus salinus*, which had been eliminated by Baltic herring. The second exotic species, *Heterocope caspia*, did not establish itself. Thanks to their high euryhalinity, the successful introduced species endured the subsequent salinization of the Aral Sea and remained dominant components of the zoobenthos and zooplankton.

During the introduction of planktonic copepods, crab larvae of the species *Rhithropanopeus harrisii tridentata* were unintentionally brought into the ecosystem. Since 1976, this benthic crustacean has become widely distributed across the southern region of the Aral Sea (Andreev and Andreeva 1988).

The adaptation of euryhaline planktonic and benthic invertebrates can be viewed as a carefully planned and effective introduction. These euryhaline species managed to preserve the nutritional quality of zooplankton and benthos during the anthropogenic salinization of the Aral Sea. The beneficial impact of euryhaline acclimatizants became particularly evident from the start and middle of the 1970s, when the Aral Sea's salinity rose above 12–14‰, leading to the decline of freshwater and brackish-water plankton and benthos organisms.

However, increasing salinity levels had a detrimental impact on the fish populations. The early developmental phases of freshwater fish species, which were initially dominant in the Aral Sea, proved especially sensitive. The survival rates of larvae and juvenile fish started to decline rapidly, even with salinity increases of just 1–2‰ above the typical range of 8–10‰. Most freshwater fish and invertebrates in the Aral Sea were already living at the upper edge of their salinity tolerance, which accounts for their rapid disappearance (Karpevich 1975; Aladin and Kotov 1989; Plotnikov et al. 1991). Within a single decade after the onset of human-induced desiccation, over 50–70% of fish and free-living invertebrate species in the Aral Sea had gone extinct.

Under such extreme environmental circumstances, the successful adaptation of euryhaline invertebrates led to the proposal of introducing commercially valuable euryhaline fish species. In the late 1970s, the flat-fish *Platichthys flesus* from the Sea of Azov was introduced. By 1981, this commercial species had become dominant in local catches (Lim and Markova 1981). Of the 20 native fish species originally inhabiting the Aral Sea, only the euryhaline stickleback *Pungitius platygaster aralensis* managed to survive. All other indigenous fish vanished as salinity increased, with only a few persisting in the deltas and deltaic water bodies of the Amu Darya and Syr Darya rivers. The successful establishment of plaice enabled the continuation of fishing activities in the Aral Sea. In the early 1980s, aside from these two fish, an accidentally introduced atherine species and two to three species of gobies were also present.

During the late 1970s and early 1980s, a final effort was made to introduce sturgeon species into the Aral Sea. Russian sturgeon, *Acipenser gueldenstaedtii*, was the species selected for this initiative (Lim and Markova 1981). However, this endeavor was unsuccessful, as the Aral Sea's salinity had increased to 18–20‰, a level far too high for the species to tolerate. In the mid-1980s, further attempts were made to introduce euryhaline invertebrates into the Aral Sea. Species such as the bivalves *Mytilus galloprovincialis* and *Mya arenaria* were brought from the Sea of Azov. These introductions also failed to establish sustainable populations.

By the end of the 1980s, the era of both intentional and unintentional species introductions into the Aral Sea came to an end. Since that time, only natural colonization processes, independent of human intervention, have taken place.

7.3.3 Fauna After the Aral Sea Division

Between 1988 and 1989, the Aral Sea's total surface area had diminished by close to 40,000 km^2 (representing 60% of its 1960 extent), its water volume had fallen by 333 km^3 (a 33% reduction from the 1960 volume), and the average salinity had climbed to 30‰. The previously brackish Aral Sea had separated into two distinct bodies—the Small and Large Aral seas—and its evolution into a multifaceted system of residual lakes with varied future trajectories was underway (Aladin and Plotnikov 2011).

7.3.4 The Small Aral Sea

Beginning in the 1990s, the reduction of Syr Darya water diverted for irrigation led to an increase in its flow into the Small Aral Sea. As the water entering the Small Aral surpassed the rate of evaporation, a slow decline in salinity began, and the excess water started to flow through the dried-up Berg Strait into the Large Aral. This development created a risk of further erosion of this channel, the gradual breakdown of the natural division between the Large and Small Aral Sea, and a renewed drop in the Small Sea's water level. Moreover, this watercourse had the potential to extend to the Syr Darya's mouth, which would divert the flow directly to the Large Aral Sea, threatening the Small Sea with complete desiccation. To avert this outcome and protect the Small Aral Sea, the former Berg Strait was blocked by a low dam in 1992. Subsequently, the Small Aral Sea's water levels started to climb, and its salinity kept falling slowly. However, due to design flaws, this dam was entirely destroyed by a storm in 1999. A new, properly engineered dam in the Berg Strait (the Kokaral Dam, Fig. 7.4) was finished in 2005. It includes flood-control outlets to discharge surplus water and keep the Small Aral at a safe level. Conversely, blocking the flow through the Berg Strait has hastened the decline of the Large Aral Sea's water level and raised its salinity.

The current salinity level of the Small Aral Sea has dropped to 14‰, and with this trend continuing, the restoration of the Small Sea into a brackish-water lake seems achievable. Wildlife began returning to the Small Aral Sea shortly after its separation and even before the dam was built. The cladoceran species *Podonevadne camptonyx*, which had disappeared in 1980, reappeared from dormant eggs in the Small Sea's plankton. In 1999, chironomid larvae (Aladin et al. 2000), not recorded since 1974, were once again found in the benthos.

The freshwater region's expanse and the population of commercially important fish species grew significantly. Fish communities broadened their spawning and feeding territories to encompass almost the entire Small Aral Sea, with the exception of Butakov Bay, where salinity levels stayed excessively elevated. The relative stabilization of the hydrological conditions and the freshening of the Small Aral Sea led to commercially viable populations of valuable food fish, such as carp, bream, and zander.

It is anticipated that the salinity levels in the Small Aral Sea will continue to decline, allowing it to revert to a brackish-water ecosystem. As a result of this transformation, a rise in the population of native freshwater fish species and a restoration of the area's fishery significance are likely for this section of the Aral Sea.

7.3.5 The Large Aral Sea

Following the separation of the Aral Sea and the building of the dam in the Berg Strait, the outflow from the Small Aral Sea diminished. This led to a significant decline in the water level of the Large Aral Sea and hastened its salinization process. The Amu Darya's flow now only sporadically and minimally reaches the Large Sea from its southern side.

Following the significant drop in the Large Aral Sea's water level, Vozrozhdeniya Island expanded considerably and, by 2001, had merged with the southern mainland to create a peninsula. Consequently, in 2001, the Large Aral Sea was split into distinct Western and Eastern sections, linked only by a northern channel. This separation caused the salinity in the shallower eastern basin to increase much more rapidly than in the deeper western part. By the close of 2009, the Eastern Large Aral had almost completely desiccated.

The swift increase in salinity within the Large Aral Sea caused a rapid transformation in its biological community. By the close of the 1990s, the isolated Large Aral Sea had evolved into a hypersaline environment, hosting a fauna that exhibited clear signs of adaptation to the altered ecological circumstances.

By 1997, a salinity level of 57‰ enabled the zooplankton *Calanipeda aquaedulcis* to become dominant, while rotifers from the genus *Synchaeta* vanished. During the latter part of the 1990s, most of the few remaining rotifer species also disappeared, as did the native copepod *Halicyclops rotundipes aralensis*, the shrimp *Palaemon elegans*, and the crabs.

By 2001, a salinity level of 67‰ had led to a major shift in the composition of zoobenthos species. The polychaete worm *Hediste diversicolor* and the bivalve mollusk *Cerastoderma isthmicum* were no longer present. By 2002, the bivalve mollusk *Syndosmya segmentum* had also vanished.

According to certain researchers, the benthic fauna of the Western Large Aral Sea—specifically the gastropods *Caspiohydrobia* spp. and the ostracod *Cyprideis torosa*—continue to persist (Zavyalov et al. 2006).

In 1996, the species *Moina mongolica*, which had vanished in 1973, reappeared in the Large Aral. Concurrently, the brine shrimp *Artemia parthenogenetica* (Bowen and Sterling) emerged and established itself as the dominant zooplankton. By 2004, the halophilous copepod *Apocyclops dengizicus* (Lepeschkin) was observed in the Western Large Aral (Mirabdullayev et al. 2007). Two native rotifer species, *Hexarthra fennica* (Levander) and *Brachionus plicatilis*, are now abundant, whereas they were previously scarce. During summer, the infusorium *Fabrea salina* (Henneguy) is a frequent zooplankton species. Additionally, the halophilous ostracod *Eucypris inflata* (Sars) and the native *Cyprideis torosa* are present in the Western Large Aral (Aladin and Plotnikov 2008).

The benthic community of the Western Large Aral includes a significant population of larvae from the halophilous chironomid *Chironomus salinarius* (Mirabdullayev et al. 2007), the euryhaline indigenous turbellarian worm *Mecynostomum agile* (Jensen), and the large infusorian *Frontonia marina* (Fabre-Domergue). In contrast, the Eastern Large Aral is inhabited solely by *Artemia parthenogenetica* (Aladin and Plotnikov 2008). By the late 1990s, as the salinity of the Large Aral Sea increased to 60–70‰, the introduced flounder had also disappeared, leaving no remaining fish fauna.

Generally, the decline of the Aral Sea was mainly triggered by the unwise redirection of water for agricultural use, which failed to account for long-term environmental consequences. This led to a multifaceted crisis that catastrophically disrupted the area's entire water systems, wildlife, and even weather patterns while also profoundly impacting the health and economic stability of local communities.

References

Aladin NV, Kotov SV (1989) Natural condition of the Aral Sea ecosystem and the effect of manmade pollution. Proc Zool Inst Acad Sci USSR 199:110–114. (in Russian)

Aladin NV, Plotnikov IS (1995) Changes in the Aral Sea level: palaeolimnological and archaeological evidences. Proc Zool Inst RAS 262:17–46. (in Russian)

Aladin NV, Plotnikov IS (2008) Modern fauna of residual water bodies formed on the place of the former Aral Sea. Proc Zool Inst RAS 312(1/2):145–154. (in Russian)

Aladin NV, Plotnikov IS (2011) An overview of hybrid marine and lacustrine seas and saline lakes of the world. Lakes Reserv Res Manag 16:97–108

Aladin NV, Filippov AA, Plotnikov IS, Egorov AN, Piriulin DD, Smurov AO (2000) Modern ecological state of the small Aral Sea. In: Ecological research and monitoring of the Aral Sea deltas. A basis for restoration. Book 2. UNESCO Aral sea project. 1997–1999 final scientific report, pp 73–81

Aladin N, Plotnikov I, Letolle R (2004) Hydrobiology of the Aral Sea. In: Dying and dead seas: climatic versus anthropic causes, NATO science series. IV: Earth and environmental sciences, vol 36, pp 125–158. https://doi.org/10.1007/978-94-007-0967-6_6

Andreev NI (1978) Results of Calanipeda aquea-dulcis Kritschagin acclimatization in the Aral Sea. Biol. osnovy rybn. khoz-va vodoemov Credney Azii I Kazakhstana, Ilim, Frunze:6–9. (in Russian)

Andreev NI (1980) On the issue of planktonic invertebrates' acclimatization in the Aral Sea. In: Itogi I perspektivy akklimatizatsii ryb I bespozvonochnykh v vodoemakh SSSR. Tezisy dokladov (Mahachkala, 23–25 sent. 1980), pp 130–132 (in Russian)

Andreev NI (1989) Zooplankton of the Aral Sea in the initial period of its salinization. Proc Zool Inst Acad Sci USSR 199:26–52. (in Russian)

Andreev NI, Andreeva SI (1988) Crab Rhithropanopeus harrisii tridentatus (Decapoda, Xanthidae) in the Aral Sea. Zool Zhurnal 67(1):131–132. (in Russian)

Andrulionis NY, Izhitskiy AS, Savvichev AS, Zavialov PO (2025) Recent changes in physical and biogeochemical state in residual basins of the Aral Sea. Aquat Geochem 31:1. https://doi.org/10.1007/s10498-024-09436-5

Averintsev SV (1936) About extension of food reserve for commercial fishes of the Aral Sea (as a basis for discussion). Rybnoe khoz'yastvo SSSR 8:54–55. (in Russian)

Behning AL (1936) Invasion of new animal species to the Caspian Sea. Природа 25(4):107–108. (in Russian)

Bekmurzaev B (1974) Some data on the biology of mysids acclimatized in the south of the Aral Sea. In: Biologicheskie osnovy rybnogo khozyaistva vodoemov Sredney Azii I Kazakhstana. Tezisy dokladov, Ashkhabad, pp 41–42. (in Russian)

Belyanina SI, Konstantinov AS (1974) Order Diptera. In: Mordukhai-Boltovskoi FD (ed) Atlas of the Aral Sea invertebrates. Pischevaya Promyshlennost, Moscow, pp 228–236. (in Russian)

Bervald EA (1964) Ways of organizing rational fish economy on inland water bodies. Rostov-on-Don: Izd. Rostovskogo Gos. Universiteta., 148 pp (in Russian)

Boomer I, Aladin NV, Plotnikov IS, Whatley R (2000) The Palaeolimnology of the Aral Sea: a review. Quat Sci Rev 19:1259–1278

Borutzkiy EV (1974) Subcalass Copepoda. In: Mordukhai-Boltovskoi FD (ed) Atlas of the Aral Sea invertebrates. Pischevaya Promyshlennost, Moscow, pp 134–178. (in Russian)
Bykov NE (1964) About incidental introduction of gobies and atherine in the Aral Sea and their interrelation with acclimatized and aboriginal fishes. Bulletin MOIP, Otd Biol 69(1):51–58. (in Russian)
Dengina RS (1959) Benthos of Karabaili archipelago of the Aral Sea. Trudy Laboratorii Ozerovedeniya 8:234–255. (in Russian)
Doroshev SI (1968) Abundance of fishes acclimatized in the Aral Sea and modification of commercial catches. In: Acclimatization of fishes and invertebrates in water bodies of the USSR. VNIRO, Moscow, pp 105–112. (in Russian)
Gavrilov GB (1970) Some acclimatized invertebrates in the food of Aral Sea commercial fishes. Proc VNIRO 76(3):207–211. (in Russian)
Husainova, NZ (1958) Biological properties of some mass bottom food invertebrates of the Aral Sea. Publ. Kazachstan State Univ., Alma-Ata: 1–116. (in Russian)
Husainova NZ (1971) The newest changes of biological processes in the Aral Sea. Biol Nauk 1: 176–190. (in Russian)
Ilyin BS (1954) Acclimatization of fishes in the connection with construction of hydraulic structures. In: Trudy soveschaniya po probleme akklimatizatsiyi ryb I kormovykh bespozvonochykh, Nauka, Moscow, pp 27–31. (in Russian)
Karpevich AF (1947) Prerequisites for acclimatization of new forms in the Aral Sea. Doklady VNIRO 6:13–17. (in Russian)
Karpevich AF (1948) Results and perspectives of acclimatization of fishes and invertebrates in the USSR. Zool Zhurnal 27(6):469–480. (in Russian)
Karpevich AF (1953) State of forage reserve in the southern seas after flow regulation of rivers. Trudy soveschaniya ihtiologicheskoy komissii AN SSSR 1:124–150. (in Russian)
Karpevich AF (1960) Basing of aquatic organisms acclimatization in the Aral Sea. Trudy VNIRO 43(1):76–115. (in Russian)
Karpevich AF (1964) Features of reproduction and growth of bivalve mollusks from brackish-water seas of the USSR. In: Ekologiya baspozvonochykh yuzhykh morey SSSR, Moscow, pp 3–61. (in Russian)
Karpevich AF (1975) Theory and practice of aquatic organisms acclimatization. Pischevaya pomyshlennost, Moscow. (in Russian)
Karpevich AF (1986) Tendencies of changes in regime, biota composition and bioproductivity of the Aral and Caspian seas. In: Antropogennoe vosdeistvie na pribrezhno-morskie ekosistemy, Nauka, Moscow, pp 12–33. (in Russian)
Karpevich AF, Bokova EN (1970) Effect of climate and biotechnology on acclimatization of mysids of Caspian complex. Proc VNIRO 76(3):163–178. (in Russian)
Kazakhbaev (1974) Calanipeda in the southern part of the Aral Sea. Gidrobiol Zh 10(1):89–91
Konovalov PM, Markova EA, Binting EA (1958) Acclimatization of Baltic herring in the Aral Sea. Rybnoe khoz'yastvo 3:12–16. (in Russian)
Kortunova TA (1970) Some data on food organisms introduced to the Aral Sea. Proc VNIRO 76(3): 178–184. (in Russian)
Kortunova TA (1975) On the changes in the Aral Sea zooplankton in 1959–1968. Zool Zhurnal 54(5):567–669. (in Russian)
Kortunova TA, Lukonina NK (1970) Quantitative characteristic of the Aral Sea zooplankton. In: Rybnye resursy vodoemov Kazakhstana i ikh ispolzovanie, vol 6. Nauka, Alma-Ata, pp 52–60. (in Russian)
Kostianoy A, Kosarev N (eds) (2010) The Aral Sea Environment. Springer-Verlag, Berlin, Heidelberg, New York, p 332. https://doi.org/10.1007/978-3-540-88277-0. eBook ISBN 978-3-540-88277-0
Kutikova LA (1970) Rotifers of the USSR fauna. Nauka, Leningrad. (in Russian)
Lim RM, Markova EA (1981) Results of introducing sturgeons and plaice to the Aral Sea. Rybnoe Khoz-vo 9:26–26. (in Russian)

Lukonina NK (1960) Zooplankton of the Aral Sea. Proc VNIRO 43(1):177–197. (in Russian)
Malinovskaya AS (1961) On the biology of srimps acclimatized in the Aral Sea. In: Sbornik rabot po ihtiologii i gidrobiologii, Issue 3, Nauka, Alma-Ata, pp 113–124. (in Russian)
Markova EL (1972) Results of acclimatizations into the Aral Sea during period from 1963 till 1971. In: Akklimatizatsiya ryb I besposvonochnykh v vodoemah SSSR: Tezisy dokladov, Sept. 1972. Ilim, Frunze, pp 56–58. (in Russian)
Mirabdullayev I, Abdullaeva L, Musaev A, Zholdasova I, Mustafaeva Z, Jumaniezova N (2007) Sharp fluctuations in ecosystem parameters of the East Big Aral. Geophys Res Abstr 9:772
Nikolsky GV (1940) Fishes of the Aral Sea. MOIP, Moscow, pp 83–111. (in Russian)
Plotnikov IS, Aladin NV, Filippov AA (1991) The past and present of the Aral Sea fauna. Zool Zh 70(4):5–15. (in Russian)
Plotnikov I, Aladin N, Zhakova L, Mossin J, Høeg J (2023) Past, present and future of the Aral Sea -a review of its Fauna and Flora before and during the regression crisis. Zool Stud 62:e19. https://doi.org/10.6620/ZS.2023.62-19
Yablonskaya EA (1960) Food base of the Aral Sea and its use. Proc VNIRO 43(1):150–176. (in Russian)
Yablonskaya EA (1974) History of studies and composition of invertebrate fauna of the Aral Sea. In: Mordukhai-Boltovskoi FD (ed) Atlas of the Aral Sea invertebrates. Pischevaya Promyshlennost, Moscow, pp 3–8. (in Russian)
Yablonskaya EA, Lukonina NK (1962) On the problem of the Aral Sea productivity. Okeanologiya 2(2):298–304. (in Russian)
Yablonskaya EA, Kortunova TA, Gavrilov GB (1973) Long-term changes in the Aral Sea benthos. Proc VNIRO 80(3):147–158. (in Russian)
Zavyalov PO, Arashkevich AG, Grabovsky AB, Dikarev SN, Dzhalilov G, Evdokimov YV, Kudyshkin TV, Kurbaniyazov AK, Matchanov AT, Ni AA, Sapozhnikov FV, Tomashevskaya IG (2006) Quasisynoptic field studies in the western and eastern basins of the Aral Sea (October 2005). Okeanologiya 46(5):750–754. (in Russian)
Zenkevich LA (1947) Fuana and biological productivity of sea. In: Seas of the USSR, vol 2. Academy of Sciences of the USSR, Moscow-Leningrad. (in Russian)
Zenkevich LA (1963) Biology of the seas of the USSR. Academy of Sciences of the USSR, Moscow. (in Russian)
Zenkevich LA, Birshtein YA (1934) About possible measures to increase productivity of the Caspian and Aral. Rybnoe Khoz'yaistvo of the USSR 3:38–40. (in Russian)
Zenkevich LA, Birshtein YA (1937) Against proposal to acclimatize Chinese crab. Rybnoe Khoz'yaistvo of the USSR 6:33–34. (in Russian)

Chapter 8
Impact of Abiotic Environmental Factors and Climate Change on Non-native Species

Abstract The influence of abiotic environmental conditions and global climate change on non-native species is examined. The significance of multidecadal, interannual, and seasonal variability in abiotic factors for invasive ctenophore population dynamics is analyzed. This is essential for comprehending the role of the abiotic component in ecosystem health and for evaluating the anticipated effects of climate change. The ranges of key environmental parameters—SST, SSS, and Chl concentration—that are conducive to the occurrence and reproduction of *M. leidyi* and *B. ovata* have been identified. Regions where all three parameters are currently favorable at the same time have been mapped on a global scale and specifically for the Ponto-Caspian Seas using climatological data. Various global warming scenarios were employed to predict future changes in ctenophore habitats. A key finding from these projections is the significant expansion of *M. leidyi* reproduction zones: northward into the southern Norwegian Sea and westward into the open North Atlantic. Extensive areas along the Arctic coasts of Eurasia may become suitable for *M. leidyi* establishment and reproduction. Detailed forecasts of habitat changes for *M. leidyi* and *B. ovata* in the Black, Azov, and Caspian Seas are also provided.

Abiotic environmental conditions, such as air and water temperature, wind, and atmospheric humidity, have a powerful influence on the ecosystem. This section tackles the essential issue of examining the variability of fundamental abiotic environmental factors that could influence the invasions of harmful ctenophores in the Ponto-Caspian Seas. Furthermore, projections regarding anticipated shifts in the habitats of ctenophores, linked to global warming, are also examined.

T. A. Shiganova et al., *Non-Native Species in the Ponto-Caspian Seas*,
Monographiae Biologicae 96, https://doi.org/10.1007/978-3-032-18640-9_8

8.1 Favorable Environmental Conditions for Ctenophore Occurrence and Reproduction

The ranges of essential abiotic environmental variables—sea surface temperature (SST) and sea surface salinity (SSS)—necessary for the occurrence and reproduction of *M. leidyi* and *B. ovata* have been established through a synthesis of published literature and our own field data. Regions classified as having "occurrence conditions" are considered habitats capable of supporting self-sustaining populations that can reproduce during specific seasons, as well as areas where these species can disperse and persist even when reproduction is not possible during unfavorable periods. Meanwhile, areas identified with "reproduction conditions" represent habitats that are conducive to both reproduction and the development of self-sustaining populations. However, mentioned abiotic factors along are not sufficient for ctenophore occurrence and reproduction. Vital limiting biotic factor is food availability. Juvenile and adult ctenophores consume a diverse diet that includes mesozooplankton, meroplankton, small pelagic fish, and fish eggs (Waggett and Sullivan 2006; Stanlaw et al. 1981). Surface chlorophyll concentration (Chl) may be used as a comprehensive estimate of ecosystem productivity. Ranges of suitable conditions for *M. leidyi* and *B. ovata* occurrence and reproduction, compiled from published sources (Shiganova 2009; Costello et al. 2012; Javidpour et al. 2009; Riisgård and Goldstein 2014; Jaspers et al. 2013; Kremer 1994; Sarpe et al. 2007; Shiganova et al. 2014b, 2019a, b, 2024; Purcell et al. 2001; Rapoza et al. 2005; Sullivan and Gifford 2004, 2007; Kremer and Reeve 1989) and our own field and laboratory observations, are summarized in Table 8.1.

The parameters outlined in Table 8.1 were applied to determine the global climatology of essential environmental variables (SST, SSS, and Chl concentration) that support *M. leidyi* reproduction (Shiganova et al. 2019a, b). Worldwide climatological SST and SSS datasets were sourced from the World Ocean Atlas 2013 (WOA2013; Mishonov and National Center for Atmospheric Research Staff 2016). Global surface chlorophyll concentration climatology was derived from SeaWiFS ocean color scanner observations (NASA 2018). The findings are illustrated in Fig. 8.1.

Table 8.1 Occurrence and reproduction requirements for *M. leidyi* and *B. ovata*

Species	Parameter		Occurrence	Reproduction
M. leidyi	T, °C	Northern seas	1–24	10–24
		Southern seas	3–30	12–27
	S, ‰	Northern seas	4.5–35	N/A
		Southern seas	3.5–45	6–40
	Chl, $mg.m^{-3}$		>0.1	>0.5
B. ovata	T, °C		N/A	10–27
	S, ‰		N/A	10–40
	Chl, $mg.m^{-3}$		>0	>0

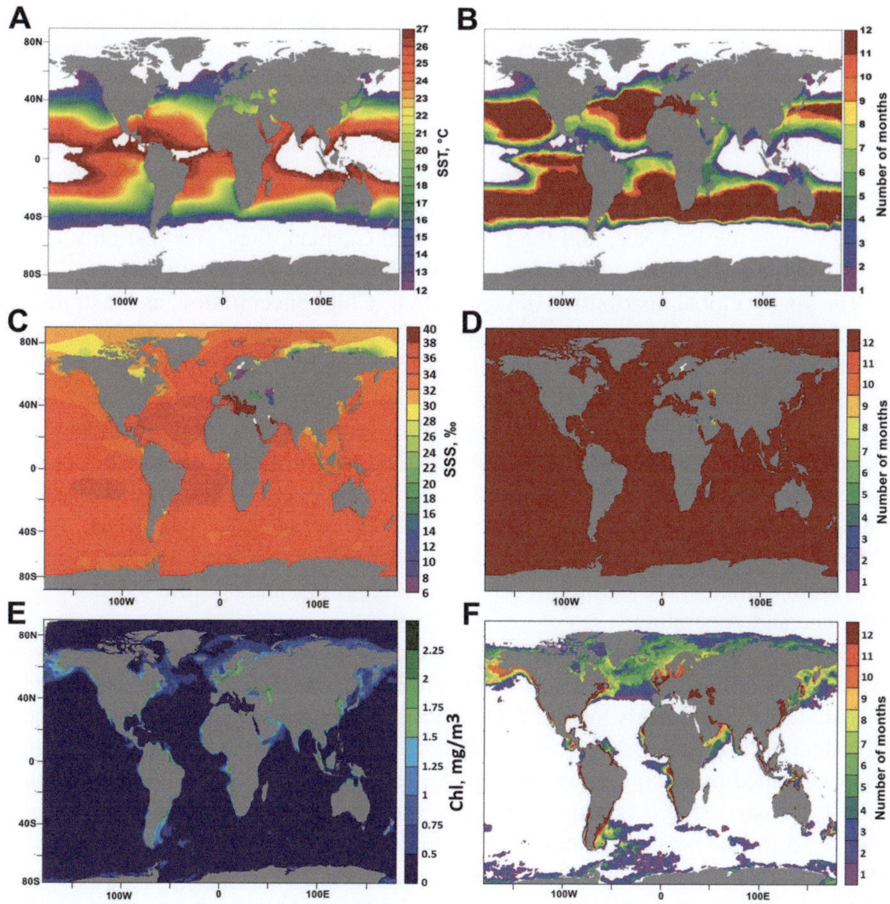

Fig. 8.1 Annual climatology of SST (**a**) and duration of conditions with 12 °C < SST < 27 °C (**b**), SSS (**c**) and duration of conditions with 6‰ < SSS < 40‰ (**d**) and Chl concentration (**e**) and duration (in months) of conditions with Chl concentration exceeding 0.5 mg.m^{-3} (**f**). Blank areas indicate conditions, unsuitable for *M. leidyi* reproduction

Regions with SST conducive to *M. leidyi* reproduction are found in the temperate and subtropical zones of the World Ocean (Fig. 8.1a). The blank zones represent equatorial, subpolar, and polar waters where temperatures are either too high or too low for *M. leidyi* reproduction. Nevertheless, the length of the favorable SST period differs across these regions. The annual duration of SST conditions suitable for *M. leidyi* reproduction is illustrated in Fig. 8.1b. In the Mediterranean Sea, these favorable conditions persist for 8–12 months, in the Black and Azov Seas for 6–7 months, in the North and Baltic Seas for 3–4 months, and for just 1 month in the Bothnian Bay. The blank areas in Fig. 8.1b indicate locations where suitable conditions for *M. leidyi* reproduction are absent year-round.

When it comes to salinity levels, the conditions are generally suitable for *M. leidyi* reproduction across most of the World Ocean, with a few limited but significant exceptions (Fig. 8.1c). In the northern Persian Gulf and the Red Sea, salinity surpasses the upper threshold, whereas in the northern Baltic Sea and the northeastern Caspian, it falls below the favorable range. The annual duration of suitable salinity conditions (Fig. 8.1d) reinforces these exceptions: favorable conditions are entirely absent in the northern Baltic Sea (shown as a blank area), and in the northern Persian Gulf, the Red Sea, and the northeastern Caspian, they may last only a few months—too short a period to support the establishment of a population.

The yearly global distribution of chlorophyll (Chl) concentration, as illustrated in Fig. 8.1e, confirms that coastal waters are the most biologically productive regions and are particularly conducive to *M. leidyi* reproduction. The most elevated Chl levels are found in the Black, Azov, and Caspian seas, the Persian Gulf, the Yellow Sea, and the North and Baltic seas, along with the eastern coasts of the Americas, the western coast of Africa, and the Bering Sea. The period during which Chl concentration remains above 0.5 $mg.m^{-3}$ (Fig. 8.1f) identifies zones with consistently high Chl levels, adequate for *M. leidyi* reproduction. These encompass the Black, Azov, Caspian, North, and Yellow seas, the Atlantic Ocean coastal areas bordering North and South America, the northern Pacific coastal regions, and specific locations near Malaysia and Australia.

Therefore, focusing on a single environmental factor reveals extensive regions that could support *M. leidyi* reproduction. However, integrating all three parameters is essential to accurately identify the specific locations favorable for both the reproduction and growth of *M. leidyi*. The chlorophyll concentration maps, which are constrained by the other two environmental variables—SST and SSS—simultaneously, illustrate the conditions for *M. leidyi* presence (Fig. 8.2a) and reproduction (Fig. 8.2b) in the western Eurasian seas, an area where established populations of *M. leidyi* already exist. Figure 8.2a suggests that *M. leidyi* could potentially be found almost anywhere. Nonetheless, in the deep green zones, which cover nearly the entire Mediterranean Sea, the Red Sea, and the offshore areas of the North Atlantic, *M. leidyi* has very low reproductive potential because of the minimal chlorophyll

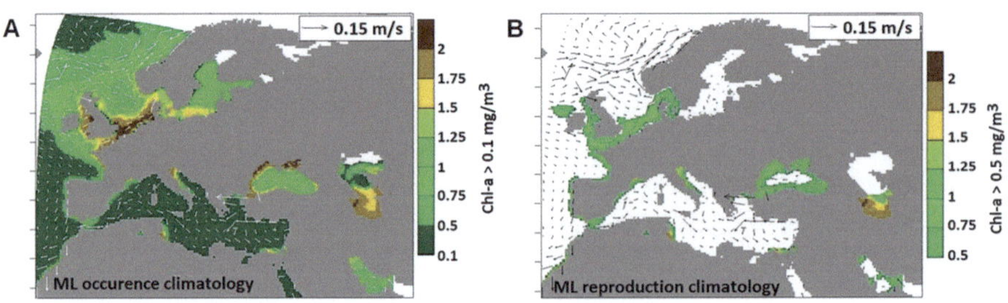

Fig. 8.2 Annual chlorophyll concentration patterns in the Eurasian seas, constrained by sea surface temperature and sea surface salinity, under environmental conditions conducive to *M. leidyi* presence (**a**) and reproduction (**b**). Blank areas indicate conditions, unsuitable for *M. leidyi* occurrence or reproduction

concentrations (0.1–0.5 mg/m^3). The most productive regions, where chlorophyll concentration surpasses 1 mg/m^3, are the coastal waters of the Black Sea, the Seas of Azov and Marmara, the northern Adriatic Sea, the northern coasts of the Aegean Sea, the Israeli and Egyptian coasts of the Levantine Sea, the Gulf of Lions, and the Catalan and Libyan coasts. In the Northern seas, the most productive areas are situated in the North Sea and along the southern coast of the Baltic Sea. The majority of these zones align with the documented locations of *M. leidyi* occurrence from field data.

The regions where *M. leidyi* successfully reproduces, which align with the areas where it has become established, are illustrated in Fig. 8.2b. Among the zones with elevated Chl concentrations (exceeding 1 mg/m^3), which correspond with the source areas confirmed by field observations, are the coastal waters of the Black Sea, the southern Caspian Sea, the bays in the Sea of Marmara, coastal waters and bays of the Aegean Sea, Israeli coasts, the extensive northern Adriatic region, the coastal waters of the Gulf of Lions, the southern Catalan coasts, and the coastal waters of the North Sea.

Figure 8.3 illustrates the seasonal variability of favorable conditions for *M. leidyi* reproduction in the western Eurasian seas. This figure is constructed similarly to Fig. 8.2 but focuses on seasonal climatology and reproduction exclusively. In the Black Sea and the Sea of Azov, suitable conditions emerge during summer and

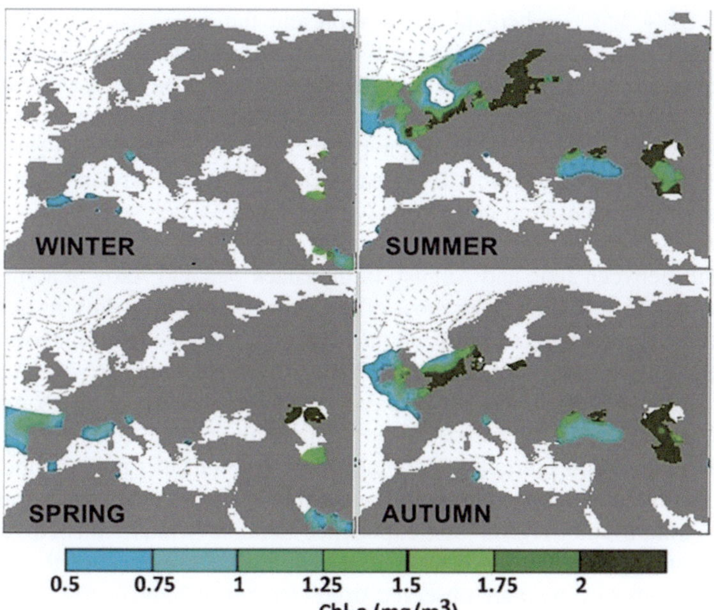

Fig. 8.3 Seasonal climatology of Chl concentration limited by SST and SSS in the seas of Eurasia favorable for *M. leidyi* reproduction. Blank areas indicate an absence of suitable conditions for *M. leidyi* reproduction

autumn. Within the productive zones of the Mediterranean Sea, favorable periods vary by region: late winter to spring in the northern Aegean Sea; year-round along the northwestern Adriatic Sea coasts; spring in the northwestern Mediterranean, such as the Gulf of Lions; winter to spring in the Alboran Sea near Gibraltar; and winter along the coasts of Libya and Algeria. In the North Sea, conditions conducive to *M. leidyi* reproduction occur in summer and autumn. Seasonal maps for the northern Baltic Sea indicate chlorophyll levels adequate for reproduction in summer, yet field observations do not detect *M. leidyi*, likely due to low salinity. Along the Atlantic coasts, favorable reproduction conditions are noted from spring to autumn off the coast of France and in spring along the coasts of Spain and Portugal.

The majority of the climatologically projected reproduction zones align with field data. Nonetheless, *M. leidyi* has not been observed in several regions despite suitable environmental conditions, including the coasts of Libya and Algeria, the Alboran Sea, the Atlantic coasts of Spain and Portugal, the Gibraltar Strait area, and the northern Baltic Sea.

Physical mechanisms that may facilitate the dispersal of *M. leidyi* encompass turbulent mixing, both vertical and horizontal, as well as transport by currents.

Vertical turbulent mixing has the potential to influence the population dynamics of *M. leidyi* (Shiganova 2009; Shiganova et al. 2014a, 2018). This process is especially significant during the formation of the upper mixed layer and the seasonal thermocline, as the food concentration available to *M. leidyi* is linked to the thickness of the mixed layer. Generally, this thickness is proportional to wind speed, expressed as $W = (U^2 + V^2)^{0.5}$, where U and V represent the zonal and meridional wind components. Consequently, areas with lower wind speeds, which result in a shallower mixed layer, are typically more advantageous for *M. leidyi* feeding and survival. Seasonal wind speed variations in the western Eurasian seas were analyzed by Shiganova et al. (2019c). In the northern habitats of *M. leidyi*—such as the English Channel, the northwest European coast, the Scandinavian coast, and the Baltic Sea—favorable wind conditions (with a mean W below 2 m/s) occur from February to September. An exception is the Skagerrak area, where wind speeds can reach 4 m/s between June and August. In the Black Sea and Azov Sea, mean wind speeds remain low (<2 m/s) for most of the year, except from July to September, when speeds in open waters rise to 3–4 m/s. Nevertheless, even during these months, coastal areas typically experience wind speeds not exceeding 2 m/s. Across the Caspian Sea, wind speeds do not surpass 2 m/s from January through April. Key *M. leidyi* habitats in the Mediterranean Sea—including coastal zones of the Adriatic Sea, Tyrrhenian Sea, Ligurian Sea, the northwestern Gulf of Lion, and the Catalan Coast—also maintain favorable wind conditions (<2 m/s) nearly year-round. The Sea of Marmara and Aegean Sea experience suitable winds from November to December and again from April to June. Along the eastern coast of the Levantine Sea, weak winds (<2 m/s) are observed from October to May.

The movement of *M. leidyi* is also influenced by advective transport through local marine currents. Table 8.2 outlines the typical current characteristics in the primary habitats of *M. leidyi*, offering a fundamental understanding of the potential spatial and temporal dimensions of its advective dispersal (Dobrovolskii and Zalogin 1992;

Table 8.2 Common features of water movement in the western Eurasian marine regions

1	2	3	4
Area	Specific conditions	Current speed, (m/s)/suspended sediment transport rate (km/day)	Duration required to cover the distance specified in column (2), days
Southern seas			
Black Sea	Cyclonic transport along the coast with the Rim Current	0.1–0.25/9–22	N/A
The Azov Sea	Unstable circulation patterns and vigorous wind mixing	0.1–0.2/9–17	N/A
The Caspian Sea	Cyclonic circulation; northward movement along the eastern shoreline from the southern to the northern region over a span of 700 km	0.1–0.3/9–26	80–28
The Mediterranean Sea	Generally westward transport along northern coast	0.01–0.07/1–6	N/A
Northern seas			
The Baltic Sea	Cyclonic circulation patterns drive northward and southward transport along the eastern and western coastlines	0.03–0.04/2.5–3.5	N/A
The North Sea, the European coastline	Coastal maritime transport spanning 1000 km from the English Channel to the Skagerrak	0.03–0.07/2.5–6	400–167
The North Sea, Scandinavian coastline	Transportation covering 600 km from the southernmost point of Scandinavia in northerly direction	0.01–0.05/1–5	694–138

Jaspers et al. 2018; Shiganova et al. 2019c). By comparing the time needed to travel a given distance (detailed in columns 4 and 2) with *M. leidyi*'s approximately 7-month (~200-day) growth period, it is possible to assess the likelihood of its expansion via this method. In certain regions, such as the Caspian and North seas, the maximum current speeds allow the organism to traverse the distance between suitable habitats in less time than its reproductive cycle. Conversely, in the North Sea, the migration duration at minimal current velocities considerably exceeds the reproduction period, which could restrict *M. leidyi*'s capacity to colonize new locations.

An evaluation provided earlier suggests that localized currents are capable of transporting *M. leidyi* only over relatively short distances in size-restricted enclosed seas. In open marine environments, such as the European and Scandinavian coastlines of the North Sea, the potential travel range for *M. leidyi* could be significantly greater. Nevertheless, the duration required to traverse such distances surpasses the *M. leidyi* life cycle, thereby reducing the likelihood of the species establishing populations far from their native habitats. Consequently, global maritime traffic, with its growing frequency and reach, continues to be the primary mechanism for species invasions.

8.2 Seasonal, Interannual, and Multidecadal Variability of Environmental Conditions and Ctenophore Population Dynamics

The population cycles of the ctenophores *M. leidyi* and *B. ovata* are primarily regulated by the seasonal cycle of water temperature within the upper mixed layer (Shiganova et al. 2014a, b, c, 2018, 2024). Given that this upper layer is nearly uniform because of strong mixing, SST serves as an adequate indicator for the temperature throughout the entire layer. The seasonal SST patterns observed in the Ponto-Caspian basins—the Black, Azov, and Caspian seas—are alike, generally featuring a period of spring warming, a summer temperature peak, and autumn cooling. Shiganova et al. (2024) have supplied field data detailing the seasonal population fluctuations of *M. leidyi* and *B. ovata* in these seas. Quantitative assessments were carried out by graphing the monthly average ctenophore abundance over the full observation periods (1999–2020 for the Black Sea and 2000–2015 for the Caspian Sea) against the corresponding seasonal SST cycle (Fig. 8.4).

A statistical evaluation of the time series data, presented in Fig. 8.4, indicates the following key findings (Table 8.3): (1) a presence of strong, statistically meaningful relationships between ctenophore abundance and SST variability at the seasonal timescale; (2) an observed increase in the 1-month time-lag correlation for *B. ovata* in the Black Sea, suggesting its peak abundance follows the SST and *M. leidyi* abundance peaks by 1 month; and (3) a rise in the 1-month time-lag correlation for *M. leidyi* in the Caspian Sea, showing that its highest abundance happens 1 month after the SST maximums.

Year-to-year population fluctuations of *M. leidyi* and *B. ovata* generally align with interannual SST variability. In the Black Sea between 1999 and 2020, peak abundances of both ctenophore species occurred during SST increases in 2001–2003 and 2004–2008 (Fig. 8.5a). These temperature rises fell within ranges that support ctenophore development and likely enhanced population growth. However, a sharp SST increase to 27.5 °C in 2010—the critical upper survival threshold for these species—led to a rapid decline in both ctenophore populations (Fig. 8.5a). Statistical evaluation of the time series in Fig. 8.5a reveals that from 1999 to 2009, *M. leidyi* and *B. ovata* abundances were strongly correlated with SST on an interannual scale. Correlation coefficients were 0.59 ($p < 0.05$) for *M. leidyi* and 0.72 ($p < 0.02$) for *B. ovata*. Nevertheless, the extreme temperature spike in 2010 likely caused such severe population damage that recovery was not possible even when temperatures later returned to favorable levels, and no significant correlation was detected after 2010.

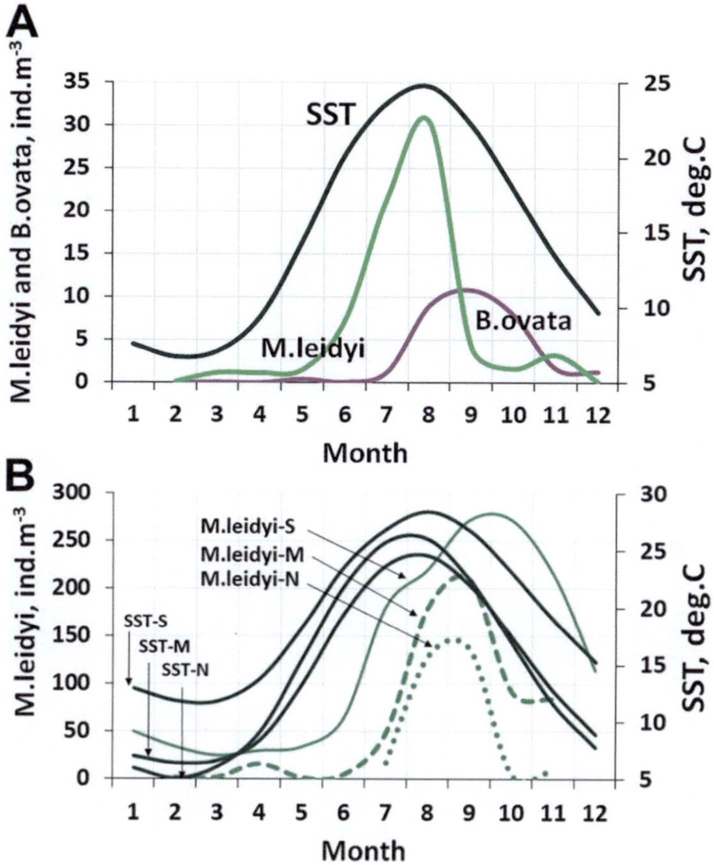

Fig. 8.4 (**a**) seasonal fluctuations in *M. leidyi* (green line) and *B. ovata* (purple line) abundance, along with the seasonal SST cycle (black line) in the Black Sea. (**b**) seasonal variations in *M. leidyi* (green lines) abundance and the seasonal SST pattern (black lines) in the Caspian Sea. N, M, and S denote Northwestern, Middle, and Southern Caspian

Table 8.3 Correlation coefficients (R) and their associated probability values (p) for the relationship between SST and ctenophore abundance at seasonal timescale

	R/p		
Species	Black Sea	Caspian Sea	
M. leidyi	0.75/$p < 0.01$	Northwestern	0.60/$p < 0.05$ 0.63/$p < 0.05$ (1 month time lag)
		Middle	0.66/$p < 0.02$ 0.88/$p < 0.01$ (1 month time lag)
		Southern	0.76/$p < 0.01$ 0.97/$p < 0.01$ (1 month time lag)
B. ovata	0.63/$p < 0.05$ 0.84/$p < 0.01$ (1 month time lag)	N/A	

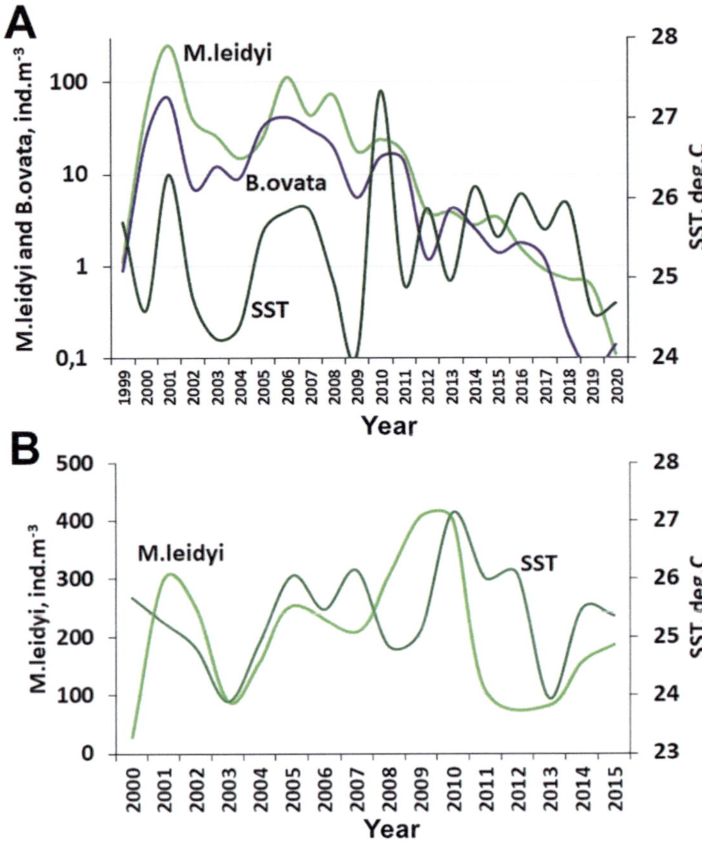

Fig. 8.5 (**a**) Interannual variability of SST (deep green line) and *M. leidyi* (green line) and *B. ovata* (purple line) abundance in the Black Sea; (**b**) Interannual variability of SST (deep green line) and *M. leidyi* (green line) abundance in the Middle Caspian

In the Caspian Sea, the year-to-year population dynamics of *M. leidyi* generally align with SST variations, as increases or decreases in abundance correspond with rises or falls in SST (Fig. 8.5b). This pattern, however, is observed only when temperatures remain within favorable bounds. Once these limits are surpassed, a rapid population decline takes place.

Although ctenophore invasions in the Ponto-Caspian region only started in the 1980s, it is valuable to illustrate the timing of their introductions against the backdrop of long-term, multidecadal patterns in abiotic parameter variability. Recent analyses of hydrometeorological parameters—including thermal, dynamic, and atmospheric moisture content—in the Black and Caspian seas have been conducted for the period from 1948 to 2020 (Kazmin 2021; Kazmin and Shiganova 2022, 2024). These two basins, while geographically close, are significantly distinct and exhibit different patterns of multidecadal variability. In the Black Sea, two time intervals with average durations of 20 and 40 years, characterized by opposing

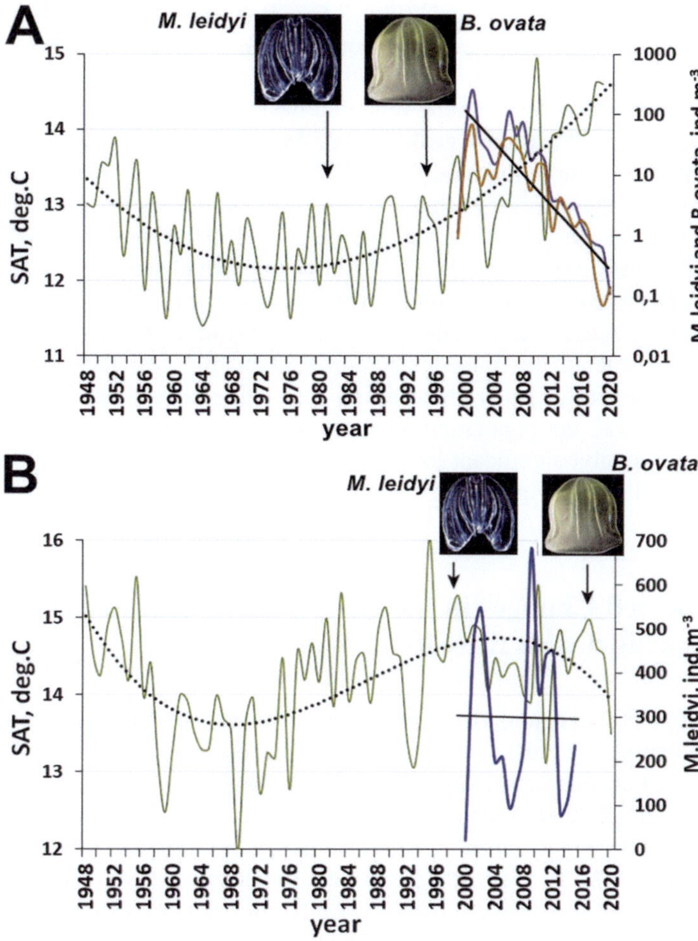

Fig. 8.6 (**a**) Long-term fluctuations in surface air temperature (SAT) and changes in the abundance (ind.m^{-3}) of *M. leidyi* (blue line) and *B. ovata* (red line) in the Black Sea; (**b**) multidecadal SAT variability and shifts in *M. leidyi* (blue line) abundance (ind.m^{-3}) in the Caspian Sea. Included are images and the timing of invasions (indicated by downward arrows) for *M. leidyi* and *B. ovata*. Dotted lines represent polynomial approximations of SAT. Solid black lines indicate linear trends in ctenophore abundance variability

trends in hydrometeorological parameters, have been identified. Conversely, the Caspian Sea shows three such intervals, each lasting between 10 and 25 years (Fig. 8.6).

The invasive ctenophore *M. leidyi* was first introduced to the Black Sea in the early 1980s, arriving via ballast water from the Northeastern coast of America (Vinogradov et al. 1989; Shiganova et al. 2004; Oguz et al. 2008; Ghabooli et al. 2011). Roughly a decade later, during the 1990s, its predator *B. ovata* also invaded the region (Shiganova and Abusova 2021). Both species became established during a period of multidecadal warming (Fig. 8.6a). Between 1999 and 2020, a clear downward trend in the populations of both *M. leidyi* and *B. ovata* was recorded (solid black line in Fig. 8.6a), coinciding with a sharp rise in temperature. This decline is likely due to deteriorating reproductive conditions for *M. leidyi* linked to the intense warming (Shiganova et al. 2014a). As a result, the population of *B. ovata* also decreased because of the reduced availability of its primary prey, *M. leidyi*.

Subsequently, these species expanded from their original point of invasion in the Black Sea into the Caspian Sea (Studenikina et al. 1991; Shiganova 1993; Ivanov et al. 2000; Shiganova et al. 2001). The first documented observation of *M. leidyi* in the Caspian occurred in 1999, while *B. ovata* was recorded in 2019 (Fig. 8.6b). The introduction and successful establishment of *M. leidyi* in the Caspian Sea happened during a phase of relatively stable temperatures, coinciding with the transition of the multidecadal climate phase from warming to cooling (Fig. 8.6b). Consequently, during this interval, the population levels of *M. leidyi* (averaged across the entire basin) did not exhibit significant long-term fluctuations, as indicated by the nearly horizontal trend line in Fig. 8.6b.

8.3 Expected Changes in the Ctenophore's Habitats, Associated with the Global Warming

Pinpointing potential regions where *M. leidyi* may establish and reproduce as a result of global warming is a crucial concern. As an initial estimate, the spread of *M. leidyi* in the western Eurasian seas was evaluated by increasing the WOA13 climatological SST by 2 °C and 4 °C (Shiganova et al. 2019c) (Fig. 8.7). A key finding from this projection was the significant enlargement of the species' reproductive zones (shown in purple in Fig. 8.7). To the north, this zone is projected to stretch into the southern Norwegian Sea and further west into the open North Atlantic, with the extent of this spread varying according to the level of SST increase (Fig. 8.7b, c). In the southern seas of western Eurasia, the entire Black Sea is anticipated to become suitable for *M. leidyi* reproduction. Conversely, in the Mediterranean Sea, the existing reproductive areas are not predicted to expand with rising SST. Concerning areas of occurrence, the southern Levantine Sea and the Libyan coast are forecast to become unsuitable for *M. leidyi* presence if SST increases by 4 °C (Fig. 8.7c, blank areas).

Furthermore, the anticipated shifts in favorable conditions for *M. leidyi* reproduction were projected on a global scale using SST forecasts from the CMIP5 (Coupled Model Intercomparison Project Phase 5; IPCC 2013) (Shiganova et al. 2019c). The SST increase predicted under the RCP8.5 scenario (Representative Concentration Pathways; a scenario reflecting relatively high greenhouse gas emissions) for the year 2100 was applied (Fig. 8.8). This scenario indicates an accelerated rise in SST across the Northern Hemisphere. Consequently, marine regions that were previously too cold for *M. leidyi* reproduction may become suitable. During summer and autumn, these favorable zones are projected to expand significantly and extend

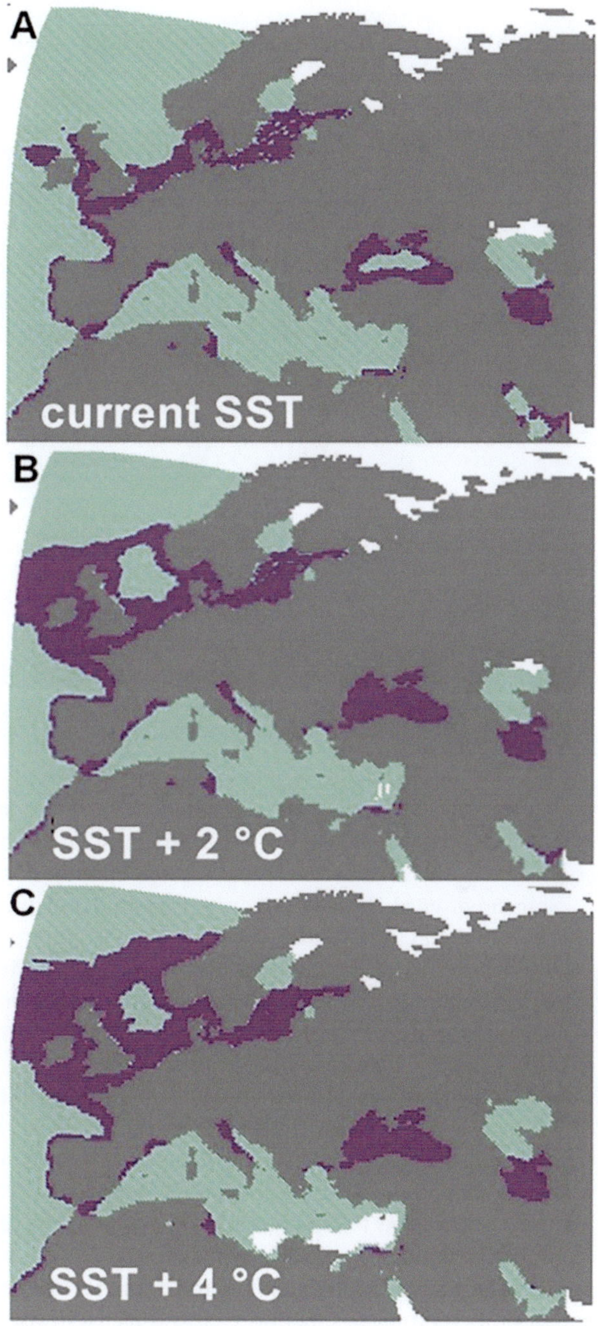

Fig. 8.7 Reproduction (purple) and occurrence (light green) areas of *M. leidyi* in the western Eurasia seas under the current temperature conditions (**a**) and after an increase of temperature by 2 °C (**b**) and 4 °C (**c**)

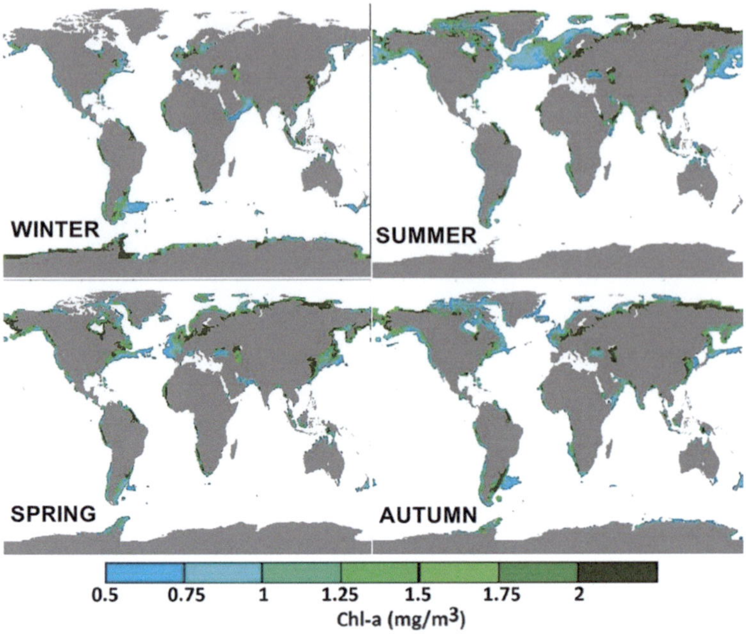

Fig. 8.8 Global seasonal patterns of chlorophyll concentration (mg/m^3), displayed for the preferred temperature (12 < SST < 27 °C) and salinity (6 < SSS < 40) ranges supporting *M. leidyi* reproduction, based on model projections (using the RCP8.5 scenario) for the year 2100

farther toward the North Pole compared to present conditions (Fig. 8.8). Notably, extensive coastal areas along the Eurasian Arctic are expected to become habitats conducive to the establishment and reproduction of *M. leidyi*.

Shiganova et al. (2024) conducted an in-depth investigation into the possible effects of global SST shifts on the habitats and reproductive environments of *M. leidyi* and *B. ovata* in the Black and Caspian seas. Multiple SST change scenarios were examined: (1) a consistent annual SST increase or decrease of 2 °C and (2) an isolated spring temperature rise or fall of 2 °C, as warmer spring conditions could encourage earlier ctenophore blooms.

Figure 8.9 illustrates the effects of a 2 °C SST warming on the reproductive zones of *M. leidyi* and *B. ovata* in the Black and Caspian seas. In the Black Sea, the reproductive range of *M. leidyi* is projected to extend further offshore under global warming (Fig. 8.9c) relative to current conditions (Fig. 8.9a). The reproductive period is anticipated to lengthen by 1–2 months. However, in the Middle Caspian, reproductive conditions for *M. leidyi* are expected to vanish, while the duration of favorable reproduction in the Southern Caspian is likely to increase. For *B. ovata*, the length of its reproductive period is also forecast to rise (Fig. 8.9d) compared to the present scenario (Fig. 8.9b). In the Caspian Sea, the region with the longest reproduction period is predicted to move from the Southern Caspian to the Middle Caspian. Consequently, global warming is likely to result in extended reproductive

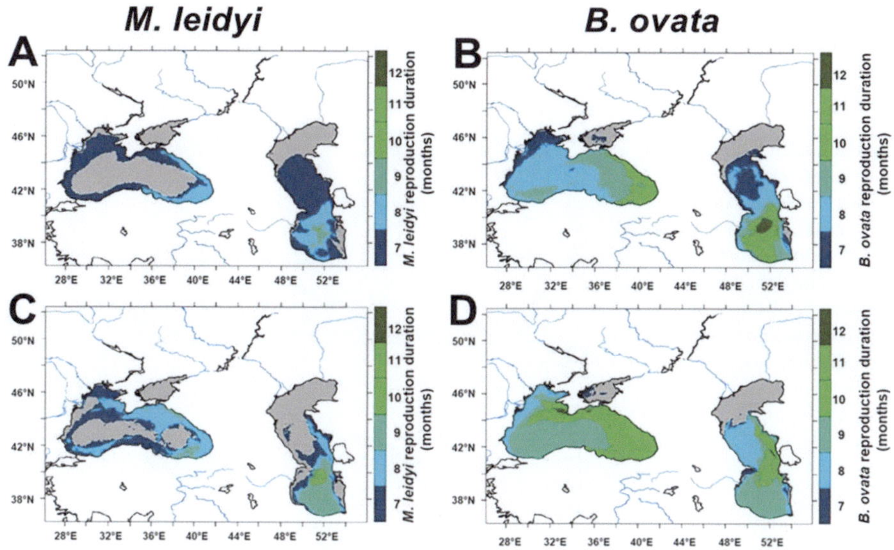

Fig. 8.9 Duration (in months) of environmental conditions suitable for *M. leidyi* (**a**) and *B. ovata* (**b**) reproduction under present-day sea surface temperatures and duration of favorable conditions for *M. leidyi* (**c**) and *B. ovata* (**d**) reproduction with a 2 °C rise in SST. Regions where data exists but the duration is below the threshold (<7 months) are shown in gray

durations for both ctenophores, and it is projected that the Middle Caspian could become a potential permanent reproduction area for B. ovata.

In recent times, ctenophores have frequently been observed in the Black and Caspian seas starting in April, suggesting their earlier developmental stages, likely as a result of warmer spring conditions. The responsiveness of ctenophores to spring sea surface temperature fluctuations has been examined with regard to their reproductive period (Shiganova et al. 2024).

Warmer spring conditions may lead to an earlier onset of *M. leidyi* reproduction in the Black Sea coastal regions and the transitional area between the Southern and Middle Caspian (Fig. 8.10a). In contrast, a later start to reproduction during colder springs is projected to take place solely in the Southern Caspian (Fig. 8.11a), with the Black Sea remaining unaffected.

In the Black Sea, *B. ovata* appears more responsive to spring sea surface temperature fluctuations, whether they rise or fall. Warmer spring conditions lead to a longer reproductive period in the western region (Fig. 8.10b), whereas colder springs result in a shorter duration in the eastern part (Fig. 8.11b). For the Caspian Sea, projections indicate that *B. ovata* will prolong its reproductive phase in the Middle Caspian during warmer springs (Fig. 8.10b) and shorten it in the area between the Southern and Middle Caspian under colder spring conditions (Fig. 8.11b).

Overall, the multidecadal, interannual, and seasonal variations in abiotic environmental conditions, along with the effects of climate change on invasive

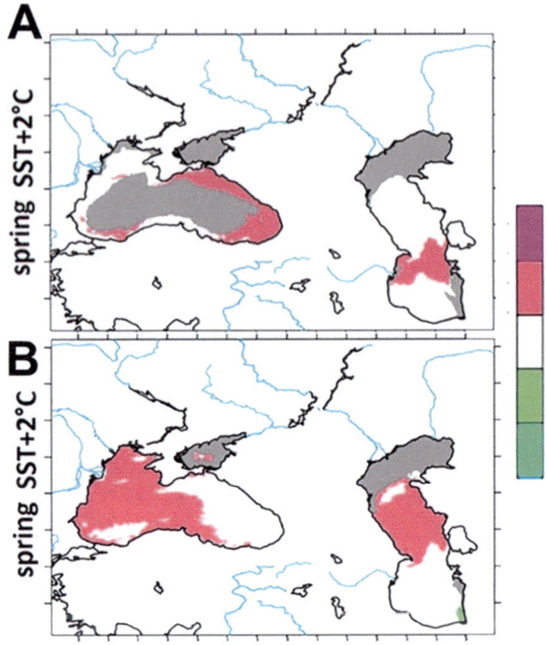

Fig. 8.10 Reproductive sensitivity to warmer spring conditions (SST + 2 °C) for *M. leidyi* (**a**) and *B. ovata* (**b**). The difference between the climate change scenario and current climate is shown. Positive values indicate an extension of the favorable reproductive period. Gray shading denotes the absence of suitable conditions for reproduction

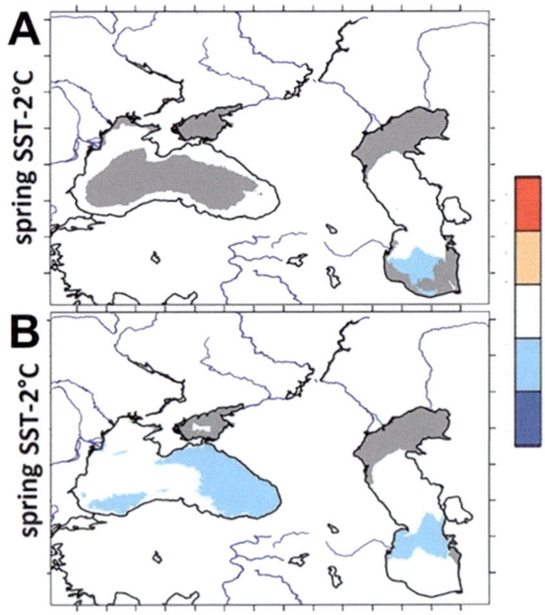

Fig. 8.11 Sensitivity of reproductive conditions to a cooler spring (SST-2 °C) for *M. leidyi* (**a**) and *B. ovata* (**b**). The difference between the climate change scenario and the current climate is shown. Negative values indicate a shorter favorable period for reproduction. Areas shaded in gray represent the lack of suitable conditions for reproduction

ctenophore populations discussed in this chapter, are essential for recognizing the importance of the abiotic component in the ecosystem's condition and for evaluating the anticipated outcomes of climate change.

References

Costello JH, Bayha KM, Mianzan HW, Shiganova TA, Purcell JE (2012) Transitions of *Mnemiopsis leidyi* (Ctenophora: Lobata) from a native to an exotic species: a review. Hydrobiologia 690(1):21–46. https://doi.org/10.1007/s10750-012-1037-9

Dobrovolskii AD, Zalogin BS (1992) Seas of the USSR. Moscow State University Publishing, Moscow. 192 p. (in Russian)

Ghabooli S, Shiganova TA, Zhan A, Cristescu M, Eghtesadi-Araghi P, MacIsaac H, Kamakin (2011) Multiple introductions and invasion pathways for the invasive ctenophore Mnemiopsis leidyi in Eurasia. Biol Invasions 13:679–690

IPCC (2013) In: Stocker TF, Qin D, Plattner G-K, Tignor M, Allen SK, Boschung J, Nauels A, Xia Y, Bex V, Midgley PM (eds) Climate change 2013: the physical science basis. Contribution of working group I to the fifth assessment report of the intergovernmental panel on climate change. Cambridge University Press, Cambridge, New York. 1535 pp

Ivanov VP, Kamakin AM, Ushivtsev VB, Shiganova TA, Zhukova OP, Aladin N, Wilson SI, Harbison GR, Dumont HJ (2000) Invasion of the Caspian Sea by the comb jellyfish Mnemiopsis leidyi (ctenophora). Biol Invasions 2000(2):255–258

Jaspers C, Haraldsson M, Lombard F, Bolte S, Kiørboe T (2013) Seasonal dynamics of early life stages of invasive and native ctenophores give clues to invasion and bloom potential in the Baltic Sea. J Plankton Res 35:582–594

Jaspers C, Huwer B, Antajan E, Hosia A, Hinrichsen HH, Biastoch A et al (2018) Ocean current connectivity propelling the secondary spread of a marine invasive comb jelly across western Eurasia. Glob Ecol Biogeogr 2018:1–14. https://doi.org/10.1111/geb.12742

Javidpour J, Molinero JC, Peschutter J, Sommer U, Shiganova T (2009) Seasonal changes and population dynamics of the ctenophore Mnemiopsis leidyi after its first year of invasion in the Kiel Fjord, Western Baltic Sea. Biol Invasions 11:873–882. https://doi.org/10.1007/s10530-008-9300-8

Kazmin AS (2021) multidecadal variability of the hydrometeorological parameters in the Caspian Sea. Estuar Coast Shelf Sci 250(2021):107150. https://doi.org/10.1016/j.ecss.2020.107150

Kazmin AS, Shiganova TA (2022) Comparative analysis of multidecadal variability of abiotic parameters in the Ponto-Caspian seas. J Oceanol Res 50(3):51–69. https://doi.org/10.29006/1564-2291.JOR-2022.50(3).3

Kazmin AS, Shiganova TA (2024) ctenophore invasions in the Ponto-Caspian seas: role of abiotic factors variability. Biol Invasions 26:1381–1397. https://doi.org/10.1007/s10530-024-03252-2

Kremer P (1994) Patterns of abundance for *Mnemiopsis* in US coastal waters - a comparative overview. ICES J Mar Sci 51:347–354. https://doi.org/10.1006/jmsc.1994.1036

Kremer P, Reeve R (1989) Growth dynamics of a ctenophore (*Mnemiopsis*) in relation to variable food supply. 11. Carbon budgets and growth model. LPlank Res 11:553–574

Mishonov A, National Center for Atmospheric Research Staff (eds). Last modified 13 Oct 2016. The Climate Data Guide: World Ocean Atlas 2013 (WOA13). Retrieved from https://climatedataguide.ucar.edu/climate-data/world-ocean-atlas-2013-woa13

NASA Goddard Space Flight Center, Ocean Ecology Laboratory, Ocean Biology Processing Group. Sea-viewing Wide Field-of-view Sensor (SeaWiFS) Chlorophyll Data (2018) Reprocessing. NASA OB. DAAC, Greenbelt. https://doi.org/10.5067/ORBVIEW-2/SEAWIFS/L3B/CHL/2018

Oguz T, Salihoglu B, Fach B (2008) A coupled plankton–anchovy population dynamics model assessing nonlinear controls of anchovy and gelatinous biomass in the Black Sea. Mar Ecol Prog Ser 369:229–256. https://doi.org/10.3354/meps07540

Purcell JE, Shiganova TA, Decker MB, Houde ED (2001) The ctenophore *Mnemiopsis* in native and exotic habitats: U.S. estuaries versus the Black Sea basin. Hydrobiologia 451:145–176

Rapoza R, Novak D, Costello JH (2005) Life-stage dependent, in situ dietary patterns of the lobate ctenophore Mnemiopsis leidyi Agassiz 1865. J Plankton Res 27:951–956

Riisgård HU, Goldstein J (2014) Jellyfish and ctenophores in Limfjorden (Denmark) mini-review, with recent new observations. J Mar Sci Eng 2:593–615. https://doi.org/10.3390/jmse2040593

Sarpe D, Grosskopf T, Javidpour J (2007) Mnemiopsis leidyi: analysis of an invader in the Kiel Fjord with focus on respiration and reproduction rate. In: 42nd European marine biology symposium, 27–32 August 2007. Kiel

Shiganova TA (1993) Ctenophore Mnemiopsis leidyi and ichthyoplankton in the sea of Marmara in October of 1992. Oceanology 33:900–903

Shiganova TA (2009) Non-native species in the seas of Eurasia. DS thesis. A. N. Severtsov Institute of Ecology and Evolution, Moscow. Available at: http://www.sevin.ru/dissertations/gidrobiol/49.pdf

Shiganova TA, Abusova GA (2021) Revision of Beroidae Eschscholtz, 1825 systematics and distribution in the southern seas of Europe based on genetic and morphological approaches. Zool J Linnean Soc 194(1):297–322. https://doi.org/10.1093/zoolinnean/zlab021

Shiganova TA, Bulgakova YV, Volovik SP, Mirzoyan ZA, Dudkin SI (2001) The new invader *Beroe ovata* Mayer 1912 and its effect on the ecosystem in the northeastern Black Sea. Hydrobiologia 451:187–199

Shiganova TA, Dumont HJ, Mikaelyan A, Glazov DM, Bulgakova YV, Musaeva EI et al (2004) Interaction between the invading ctenophore Mnemiopsis leidyi A. Agassiz and Beroe ovata Mayer 1912, and their Influence on the pelagic ecosystem of the northeastern Black Sea. In: Dumont H, Shiganova TA, Niermann U (eds) The Ctenophore Mnemiopsis leidyi in the Black, Caspian and Mediterranean Seas and Other Aquatic Invasions, NATO science series: IV. Earth and environmental sciences, vol 35. Kluwer Academic Publishers, Dordrecht, pp 33–70

Shiganova TA, Legendre L, Kazmin AS, Nival P (2014a) Interactions between invasive ctenophores in the Black Sea: assessment of control mechanisms based on long-term observations. Mar Ecol Prog Ser 507:111–123. https://doi.org/10.3354/meps10806

Shiganova TA, Sommer U, Javidpour J, Molinero JC et al (2014b) Adaptive strategy of the invader *Mnemiopsis leidyi* and *Beroe spp.* in the Eurasian seas. In: Pitois S, Shiganova T (eds) ICES/CIESM workshop A Coruna. 18–20 September 2014, pp 4–11

Shiganova TA, Alekseenko E, Moskolenko L, Nival P (2018) Modelling assessment of interactions in the Black Sea of the invasive ctenophore Mnemiopsis leidyi and Beroe ovata. Ecol Model 376:1–14. https://doi.org/10.1016/j.ecolmodel.2018.02.008

Shiganova TA, Mikaelyan AS, Moncheva S, Stefanova K, Chasovnikov VK, Mosharov SA, Mosharova IN, Slabakova N, Mavrodieva R, Stefanova E, Zasko DN, Dzhurova B (2019a) Effect of invasive ctenophores *Mnemiopsis leidyi* and *Beroe ovata* on low trophic webs of the Black Sea ecosystem. Mar Pollut Bull 141(2019):434–447. https://doi.org/10.1016/j.marpolbul.2019.02.049

Shiganova TA, Sommer U, Javidpour J, Molinero JC, Malej A, Kazmin AS, Isinibilir M, Christou E, Siokou- Frangou I, Marambio M, Fuentesf V, Mirsoyang ZA, Gülsahinh N, Lombardi F, Lilleyi MKS, j, Angelk DL, Galill BS, Bonnetm D, Delpy F (2019b) Patterns of invasive ctenophore *Mnemiopsis leidyi* distribution and variability in different recipient environments of the Eurasian seas: a review. Mar Environ Res. https://doi.org/10.1016/j.marenvres.2019.104791

Shiganova TA, Alekseenko E, Kazmin AS (2019c) Predicting range expansion of invasive ctenophore Mnemiopsis leidyi A. agassiz 1865 under current environmental conditions and future climate change scenarios. Estuar Coast Shelf Sci 227:106347. https://doi.org/10.1016/j.ecss.2019.106347

Shiganova TA, Alekseenko EV, Mirzoyan ZA, Kazmin AS (2024) Prediction of ctenophore Beroe ovata Bruguiere, 1789 habitat expansion and control of invading Mnemiopsis leidyi Agassiz, 1865 in the Ponto-Caspian basin associated with climate change. Mar Environ Res 194(2024): 106315. https://doi.org/10.1016/j.marenvres.2023.106315

Stanlaw KA, Reeve MR, Walter MA (1981) Growth, food, and vulnerability to damage of the ctenophore *Mnemiopsis mccradyi* in its early life history stages. Limnol Oceanogr 26:224–234

Studenikina EI, Volovik SP, Miryozan IA, Luts GI (1991) The ctenophore Mnemiopsis leidyi in the sea of Azov. Oceanology 3:722–725

Sullivan LJ, Gifford DJ (2004) Diet of the larval ctenophore *Mnemiopsis leidyi* A. Agassiz (Ctenophora, Lobata). J Plankton Res 26:417–431

Sullivan LJ, Gifford DJ (2007) Growth and feeding rates of the newly hatched larval ctenophore *Mnemiopsis leidyi* A. Agassiz (Ctenophora, Lobata). J Plankton Res 29:949–965

Vinogradov ME, Shushkina EA, Musaeva IA, Sorokin PY (1989) Ctenophore Mnemiopsis leidyi (A. Agassiz) (ctenophora: lobata) – new settlers in the Black Sea. Oceanology 29:293–298

Waggett RJ, Sullivan LJ (2006) Feeding efficiency of the larval ctenophore *Mnemiopsis leidyi* A. Agassiz (Ctenophora, Lobata). J Plankton Res 28:719–723

Summary and Conclusions

Key Role of the Black Sea as a Recipient and Donor Basin

Using the southern seas of Eurasia as a case study, we examined the pathways of species invasion, their establishment, subsequent dispersal to other marine areas, and the resulting impacts on local ecosystems. Our analysis of the dynamics within these basins was grounded in key data: the origins and evolution of marine life, current levels of biodiversity and productivity, and the overall state and operational health of the ecosystems.

While the southern seas of Eurasia—the Black Sea, the Sea of Azov, the Caspian and Aral Seas, and the Mediterranean Sea—share a common geological origin, they have evolved into distinct basins in the modern era. These bodies of water now exhibit varying temperature and salinity patterns, climatic influences, levels of biological productivity and diversity, and differing susceptibilities to the introduction and naturalization of non-native species. These distinctions are further accentuated by varying degrees of human-induced disruption and the resulting alterations to the functioning of their ecosystems.

The Black Sea, the Sea of Azov, and the Caspian Sea—and at certain geological periods, the Aral Sea—share the closest similarities in both origin and features. Historically, they formed the Ponto-Caspian basin, and today, they exist as brackish-water bodies characterized by limited biodiversity and elevated productivity.

Currently, the Caspian Sea and the Aral Sea are isolated from one another and also from the Black Sea and the Sea of Azov. The latter two are part of the Mediterranean basin and are linked to it through various straits. However, following the completion of the Volga-Don Canal in 1952, the Black Sea and the Sea of Azov were reconnected with the Caspian Sea. This new shipping lane enabled the movement of plant and animal species into the Caspian Sea via vessels.

Previously, the Suez Canal was built, establishing a connection between the Mediterranean Sea and the Red Sea, and thereby with the entire Indo-Pacific region.

T. A. Shiganova et al., *Non-Native Species in the Ponto-Caspian Seas*, Monographiae Biologicae 96, https://doi.org/10.1007/978-3-032-18640-9

This created a pathway through which plant and animal species can migrate directly or be carried by currents from the Red Sea into the Mediterranean Sea. Shipping routes were established, allowing vessels to transport organisms—either as fouling communities or in ballast water—from the Indian and Pacific oceans to different parts of the Atlantic Ocean and the Mediterranean basin, including the Black Sea.

Furthermore, during the twentieth century, dams and reservoirs were built on European rivers that flow into the Black Sea, the Sea of Azov, and the Caspian Sea. New freshwater canals also linked the basins of these seas with freshwater systems in Northern, Central, and Southern Europe. These newly established pathways enabled freshwater and brackish-water Ponto-Caspian species to disperse from the Sea of Azov and the Caspian Sea across Europe, entering rivers, lakes, the Baltic Sea, and the North Sea, and later traveling by ship to the Great Lakes of North America. Movement in the reverse direction also took place: some of the most invasive freshwater and brackish-water species spread along European coasts and then entered the less saline areas of the Black Sea, the Sea of Azov, and the Caspian Sea.

To fully grasp the mechanisms behind the penetration, naturalization, and proliferation of non-native species from the Ponto-Caspian basin into other marine environments, it is essential to analyze them as a unified phenomenon, as this currently represents a collective process across all three seas.

Three key phases mark the introduction of non-native species into the Ponto-Caspian basin's marine environments. The earliest non-native organisms emerged spontaneously prior to the nineteenth century, though their numbers were limited and their ecological influence remained minimal. Subsequently, from the 1930s through the late 1970s, the USSR undertook a systematic program to introduce commercially valuable fish and edible invertebrates into the Black Sea, the Sea of Azov, and the Caspian and Aral seas. During this era, deliberate acclimatization represented the main pathway for the arrival of non-native species. Alongside the targeted invaders, undesirable animal and plant species, as well as parasites, were also inadvertently introduced. In the Caspian Sea, and typically in the Aral Sea too, only species originating from the Azov-Black Sea region successfully established themselves, despite the broad geographic range of the species that were introduced.

Following the establishment of canals that linked previously separate basins, a new and significant pathway for species introduction emerged—direct transfer through rivers and canals. As navigation activity increased, organisms began to spread from different regions within fouling communities. By the early 1980s, after the adoption of isolated tanks for ship ballast water, a major new vector for organism transport developed through ballast water and residual material in tanks and chambers. The movement of plants and animals thereby took on a worldwide scope.

In examining the pathways and mechanisms for the introduction, establishment, and dispersal of non-native species across Eurasia's inland seas, it is clear that the Black Sea became a major recipient basin for numerous alien species during the latter half of the twentieth century. A significant number of these species successfully established breeding populations and began to dominate the ecological communities they colonized. Subsequently, their range expanded further—initially into the Sea of Azov through the Kerch Strait, and later into the Caspian Sea, transported via fouling communities and in ballast water. Euryhaline species also moved in a southern direction, reaching the Sea of Marmara and even extending into the Mediterranean Sea.

So, what were the factors that favored this process?

The Black Sea stands as the sole body of water within the Ponto-Caspian basin that maintains a direct link to the global ocean and supports heavy maritime traffic, connecting it to numerous parts of the World Ocean. As a result, during the latter half of the twentieth century, the sea became a receiving area for a multitude of marine and brackish-water organisms, which are continuously introduced by vessels arriving at its various ports. It is important to recognize that both the rate of invasions and the level of propagule pressure are essential prerequisites for a species to establish and become naturalized in the receiving environment. At the same time, several other factors or filters are also necessary for a species to achieve naturalization:

1	2	3	4
The presence of pathways for the introduction of non-native species, the frequent occurrence of their invasions, and the substantial abundance of the organisms introduced	The presence of a highly resilient species that boasts a broad dietary range, strong competitive traits, and high reproductive capacity	The presence of a suitable recipient area, characterized by conditions that support colonization by the donor species, is essential. These conditions include minimal competitive, predatory, and parasitic pressures, an adequate supply of food resources, and disrupted natural biocoenotic relationships within the ecosystem	Availability of conditions favorable for maintaining a reproductive population

Currently, the Black Sea functions as a semi-enclosed, nontidal, meromictic basin with relatively low salinity and a disturbed, deformed ecosystem. It possesses a very thin active layer, measuring 60–200 m, which contains oxygen and supports living organisms. The contemporary Black Sea's flora and fauna are notable for the low species diversity across most taxonomic groups and the total absence of certain groups. Coastal waters exhibit high productivity, a large abundance of common species, and a wealth of fish resources. Since the 1960s, an anthropogenic factor—eutrophication—has compounded the basin's natural deformation. This was triggered by increased inputs of phosphates, nitrates, and organic matter from major rivers in the sea's northwestern area, along with intensified fishing activity. Eutrophication has, on one hand, altered the plankton community's species composition, leading to a marked decline in several valuable plankton species and a rise in certain edible plankton types that thrive in nutrient-rich waters. Conversely, it has encouraged outbreaks of harmful algae and gelatinous plankton, including the dinoflagellate *Noctiluca scintillans* and the scyphomedusae *Aurelia aurita* and *Rhizostoma pulmo*.

Consequently, by the start of the 1980s, the Black Sea had transformed into a mesotrophic body of water, with the exception of its open-sea regions. In contrast, the northwestern section had developed into a eutrophic basin, experiencing a decline in predatory fish and dolphin populations—and subsequently, a reduction in small pelagic fish numbers. This area, however, was marked by increased productivity and a high biomass of certain edible plankton species. The conditions that supported the establishment of numerous non-native species include:

- Lack of natural predators
- Minimal presence of food competitors
- Elevated densities of food sources in both planktonic and benthic ecosystems
- Accessible and recently created unoccupied niches

This set should be supplemented by the climatic element that has emerged over the past ten years—the warming of the Black Sea's upper mixed and cold intermediate layers, which has enabled the establishment of non-native thermophilic species.

Principal Routes of Invasion

An examination of the makeup of established non-native species and their geographic origins reveals several key invasion pathways that enabled organisms to successfully enter the Black Sea, either spontaneously or intentionally.

A notable portion of these established invaders (11%) were primarily introduced via ships from the Atlantic coastal waters of North America between the 1950s and 1990s, although some species had arrived earlier. All non-native species in this category are marine neritic organisms that are eurythermal and euryhaline, commonly found in coastal regions across the World Ocean. Among them, the most invasive include *Mnemiopsis leidyi*, *Beroe ovata*, *Amphibalanus improvisus*, *Acartia tonsa*, and *Rhithropanopeus harrisii*.

A comparable invasion pattern was observed among the inhabitants of the nearby brackish-water bays and estuaries in this region. The most aggressive invaders are the hydromedusas *Blackfordia virginica* and *Calyptospadix cerulea*. These species have expanded across the desalinated western and northwestern parts of the Black Sea. Certain freshwater fish species were deliberately introduced to boost the area's fish productivity, yet they either failed to establish populations or remain extremely uncommon.

Another category of non-native organisms that naturally entered the Black Sea also originates from the North Atlantic, specifically from its eastern sector—the coastal waters of Europe (5%). While shipping has introduced the majority of species from this region, others have arrived through rivers and canals. Among these, the most invasive are the crab *Eriocheir sinensis*, which was transported to the North Sea from eastern Asian estuaries, subsequently spread widely across Europe, and entered the freshened zones of the Black Sea's northwestern area, and the kingfish *Lepomis gibbosus*, introduced to Europe from North America in the eighteenth century before moving into the Danube and reaching its delta.

A number of species entered from the northern Atlantic Ocean (10%), though their precise origin—eastern or western—remains unclear. The most invasive among these are the diatom alga *Pseudosolenia calcar-avis* and the bivalve mollusk *Mya arenaria*, both of which have expanded widely and achieved significant population densities.

The third category consists of non-native species from the Mediterranean Sea, which make up a substantial portion (44%). Their numbers have risen notably in recent years, largely due to increasing water temperatures. However, none of these species has become widespread across the entire sea, nor have any moved into other seas. These organisms were transported by currents and vessels, primarily via ballast water. They include representatives of phyto- and zooplankton, macrophytes, benthos, demersal plankton, and fish.

In contrast, the small percentage of species introduced from the Adriatic Sea (2%) not only successfully established themselves and formed breeding populations in the Black Sea but also became abundant and spread to other marine areas. Notable invasive species among them include *Anadara inaequivalvis*, which was first transported to the Adriatic Sea from the coastal waters of the Philippine Islands in the Pacific Ocean, and *Gambusia holbrooki*, which was deliberately introduced from the Adriatic Sea basin and the freshwater systems of Italy, where it had previously been brought from Central America. The fouling polychaete *Ficopomatus enigmaticus* (Fauvel, 1923) (formerly *Mercierella enigmatica*) is also believed to have likely entered from the Adriatic Sea. The oyster *Magallana gigas* (Thunberg 1793) (previously *Crassostrea gigas*) was initially carried by ships to the Black Sea from the Adriatic Sea, having originally arrived there from the Sea of Japan. Later cultivation attempts were made, though this oyster is still only encountered as scattered individuals. The successful establishment of most species originating from the Adriatic Sea can be attributed to the lower salinity levels of the Adriatic compared to the Mediterranean Sea, as well as its cooler summer and winter water temperatures, which are similar to those in the Black Sea. These conditions help Adriatic invaders endure the Black Sea's winter cooling. Furthermore, intensive shipping traffic between the ports of the Black and Adriatic Seas increases the likelihood of species exchange between these two regions.

Another pathway for non-native species entering the Black Sea originates from the Sea of Japan, along with its estuaries and rivers, from which several species have been introduced either accidentally or deliberately (6%). These include the mollusk *Rapana venosa*, which was unintentionally transported as part of fouling communities; the gray mullet *Planiliza haematocheilus* (Temminck and Schlegel 1845), previously known as *Liza haematochilos* (*Mugil soiuy*), which was intentionally introduced; and the medaka *Oryzias latipes*, a freshwater species from Japanese basins that has occasionally been released by aquarium hobbyists (2%).

The accidental introduction of non-native species from the Sea of Japan via shipping became feasible after the opening of the Suez Canal, which created new maritime routes. However, until the final decade of the twentieth century, no species other than *Rapana* had successfully established itself in the Black Sea through spontaneous invasion. During the 1990s and, more notably, the 2000s, there was an increase in observations of Indian-Pacific species in the Black Sea, although their long-term establishment remains uncertain.

To boost fish yields, several freshwater fish varieties were introduced from the estuaries of Northeast Asia (including the Sea of Japan and the Amur and other rivers) into freshwater systems, primarily ponds in the Black Sea area. Among these, silver carp and amur became abundant species, inhabiting freshwater lagoons, rivers, and deltas of the Black Sea, although their numbers are maintained through artificial breeding. Additionally, two types of ferns and strains of *Vibrio cholerae* (6%) were introduced from the same area.

As indicated by the entry pathways mentioned earlier, certain species are not native to the regions from which they reached the Black Sea. It is likely that the successful establishment of some of these species in the Black Sea occurred as a secondary invasion, facilitated by the gradual environmental changes within the basin.

Biota Richness

Within the Ponto-Caspian region, the Black Sea hosts the greatest number of both native and non-native species, followed by the Caspian Sea and then the Sea of Azov. The lower combined totals of native and non-native species in the Azov and Caspian Seas may be attributed to their salinity falling below the mesohaline threshold (approximately 15‰), a crucial biotic limit.

The most substantial constraints on species establishment are the low winter temperatures that hinder overwintering, as seen in the Northern Caspian Sea and the Sea of Azov. Consequently, warmer regions like the Mediterranean Sea exhibit a greater taxonomic capacity. The lowest primary productivity, measured by Chl "a" concentration, is found in the Mediterranean Sea, while the highest levels are in the Sea of Azov. Nevertheless, the immense size of the Mediterranean Sea permits species to migrate and locate favorable conditions, thereby supporting higher biodiversity.

Summary of Ctenophore Occurrence, Development, and Variability

Research on the proliferation of non-native gelatinous ctenophores, specifically *M. leidyi* and *B. ovata*, in Eurasian seas—particularly the Caspian Sea—has greatly enhanced our comprehension of how invasive species can profoundly influence marine environments. The underlying processes that enable these relatively simple gelatinous organisms to exert such extensive ecosystem-wide effects have been substantially clarified. For instance, one of these invaders, *M. leidyi*, succeeded in nearly overwhelming several highly productive ecosystems across Eurasian seas. In contrast, the arrival of *B. ovata* has supported the restoration of the Black Sea ecosystem and holds potential for aiding the recovery of the Caspian Sea ecosystem in the future.

We contend that invasive species represent a significant danger to biodiversity and ecosystem operations: within the Ponto-Caspian seas, they have caused substantial harm to the ecosystem, its distinctive endemic life forms, and the regional economy due to a fisheries collapse. Fortunately, restoration of the ecosystem remains achievable, at least in terms of reducing the influence of *M. leidyi* following the establishment of *B. ovata*.

The harmful ctenophore *M. leidyi* first invaded the Ponto-Caspian seas, entering the Black Sea via ballast waters in the early 1980s. This was followed by the arrival of its predator, *B. ovata*, during the 1990s. The introduction of both species into the Black Sea took place amid a period of general multi-decadal temperature increase. From 1999 to 2020, populations of both *M. leidyi* and *B. ovata* progressively declined, a trend that coincided with a marked rise in temperature.

In the Black Sea, prior to the introduction of *B. ovata*, *M. leidyi* typically emerged in the spring, commenced reproduction in July, and reached its reproductive peak in August. This peak aligned with the highest water temperatures, and reproduction often continued into early autumn and occasionally later, depending on thermal conditions. The seasonal pattern of *B. ovata* in the Black Sea is directly influenced by the seasonal cycle of its prey, *M. leidyi*. *B. ovata* appears once *M. leidyi* reaches its developmental peak, and it begins to reproduce when *M. leidyi* is at its reproductive maximum, largely because *B. ovata* larvae feed on *M. leidyi* larvae. *B. ovata* persists as long as its prey is available and vanishes once the prey population is depleted. In recent years, with rising temperatures, both ctenophore species have been observed in small quantities even into December. The seasonal dynamics of both ctenophores in the Azov Sea, which acts as a sink area, mirror the patterns seen in the Black Sea.

In the Caspian Sea, *M. leidyi* exhibits a distinct distribution pattern characterized by source and sink regions. The primary habitat, or source area, is the Southern Caspian, where the species persists year-round due to temperatures never falling below 10°C, with its highest densities observed from August to December. From this region, it spreads into other parts of the Caspian Sea, which serve as sink areas. *M. leidyi* initiates growth with the onset of warming in late winter to early spring in the Southern Caspian and extends northward into the Middle and Northwestern Caspian, and more recently into the Northeastern Caspian, where environmental conditions are suitable. It inhabits, develops, and reproduces in these areas during the warmer months and disappears with the onset of winter cooling.

The interannual population dynamics of *M. leidyi* in the Caspian Sea generally align with sea surface temperature (SST) fluctuations, where increases or decreases in abundance correspond with rises or falls in SST. However, this relationship holds only when temperatures remain within favorable bounds. When these limits are surpassed, a significant population decline has occurred.

B. ovata as a Biocontroller of *M. leidyi* Population

Our evaluations and modeling indicate that *B. ovata* serves as a significant biocontrol organism for *M. leidyi* and poses no threat to the ecosystem, as it feeds almost exclusively on *M. leidyi*. *B. ovata* preys on M. l*eidyi* across all life stages: adult *B. ovata* primarily consume adult *M. leidyi*, while *B. ovata* larvae feed on *M. leidyi* larvae. An additional physiological benefit of *B. ovata* is its capacity for rapid reproduction; as a hermaphrodite, it can self-fertilize, exhibits high fecundity, and matures quickly. Under suitable environmental conditions and with adequate food supply (*M. leidyi*), only a small number of individuals are needed to establish a large population within a month. Following the introduction of *B. ovata* into the Black Sea, it successfully managed the *M. leidyi* population, contributing to the gradual restoration of the Black Sea ecosystem. A similar outcome has been noted in the Caspian Sea since 2022. We anticipate that *B. ovata* could be employed as a biocontrol agent in other regions affected by *M. leidyi* outbreaks. One reason for this expectation is the experiments conducted by Dr. T. Shiganova in Iran (the Caspian Sea), which sought to determine if *Beroe ovata* could consume anything other than *M. leidyi*. The findings provided a clear negative answer. This outcome further supports the conclusion that *B. ovata* is harmless to other elements of the marine ecosystem.

Harmful and Beneficial Invaders, Examined in Our Research

Microplankton

The El Tor strain of *Vibrio cholerae* presents a significant threat to human populations. It has the ability to survive in coastal marine environments and subsequently infect individuals in nearby coastal communities. The initial epidemic outbreaks of cholera attributed to the El Tor strain were documented in 1970 along the Black Sea coastline, specifically in areas adjacent to the Kerch and Odessa regions. From that period onward, El Tor outbreaks have become a recurrent phenomenon in numerous cities throughout the Azov-Black Sea basin.

Phytoplankton

Certain toxic algae varieties, which can form cysts, rank among the most hazardous invasive organisms. The ability to produce cysts enables them to endure unfavorable environments, both during transit and after introduction to new habitats. The toxic dinoflagellate *Alexandrium monilatum* impacts shellfish through toxic strains that release hemolytic and neurotoxic endotoxins, potentially leading to mortality in fish and invertebrates. Although *Alexandrium tamarense*, *A. affine*, and *A. catenella* were not detected in plankton samples, researchers have identified them as potential invaders because their cysts have been recorded in port regions. In Bulgarian waters, the potentially toxic dinoflagellates *Alexandrium monilatum* and *Mantoniella squamata* have been observed.

During 2016 and 2017, a new nonindigenous macroalga, *Bonnemaisonia hamifera* (Hariot 1891), was identified along the Northern Caucasus coastline. It is believed to have been introduced from the Atlantic, where it is also non-native, originally hailing from the Pacific region. In the Black Sea, the invasive and harmful *B. hamifera* outcompetes other species, establishing dominance, altering the visual and structural characteristics of plant communities, contributing to their decline, and releasing substances that inhibit colonization by other algal types.

Green Bottom Algae

A prominent invasive species in the Mediterranean Sea is the green bottom alga *Caulerpa taxifolia*, which has drastically changed the composition of the seabed and had a major impact on the marine ecosystem. However, the expansion of *C. taxifolia* in the eastern Mediterranean is hindered by another alga, *Posidonia oceanica*, which acts as a natural blockade. It creates extensive underwater meadows that restrict the spread of the damaging *C. taxifolia*.

Macroplankton (Gelatinous Plankton)

Besides *M. leidyi* and *B. ovata*, it is also important to mention another gelatinous species, *Blackfordia virginica* (Mayer, 1910), which has invaded the Black Sea, the Sea of Azov, and the Caspian Sea. *B. virginica* impacts zooplankton and fish larvae populations, competes with pelagic fish by preying directly on early life stages, and reduces the abundance of commercially valuable species through both predation and competition.

Benthos

The common shipworm, *Teredo navalis*, is an invasive species in the Black Sea and ranks among the most globally distributed marine wood-boring bivalves, as well as one of the most destructive wood-consuming invertebrates in marine environments.

Amphibalanus improvisus possesses a cosmopolitan range, inhabiting temperate and tropical zones of the Atlantic Ocean, the Arctic Ocean, the Baltic Sea, the North Sea, the Mediterranean Sea, and the Black Sea. This species can colonize ship hulls, which decreases vessel speed, harms fishing equipment and industrial machinery, and competes with indigenous species for both food and habitat.

The gastropod *Rapana venosa* (Valenciennes, 1846) was first recorded in the Black Sea in 1946. Its presence led to considerable ecological shifts among benthic organisms, and it is regarded as a significant marine pest in the region. In recent times, however, *R. venosa* has become a valuable product in aquaculture.

Both the red-claw crayfish *Cherax quadricarinatus* and the long-fingered crayfish *Pontastacus leptodactylus* are raised at an aquaculture facility in the Sea of Azov (AzNIIRH, Azov Scientific Research Institute of Fishery, Russian Federation), where juveniles achieve market size rapidly. The long-fingered crayfish has also been accidentally introduced into Lake Sevan in Armenia.

Mariculture of Mollusks

The Mediterranean-Black Sea mussel *Mytilus galloprovincialis* is farmed at coastal installations close to Sevastopol on the Crimean Peninsula.

The Pacific giant oyster *Magallana gigas* (Thunberg, 1793), previously known as *Crassostrea gigas*, is raised at an oyster farm near the Bolshoy Utrish settlement along the northwestern Black Sea coast.

Mytilus edulis (Linnaeus 1758), commonly referred to as the common mussel or "true mussel," is a medium-sized edible marine bivalve mollusk belonging to the family Mytilidae, the only surviving family in the order Mytilida, and is the focus of intensive fishing efforts.

Fishes

In the Black Sea, periodic migrations of two harmful Indo-Pacific fish species, known as Lessepsian migrants, occur from the Sea of Marmara. These are the red barracuda, *Sphyraena pinguis* (Günther 1874), and the tetraodon, Lagocephalus sceleratus.

The Pilengas, *Planiliza haematocheilus* (Temminck and Schlegel 1845), previously identified as *Mugil soiuy*, is a commercially significant fish. Initially introduced into the Sea of Azov, it successfully established a population, reproduced, and expanded its range throughout the Azov and Black Seas, eventually reaching the Marmara and Mediterranean seas.

The Dorado, *Salminus brasiliensis*, has been regularly observed since 1999 in small groups or as solitary individuals along the Crimean coast, particularly in Balaklava Bay and nearby areas. Young fish tend to remain in shallow coastal waters, while adults inhabit deeper zones—around 30 meters and up to 150 meters during spawning. There is potential for cultivating this globally popular species in the Black Sea in the future.

Effects of Present Abiotic Environmental Factors and Projected Climate Shifts on the Spread of Invasive Ctenophores

This research also addresses the critical issue of how favorable environmental conditions for invasive ctenophores are geographically distributed and how this distribution is projected to shift due to climate change.

Based on a combination of field observations and experimental data, the study established the specific ranges for key environmental parameters—sea surface temperature (SST), sea surface salinity (SSS), and chlorophyll concentration—that are conducive to the occurrence, reproduction, and population growth of *M. leidyi* and *B. ovata*. Subsequently, by applying climatological data, the geographical areas where all three of these parameters are simultaneously favorable were identified, both on a global scale and with a specific focus on the Ponto-Caspian Seas.

On a global level, the climatology-based evaluation of current conditions not only confirmed the known native habitats of *M. leidyi* along the eastern coasts of the Americas and its established recipient regions in Eurasian seas but also highlighted prospective areas that are environmentally suitable for *M. leidyi* establishment where it is not yet found. These prospective zones include:

- In the Atlantic: from the Iberian coast to Great Britain and the western coasts of Africa
- In the Pacific: the western coasts of North and South America, the eastern coasts of Asia, and the waters surrounding New Zealand
- In the Indian Ocean: the Persian Gulf, the Arabian Sea, and the northeastern coast of Africa

A significant finding is that all habitats of *M. leidyi*, encompassing native, recipient, and prospective areas, are located in shallow coastal zones and inland water bodies such as semi-enclosed and closed seas, bays, lagoons, and fjords.

In the Ponto-Caspian Seas, the full expanse of the Black Sea and the Southern and Middle Caspian regions offer year-round conditions suitable for the presence of *M. leidyi*. In contrast, the Sea of Azov and the Northern Caspian experience shorter favorable periods because of their colder winter climates. The northwestern coastal area of the Black Sea typically supports a faster population growth rate for *M. leidyi*, attributed to greater food supply, which is also anticipated to promote the expansion of *B. ovata* populations.

Active reproduction of *M. leidyi* is projected along the entire Black Sea coast. Within the Caspian Sea, reproduction is expected in the southern, middle, and northwestern zones. The reproductive range of *B. ovata* is considerably broader, encompassing the whole Black Sea basin as well as the Southern and Middle Caspian. The prolonged reproductive season of *B. ovata* in the Southern Caspian suggests the potential for suitable conditions all year. Nevertheless, successful reproduction for *B. ovata* is dependent on the presence of its prey, specifically *M. leidyi*. In the Sea of Azov, both *M. leidyi* and *B. ovata* experience interrupted reproduction cycles as a result of low winter temperatures.

A key priority involves identifying potential regions where *M. leidyi* may emerge and reproduce as a consequence of global warming. Initially, the spread of *M. leidyi* was evaluated under scenarios where sea surface temperature (SST) increases by 2° C and 4°C above current climatological levels. The most striking outcome of this projection is a substantial enlargement of the species' reproductive zones. In northern waters, these areas are anticipated to stretch northward into the southern Norwegian Sea and westward into the open North Atlantic.

Subsequently, projected shifts in favorable reproductive conditions for *M. leidyi* were analyzed using SST forecasts from models that simulate the RCP8.5 scenario for the year 2100. This scenario anticipates a rapid increase in SST across the Northern Hemisphere. Consequently, marine regions that were formerly too cold for *M. leidyi* reproduction may become suitable. During summer and autumn, these zones are predicted to be significantly more extensive and to reach closer to the North Pole compared to present conditions. It is especially noteworthy that extensive coastal areas along the Eurasian Arctic could transform into habitats conducive to the establishment and reproduction of *M. leidyi*.

Within the Ponto-Caspian basin, under a uniform global warming of 2°C, the reproductive range of *M. leidyi* is projected to extend further offshore in the Black Sea, with the reproductive season expected to lengthen by 1 to 2 months. Reproduction in the Middle Caspian is forecast to cease, while its duration in the Southern Caspian is set to increase.

For *B. ovata*, the reproductive period is also projected to lengthen relative to current conditions. In the Caspian Sea, the zone with the longest reproduction duration is anticipated to move from the Southern Caspian to the Middle Caspian. This suggests that global warming leads to extended reproductive seasons for both ctenophore species, with the Middle Caspian emerging as a likely permanent reproduction zone for *B. ovata*.

Additional estimates were made for scenarios where only spring temperatures rise or fall by 2°C, simulating periodically occurring warmer or colder springs. In the Black Sea's coastal zones and the transitional area between the Southern and Middle Caspian, *M. leidyi* reproduction is expected to begin earlier during warmer springs. Conversely, colder springs would delay the start of reproduction only in the

Southern Caspian. For *B. ovata*, the entire Black Sea is sensitive to variations in spring sea surface temperatures, whether increases or decreases. Warmer springs lead to a longer reproduction period in the western Black Sea, while colder springs shorten the duration in the eastern part. In the Caspian Sea, *B. ovata*'s reproduction period lengthens in the central region during warmer springs and contracts in the area between the Southern and Middle Caspian during colder springs.

Final Remarks

Throughout all historical periods, the movement of living organisms between marine basins has occurred. This migration was driven by geological events that either separated or reconnected these basins, as well as by shifts in climate that altered habitat conditions. For a long time, the geographical separation of seas helped preserve the unique species diversity that had evolved through each region's specific geological history. Although non-native species of animals or plants were occasionally documented, these were rare, isolated incidents that did not seriously harm local ecosystems.

However, during the twentieth century—particularly in the latter half—there was a marked surge in the number of non-native species introduced either deliberately or accidentally into marine and coastal zones. Starting in the 1980s, with the widespread adoption of ballast water tanks on ships, the transfer of species between regions escalated into a global phenomenon. More recently, global warming has enabled species to extend their ranges northward, colonizing regions where they could not previously survive.

Today, tens of thousands of species from various taxonomic groups—from protozoa to fish—are transported between water bodies every day. A significant number of these organisms successfully establish themselves in new biotopes, which are often already stressed by human activities such as overfishing, eutrophication, river flow regulation, and pollution. The introduction of non-native species frequently results in severe ecological and economic impacts.

This study examines the process of species invasions from both global and regional viewpoints. The primary regions of focus are the southern Eurasian seas, especially the Ponto-Caspian basin—comprising the Black, Azov, and Caspian Seas—along with the adjacent Sea of Marmara and Aegean Sea. These enclosed and semi-enclosed marine environments in Europe and Central Asia have shown particular vulnerability to the introduction and establishment of nonindigenous species. Furthermore, the Ponto-Caspian region has emerged as a source area, exporting its native species through constructed canals and reservoirs that link previously isolated basins. The Mediterranean Sea is another recipient area highly prone to the naturalization of non-native organisms. The largest influx of non-native species into the Mediterranean was documented following the opening of the Suez Canal.

Special consideration has been given to the adverse ecological effects caused by one of the most damaging invaders, *M. leidyi*, and the subsequent ecosystem recovery following the introduction of its predator, the ctenophore *B. ovata*. *B. ovata* has demonstrated an ability to regulate *M. leidyi* populations and support the restoration of the ecosystem. Similarly, the United States and Canada experienced comparable ecological and economic damage from the arrival of the Ponto-Caspian mollusk *Dreissena polymorpha*, which was introduced from the Azov-Black Sea basin into the North American Great Lakes.

During the 1990s, specialized international organizations and initiatives were created at both regional and global levels to gather and synthesize data on non-native species, formulate strategies to manage their proliferation, and safeguard the natural biodiversity of aquatic ecosystems. Notable among these international efforts are the Global Invasive Species Program and GloBallast, which focus on advancing techniques for managing ballast water discharge. An international convention addressing the control and management of ballast water and sediments has been ratified, and specialized ship coatings are under development to prevent the attachment of fouling organisms. Under the framework of the International Convention for the Conservation of Biological Diversity, established in 1992, numerous scientific and practical investigations are being carried out concerning the challenges posed by biological invasions.

Nevertheless, the movement of species continues unabated, even with the commission's control efforts. The increase in maritime traffic persists, new waterways are under construction, and the Suez Canal has undergone enlargement. It is clear that the worldwide spread of invasive species is an unavoidable outcome of advancing civilization. Still, steps can be implemented to curb the arrival and establishment of damaging organisms. Drawing from our extensive involvement in global initiatives and committees, such as GloBallast, we strongly advocate for the full implementation of all available strategies to stop the transfer of harmful plant and animal species via ship hull fouling and ballast water. Today, there are effective hull coating techniques that deter the accumulation of fouling organisms. Leading ports follow guidelines for correct ballast water handling, as outlined in the International Maritime Organization's Ballast Water Management (BWM) Convention, which was established in 2004. The current option for ballast water exchange represents a beneficial advancement, offering a method to safeguard the environment. At present, 33 certified Ballast Water Management Treatment Systems (BWMTS) are actively in use.

Our central argument is that a deeper investigation into the issues surrounding invasive species should extend beyond simply expanding the list of harmful invaders. It must also focus on characterizing the environmental conditions—both biotic and abiotic—of the basins where these species have become established. Additionally, it is important to recognize that among the non-native species, some are edible (and frequently, quite delicious) and suitable for aquaculture. Therefore, invasive species should be categorized as either harmful or beneficial.

MIX
Papier aus verantwortungsvollen Quellen
Paper from responsible sources
FSC® C105338

If you have any concerns about our products,
you can contact us on
ProductSafety@springernature.com

In case Publisher is established outside the EU,
the EU authorized representative is:
Springer Nature Customer Service Center GmbH
Europaplatz 3, 69115 Heidelberg, Germany

Printed by Libri Plureos GmbH
in Hamburg, Germany